INDUSTRIAL NOISE
AND VIBRATION CONTROL

INDUSTRIAL NOISE
AND VIBRATION CONTROL

J. D. IRWIN

and

E. R. GRAF

Department of Electrical Engineering
Auburn University, Alabama

PRENTICE-HALL, INC., *Englewood Cliffs, New Jersey 07632*

Library of Congress Cataloging in Publication Data

Irwin, J   David, 1939–
    Industrial noise and vibration control.

    Includes bibliographies and index.
    1. Industrial noise.   2. Noise control.
author.   II. Title
3. Vibration.   I. Graf, Edward R., 1931–   joint
TD892.I77   620.2′3   78-7786
ISBN 0-13-461574-3

© 1979 by Prentice-Hall, Inc., Englewood Cliffs, N.J. 07632

10 9 8 7 6 5 4 3 2 1
Printed in the United States of America

PRENTICE-HALL INTERNATIONAL, INC., *London*
PRENTICE-HALL OF AUSTRALIA PTY. LIMITED, *Sydney*
PRENTICE-HALL OF CANADA, LTD., *Toronto*
PRENTICE-HALL OF INDIA PRIVATE LIMITED, *New Delhi*
PRENTICE-HALL OF JAPAN, INC., *Tokyo*
PRENTICE-HALL OF SOUTHEAST ASIA PTE. LTD., *Singapore*
WHITEHALL BOOKS LIMITED, *Wellington, New Zealand*

This Book Is Dedicated to Our Wives,
Edie and *Trixi*

CONTENTS

ACKNOWLEDGMENTS

The authors wish to gratefully acknowledge many of their colleagues who have contributed to this text. In particular, we are especially indebted to Mr. Donald E. Bently, Bently Nevada Corporation, Minden, Nevada, and Mr. Kenneth A. Ramsey, Hewlett-Packard Corporation, Santa Clara, California, for their constructive help and criticism of the vibration material and for allowing us to use some of their work in this text. Special thanks are also expressed to Dr. Charles E. Hickman, Southern Company Services, Inc., Birmingham, Alabama, for providing an appendix and other selected material. We also extend our thanks to Mr. Robert E. Perry, Barron Air Systems Engineers, Birmingham, Alabama, who provided some examples, and Mr. Charles Jackson, Monsanto, Texas City, Texas, who reviewed the vibration material. Also, we thank the Industrial Acoustics Company for providing pictures of various pieces of noise control equipment.

Finally, we are deeply indebted to the many graduate and undergraduate students whose suggestions were extremely helpful in the review and correction of the entire text. Expecially, we wish to thank Gregory H. Williams, Theodore R. Serota, Alan H. Bissinger, and Arnold Geeslin, Jr., for their assistance with the problems and the production of the solutions manual.

PREFACE

This textbook was developed from notes used for a number of years in teaching an undergraduate course in industrial noise and vibration control. Furthermore, additional material was added to the notes as a result of numerous industrial short courses taught on the subject. The topics have been judiciously selected and every attempt has been made to present the material in a fashion that is readily usable both for an undergraduate course or for self-study by a practicing engineer.

The mathematics has intentionally been maintained at a level easily handled by college sophomores; however, juniors are probably better prepared in most instances for a rapidly moving treatment. Considerable detail has been included in selected derivations. This was done to clarify certain critical equations that are of great importance to the entire text. However, this extra detail may be skipped at the discretion of the instructor without a serious loss to practical problem solving.

Both English and metric systems of units are used throughout. A table of conversion factors is provided in an appendix to allow for quick conversion where it is desirable. No effort has been made to designate each and every equation, table, and figure in both sets of units. Only in critical instances are both given. All the remaining ones are left as exercises. Weight is given in newtons as a simplification of units.

The text is replete with examples, and drill problems appear after most sections in which an important formula or concept has been introduced. For the most part, the drill problems are straightforward and relatively simple. The answers to the drill problems appear with the problems. Additional problems at the end of each chapter are generally more difficult than the drill problems and are intended to exercise and demonstrate the new concepts presented in the chapter which they follow. The answers to the odd problems are given in at the end of the book. A complete solutions manual with all drill problems and chapter problems solved is available as an instructor's aid.

J. D. IRWIN
E. R. GRAF

Auburn University
Auburn, Alabama

INDUSTRIAL NOISE
AND VIBRATION CONTROL

1

SOUND LEVELS, DECIBELS, AND DIRECTIVITY

1.1 INTRODUCTION

Sound may be described as a propagating disturbance through a physical medium. It is perceived by the ear as a pressure wave superimposed upon the ambient air pressure at the listener. The *sound pressure* is therefore the incremental variation about the ambient atmospheric pressure.

We shall now proceed to a mathematical description of those pressure waves that we designate as sound.

1.2 SOUND WAVE CHARACTERISTICS

Sound wave characteristics are readily described upon examining the attributes of a pure tone. A *pure tone* is a sinusoidal pressure wave of a specific frequency and amplitude, propagating at a velocity determined by the temperature and pressure of the air.

Let us now consider a hypothetical sound generator, as shown in Figure 1.1. The source may be thought of as an elastic sphere that expands and contracts sinusoidally at a frequency, f. As the sphere expands, the air molecules are compressed. Then as the sphere contracts, the air molecules spread apart; that is, the gas is rarefied. The *sound wave* thus generated will have a frequency equal to the number of times per second which the sphere expands and contracts. The peak pressure amplitude is a function of the maximum excursion of the sphere.

1

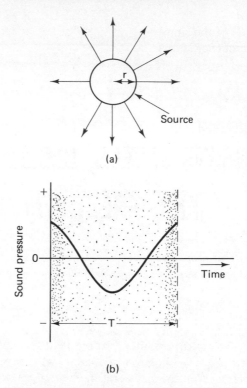

(a)

(b)

FIGURE 1.1 (a) *Spherical source oscillating at a frequency, f, to generate a spherical acoustic wave; (b) pressure vs. time for the sinusoidal wave with the compression and rarefaction of the gas depicted*

1.2.1 Frequency, Period, and Wavelength

The concept of *frequency* is common to both electrical and mechanical oscillations. The frequency, f, of an oscillating disturbance is equal to the number of times per second that the disturbance passes through both its positive and negative excursions. The number of cycles per second is termed *hertz* (Hz). For example, in the United States our electrical utility systems supply electrical power at 60 Hz, which simply means that a voltage is delivered with a sinusoidal waveform varying at 60 cycles per second.

The frequency of a simple pure-tone sound wave is recognized as the *pitch* of the tone. The human ear responds to a range of frequencies from approximately 20 to 16,000 Hz, with a maximum sensitivity at about 3,000 Hz.

In the area of industrial noise reduction we shall be interested primarily in the range of frequencies from about 63 Hz to 16,000 Hz. This is because the sensitivity of the human ear is greatly reduced below 63 Hz and above 16,000 Hz compared to its peak sensitivity.

The *period*, T, of the sinusoidal wave is depicted in Figure 1.1. Period is related to the frequency, f, by

$$T = \frac{1}{f} \quad \text{s} \tag{1.1}$$

We note that the period is the time required for one complete cycle.

The *wavelength*, λ, is the distance between like points on two successive waves. Wavelength is related to the frequency and velocity of propagation by

$$\lambda = \frac{c}{f} = cT \quad \text{ft} \quad \text{or} \quad \text{m} \tag{1.2}$$

where the *velocity of propagation*, c, is in turn a function of the characteristics of the propagation supporting medium.

1.2.2 Velocity of Sound

The speed of sound in air is given by

$$c = \sqrt{\frac{\gamma p_0}{\rho}} \quad \text{ft/s} \quad \text{or} \quad \text{m/s} \tag{1.3}$$

where $\gamma = \dfrac{\text{specific heat (constant pressure)}}{\text{specific heat (constant volume)}}$

p_0 = ambient or equilibrium pressure

ρ = ambient or equilibrium density

In the case of air, within the range of conditions of interest, γ is taken as 1.4. Equation (1.3) then becomes

$$c = \sqrt{\frac{1.4 p_0}{\rho}} \quad \text{ft/s} \quad \text{or} \quad \text{m/s} \tag{1.4}$$

which can be further simplified by taking advantage of the fact that the ratio p_0/ρ is related to the temperature of the gas. Upon assuming that the air behaves virtually as an ideal gas, the velocity, c, is related to the absolute temperature in degrees Rankine by

$$c = 49.03\sqrt{R} \quad \text{ft/s} \tag{1.5}$$

where R, the temperature in degrees Rankine, is

$$R = [459.7° + (\text{degree Fahrenheit})] \quad \text{degrees} \tag{1.6}$$

The velocity as related to degrees Kelvin is

$$\boxed{c = 20.05\sqrt{T} \quad \text{m/s}} \tag{1.7}$$

where T, the temperature in degrees Kelvin, is

$$T = [273.2° + (\text{degree Celsius})] \quad \text{degrees} \tag{1.8}$$

EXAMPLE 1.1 Calculate the velocity of sound in air at 70°F (21.1°C) in both English and metric units. Then determine the wavelength of a 1000-Hz tone at the same temperature.

SOLUTION

$$R = (459.7° + 70°) = 529.7° \text{ Rankine}$$
$$c = 49.03\sqrt{529.7} = 1128 \text{ ft/s}$$
$$T = (273.2° + 21.1°) = 294.3° \text{ Kelvin}$$
$$c = 20.05\sqrt{294.3} = 344 \text{ m/s}$$
$$\lambda = \frac{c}{f} = \frac{1128}{1000} = 1.128 \text{ ft} \quad \text{at 1 kHz}$$
$$\lambda = \frac{c}{f} = \frac{344}{1000} = 0.344 \text{ m} \quad \text{at 1 kHz}$$

It is also important to note that the velocity of sound in common building materials is generally quite different from that for air. This, in turn, means that the wavelength in these materials is proportionately different from that in air. This becomes of particular importance to us when we consider the isolation of low-frequency sounds. Table 1.1 lists the approximate veloc-

TABLE 1.1

*Approximate velocity of sound in certain
common media at room temperature (70°F or 21.1°C)*

Material	Velocity of Sound ft/s	m/s
Air	1,128	344
Water	4,500	1,372
Concrete	10,000	3,048
Glass	12,000	3,658
Iron	17,000	5,182
Lead	4,000	1,219
Steel	17,000	5,182
Wood (hard)	14,000	4,267
Wood (soft)	11,000	3,353

ity of sound in some common building materials, along with air and water for comparison, all taken at room temperature.

EXAMPLE 1.2 Calculate and compare the wavelength of a 1000-Hz tone in steel to that in air.

SOLUTION In steel, $c = 17,000$ ft/s or 5182 m/s.

$$\lambda = \frac{17,000}{1000} = 17 \text{ ft in steel}$$

or

$$\lambda = \frac{5182}{1000} = 5.182 \text{ m in steel}$$

In air, $c = 1128$ ft/s or 344 m/s.

$$\lambda = \frac{1128}{1000} = 1.128 \text{ ft in air}$$

or

$$\lambda = \frac{344}{1000} = 0.344 \text{ m in air}$$

Comparing the two wavelengths,

$$\frac{\lambda(\text{steel})}{\lambda(\text{air})} = \frac{17}{1.128} = 15.1$$

The wavelength in steel is 15.1 times longer than in air.

D1.1 What is the frequency of a wave with a wavelength of 0.025 m at 0°C in air?
Ans. $f = 13.26$ kHz.

D1.2 What is the approximate wavelength of a 5.5-kHz sound wave propagating at room temperature in (a) water (b) glass (c) lead?
Ans. (a) 0.249 m   (b) 0.665 m   (c) 0.222 m.

1.3 LEVELS AND DECIBELS

Sound pressures and powers are commonly expressed in terms of *decibel levels*. This allows us to use a logarithmic rather than a linear scale. It has the distinct advantage of allowing us to do our calculations within a scale of small numbers rather than over an extremely large scale of numbers.

The decibel was originated by electrical engineers in the area of telecommunications. However, it is well suited to acoustics by simply substituting pressures and acoustic powers for voltages and electrical powers respectively.

Furthermore, since the *decibel* is the logarithm to the base 10 of the ratio of the quantity in question to an arbitrarily chosen reference quantity, the argument of the logarithm is dimensionless. We therefore refer to the designation in decibels as the level of the quantity referenced to the chosen reference quantity. Expressed mathematically, we have

$$\text{level} = 10 \log \left(\frac{W}{W_0} \right) \quad \text{dB} \tag{1.9}$$

where W = power in question

$\quad\quad$ W_0 = chosen reference power

EXAMPLE 1.3   Calculate the level of a power of 100 W referenced to 1 W.

SOLUTION

$$\text{level} = 10 \log \left(\frac{100}{1} \right) \quad \text{dB}$$

$$= 10 \log (100) = 10 \times 2 = 20 \, \text{dB}$$

$$= 20 \, \text{dB} \quad \text{referenced to 1 W}$$

1.3.1   Sound Power Level

Sound power level describes the acoustical power radiated by a given source with respect to the international reference of 10^{-12} W. (In certain older literature, 10^{-13} W was used as a reference.) The sound power level, L_W, is defined as

$$L_W = 10 \log \left(\frac{W}{W_{\text{re}}} \right) \quad \text{dB} \tag{1.10}$$

where W = sound power in question

$\quad\quad$ W_{re} = 10^{-12} W (reference)

EXAMPLE 1.4   Determine the sound power level of a small siren that generates 0.1 W of sound power.

SOLUTION

$$L_W = 10 \log \left(\frac{W}{W_{\text{re}}} \right) \quad \text{dB}$$

$$= 10 \log \left(\frac{0.1}{10^{-12}} \right) = 10 \log (10^{11})$$

$$= 10 \times 11 = 110 \, \text{dB}$$

$$= 110 \, \text{dB}$$

Upon a moment's reflection, we note that a very small amount of acoustical power, 0.1 W, is associated with a loud sound source referenced to the sensitivity of the human ear.

In certain instances it is desirable to determine the absolute acoustic power from the power level. This is simply accomplished by solving for W in equation (1.10); we obtain

$$W = W_{re} \text{ antilog} \left(\frac{L_W}{10}\right) \quad W \tag{1.11}$$

or

$$W = W_{re} \times 10^{L_W/10} \quad W \tag{1.12}$$

EXAMPLE 1.5 Determine the sound power of a machine whose specified sound power level is 125 dB.

SOLUTION

$$W = W_{re} \times 10^{L_W/10} \quad W$$
$$= 10^{-12} \times 10^{125/10} = 10^{-12} \times 10^{12.5}$$
$$= 10^{-12} \times 10^{12} \times 10^{0.5} = 10^{0.5} = 3.2$$
$$= 3.2 \text{ W}$$

It is interesting to note that 3.2 W represents only 0.004 hp. Certainly from the efficiency point of view, the power lost by the device in the form of acoustical power is virtually zero, and yet with regard to the noise to the operator, it is very loud.

Representative sound sources with typical sound powers and sound power levels are given in Table 1.2. We note that as the power goes from 10^{-7} W to 30×10^6 W, the power level ranges from 50 dB to only 195 dB.

TABLE 1.2

Typical powers and power levels for certain representative sound sources

Source	Power (W)	Power level (dB referenced to 10^{-12} W)
Whisper	10^{-7}	50
Voice (conversational)	10^{-5}	70
Voice (shouting average)	10^{-3}	90
Record player (loud)	10^{-2}	100
Truck horn	10^{-1}	110
Airplane engine (propeller)	1	120
Pipe organ (peak)	10	130
Airliner (four-propeller)	100	140
Saturn rocket	30×10^6	195

D1.3 How many watts are represented by a sound power level of (a) 120 dB (b) 123 dB   (c) 129 dB   (d) 125 dB?

$\qquad$ *Ans.*   (a) 1 W   (b) 2 W   (c) 8 W   (d) 3.16 W.

D1.4 What is the sound power level of a source that generates (a) 2.2 W   (b) 1.5 W (c) 3 W   (d) 0.5 W?

$\qquad$ *Ans.*   (a) 123.4 dB   (b) 121.76 dB   (c) 124.77 dB   (d) 117 dB.

1.3.2 Sound Pressure Level

Sound pressure levels are expressed in decibels just as are sound power levels. The sound pressure level, L_p, is defined as

$$\boxed{L_p = 10 \log \left(\frac{p^2}{p_{\text{re}}^2}\right) \quad \text{dB}} \tag{1.13}$$

or

$$L_p = 20 \log \left(\frac{p}{p_{\text{re}}}\right) \quad \text{dB} \tag{1.14}$$

where p = root-mean-square (rms) sound pressure in question (Pa or N/m²)

p_{re} = international reference pressure of 20×10^{-6} Pa (0.0002 μbar)

The pressure of 20×10^{-6} Pa has been chosen as a reference because it has been found that the average young adult can just perceive a 1000-Hz tone at this pressure. Therefore, this reference is often referred to as the threshold of hearing at 1000 Hz.

It is also noteworthy that the sound pressure level is proportional to the logarithm of the ratio of pressures squared. This is significant in that the pressure squared is proportional to some sound power; thus, both the sound power level and the sound pressure level are associated with power. In noise abatement we are almost exclusively interested in relatively broad frequency bands of noise with random amplitude and phase relationships in both time and space. This requires that in adding and subtracting sound pressure levels, pressure squared rather than simple pressure ratios be used in calculations.

EXAMPLE 1.6 Calculate the sound pressure level for a sound with an rms acoustic pressure of 2.5 Pa.

SOLUTION

$$L_p = 10 \log \left(\frac{p}{p_{\text{re}}}\right)^2 \quad \text{dB}$$

and

$$p_{\text{re}} = 20 \times 10^{-6} \text{ Pa}$$

Therefore,

$$L_p = 10 \log \left(\frac{2.5}{20 \times 10^{-6}}\right)^2$$

$$= 10 \log (12.5 \times 10^4)^2 = 20(\log 12.5 + \log 10^4)$$

$$= 20(1.096 + 4) = 20(5.096) = 101.9$$

$$= 101.9 \text{ dB}$$

D1.5 What is the rms pressure of a noise with a sound pressure level of (a) 110 dB
(b) 115 dB   (c) 112 dB   (d) 102 dB?

   Ans.   (a) 6.32 N/m²   (b) 11.25 N/m²   (c) 7.96 N/m²   (d) 2.52 N/m².

D1.6 Determine the sound pressure level associated with the following acoustic
pressures: (a) 0.3320 Pa   (b) 0.106 Pa   (c) 2.97 Pa

   Ans.   (a) 84.4 dB   (b) 74.5 dB   (c) 103 dB.

1.3.3 Adding, Subtracting, and Averaging Decibels

The solution of industrial noise problems commonly demands that pressures and powers be manipulated by means of decibel additions and subtrac-
tions. Furthermore, sound pressure levels are averaged, again in decibels,
in the calculation of source directivity. Mathematical procedures, of course,
as well as charts and approximations, are available for adding, subtracting,
and averaging decibels.

1.3.3.1 Adding decibels

Rare, indeed, is the worker position affected by the noise of only a single
source. Even in those few cases, a background or ambient noise level is
always present. We are, therefore, generally concerned with the sound pressure level due to several noise sources.

Furthermore, it is general practice to determine individually the sound
pressure levels from a single source in the various frequency bands in question. We may then determine the total sound pressure level at a given position due to the source by the decibel addition of the band levels in question.
More often than not, a single measurement is also made to determine the
total sound pressure level at the position in question. Obviously, the calculated value, that obtained by means of the decibel summation, and the
measured value may be compared as an accuracy check.

Another convenient check is obtained by converting each of the band
sound pressure levels to a sound level and summing for the total. This value
may then be compared to a single measured value for an additional accuracy
check.

Sound power levels are most commonly added in the determination of
the total sound power level of a source. The sound power level in the several

bands of interest may be determined by measurements and calculations. Finally, the total sound power level is the decibel summation of the contribution of each of the frequency bands of interest.

We shall now develop the procedure for adding decibels. Note in the development that we always combine on an energy basis. This is so because in our analysis we are considering the noise to be random with respect to phase.

First, let us assume that we wish to add the sound pressure levels L_{p_1}, L_{p_2}, L_{p_3}, through L_{p_i}. Recall that the sound pressure level, L_{p_i}, by definition, is

$$L_{p_i} = 10 \log \left(\frac{p}{p_{re}}\right)^2_i \quad \text{dB} \tag{1.15}$$

We now proceed by determining the square of the pressure ratio, that is,

$$\left(\frac{p}{p_{re}}\right)^2_i = \text{antilog} \left(\frac{L_{p_i}}{10}\right) \tag{1.16}$$

The total sound pressure level, L_{p_t}, is simply

$$L_{p_t} = 10 \log \left[\sum_{i=1}^{n} \left(\frac{p}{p_{re}}\right)^2_i\right] \quad \text{dB} \tag{1.17}$$

or, in terms of the sound pressure levels,

$$L_{p_t} = 10 \log \left[\sum_{i=1}^{n} \text{antilog} \left(\frac{L_{p_i}}{10}\right)\right] \quad \text{dB} \tag{1.18}$$

Upon further simplification, one obtains

$$L_{p_t} = 10 \log \left(\sum_{i=1}^{n} 10^{L_{p_i}/10}\right) \quad \text{dB} \tag{1.19}$$

This final expression clearly indicates the ease by which L_{p_t} may be calculated.

In an analogous fashion, we find that the summation of sound power levels may be expressed as

$$L_{w_t} = 10 \log \left(\sum_{i=1}^{n} 10^{L_{w_i}/10}\right) \quad \text{dB} \tag{1.20}$$

where L_{w_t} = total sound power level (dB)

L_{w_i} = ith sound power level (dB)

EXAMPLE 1.7 Determine the total sound pressure level due to $L_{p_1} = 90$ dB, $L_{p_2} = 95$ dB, and $L_{p_3} = 88$ dB.

SOLUTION According to equation (1.19), in this case

$$L_{p_t} = 10 \log \left(\sum_{i=1}^{3} 10^{L_{p_i}/10} \right) \quad \text{dB}$$

which may be expanded as

$$L_{p_t} = 10 \log (10^{90/10} + 10^{95/10} + 10^{88/10}) \quad \text{dB}$$

Upon simplification, we obtain

$$L_{p_t} = 10 \log (10^9 + 10^{9.5} + 10^{8.8}) \quad \text{dB}$$

which reduces to

$$L_{p_t} = 10 \log (4.79 \times 10^9) \quad \text{dB}$$

Finally, we obtain

$$L_{p_t} = 96.8 \text{ dB}$$

EXAMPLE 1.8 Determine the total sound power level due to the contributions from three octave bands, designated as $L_{w_1} = 100$ dB, $L_{w_2} = 103$ dB, and $L_{w_3} = 106$ dB.

SOLUTION We recall from equation (1.20) that

$$L_{w_t} = 10 \log \left(\sum_{i=1}^{n} 10^{L_{w_i}/10} \right) \quad \text{dB}$$

In this example we have

$$L_{w_t} = 10 \log (10^{100/10} + 10^{103/10} + 10^{106/10}) \quad \text{dB}$$

which reduces to

$$L_{w_t} = 10 \log (10^{10} + 10^{10.3} + 10^{10.6}) \quad \text{dB}$$

Upon performing the indicated operations,

$$L_{w_t} = 108.4 \text{ dB}$$

An alternative, but less accurate, means to add decibels is to employ a decibel addition chart such as that given in Figure 1.2. Before we proceed to an example of this method, let us briefly examine the chart. Two impor-

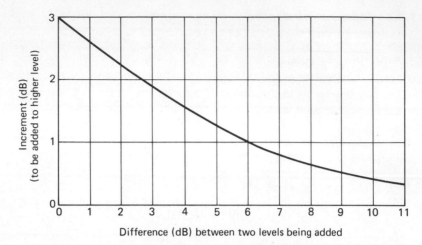

Difference (dB) between two levels being added

FIGURE 1.2 *Chart for adding decibels*

tant attributes of decibel additions are readily noted. First, one observes that when two equal decibel levels are added, the sum is equal to 3 dB above the single original level. That is, if x dB and x dB are added, the sum is $x + 3$ dB. Second, we observe that if two dB levels are separated by 10 dB or more, the sum is less than 0.5 dB more than the larger of the two levels being added. This is very important: it clearly means that if the lower dB level were eliminated, the total decibel level would remain essentially unchanged.

EXAMPLE 1.9 Determine the total sound pressure level due to $L_{p_1} = 90$ dB, $L_{p_2} = 95$ dB, and $L_{p_3} = 88$ dB. Use the chart in Figure 1.2 and compare the result obtained for this same addition in Example 1.7.

SOLUTION No particular order of addition is required; we will begin with the lowest level and proceed to higher-dB levels.

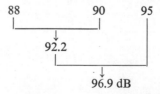

For this summation in Example 1.7 we obtained $L_{p_t} = 96.8$ dB. It should be noted, however, that in practical applications we are rarely interested in a fraction of a decibel. Variations in the noise source, position of the worker, and simply the instrumentation reduce fractional-dB values to sheer nonsense in most problem areas.

1.3.3.2 Subtracting decibels

In certain instances it is desirable to subtract an ambient or background sound pressure level from a total measured level. This allows one to determine the sound pressure level produced by a particular source. However, in general, it is not possible to make meaningful measurements unless the background sound pressure level is at least 3 dB below the level of the source under consideration.

The procedure for subtracting decibels is analogous to their addition. Upon reference to equations (1.15) and (1.16), we obtain for the total sound pressure level, L_{p_t}, the expression

$$L_{p_t} = 10 \log \left(\frac{p}{p_\text{re}} \right)_t^2 \quad \text{dB} \tag{1.21}$$

or, in terms of the mean-square pressure ratio,

$$\left(\frac{p}{p_\text{re}} \right)_t^2 = \text{antilog} \left(\frac{L_{p_t}}{10} \right) = 10^{L_{p_t}/10} \tag{1.22}$$

Similarly, one may express the background or ambient noise as

$$L_{p_B} = 10 \log \left(\frac{p}{p_\text{re}} \right)_B^2 \quad \text{dB} \tag{1.23}$$

or

$$\left(\frac{p}{p_\text{re}} \right)_B^2 = \text{antilog} \left(\frac{L_{p_B}}{10} \right) = 10^{L_{p_B}/10} \tag{1.24}$$

where L_{p_B} = sound pressure level of the ambient or background noise (dB)

The sought-after sound pressure level of the source, L_{p_s}, is, therefore,

$$L_{p_s} = 10 \log \left[\left(\frac{p}{p_\text{re}} \right)_t^2 - \left(\frac{p}{p_\text{re}} \right)_B^2 \right] \quad \text{dB} \tag{1.25}$$

or, more conveniently,

$$L_{p_s} = 10 \log \left(10^{L_{p_t}/10} - 10^{L_{p_B}/10} \right) \quad \text{dB} \tag{1.26}$$

Finally, we note that in practical applications it is necessary to make two measurements: (1) L_{p_t}, the sound pressure level with the machine running in its desired noise environment, and (2) L_{p_B}, the sound pressure level of the noise environment with the machine under study remaining silent.

13

EXAMPLE 1.10 Determine the sound pressure level at a point due to a particular machine if, at the point, $L_p = 85$ dB with the machine "off" and 94 dB with the machine operating.

SOLUTION According to equation (1.26), we have

$$L_{p_s} = 10 \log (10^{L_{p_t}/10} - 10^{L_{p_B}/10}) \quad \text{dB}$$

In this case, $L_{p_t} = 94$ dB and $L_{p_B} = 85$ dB. Therefore,

$$L_{p_s} = 10 \log (10^{94/10} - 10^{85/10}) \quad \text{dB}$$

which reduces to

$$L_{p_s} = 10 \log (2.196 \times 10^9) \quad \text{dB}$$

and finally, for the machine,

$$L_{p_s} = 93.4 \text{ dB}$$

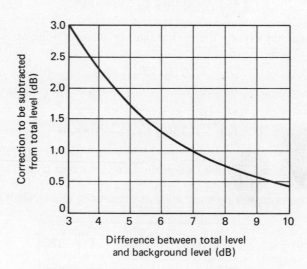

FIGURE 1.3 *Chart for subtracting decibels*

Just as for the addition of decibels, we may also use a chart for subtracting decibels. Such a chart is given in Figure 1.3. Obviously, it is not as accurate a method as that of calculating, but in a majority of practical situations, it is completely adequate.

EXAMPLE 1.11 Repeat Example 1.10, but solve using the chart of Figure 1.3.

SOLUTION Again, $L_{p_t} = 94$ dB and $L_{p_B} = 85$ dB. To use the chart, we obtain

$$L_{p_d} = L_{p_t} - L_{p_B} = 94 - 85 = 9 \text{ dB}$$

that is, the difference between the total level and the background level, $L_{p_d} = 9$ dB. Entering the chart at 9 dB on the abscissa, we find that 0.6 dB is to be subtracted from L_{p_t}. Therefore,

$$L_{p_s} = 94 - 0.6 \text{ dB}$$

and

$$L_{p_s} = 93.4 \text{ dB}$$

It is interesting to note that this result is the same as that calculated in Example 1.10, to the nearest tenth of 1 dB.

D1.7 An operator in a textile mill is operating five machines. The sound pressure levels of the machines at his position are 95, 90, 92, 88, and 82 dB, respectively. (a) What is the total sound pressure level at his position due to all five machines? (b) What would be the total if the sound pressure levels were 91 dB, 86 dB, 83 dB, 80 dB, and 87 dB? Check your answers using Figure 1.2.

Ans. (a) 98 dB   (b) 94 dB.

D1.8 Determine the sound pressure level at the operator's position in part (a) of drill problem D1.7 if (a) machine 1 is turned off. (b) machines 2 and 3 are turned off.

Ans. (a) 95 dB   (b) 96 dB.

D1.9 Determine the sound pressure level at a point due only to three machines if the sound pressure level at the point due only to machine 1 and the background noise is 88 dB, only machine 2 and the background noise is 90 dB, and only machine 3 and the background noise is 87 dB. When the machines are off, the sound pressure level at the point is 86 dB.

Ans. 90 dB.

1.3.3.3 Averaging decibels

A requirement to average decibels presents itself from time to time. In particular, we are required to average decibels in the calculation of the average sound pressure level, $\bar{L}_p$, about a source for directivity determinations. In other instances we may desire to measure the sound pressure level at a single position several times and then take an average value for calculation purposes.

The procedure for averaging decibels follows directly from that for the summation. The addition of decibels was expressed by equation (1.19) as

$$L_{p_t} = 10 \log \left(\sum_{i=1}^{n} 10^{L_{p_i}/10} \right) \quad \text{dB} \tag{1.27}$$

We may determine the average decibel level, $\bar{L}_p$, by dividing the sum by the number of levels, that is,

$$\boxed{\bar{L}_p = 10 \log \left(\frac{1}{n} \sum_{i=1}^{n} 10^{L_{p_i}/10} \right) \quad \text{dB}} \tag{1.28}$$

This equation is a convenient expression for $\bar{L}_p$ and will be used in the determination of the directivity factor, Q.

EXAMPLE 1.12   Determine the average sound pressure level, $\bar{L}_p$, of the following set of measured values: $L_{p_1} = 96$ dB, $L_{p_2} = 100$ dB, $L_{p_3} = 90$ dB, and $L_{p_4} = 97$ dB.

SOLUTION   From equation (1.28) we have

$$\bar{L}_p = 10 \log \left(\frac{1}{n} \sum_{i=1}^{n} 10^{L_{p_i}/10} \right) \quad \text{dB}$$

which in this case yields

$$\bar{L}_p = 10 \log \left[\tfrac{1}{4} (10^{96/10} + 10^{100/10} + 10^{90/10} + 10^{97/10}) \right] \quad \text{dB}$$

Upon simplification,

$$\bar{L}_p = 10 \log \left[\tfrac{1}{4} (10^{9.6} + 10^{10} + 10^{9} + 10^{9.7}) \right] \quad \text{dB}$$

and finally,

$$\bar{L}_p = 97 \text{ dB}$$

1.3.3.4 Approximation of $\bar{L}_p$

In practical applications we often find that the variation in the sound pressure levels from measuring point to measuring point is 10 dB or less. In those instances in which the variation is 5 dB or less, the average value, $\bar{L}_p$, may be approximated by averaging arithmetically. That is,

$$\boxed{\bar{L}_p = \frac{1}{n} \sum_{i=1}^{n} L_{p_i} \quad \text{dB}} \tag{1.29}$$

under the condition that

$$L_{p_{i_{max}}} - L_{p_{i_{min}}} \leq 5 \text{ dB}$$

where n = number of measured values

      L_{p_i} = ith measured sound pressure level (dB)

Then in those cases where the maximum variation is between 5 and 10 dB, an approximate value is found by averaging arithmetically and adding 1 dB. Simply expressed,

$$\boxed{\bar{L}_p = \left(\frac{1}{n}\sum_{i=1}^{n} L_{p_i}\right) + 1 \quad \text{dB}} \tag{1.30}$$

under the condition that

$$5 \text{ dB} < L_{p_{i_{max}}} - L_{p_{i_{min}}} \leq 10 \text{ dB}$$

Equations (1.29) and (1.30) are widely used in calculations involving industrial problems.

EXAMPLE 1.13 Determine the average sound pressure level, $\bar{L}_p$, of the following set of measured values: L_{p_1} = 94 dB, L_{p_2} = 89 dB, L_{p_3} = 99 dB, and L_{p_4} = 90 dB. Solve by both the approximate and the exact methods and compare the results.

SOLUTION We find the variation

$$L_{p_{i_{max}}} - L_{p_{i_{min}}} = 99 - 89 = 10 \text{ dB}$$

For an approximate solution, we then use equation (1.30),

$$\bar{L}_p = \left(\frac{1}{n}\sum_{i=1}^{n} L_{p_i}\right) + 1 \text{ dB}$$

to obtain

$$\bar{L}_p = \tfrac{1}{4}(94 + 89 + 99 + 90) + 1 \text{ dB}$$
$$= 94 \text{ dB} \quad \text{(approximate value)}$$

By the exact method using equation (1.28), we obtain

$$\bar{L}_p = 10 \log \left[\tfrac{1}{4}(10^{94/10} + 10^{89/10} + 10^{99/10} + 10^{90/10})\right] \quad \text{dB}$$

which yields

$$\bar{L}_p = 94.9 \text{ dB}$$

We note that the two solutions yield approximately the same result.

D1.10 Three sound-pressure-level measurements were taken: 88 dB, 90 dB, and 87 dB. Determine the average by exact and approximate methods.

Ans. exact method 88.5 dB; approximate method 88.33 dB.

D1.11 The sound pressure level due to a machine in a room is measured at five different positions in the room and found to be 83 dB, 86 dB, 82 dB, 85 dB, and 83 dB. The sound pressure level due to a second machine is also measured at five positions and found to be 83 dB, 86 dB, 88 dB, 81 dB, and 84 dB. What is the average sound pressure level of (a) the first machine? (b) the second machine? (Use the approximate method.)

Ans. (a) 84 dB   (b) 85 dB.

D1.12 Determine the average of the following three sound pressure levels: 84 dB, 85 dB, and 80 dB.

Ans. 83 dB.

1.4 DIRECTIVITY

In general, we are concerned with noise sources that do not radiate uniformly in all directions. Indeed, it is not a simple matter to obtain a virtual isotropic source for laboratory experimentation. One needs only to consider the directional attributes of an entertainment loudspeaker to note that the sound is not radiated uniformly, but rather directed as exemplified in Figure 1.4. The pattern shown in Figure 1.4 is for the xy plane only; obviously, the radiation takes place in a three-dimensional space, and therefore an infinite number of such planes could be designated and polar patterns plotted. Furthermore, the pattern was determined at a specific frequency and may in general not be assumed to be constant over a very large range of frequencies; that is, an octave change can change the pattern considerably.

In the case of industrial noise sources, the same general characteristics apply except that usually the problem is much more complicated, owing to the multiplicity of noise sources within a single machine as well as the proximity of other noisy machinery. Nonetheless, it is absolutely necessary to estimate at least the directional characteristics of the noise source in question. This requirement is met by an estimation of the directivity factor, Q, directly or by an approximate calculation using the directivity index, DI.

1.4.1 Directivity Factor, *Q*

The directivity factor, Q, is defined as the ratio of the intensity (W/m²) at some distance and angle from the source to the intensity at the same distance, if the total power from the source were radiated uniformly in all directions. As derived and discussed in Chapter 6, the intensity, I, of a spher-

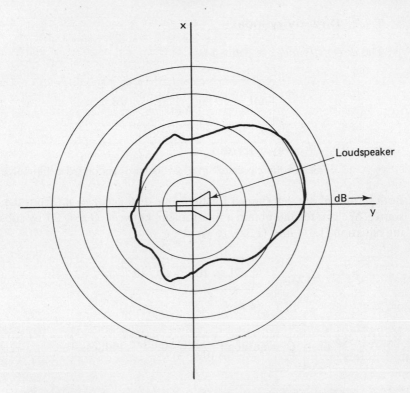

FIGURE 1.4 *Sound-pressure-level polar plot of the radiation pattern of a particular loudspeaker at 1 kHz in the xy plane*

ical acoustic wave, many wavelengths from the source, is

$$I = \frac{p^2}{\rho_0 C} \quad \text{W/m}^2$$

where I = sound intensity (W/m²)

p = rms sound pressure (Pa)

$\rho_0 C$ = acoustic characteristic impedance

Thus,

$$Q = \left(\frac{I_\theta}{\bar{I}}\right)\bigg|_{r=r_1} = \left(\frac{p_\theta^2}{\bar{p}^2}\right)\bigg|_{r=r_1} \quad \text{unitless} \tag{1.31}$$

where $\bar{I}$ = average intensity over a spherical surface at the distance r_1 (W/m²)

$\bar{p}$ = average rms pressure over a spherical surface at the distance r_1 (Pa)

Since we measure the sound pressure level, it is most convenient to continue our discussion in terms of sound pressures and sound pressure levels.

1.4.2 Directivity Index

The *directivity index* is defined as

$$\mathrm{DI} = 10 \log \left(\frac{p_\theta^2}{\bar{p}^2} \right) \Bigg|_{r=r_1} \quad \mathrm{dB} \qquad (1.32)$$

where   DI = directivity index (dB)

p_θ^2 = mean-square pressure in a particular direction from the source (Pa²)

Both p_θ^2 and $\bar{p}^2$ are determined for the same distance, $r = r_1$, from the noise source. We can readily obtain a relationship between Q and DI by substituting equation (1.31) into (1.32), to yield

$$\mathrm{DI} = 10 \log Q \quad \mathrm{dB} \qquad (1.33)$$

or

$$Q = \mathrm{antilog} \left(\frac{\mathrm{DI}}{10} \right) = 10^{\mathrm{DI}/10} \quad \mathrm{unitless} \qquad (1.34)$$

For practical applications it is much more convenient to express DI in terms of the sound pressure level, L_{p_θ}, and the average sound pressure level, $\bar{L}_p$. This can be easily accomplished by the simple substitution of the defining expression for L_p into equation (1.32). Substitution yields

$$\mathrm{DI} = 10 \log \left[\frac{\left(\dfrac{p_\theta}{p_{\mathrm{re}}} \right)^2}{\left(\dfrac{\bar{p}}{p_{\mathrm{re}}} \right)^2} \right] \Bigg|_{r=r_1} \quad \mathrm{dB} \qquad (1.35)$$

which may be written as

$$\mathrm{DI} = 10 \log \left(\frac{p_\theta}{p_{\mathrm{re}}} \right)^2 - 10 \log \left(\frac{\bar{p}}{p_{\mathrm{re}}} \right)^2 \quad \mathrm{dB} \qquad (1.36)$$

or simply

$$\mathrm{DI} = L_{p_\theta} - \bar{L}_p \quad \mathrm{dB} \qquad (1.37)$$

where both L_{p_θ} and $\bar{L}_p$ are determined for the same fixed distance from the noise source.

Furthermore, Q, which is required in most room acoustic calculations, is given by

$$Q = \text{antilog}\left(\frac{L_{p_\theta} - \bar{L}_p}{10}\right) \quad \text{unitless} \tag{1.38}$$

We note that in practical situations the determination of $\bar{L}_p$, the average sound pressure level about the source, is required.

EXAMPLE 1.14   Calculate the directivity factor, Q, for a directivity index of 3.5 dB.

SOLUTION

$$Q = \text{antilog}\left(\frac{\text{DI}}{10}\right)$$

$$= 10^{3.5/10} = 2.24$$

D1.13   Determine the directivity index if (a) $p_\theta = 5450$ Pa and $\bar{p} = 3450$ Pa (b) $p_\theta = 437$ Pa and $\bar{p} = 275$ Pa   (c) $p_\theta = 1120$ Pa and $\bar{p} = 912$ Pa
 Ans.   (a) 3.97 dB   (b) 4.02 dB   (c) 1.78 dB.

D1.14   The average sound pressure level 10 ft from a source is 102 dB. In a specified direction the sound pressure level 10 ft from the source is 104 dB. What is the directivity factor?
 Ans. $Q = 1.58$.

1.4.3   Determination of $\bar{L}_p$

Inasmuch as we are concerned with the determination of the space average of the sound pressure level about the sound source, it is necessary to develop a technique whereby a few discrete measurements will be satisfactory. If one has access to a sufficiently large anechoic chamber, the measurements can be made on a hypothetical spherical surface about the source, where the center of the sound source is the center of the sphere. Such a technique is clearly not applicable to large industrial machines, but is suited for instance to small appliances, loudspeakers, and the like.

Measurements about large machines are usually performed with the machine on a large, flat reflecting surface, a concrete floor, for example, in a large room with the walls and ceiling covered with a suitable anechoic material. Numerous measurement laboratories employ large rooms covered with thick, wedge-shaped layers of mineral wool insulation. Another method is simply to place the machine outdoors on a large flat concrete or asphalt pad for measurement.

In those cases where we place the machine on a reflecting surface, measurements are made over a hemispherical surface rather than a spherical surface. The success of this method depends on the assumption that the sound source is effectively at the center of the hemisphere and in the plane of the reflecting surface.

1.4.3.1 Measurements in anechoic full space

As in most measurements, the more accurately we measure the total power radiated by the source, the more data points are required and the more costly the determination. In an effort to make measurements in an orderly and readily describable manner, let us assume the sound source to be located at the center of a polyhedron of 12 equal sides, as depicted in Figure 1.5. The coordinates of the center points are given in Table 1.3 for the 8-, 12-, and 20-sided polyhedrons shown in Figure 1.6. One measurement is made on each face at the center point of each, with the assumption that the same sound pressure exists over that entire face. Obviously, the accuracy of this method is increased with an increase in the number of sides of the polyhedron chosen.

TABLE 1.3

Coordinates of midpoints of faces

Position	20-SIDED			12-SIDED			8-SIDED		
	x/r	y/r	z/r	x/r	y/r	z/r	x/r	y/r	z/r
1	0.00	0.93	0.36	0.00	0.89	0.45	0.00	0.82	0.58
2	0.00	0.93	−0.36	0.00	−0.89	−0.45	0.00	0.82	−0.58
3	0.36	0.00	0.93	0.53	0.72	−0.45	0.82	0.00	0.58
4	0.36	0.00	−0.93	0.53	−0.72	0.45	0.82	0.00	−0.58
5	0.58	0.58	0.58	0.85	0.28	0.45	0.00	−0.82	0.58
6	0.58	0.58	−0.58	0.85	−0.28	−0.45	0.00	−0.82	−0.58
7	0.58	−0.58	0.58	0.00	0.00	1.00	−0.82	0.00	0.58
8	0.58	−0.58	−0.58	0.00	0.00	−1.00	−0.82	0.00	−0.58
9	0.93	0.36	0.00	−0.53	−0.72	0.45			
10	0.93	−0.36	0.00	−0.53	0.72	−0.45			
11	0.00	−0.93	0.36	−0.85	−0.28	−0.45			
12	0.00	−0.93	−0.36	−0.85	0.28	0.45			
13	−0.36	0.00	0.93						
14	−0.36	0.00	−0.93						
15	−0.58	−0.58	0.58						
16	−0.58	−0.58	−0.58						
17	−0.58	0.58	0.58						
18	−0.58	0.58	−0.58						
19	−0.93	−0.36	0.00						
20	−0.93	0.36	0.00						

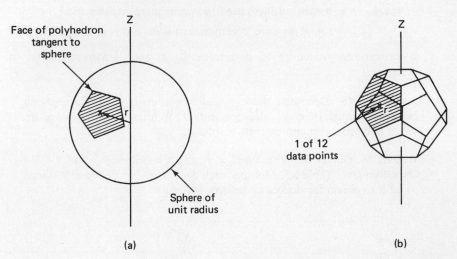

FIGURE 1.5 *Twelve-sided polyhedron with the center of each face tangent to a sphere centered inside the polyhedron*

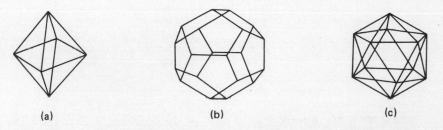

FIGURE 1.6 *Geometry of (a) 8-sided, (b) 12-sided, and (c) 20-sided polyhedron*

Furthermore, one should choose a radial distance from the sound source which is at least twice the largest dimension of the source to assure a valid measurement.

The following step-by-step procedure is used to obtain $\bar{L}_p$:

1. Choose the polyhedron to be used and the distance the measurements are to be taken from the sound source.

2. Look up the normalized coordinates of the midpoints of the faces and multiply each by the distance chosen for the measurements.

3. Measure the sound pressure level at the midpoint of each face in dB.

4. Determine the average sound pressure level by means of equation (1.28):

$$\bar{L}_p = 10 \log \left(\frac{1}{n} \sum_{i=1}^{n} 10^{L_{p_i}/10} \right) \quad \text{dB} \qquad (1.28)$$

where $n =$ number of faces used to approximate a sphere

$L_{p_i} =$ sound pressure level measured at the center of the ith face (dB)

5. Repeat the procedure to determine $\bar{L}_p$ in each frequency band of interest.

EXAMPLE 1.15   Determine the x, y, and z coordinates of the microphone positions required to make measurements 12 ft from a source using an 8-sided polyhedron to approximate a sphere.

SOLUTION   Obtain the normalized x, y, and z coordinates of an 8-sided polyhedron from Table 1.3. Multiply each coordinate by the radial distance $r = 12$ ft to obtain the desired coordinate values.

Position	x ft	y ft	z ft
1	0.00	9.84	6.96
2	0.00	9.84	−6.96
3	9.84	0.00	6.96
4	9.84	0.00	−6.96
5	0.00	−9.84	6.96
6	0.00	−9.84	−6.96
7	−9.84	0.00	6.96
8	−9.84	0.00	−6.96

EXAMPLE 1.16   Assume that at each of the positions of Example 1.15 a sound-pressure-level measurement was made in the 1000-Hz octave band (directivity is, in general, a function of frequency) and that the values were $L_{p_1} = 95$ dB, $L_{p_2} = 90$ dB, $L_{p_3} = 94$ dB, $L_{p_4} = 94$ dB, $L_{p_5} = 90$ dB, $L_{p_6} = 93$ dB, $L_{p_7} = 90$ dB, and $L_{p_8} = 89$ dB. Determine the directivity index, DI, and the directivity factor, Q, for each of the 8 directions associated with the sound-pressure-level measurements.

SOLUTION   Calculate $\bar{L}_p$ by means of equation (1.28),

$$\bar{L}_p = 10 \log \left(\frac{1}{n} \sum_{i=1}^{n} 10^{L_{p_i}/10} \right) \quad \text{dB}$$

which becomes

$$\bar{L}_p = 10 \log \left[\tfrac{1}{8} (10^{95/10} + 10^{90/10} + 10^{94/10} + 10^{94/10} \right.$$
$$\left. + 10^{90/10} + 10^{93/10} + 10^{90/10} + 10^{89/10}) \right] \quad \text{dB}$$

Upon performing the indicated operations,

$$\bar{L}_p = 92.4 \text{ dB}$$

Since the maximum variation in L_p is

$$95 \text{ dB} - 89 \text{ dB} = 6 \text{ dB}$$

we could have used the approximate method, equation (1.30), to determine $\bar{L}_p$. The approximation is

$$\bar{L}_p = \left(\frac{1}{n}\sum_{i=1}^{n} L_{p_i}\right) + 1 \text{ dB}$$

which upon application to this problem yields

$$\bar{L}_p = \tfrac{1}{8}(95 + 90 + 94 + 94 + 90 + 93 + 90 + 89) + 1 \text{ dB}$$

which becomes

$$\bar{L}_p = 92.9 \text{ dB}$$

We note that the exact and approximate methods yielded results within 0.5 dB of each other. For further calculations we shall use $\bar{L}_p = 92$ dB. Next, we calculate the eight values for DI and Q by means of equations (1.37) and (1.38), respectively. We have

$$\text{DI} = L_p - \bar{L}_p \quad \text{dB}$$

and

$$Q = \text{antilog}\left(\frac{L_{p_\theta} - \bar{L}_p}{10}\right)$$

These two equations, in terms of this problem, may be more conveniently expressed by

$$\text{DI}_i = L_{p_i} - \bar{L}_p \quad \text{dB}$$

and

$$Q_i = \text{antilog}\left(\frac{L_{p_i} - \bar{L}_p}{10}\right)$$

where the subscript i indicates the direction of the center of the ith face of the 8-sided polyhedron which was used as a basis for measurements. Using these two equations and tabulating the results, we obtain

Position	DI (dB)	Q
1	+3	2.00
2	−2	0.63
3	+2	1.58
4	+2	1.58
5	−2	0.63
6	+1	1.26
7	−2	0.63
8	−3	0.50

Note that, in general, both DI and Q are functions of direction with respect to the center of the sound source.

D1.15 A study was made of a device in an anechoic chamber. It was decided that 12 measurements would be taken 15 ft from the source in the 2000-Hz band. The results of the test were as follows: $L_{p_1} = 87$ dB, $L_{p_2} = 90$ dB, $L_{p_3} = 92$ dB, $L_{p_4} = 88$ dB, $L_{p_5} = 91$ dB, $L_{p_6} = 90$ dB, $L_{p_7} = 93$ dB, $L_{p_8} = 87$ dB, $L_{p_9} = 91$ dB, $L_{p_{10}} = 90$ dB, $L_{p_{11}} = 92$ dB, and $L_{p_{12}} = 91$ dB. (a) Determine the average sound pressure level 15 ft from the source. (b) Determine the directivity index and the directivity factor at positions 2 and 7.

$$Ans. \quad (a) \ \bar{L}_p = 91 \text{ dB} \quad (b) \ Q_2 = 0.794, \ Q_7 = 1.58.$$

1.4.3.2 Measurements in anechoic hemispherical space

As has been pointed out, measurements about large machines and equipment are made almost exclusively over large reflecting surfaces, outdoors or in large anechoic chambers with reflecting rather than absorbing floors.

Of course, the coordinates listed in Table 1.3 can be used for hemispherical measurements except that we will obviously use only those coordinates for which $z \geq 0$. In certain instances for which $z = 0$, we use only half a face; thus, 3 dB must be subtracted from the sound pressure level for each of these positions.

There is available an array of coordinates that are derived from the midpoints of the 20-sided polyhedron, which has been tilted by 20.9°. This places two of the otherwise halved faces totally above the reflecting plane, and the other two are eliminated. The coordinates for the 10-point array thus derived is given in Table 1.4. This array is convenient for large machines.

TABLE 1.4

Coordinates of midpoints for a 10-point hemispherical array

Position	x/r	y/r	z/r
1	0	0.00	1.00
2	0.13	0.93	0.33
3	0.75	0.58	0.33
4	0.75	−0.58	0.33
5	0.13	−0.93	0.33
6	−0.87	0.36	0.33
7	−0.87	−0.36	0.33
8	−0.33	0.58	0.75
9	0.67	0.00	0.75
10	−0.33	−0.58	0.75

Figure 1.7 depicts a machine, a possible orientation of the coordinate system, and three measurement points as given in Table 1.4 for the adjusted 10-point hemispherical array. Note that the coordinates of each point have been multiplied by the radius; furthermore, we have chosen a radius that is twice the largest dimension of the machine. Inasmuch as we have located the origin of the coordinate system on the floor, it follows that we are assuming the center of the noise source to be at the same location. This is, of course, as noted, only an assumption, but it is one that has been proved to be valid in large numbers of cases. In those instances where this approximation is grossly in error, one must adjust the polyhedron used accordingly.

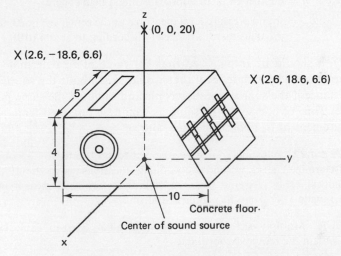

FIGURE 1.7 *Machine with a possible orientation of the coordinate system with measuring points from Table 1.4 ($r = 20$)*

Another problem in many industrial situations is that it will simply not be physically possible to make measurements at each of the prescribed points. Under such conditions, approximations must be made with the best accuracy possible within the practical constraints.

The following step-by-step procedure may be used to obtain $\bar{L}_p$ when a hemispherical configuration is employed:

1. Choose the polyhedron or hemispherical array of points and the distance from the source at which the measurements are to be taken.

2. Multiply each of the normalized coordinates of the midpoints of the faces by the distance chosen for the measurements.

3. Measure the sound pressure level at the midpoint of each face.

4. In those special cases in which a midpoint falls in a half-face at the perimeter of the hemisphere, subtract 3 dB from those measured values of L_{p_i}.

5. Determine the average sound pressure level, $\bar{L}_p$, just as for the spherical case, except now subtract 3 dB, since the measurements were taken only in hemispherical space. That is,

$$\bar{L}_p = \left[10 \log \left(\frac{1}{n} \sum_{i=1}^{n} 10^{L_{p_i}/10} \right) \right] - 3 \quad \text{dB} \qquad (1.39)$$

where n = number of faces approximating a hemisphere

L_{p_i} = sound pressure level measured at the center of the ith face with an adjustment if necessary according to step 4 (dB)

Finally, it should be emphasized that the value of $\bar{L}_p$ determined is the space-average sound pressure level that would exist about the source at the chosen distance if the source were isotropic and radiated the same power as the source. This, of course, is the same as would be obtained from a spherical measurement of $\bar{L}_p$ about the same source.

EXAMPLE 1.17   A sound source is assumed to be located on a concrete floor. Determine the x-, y-, and z-coordinate locations using the 10-point hemispherical array. Assume that the sound-pressure-level measurements are to be made 15 ft from the center of the source.

SOLUTION   Multiply each of the normalized values given in Table 1.4 by 15 ft to obtain the required coordinates.

Position	x ft	y ft	z ft
1	0.00	0.00	15.00
2	1.95	13.95	4.95
3	11.25	8.70	4.95
4	11.25	−8.70	4.95
5	1.95	−13.95	4.95
6	−13.05	5.40	4.95
7	−13.05	−5.40	4.95
8	−4.95	8.70	11.25
9	10.05	0.00	11.25
10	−4.95	−8.70	11.25

EXAMPLE 1.18   Assume that about the source the following measurements were made at the positions determined in Example 1.17: $L_{p_1} = 94$ dB,

$L_{p_2} = 96$ dB, $L_{p_3} = 93$ dB, $L_{p_4} = 98$ dB, $L_{p_5} = 105$ dB, $L_{p_6} = 102$ dB, $L_{p_7} = 98$ dB, $L_{p_8} = 93$ dB, $L_{p_9} = 90$ dB, and $L_{p_{10}} = 91$ dB. Determine the directivity factor, Q, in the direction of the measurement of L_{p_5}.

SOLUTION Calculate $\bar{L}_p$ for hemispherical measurements by

$$\bar{L}_p = \left[10 \log \left(\frac{1}{n} \sum_{i=1}^{n} 10^{L_{pi}/10} \right) \right] - 3 \quad \text{dB} \qquad (1.39)$$

We obtain

$$\bar{L}_p = 10 \log \left[\tfrac{1}{10} (10^{94/10} + 10^{96/10} + 10^{93/10} + 10^{98/10} + 10^{105/10} \right.$$
$$\left. + 10^{102/10} + 10^{98/10} + 10^{93/10} + 10^{90/10} + 10^{91/10}) \right] - 3 \quad \text{dB}$$

which yields

$$\bar{L}_p = 95.6 \approx 96 \text{ dB}$$

We next find the directivity index, DI, in the direction of L_{p_5} to be

$$\text{DI}_5 = L_{p_5} - \bar{L}_p \quad \text{dB} \qquad (1.37)$$
$$= 105 - 96 = 9 \text{ dB}$$

Thus,

$$\text{DI}_5 = 9 \text{ dB}$$

The directivity factor, Q, in this same direction is

$$Q_5 = \text{antilog} \left(\frac{\text{DI}_5}{10} \right) \qquad (1.34)$$

or

$$Q_5 = 10^{\text{DI}_5/10}$$

which becomes

$$Q_5 = 10^{9/10}$$

and yields

$$Q_5 = 7.94$$

Note that once $\bar{L}_p$ is determined, DI and Q are calculated just as for the spherical case.

D1.16 A machine is in a large anechoic chamber that has a reflecting floor. A 10-point hemispherical array is used in determining the average sound pressure level and the directivity at each position. The sound pressure levels measured at each position on the hemisphere are as follows: $L_{p_1} = 95$ dB, $L_{p_2} = 93$ dB, $L_{p_3} = 91$ dB, $L_{p_4} = 92$ dB, $L_{p_5} = 90$ dB, $L_{p_6} = 93$ dB, $L_{p_7} = 91$ dB, $L_{p_8} = 89$ dB, $L_{p_9} = 90$ dB, and $L_{p_{10}} = 91$ dB.

What is (a) the average sound pressure level? (b) the directivity factor in the direction of L_{p_1}? (c) the directivity factor in the direction of L_{p_3}?

Ans. (a) $\bar{L}_p = 89$ dB (b) $Q_1 = 3.98$ (c) $Q_3 = 1.58$.

PROBLEMS

1.1 A certain pure tone sound with a wavelength of 0.2 m produced a sound pressure level of 87 dB. Determine:

(a) the frequency of the tone

(b) its rms pressure (Assume that $c = 344$ m/s.)

1.2 An octave-band analysis was done on an automatic wood lathe in operation. It was found that the octave-band sound pressure levels were 93 dB at 250 Hz, 94 dB at 500 Hz, 96 dB at 1000 Hz, 95 dB at 2000 Hz, 94 dB at 4000 Hz, and 93 dB at 8000 Hz. What is the total mean-square pressure?

1.3 At an operator's position the sound pressure level in the 1-kHz band was found to be 104 dB and 98 dB with the machine "on" and "off", respectively. Determine the rms sound pressure at the operator's position which was produced by the machine only.

1.4 It was found that in free space, $\bar{L}_p = 96$ dB at a distance of 13 ft from a source. It was also determined that $L_{p_{\theta_1}}$ was 102 dB at a distance of 13 ft from the source. What is Q_{θ_1}?

1.5 A particular loudspeaker creates an average sound pressure level of 75.6 dB measured at a distance of 15 ft. The sound pressure levels at three different positions 15 ft from the loudspeaker are measured. What is the directivity factor at each position if the measured levels are:

(a) 78.3 dB?

(b) 80.1 dB?

(c) 83.2 dB?

1.6 In the same area of a warehouse there are four large machines. Machine 1 produces a sound power of 1 W. Machines 2, 3, and 4 produce an acoustical power of 0.5 W, 0.75 W, and 1.25 W, respectively. What is the total sound power level generated in the area by the four machines?

1.7 A particular machine produced the following sound power levels in the octave bands of interest: 78 dB, 84 dB, 94 dB, 96 dB, 95 dB, and 91 dB. What is the total sound power produced by the machine?

1.8 A machine operator in a factory is surrounded by five machines. The machines produce sound pressure levels of 95 dB, 87 dB, 90 dB, 93 dB, and 88 dB, respectively, at the operator's position, not including ambient noise. When the machines are "off", the sound pressure level at his position is 88 dB.

Determine the total sound pressure level at his position due to both the machines and the ambient noise.

1.9 A detailed study was done on a loudspeaker producing a tone at 10 kHz. Twenty points were measured around the speaker, 9 ft away, at positions corresponding to the icosahedron in Figure 1.6(c). The sound pressure levels measured at positions 1–20 were $L_{p_1} = 84$ dB, $L_{p_2} = 82$ dB, $L_{p_3} = 80$ dB, $L_{p_4} = 86$ dB, $L_{p_5} = 84$ dB, $L_{p_6} = 83$ dB, $L_{p_7} = 87$ dB, $L_{p_8} = 86$ dB, $L_{p_9} = 90$ dB, $L_{p_{10}} = 88$ dB, $L_{p_{11}} = 89$ dB, $L_{p_{12}} = 89$ dB, $L_{p_{13}} = 90$ dB, $L_{p_{14}} = 86$ dB, $L_{p_{15}} = 87$ dB, $L_{p_{16}} = 85$ dB, $L_{p_{17}} = 85$ dB, $L_{p_{18}} = 84$ dB, $L_{p_{19}} = 83$ dB, and $L_{p_{20}} = 81$ dB, respectively. Make a table of:

    (a) the microphone positions required to make the measurements

    (b) the directivity index at each position

    (c) the directivity factor at each position

1.10 Directivity measurements were made about a noise source situated on a reflecting pad outdoors. The following 10 sound pressure levels were measured, all with equal weighting, for the hemispherical directivity calculation in the 1-kHz band at a distance of 3 m from the source: 96 dB, 94 dB, 103 dB, 95 dB, 99 dB, 101 dB, 104 dB, 97 dB, 102 dB, and 98 dB. In the θ_1 direction, also at 3 m from the source, $L_{p_{\theta_1}} = 103$ dB. Determine:

    (a) the directivity index

    (b) the directivity factor in the θ_1 direction

BIBLIOGRAPHY

BERANEK, L. L. *Acoustics*, McGraw-Hill Book Company, New York, 1954.

BERANEK, L. L., ed. *Noise and Vibration Control*, McGraw-Hill Book Company, New York, 1971.

DIEHL, G. M. *Machinery Acoustics*, John Wiley & Sons, Inc., New York, 1973.

MAGRAB, E. B. *Environmental Noise Control*, John Wiley & Sons, Inc., New York, 1975.

2

HEARING, HEARING LOSS, AND PSYCHOLOGICAL EFFECTS OF NOISE

2.1 INTRODUCTION

In the course of the study of noise control, we need to briefly consider the organ of hearing, the ear. The perception of sound by the human ear is a complicated process, dependent both on the frequency and the pressure amplitude of the sound. In this chapter we shall consider the structure of the ear and the basic hearing mechanism for the purpose of understanding how excessive noise can cause a hearing loss. Furthermore, we shall discuss various means of measuring the psychological effects of noise by one or more *levels* or other single-number designations.

2.2 THE HUMAN EAR

The human ear, which is no less complicated than the eye, is the transducer that enables man to sense sound waves and thus hear. The main components of the anatomy of the ear and its frequency and amplitude response to sound waves will be discussed briefly.

2.2.1 Anatomy

The main components of the human ear are depicted in Figure 2.1. The ear is commonly divided into three main components: (1) the outer ear, (2) the middle ear, and (3) the inner ear.

The visible portion of the ear, because of its small size compared to the primary wavelengths that we hear, serves only to produce a small enhance-

32

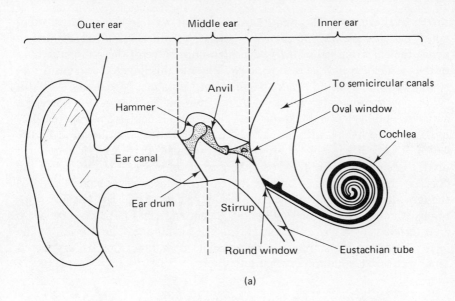

Outer ear   Middle ear   Inner ear

Anvil

To semicircular canals

Hammer

Oval window

Cochlea

Ear canal

Ear drum

Stirrup

Round window   Eustachian tube

(a)

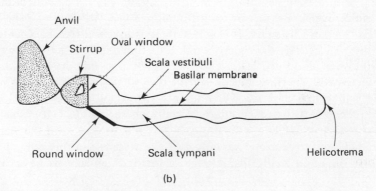

Anvil

Stirrup   Oval window

Scala vestibuli

Basilar membrane

Round window   Scala tympani   Helicotrema

(b)

FIGURE 2.1 (a) *Main components of the human ear;* (b) *the cochlea unwound*

ment of the sounds that arrive from the front of the listener as compared to those which arrive from behind; that is, man's sound reception system has a small frontal directivity. The remainder of the outer ear, which consists of the ear canal terminated in the ear drum, forms a resonant cavity at about 3 kHz, the midrange frequency of hearing. This resonant or near-resonant condition allows for a nearly reflection-free termination of the ear canal and thus a good impedance match of the ear drum to the air in which the sound wave was propagated.

The middle ear consists of three small bones, the hammer, anvil, and stirrup. The *hammer* is attached to the eardrum and the *stirrup* is attached to the oval window. The middle ear serves as an impedance transformer which matches the low impedance of the air in which sound travels and in which the ear drum is located to the high impedance of the lymphatic fluid of the cochlea beyond the oval window. Without this impedance transformation, a mismatch would occur, resulting in a loss of approximately 30 dB.

It is in the inner ear, whose main component is the *cochlea*, where the actual reception of sound takes place. The cochlea is depicted in Figure 2.1(a) in its coiled form as it is in the ear; the unwound cochlea is shown in Figure 2.1(b). The cochlea, which is located in extremely hard temporal bone, is divided almost its entire length by the *basilar membrane*. At the end of the cochlea, the two canals are connected by the *helicotrema*, which allows for the flow of the lymphatic fluid between the two sections. The basilar membrane, which is about 3 cm long and 0.02 cm wide, has about 24,000 nerve ends terminated in hair cells located on the membrane. The motion of the oval window is transmitted to the basilar membrane and its associated sensing cells. This motion is sensed as sound.

2.2.2 Frequency and Loudness Response

The human ear has great sensitivity, as well as a tremendous dynamic range over which it normally functions. The *threshold* of hearing, defined for binaural listening, is that sound pressure in a free field which one can just still hear as the signal is reduced. Looking at it another way, it is the pressure associated with a signal which one first hears as the signal is increased. The threshold of hearing, for what is considered normal hearing, is shown in Figure 2.2. The reference rms sound pressure level, as discussed in Chapter 1, has been internationally chosen as 20×10^{-6} Pa (N/m^2). As seen from the curve, human hearing is most sensitive in the range 2000 to 5000 Hz; furthermore, we note that the response in this range is very close to 0 dB, or 20×10^{-6} Pa. At the other end of the scale, we have the threshold of pain, which is usually taken as about 135 to 140 dB. Thus, we note a dynamic range of normal hearing of approximately 140 dB. Stated another way, our hearing ranges in sensitivity from just not being able to constantly hear thermal noise to being able to accept pressure amplitudes just short of producing physical damage to the hearing mechanism itself.

One also readily notes from the curves of Figure 2.2 that the threshold of hearing is a function of frequency. For example, with normal hearing, one would just be able to hear a 2000-Hz tone at a 0-dB level. However, one would require a pressure level of about 15 dB to be able to just hear a 200-Hz tone. Thus, in describing the subjective loudness of a sound, it is necessary to consider the characteristics of the human ear. This concept of loudness is quantized by the loudness level.

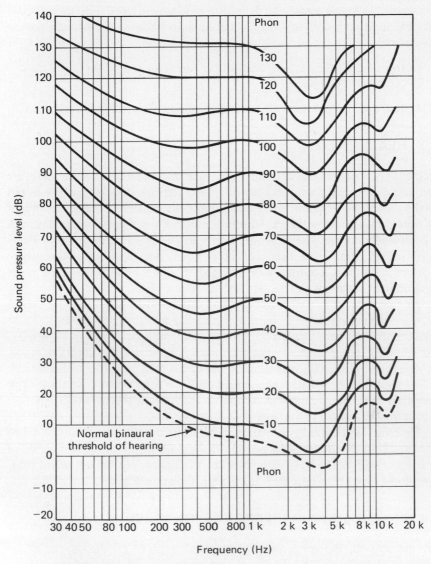

FIGURE 2.2 *Equal-loudness contours for free-field binaural listening*

The *loudness level* of a sound in question is determined by the subjective comparison of the loudness of the sound to that of a 1000-Hz pure tone. The level, measured in phons, is equal numerically to the sound pressure level, in dB, of the 1000-Hz tone, which was deemed to be of equal loudness. A set of internationally standardized equal loudness contours are plotted in

Figure 2.2. In keeping with the definition of loudness level, note that at 1000 Hz, all the equal loudness contours are equal in phons to the sound pressure level in dB.

> **EXAMPLE 2.1** Determine the sound pressure level of a 50-Hz tone with a loudness level of 20 phons.

> **SOLUTION** From Figure 2.2, we find the sound pressure level to be 50 dB.

2.3 HEARING LOSS

Regrettably, but without a doubt, excessive and prolonged noise exposure can and does cause permanent hearing loss. Various theories have been put forth in an effort to characterize and predict the possible damage that might be caused by a given exposure. Absolute proof of any theory concerning such a complex biological phenomenon is virtually impossible to achieve. However, reliable data have been collected which deal with situations where workers have been continuously exposed to more or less the same noise environment for many years.

In summary, it is well established that excessive noise exposure causes permanent hearing damage by destroying the auditory sensor cells. These cells are the hair cells located on the basilar membrane shown in Figure 2.1. Furthermore, other inner ear damage includes harm to the auditory neurons, as well as damage to the structure of the organ of Corti.

In total, the various theories and data have been taken advantage of in establishing the noise-exposure criteria set forth in the federal noise exposure regulations discussed in Chapter 3.

Hearing loss is usually measured in terms of the shift of the threshold of hearing. That is, a permanent hearing loss causes a shift in the threshold of hearing, usually selectively with respect to frequency, to higher sound pressure levels than those accepted as normal, shown in Figure 2.2.

2.4 PSYCHOLOGICAL RESPONSE TO NOISE

In this section, certain generally accepted aspects of the psychological response to noise will be discussed and quantified. Prominent throughout our discussion will be the loudness of the noise in the context of overburdening the ear, simply annoying the listener or interferring with desired speech communication.

2.4.1 Loudness Interpretation

As was discussed in conjunction with Figure 2.2, loudness level is measured in phons. The related quantity, loudness, is measured in sones. A *sone* is defined as the loudness of a 1000-Hz pure tone with a sound pressure level of 40 dB. Upon recalling the definition of loudness level, or by referring to Figure 2.2, one notes that 40 phons have a loudness equal to 1 sone. This relationship may be expressed as

$$S = 2^{(L_L - 40)/10} \quad \text{sones} \qquad (2.1)$$

where S = loudness (sones)

       L_L = loudness level (phons)

or, conversely,

$$L_L = 33.3 \log S + 40 \text{ phons} \qquad (2.2)$$

This relationship between phons and sones is also given in Figure 2.3 for convenience in converting from one to the other.

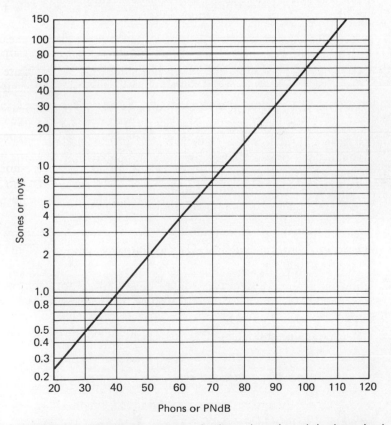

FIGURE 2.3 *Relationship between loudness (sones) and loudness level (phons) or between perceived noise (noys) and perceived noise level (PNdB)*

EXAMPLE 2.2 Make the following two conversions using the appropriate equation, (2.1) or (2.2), and the curve of Figure 2.3: (1) convert 80 phons to sones; (2) convert 100 sones to phons.

SOLUTION To convert phons to sones, use equation (2.1).

$$S = 2^{(L_L - 40)/10} \text{ sones}$$
$$= 2^{(80-40)/10}$$
$$= 2^4 = 16 \text{ sones}$$

Therefore, 80 phons = 16 sones. Note that this result checks with Figure 2.3. To convert sones to phons, use equation (2.2).

$$L_L = 33.3 \log S + 40 \text{ phons}$$
$$= 33.3 \log 100 + 40$$
$$= 66.6 + 40 = 106.6 \text{ phons}$$

Therefore, 100 sones = 106.6 phons. We also note that this result checks with Figure 2.3.

D2.1 Perform the following conversions using the appropriate equations and check your result with the curve of Figure 2.3: (a) 95 phons to sones   (b) 50 phons to sones   (c) 8 sones to phons   (d) 75 sones to phons
  Ans.   (a) 45.25 sones   (b) 2 sones   (c) 70.07 phons   (d) 102.44 phons.

Probably the most widely used method for establishing the loudness of a complex noise is that developed by Stevens [1]. The method is based upon the measurement of the 1-octave-, $\frac{1}{3}$-octave-, or $\frac{1}{2}$-octave band pressure levels. The measured band pressure levels are used in conjunction with the equal loudness index contours shown in Figure 2.4 to determine the loudness or loudness level by means of a simple calculation.

A step-by-step outline of the procedure is as follows:

1. Measure the band pressure levels (1-octave, $\frac{1}{2}$-octave, or $\frac{1}{3}$-octave) over the frequency range of interest. Usually, the range chosen is from about 50 Hz to 10,000 Hz.

2. Enter the center frequency and band pressure level for each band in the contours of Figure 2.4, and determine the loudness index for each band.

3. Calculate the total loudness, S_t, in sones by

$$\boxed{S_t = I_m(1 - K) + K \sum_{i=1}^{n} I_i \text{ sones}} \tag{2.3}$$

where S_t = total loudness (sones)
        I_m = largest of the loudness indices

I_i = each of the loudness indices, including I_m

K = weighting factor for the bands chosen: $K = 0.3$ for 1-octave bands, $K = 0.2$ for $\frac{1}{2}$-octave bands, $K = 0.15$ for $\frac{1}{3}$-octave bands

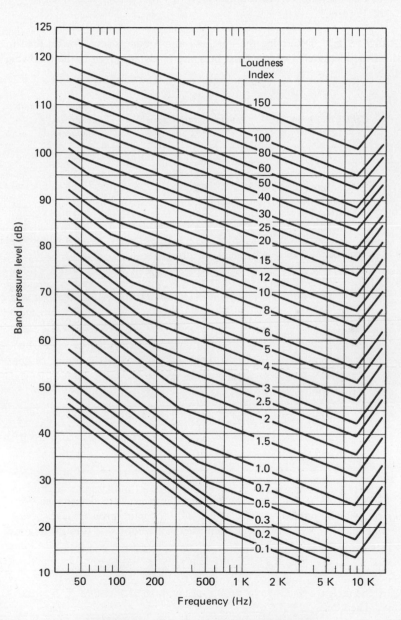

FIGURE 2.4 *Equal loudness index contours* [1]

4. If so desired, one may calculate the loudness level in phons by equation (2.2):

$$L_L = 33.3 \log S_t + 40 \text{ phons} \tag{2.2}$$

or one may convert to loudness level by means of the conversion curve of Figure 2.3.

EXAMPLE 2.3 Determine the loudness and loudness level for an area in which the measured $\frac{1}{3}$-octave band pressure levels are given in the following table.

Center frequency (Hz)	Band pressure level (dB)	Loudness index (sones)
63	65	2.7
80	64	2.5
100	63	2.7
125	62	2.7
160	63	3.7
200	64	4.2
250	65	4.7
315	66	5.3
400	67	6.0
500	68	7.0
630	69	7.7
800	70	8.0
1,000	71	9.0
1,250	72	10.0
1,600	74	13.0
2,000	75	15.0
2,500	77	18.0
3,150	79	22.0
4,000	80	25.0
5,000	82	30.0
6,300	81	30.0
8,000	79	29.0
10,000	77	23.0

SOLUTION As a first step, the loudness indices were determined from Figure 2.4 and recorded in tabular form with the band pressure levels. Next, we note that $\frac{1}{3}$-octave bands have been used. Therefore $K = 0.15$ in equation (2.3), and

$$S_t = I_m(1 - 0.15) + 0.15 \sum_{i=1}^{23} I_i \quad \text{sones}$$

From the table we find that $I_m = 30$ sones and, summing, find $\sum I_i = 281.2$ sones. Therefore,

$$S_t = 30(1 - 0.15) + 0.15(281.2) = 67.68 \text{ sones}$$

We find that the loudness, S_t, is 67.68 sones. The loudness level may now be calculated by means of equation (2.2):

$$L_L = 33.3 \log S_t + 40 \text{ phons}$$
$$= 33.3 \log (67.68) + 40 \text{ phons}$$

Therefore, L_L, the loudness level, is 100.95 phons. Note that one obtains the same value using Figure 2.3. Upon comparing, we find that for this particular complex noise, the loudness is 67.68 sones and the loudness level is 100.95 phons.

D2.2 A particular complex noise was measured to yield the following 1-octave band pressures:

Center frequency (Hz)	Band pressure level (dB)
63	66
125	63
250	65
500	70
1000	73
2000	76
4000	81
8000	79

Determine: (a) the loudness   (b) the loudness level.

Ans.   (a) 50 sones   (b) 96 phons.

2.4.2 Perceived Noise Level

As is to be expected, different individuals may consider the same noise to be more or less annoying. Personal experience and environment clearly play an important role with regard to how much noise individuals will tolerate before they deem a particular noisiness unacceptable. Extensive studies have, however, arrived at certain characteristics of noise which are annoying for most individuals.

It has been found, for example, that if the noise is concentrated in a narrow bandwidth it is considered noisier than if the same energy is spread over a somewhat wider band. Furthermore, a noise with a rapid rise time is found to be noiser that the same noise with a slower rise time; in addition, if the noise is intermittent or irregular, it is deemed noisier than if it remained steady. It has also been determined that high-frequency noise (about 1.5 kHz and up) is judged to be noisier than a lower-frequency noise which is equally loud.

An endeavor has been made to quantify perceived noisiness by means of a set of equal noisiness contours, which are given in Figure 2.5. The unit of noisiness is the *noy* and 1 noy is defined as the perceived noisiness in the band from 910 to 1090 Hz, centered at 1 kHz, with a maximum sound pressure level of 40 dB. The maximum level is approached at a rate of 5 dB/s, remains at the designated 40 dB level for 2 s, and then decreases at this same rate to repeat the cycle as long as desired. Finally, a noise with a noisiness of 3 noys is perceived to be three times as noisy as a noise judged to be 1 noy.

It is often convenient and more practical to express noisiness as a *perceived noise level*, PNL, in units of dB, written as PNdB. The perceived noise level is calculated by first calculating the total noisiness in noys and then correcting to PNdB. The method is analogous to that for determining loudness in sones and converting to loudness level in phons.

The total perceived noisiness is calculated by

$$\text{PN} = N_m(1 - K) + K \sum_{i=1}^{n} N_i \quad \text{noys} \tag{2.4}$$

where   PN = total perceived noise (noys)

N_m = maximum perceived noise in the frequency bands measured (noys)

N_i = each of the band perceived noisinesses, including N_m (noys)

K = weighting factor for the bands chosen: $K = 0.3$ for 1-octave bands, $K = 0.2$ for $\frac{1}{2}$-octave bands, $K = 0.15$ for $\frac{1}{3}$-octave bands

Note that it is necessary to convert from the measured band pressure levels to perceived noise in each band by means of the curves of Figure 2.5 before applying equation (2.4).

Once the total noisiness, PN, in noys has been calculated, one may calculate the perceived noise level, PNL, by

$$\text{PNL} = 33.3 \log (\text{PN}) + 40 \text{ PNdB} \tag{2.5}$$

where   PNL = perceived noise level (PNdB)

PN = perceived noise (noys)

The curve in Figure 2.3 may also be used to make this conversion.

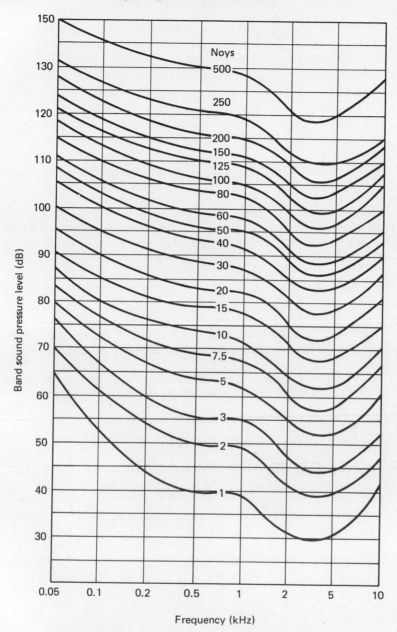

FIGURE 2.5 *Perceived noisiness contours* [2]

EXAMPLE 2.4 Determine the perceived noise and perceived noise level for an area in which the measured 1-octave band pressure levels are given in the following table.

Center frequency (Hz)	Band pressure level (dB)	Perceived noise (noys)
63	75	3
125	77	8
250	82	17
500	84	21
1000	85	26
2000	86	42
4000	81	26
8000	88	42

SOLUTION First, the curves of Figure 2.5 were used to determine the perceived noise for each band pressure level. Since the measurements were made for 1-octave bands, $K = 0.3$ is used in equation (2.4) to obtain

$$PN = N_m(1 - 0.3) + 0.3 \sum_{i=1}^{8} N_i \quad \text{noys}$$

We note from the table that $N_m = 42$ noys and that $\sum N_i = 185$ noys. Therefore,

$$PN = 42(1 - 0.3) + 0.3(185) = 84.9 \approx 85 \text{ noys}$$

Thus, we find the perceived noise, PN, to be 85 noys. The perceived noise level, PNL, in PNdB may now be calculated by equation (2.5),

$$PNL = 33.3 \log (85) + 40 \text{ PNdB}$$
$$= 104.25 \text{ PNdB}$$

The same value of PNL could also have been obtained by using Figure 2.3 to convert the perceived noise to the perceived noise level.

D2.3 A certain noise was measured in 1-octave bands to obtain the following band pressure levels:

Center frequency (Hz)	Band pressure level (dB)
63	76
125	73
250	75
500	80
1000	88
2000	86
4000	82
8000	78

Calculate:　(a) the perceived noise　(b) the perceived noise level.

Ans.　(a) 80 noys　(b) 104 PNdB.

2.4.3　Noise-Criteria Curves

One method of rating the background noise level in a room is accomplished through use of *noise-criteria* (NC) *curves*. These curves, which are shown in Figure 2.6, were established in 1957 for rating indoor noise. Each curve specifies the maximum octave-band sound pressure level for a given

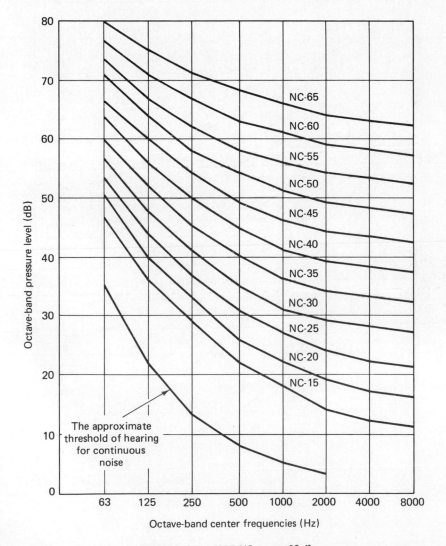

FIGURE 2.6　*1957 NC curves* [3,4]

NC rating. In addition, if the octave-band levels for a given noise spectrum are known, the rating of that noise in terms of the NC curves is given by plotting the noise spectrum on the set of NC curves to determine the point of highest penetration.

As the area of noise control progressed over the years, some objections to the NC curves led to their modification in 1971. The new curves shown in Figure 2.7 are called the *preferred noise-criteria* (PNC) *curves*. Although these curves differ from the NC curves, they are used in exactly the same manner.

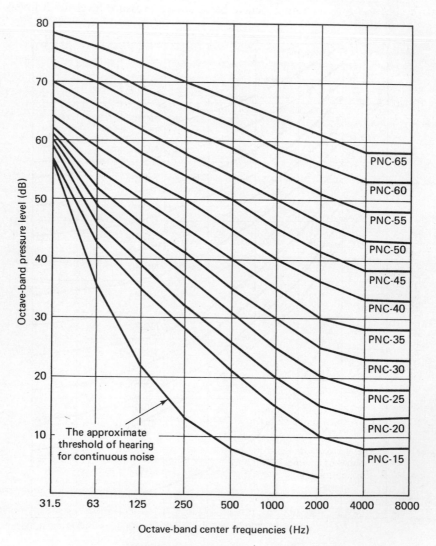

FIGURE 2.7 *1971 PNC curves* [4]

EXAMPLE 2.5　The octave-band noise spectrum in a room was measured and found to be as follows:

Center frequency (Hz)	63	125	250	500	1 k	2 k	4 k	8 k
Band pressure level (dB)	41	45	48	50	46	42	40	38

Determine the NC and PNC ratings of this noise spectrum.

SOLUTION　Plot each of the band pressure levels on the NC curves of Figure 2.6 and PNC curves of Figure 2.7. The point of highest penetration determines the rating in each case. We find that the spectrum penetrates the NC-45 curve at 500 Hz and exceeds it by about 1 dB. This yields an NC rating of NC-46. Similarly, we find that the spectrum penetrates the PNC-45 at 4 kHz and exceeds it by about 2 dB. Thus, the PNC rating is PNC-47. Recommended noise-criteria ranges for steady background noise in certain indoor areas are listed in Table 2.1.

TABLE 2.1

Recommended noise criteria range for steady indoor background noise [6]

	NC curve	PNC curve
1. Sleeping quarters	25–35	25–40
2. Living quarters	35–45	30–40
3. Office or classroom	30–35	30–40
4. Recording studio	15–20	10–20
5. Retail store or restaurant	35–50	35–45
6. Laboratory or engineering	40–45	40–50
7. Computer areas	45–60	45–55

D2.4　Determine both the NC rating and the PNC rating for the following octave-band noise spectrum:

Center frequency (Hz)	63	125	250	500	1 k	2 k	4 k	8 k
Band pressure level (dB)	50	51	48	46	50	53	57	52

Ans.　NC-59, PNC-64.

2.4.4　Sound Level

Sound levels are sound pressure levels that have been weighted according to a particular weighting curve. Weighting curves, and associated sound levels, were developed as a method to better subjectively evaluate the impact of noise upon the human ear.

Three weightings, A, B, and C, were introduced. *A-weighting* was for levels below 55 dB, *B-weighting* was for levels between 55 and 85 dB, and *C-weighting* was for levels above 85 dB. The frequency response and decibel conversions, to or from a flat response, for each of these weightings are given in Figure 2.8 and Table 2.2, respectively.

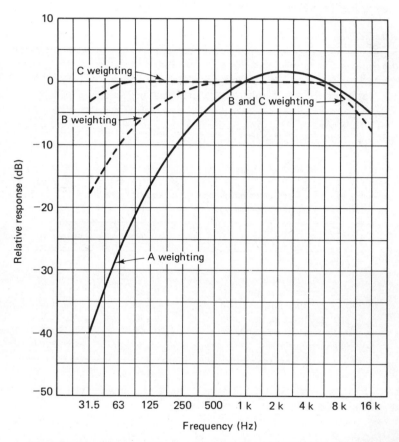

FIGURE 2.8 *Frequency response for the A, B, and C weighting networks*

The A-weighting network is now used almost exclusively in measurements that relate directly to the human response to noise, both from the viewpoint of hearing damage and of annoyance. Such measurements are referred to as *sound level measurements*. Sound level is designated by L and the designated unit is the dBA. In the event that one makes B- or C-weighted measurements, they are designated dBB or dBC, respectively.

In noise-abatement problems it is often necessary to convert calculated

TABLE 2.2

*Sound level conversion chart from flat
response to A, B, and C weightings*

Frequency (Hz)	A weighting (dB)	B weighting (dB)	C weighting (dB)
10	−70.4	−38.2	−14.3
12.5	−63.4	−33.2	−11.2
16	−56.7	−28.5	−8.5
20	−50.5	−24.2	−6.2
25	−44.7	−20.4	−4.4
31.5	−39.4	−17.1	−3.0
40	−34.6	−14.2	−2.0
50	−30.2	−11.6	−1.3
63	−26.2	−9.3	−0.8
80	−22.5	−7.4	−0.5
100	−19.1	−5.6	−0.3
125	−16.1	−4.2	−0.2
160	−13.4	−3.0	−0.1
200	−10.9	−2.0	0
250	−8.6	−1.3	0
315	−6.6	−0.8	0
400	−4.8	−0.5	0
500	−3.2	−0.3	0
630	−1.9	−0.1	0
800	−0.8	0	0
1,000	0	0	0
1,250	+0.6	0	0
1,600	+1.0	0	−0.1
2,000	+1.2	−0.1	−0.2
2,500	+1.3	−0.2	−0.3
3,150	+1.2	−0.4	−0.5
4,000	+1.0	−0.7	−0.8
5,000	+0.5	−1.2	−1.3
6,300	−0.1	−1.9	−2.0
8,000	−1.1	−2.9	−3.0
10,000	−2.5	−4.3	−4.4
12,500	−4.3	−6.1	−6.2
16,000	−6.6	−8.4	−8.5
20,000	−9.3	−11.1	−11.2

1-octave-band or $\frac{1}{3}$-octave-band sound pressure levels to a total sound level in dBA. This is usually required to ascertain if a particular abatement technique and chosen materials yield a satisfactory solution when evaluated in terms of the worker's permissible sound level.

EXAMPLE 2.6 Determine the total A-weighted sound level, L, of the following set of octave-band sound pressure levels:

Band center frequency (Hz)	Sound pressure level (dB)
31.5	74
63	66
125	71
250	61
500	60
1,000	75
2,000	82
4,000	80
8,000	87
16,000	90

SOLUTION   Refer to Table 2.2 for the dB conversion from a flat response to dBA for each of the octave bands. This yields:

$$74 \text{ dB at } 31.5 \text{ Hz} = 74 - 39.4 = 34.6 \text{ dBA}$$
$$66 \text{ dB at } 63 \text{ Hz} = 66 - 26.2 = 39.8 \text{ dBA}$$
$$71 \text{ dB at } 125 \text{ Hz} = 71 - 16.1 = 54.9 \text{ dBA}$$
$$61 \text{ dB at } 250 \text{ Hz} = 61 - 8.6 = 52.4 \text{ dBA}$$
$$60 \text{ dB at } 500 \text{ Hz} = 60 - 3.2 = 56.8 \text{ dBA}$$
$$75 \text{ dB at } 1000 \text{ Hz} = 75 + 0 = 75.0 \text{ dBA}$$
$$82 \text{ dB at } 2000 \text{ Hz} = 82 + 1.2 = 83.2 \text{ dBA}$$
$$80 \text{ dB at } 4000 \text{ Hz} = 80 + 1.0 = 81.0 \text{ dBA}$$
$$87 \text{ dB at } 8000 \text{ Hz} = 87 - 1.1 = 85.9 \text{ dBA}$$
$$90 \text{ dB at } 16,000 \text{ Hz} = 90 - 6.6 = 83.4 \text{ dBA}$$

We now sum the dBA in each of the bands for the total sound level, L, according to equation (1.19):

$$L_t = 10 \log \left(\sum_{i=1}^{n} 10^{L_i/10} \right) \quad \text{dBA}$$

(Note that L simply replaces L_p in this equation.) Expanding and substituting, we obtain

$$L_t = 10 \log (10^{34.6/10} + 10^{39.8/10} + 10^{54.9/10} + 10^{52.4/10} + 10^{56.8/10}$$
$$+ 10^{75/10} + 10^{83.2/10} + 10^{81.0/10} + 10^{85.9/10} + 10^{83.4/10}) \quad \text{dBA}$$
$$= 89.9 \text{ dBA}$$

D2.5   Determine the total A-weighted sound level from the following octave-band sound pressure levels:

Band center frequency (Hz)	Sound pressure level (dB)
31.5	79
63	76
125	84
250	84
500	92
1,000	90
2,000	96
4,000	98
8,000	81
16,000	76

Ans.   101.8 dBA.

2.4.5 Speech Interference Level

It is often necessary to determine the effect of steady background noise on speech communication in the work environment. The *speech interference level*, SIL, was established to assist in such evaluations under the condition that no speech sounds are reflected back to the listener. With the acceptance of the new designation of the octave bands, as used throughout this book, this same basic measure was adapted to the new octave bands and renamed the *preferred speech interference level*, PSIL.

By definition, the preferred speech interference level is

$$\boxed{\text{PSIL} = \frac{L_{p_{500}} + L_{p_{1000}} + L_{p_{2000}}}{3} \quad \text{dB}} \tag{2.6}$$

where $L_{p_{500}}$, $L_{p_{1000}}$, and $L_{p_{2000}}$ = the octave-band pressure levels at 500 Hz, 1000 Hz, and 2000 Hz, respectively

The curves in Figure 2.9, based on data from Beranek [4], can be used to aid in evaluating a steady noise interference with respect to a male speaker. In the case of a female speaker, one should reduce the level of the curves by 5 dB.

Another, and often more convenient, way to evaluate speech interference is to evaluate the speech communication against the A-weighted background noise level. The set of curves in Figure 2.10 rates the voice requirement for effective speech communication with various A-weighted noise background levels.

EXAMPLE 2.7   At a particular location, the following octave-band sound pressure levels were measured: $L_{p_{500}} = 45$ dB, $L_{p_{1000}} = 56$ dB, and $L_{p_{2000}}$

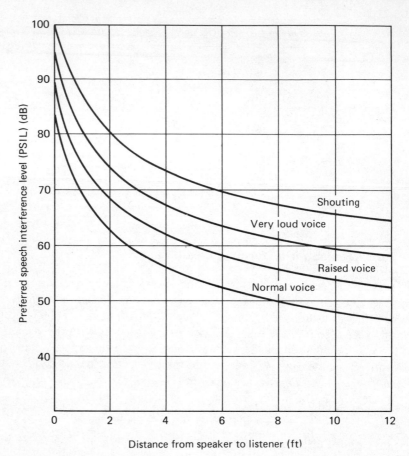

Distance from speaker to listener (ft)

FIGURE 2.9 *Expected voice levels required for speech communication (male voice) for various preferred speech interference levels and separations between speaker and listener* [4]

= 59 dB. The A-weighted total sound level was 65 dBA. A male speaker is 9 ft from the listener. Determine (a) PSIL, (b) expected voice level using Figure 2.9, and (c) expected voice level from Figure 2.10.

SOLUTION Employing equation (2.6), we obtain

$$\text{PSIL} = \frac{45 + 56 + 59}{3} = 53.3 \text{ dB}$$

Thus, PSIL $\approx$ 53 dB. From PSIL = 53 dB and Figure 2.9, we find that a raised voice is required. Now upon using $L = 65$ dBA and Figure 2.10, we note that for this same case (9 ft = 2.7 m), as is to be expected, a raised voice is necessary.

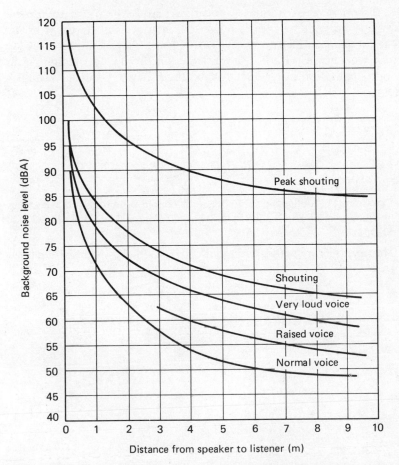

Distance from speaker to listener (m)

FIGURE 2.10 *Expected voice levels required for speech communication with various A-weighted background noise levels and separations between speaker and listener [5]*

D2.6 At a certain location the following measured data were obtained: $L_{p_{500}} = 55$ dB, $L_{p_{1000}} = 64$ dB, and $L_{p_{2000}} = 65$ dB, in the designated octave bands; furthermore, $L = 70$ dBA was the total sound level. Determine: (a) the PSIL (b) the voice level expected using Figure 2.9 and (c) Figure 2.10 if a male speaker is located 8 ft from the listener.

                    Ans.   (a) and (b) 62 dB, very loud   (c) very loud voice.

2.4.6 Masking

As discussed in the previous section, one cannot effectively use speech communication in an environment in which the background noise level is too high. That is, in such instances speech is masked by the background noise. In certain applications, however, it may be desirable to provide a background noise that serves to mask speech at a distance of 6 ft or more. This technique provides some measure of speech‚ privacy in large areas where it would otherwise not be possible. Obviously, this technique finds only rare application in production-type industries. For further details, see reference [7].

PROBLEMS

2.1   Determine the loudness level for each of the following tones, all with a sound pressure level of 80 dB: 50 Hz, 100 Hz, 200 Hz, 500 Hz, 1000 Hz, 3000 Hz, 5000 Hz.

2.2   Measurements in a particular area yielded the following $\frac{1}{3}$-octave band pressure levels:

Center frequency (Hz)	Band pressure level (dB)
63	75
80	74
100	73
125	72
160	73
200	74
250	75
315	76
400	77
500	78
630	79
800	80
1,000	81
1,250	82
1,600	84
2,000	85
2,500	87
3,150	89
4,000	90
5,000	92
6,300	91
8,000	89
10,000	87

Determine the loudness and the loudness level in the area.

2.3 Determine the perceived noise and the perceived noise level for the area investigated in Problem 2.2.

2.4 In a certain area, the following 1-octave band pressure measurements were made:

Center frequency (Hz)	Band pressure level (dB)
63	86
125	83
250	85
500	80
1,000	88
2,000	86
4,000	82
8,000	88

Determine the loudness and the loudness level.

2.5 Determine the perceived noise and the perceived noise level for the area studied in Problem 2.4.

2.6 Determine the NC and PNC ratings of the following octave-band noise spectrum:

Center frequency (Hz)	63	125	250	500	1 k	2 k	4 k	8 k
Band pressure level (dB)	51	55	58	60	56	52	50	48

2.7 The following sound pressure levels were recorded in octave bands:

Band center frequency (Hz)	Sound pressure level (dB)
31.5	90
63	92
125	94
250	88
500	86
1,000	88
2,000	92
4,000	90
8,000	81
16,000	78

From these data, determine:
- (a) The total flat response, L_p
- (b) The total A-weighted response
- (c) The total B-weighted response
- (d) The total C-weighted response

2.8 Determine the preferred speech interference level (PSIL) for each of the following noise spectrums:
- (a) Problem 2.4
- (b) Problem 2.6
- (c) Problem 2.7

Also, determine the level of male voice required (if at all possible) to communicate over a distance of 4 ft.

2.9 Determine the voice level required to communicate over a distance of 4 ft with a background noise level of
- (a) 85 dBA
- (b) 65 dBA
- (c) 75 dBA

REFERENCES

[1] "Procedure for the Computation of Loudness of Noise," *American National Standard USAS S3.4-1968*, American National Standards Institute, New York, 1968.

[2] MILLER, J. "Effects of Noise on People," *Report No. NTID 300.7*, Environmental Protection Agency, Washington, D.C., Dec. 31,1971.

[3] BERANEK, L. L. *Noise Reduction*, McGraw-Hill Book Company, New York, 1960, Chap. 20.

[4] BERANEK, L. L. *Noise and Vibration Control*, McGraw-Hill Book Company, New York, 1971, Chap. 18.

[5] "Information of Levels of Environmental Noise Requisite to Public Health and Welfare with an Adequate Margin of Safety," *Report No. 550/9-74-004*, Environmental Protection Agency, Washington, D.C., Mar. 1974.

[6] BERANEK, L. L., W. E. BLAZIER and J. J. FIGWER "Preferred Noise Criterion (PNC) Curves and Their Application to Rooms," *J. Acoust. Soc. Amer.*, Vol. 50, No. 5 (Nov. 1971), pp. 1223–1228.

[7] MEYER, E. and E. NEUMANN *Physical and Applied Acoustics*, Academic Press, Inc., New York, 1972, pp. 246–250.

BIBLIOGRAPHY

BERANEK, L. L.   *Acoustics*, McGraw-Hill Book Company, New York, 1954.

BERANEK, L. L.   *Noise and Vibration Control*, McGraw-Hill Book Company, New York, 1971.

BERANEK, L. L.   *Noise Reduction*, McGraw-Hill Book Company, New York, 1960.

DIEHL, G. M.   *Machinery Acoustics*, John Wiley & Sons, Inc., New York, 1973.

MAGRAB, E. B.   *Environmental Noise Control*, John Wiley & Sons, Inc., New York, 1975.

3

NOISE CONTROL CRITERIA AND REGULATIONS

3.1 INTRODUCTION

Over the past several years considerable progress has been made to develop and implement criteria and regulations for the control of environmental noise. The term *environmental noise* is used here in the broad sense to encompass noise in the workplace and the community. From an industrial standpoint we will consider the primary noise sources to be surface transportation, air transportation, construction/industrial equipment, and the physical presence of an industry within the community. The latter category can, of course, be related to construction/industrial equipment. In the remainder of this chapter we shall ignore air transportation as a noise source and will concern ourselves only with those aspects of surface transportation that relate to an industrial plant within a community.

The federal laws governing noise which are germane to our discussion are [1]:

- National Environmental Policy Act of 1969
- Noise Pollution and Abatement Act of 1970
- Occupational Safety and Health Act of 1970
- Noise Control Act of 1972

and the primary agencies within the federal government that are designated to issue noise control regulations under federal laws are the Environmental Protection Agency (EPA), the Department of Labor (DOL), and the Department of Transportation (DOT).

Under the National Environmental Policy Act of 1969, any construction project involving federal funds is required to prepare and submit an Environmental Impact Statement (EIS), which should assess the impact on the public of the noise to be generated by the item to be constructed.

The Noise Pollution and Abatement Act of 1970 created within the Environmental Protection Agency the Office of Noise Abatement and Control (ONAC), which has the complete responsibility for investigating the effects of noise on public health and welfare.

The Occupational Safety and Health Act of 1970 was enacted to ensure that working men and women have safe and healthful working conditions.

Perhaps the most important single piece of federal noise legislation is the Noise Control Act of 1972, which gives the Environmental Protection Agency the primary responsibility for safeguarding sound levels in the community. The act provides for a division of powers among the federal, state, and local governments. The federal government is primarily responsible for noise-source emission control. The remaining two governing bodies have the right to establish and enforce controls on environmental noise through the regulation of noise sources—their use, operation, and movement—and by establishing noise levels permitted in their environments.

In the remainder of this chapter the important aspects and ramifications of these regulations which are related to the interface between industry and the community will be discussed.

3.2 THE OCCUPATIONAL SAFETY AND HEALTH ACT OF 1970

The Department of Labor first promulgated an occupational noise exposure standard under the Walsh–Healey Public Contracts Act in 1969. In 1970, the Occupational Safety and Health Act extended the applicability of the requirements of the standard to all workers engaged in interstate commerce. The permissible noise exposures allowed under the standard are shown in Table 3.1.

It is necessary to determine if the noise dose to which a worker has been exposed is excessive. The most viable instrument for such applications, and one that will no doubt be employed extensively in industrial environments, is the portable noise-exposure meter. This device is a small unit that is attached to the worker and thus is capable of determining the noise exposure of a roving worker or one who functions in a fluctuating noise environment. Basically, this device, which consists of a sound level metering section and an integrating section, measures the noise exposure and determines the percentage of the daily allowable noise dose.

When the noise-exposure standard expressed by Table 3.1 has been exceeded, feasible administrative or engineering controls must be employed

TABLE 3.1

Permissible noise exposures*

Duration/day (h)	Sound level (dBA), slow response
16	85
8	90
6	92
4	95
3	97
2	100
1.5	102
1	105
0.5	110
0.25 or less	115

Standard for OSHA 1970

*If variations in noise level involve maxima at intervals of 1 s or less, the noise is considered to be continuous. In addition, the exposure to impulse noise should not exceed 140–dB peak sound pressure level.

to reduce the sound levels to within the values shown in Table 3.1. If such controls fail, personal protective equipment must be used to accomplish the objectives of the standard.

The daily noise exposure may be composed of more than one period of noise exposure at different levels. When this occurs, the combined effect of the various periods must be considered rather than the individual effect of each. This mixed exposure is considered to be within the limit of the standard if the value of the daily noise dose, D, defined by the following expression, does not exceed unity:

$$D = \frac{C_1}{T_1} + \frac{C_2}{T_2} + \cdots + \frac{C_n}{T_n}$$

(3.1)

where　C_n = total actual time of exposure at a specified noise level

　　　T_n = total time of exposure permitted by Table 3.1 at that level

EXAMPLE 3.1　It has been found that a group of industrial employees are exposed to noise according to the following schedule:

Exposure level (dBA)	Period of exposure (hr)
85	3
90	2
92	1
95	2

Does this daily noise dose exceed the OSHA standard?

SOLUTION   Since noise levels below 90 dBA are not part of the OSHA standard, the daily noise dose is

$$D = \tfrac{2}{8} + \tfrac{1}{6} + \tfrac{2}{4}$$
$$= 0.91667$$

D is less than unity, and therefore the employees do not exceed the noise-exposure standard. However, suppose that the standard is changed to include all the values in Table 3.1 (i.e., the addition to the standard of 85 dBA for a 16-hr period). If this were done, the value of the daily noise dose would be

$$D = \tfrac{3}{16} + \tfrac{2}{8} + \tfrac{1}{6} + \tfrac{2}{4}$$
$$= 1.10417$$

In this case, the employees would exceed the new standard.

The current standard may be deemed adequate. If it is not, however, the areas of disagreement will involve such aspects as the cost of compliance for 85 dBA versus 90 dBA, the protection afforded workers for 85 dBA versus 90 dBA for 8 hrs of continuous exposure, the use of hearing protectors rather than engineering controls as an acceptable long-term solution, and similar issues.

3.3 NOISE CONTROL ACT OF 1972

With passage of the Noise Control Act of 1972, Congress made it the policy of the United States to promote for all its citizens an environment that is free from noise that would have an adverse effect on their health, safety, and welfare. To accomplish this task, the Environmental Protection Agency was given the responsibility to approach the noise problem through a number of avenues. For example, EPA's responsibilities include the coordination of federal research in noise control, the identification of major noise sources and the development for them of noise criteria and control technology, the establishment of federal noise-emission standards for products distributed in commerce, and the development of low noise-emission products. A detailed summary of the Noise Control Act of 1972 prepared by Charles E. Hickman, Southern Company Services, Inc., Birmingham, Alabama, is given as Appendix C.

In the following section we shall examine some of the EPA programs that have a direct impact on industry.

3.4 PERFORMANCE INDICES
FOR ENVIRONMENTAL NOISE

In partial fulfillment of its obligations under the Noise Control Act of 1972, the EPA has issued two important documents on environmental noise. The first document [2], published in July 1973, deals with "criteria," and the

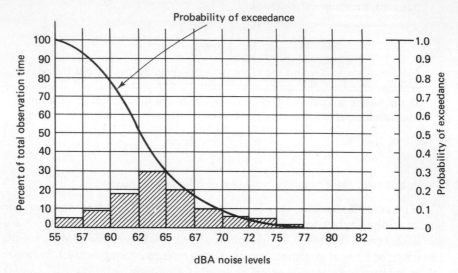

FIGURE 3.1 *Histogram of plant noise at a particular property boundary*

second document [3], published in April 1974, is concerned with "levels." Both of these publications were required by Section 5 of the Noise Control Act of 1972 as outlined in Appendix C.

The typical criteria employed to describe time-varying community noise take both level and duration into consideration. For simplicity, these performance indices are single-number criteria. In the following material four measures for quantifying noise exposures that have good correlation with human response will be presented.

The four performance indices discussed are based upon statistical data and employ A-weighted measurements. The indices are L_N, which represents noise levels exceeded N percent of the time; L_{eq}, which is the equivalent continuous sound level in dBA; L_{NP}, the noise pollution level; and L_{dn}, the day–night average sound level in dBA.

L_N is determined with an instrument that employs an amplitude distribution analyzer. In Figure 3.2 is shown a histogram derived from the output of such an instrument. The time in any particular amplitude band may be read as a percentage of the total time of observation. In addition, the probability-of-exceedance curve (cumulative distribution curve) indicates the level exceeded any chosen percentage of time. In environmental noise abatement and planning programs, criteria are often employed that require the sound levels exceeded 10, 50, and 90% of the time. These levels are usually listed as L_{10}, L_{50}, and L_{90}.

EXAMPLE 3.2 From the histogram in Figure 3.1, taken at a particular plant property line, determine the sound level that is exceeded (a) 10% of the time,

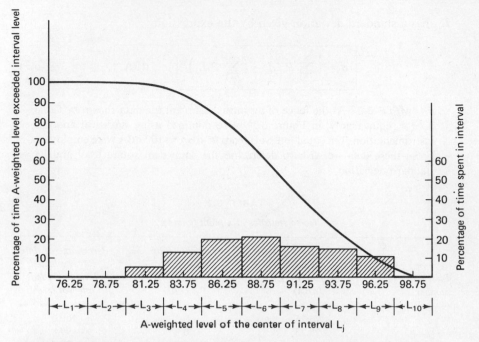

FIGURE 3.2 *Histogram and cumulative distribution function for data tabulated in Table 3.2*

(b) 50% of the time, (c) 90% of the time. Also determine the percentage of the total observation time that the sound level was between 62 and 65 dBA.

SOLUTION Refer to Figure 3.1 and read the following from the histogram using the probability-of-exceedance curve.

(a) $L_{10} \approx 69$ dBA, exceeded 10% of the time
(b) $L_{50} \approx 63$ dBA, exceeded 50% of the time
(c) $L_{90} \approx 58$ dBA, exceeded 90% of the time

From the bar-graph portion of the histogram, the amplitude band 62 to 65 dBA was recorded for 30% of the observation time.

L_{eq} is the sound energy averaged over a stated period of time; that is, it is the rms (root mean square) or mean level of the time-varying noise. It is employed by EPA as an environmental noise descriptor. This performance index is calculated from amplitude distributions measured with statistical analysis instrumentation. The equivalent continuous sound level in dBA is computed from the following equation [4]:

$$L_{eq} = 10 \log\left[\sum_{j=1}^{N} (P_j)(10^{L_j/10}) \right] \quad \text{dBA} \tag{3.2}$$

where N = number of intervals

P_j = fraction of time spent in interval j

L_j = A-weighted level of the center of interval j

L_{eq} has a standard deviation given by the expression

$$\sigma = \left[\sum_{j=1}^{N} P_j L_j^2 - \left(\sum_{j=1}^{N} P_j L_j \right)^2 \right]^{1/2} \quad \text{dBA} \qquad (3.3)$$

EXAMPLE 3.3   At the fence of an industrial plant the data shown in Table 3.2 (or, equivalently, in Figure 3.2) were obtained using statistical analysis instrumentation. Ten equal intervals from 75 dBA to 100 dBA were employed. From these data we wish to determine the equivalent sound level and its standard deviation.

TABLE 3.2

Noise statistics at a plant boundary

Interval number	A-weighted level of the center of interval, L_j (dBA)	Percentage of time spent in interval, $P_j \times 100$	Percentage of time A-weighted level exceeded interval level
1	76.25	0.2	100.0
2	78.75	0.3	99.8
3	81.25	5.0	99.5
4	83.75	12.4	94.5
5	86.25	19.6	82.1
6	88.75	21.3	62.5
7	91.25	16.7	41.2
8	93.75	13.8	24.5
9	96.25	10.2	10.7
10	98.75	0.5	0.5

SOLUTION   Using equation (3.2), the equivalent sound level can be computed as

$$L_{eq} = 10 \log [(0.002)(10^{7.625}) + (0.003)(10^{7.875})$$
$$+ (0.05)(10^{8.125}) + (0.124)(10^{8.375}) + (0.196)(10^{8.625})$$
$$+ (0.213)(10^{8.875}) + (0.167)(10^{9.125}) + (0.138)$$
$$(10^{9.375}) + (0.102)(10^{9.625}) + (0.005)(10^{9.875})]$$
$$= 10 \log (1.2963 \times 10^9)$$
$$= 91.127 \text{ dBA}$$

The standard deviation can be computed from (3.3) as

$$\sigma = (7962.938 - 7945.034)^{1/2}$$
$$= 4.231 \text{ dBA}$$

D3.1　From the data in Example 3.3, determine the L_N values L_{10}, L_{50}, and L_{90}.
　　　　　Ans.　$L_{10} = 96.25$ dBA, $L_{50} = 90.5$ dBA, $L_{90} = 85.0$ dBA.

　　The noise rating scheme which was devised to estimate the annoyance of fluctuating community noise is the noise pollution level, L_{NP}. This index is based upon two terms, one of which represents the equivalent continuous noise level and the other the annoyance due to fluctuations. The L_{NP} is defined by the following equation [4]:

$$\boxed{L_{NP} = L_{eq} + K\sigma \quad dB_{NP}}$$ (3.4)

where　$K = $ constant (normally chosen to be 2.56 because this particular value gives good correlation with available data of subjective response to noise)

　　In many community noise situations of interest, L_{NP} can be approximated by the following expression:

$$L_{NP} = L_{eq} + d \qquad dB_{NP}$$ (3.5)

or

$$L_{NP} = L_{50} + d + \frac{d^2}{60} \quad dB_{NP}$$ (3.6)

where $d = L_{10} - L_{90}$

EXAMPLE 3.4　From the data in Example 3.3, compute the L_{NP} using equation (3.4).

SOLUTION

$$L_{NP} = L_{eq} + 2.56\sigma \quad dB_{NP}$$
$$= 91.127 + (2.56)(4.231)$$
$$= 101.958 \ dB_{NP}$$

　　From the approximations (3.5) and (3.6), we obtain, using the values from D3.1,

$$L_{NP} = L_{eq} + L_{10} - L_{90} \quad dB_{NP}$$
$$= 91.127 + 96.25 - 85.0$$
$$= 102.52 \ dB_{NP}$$

and

$$L_{NP} = L_{50} + d + \frac{d^2}{60} \quad dB_{NP}$$
$$= 90.5 + 96.25 - 85.0 + \frac{(96.25 - 85.0)^2}{60}$$
$$= 103.86 \ dB_{NP}$$

respectively.

EXAMPLE 3.5 From the previous material, derive an approximate expression for L_{eq}.

SOLUTION Using equations (3.5) and (3.6), it can be seen that

$$L_{eq} \approx L_{50} + \frac{d^2}{60} \quad \text{dBA} \tag{3.7}$$

D3.2 Using equation (3.7), obtain an approximate value for L_{eq} for the data given in Example 3.3.

Ans. $L_{eq} \approx 92.61$ dBA.

Finally, L_{dn}, the day–night A-weighted average sound level introduced by the EPA [4,5] represents a 24-hr L_{eq} with a 10-dB nighttime penalty. This index is devised from the *community noise equivalent level* (CNEL). It is useful in predicting the effects on a population subjected to a long-term environmental noise. In addition, it can be used for such things as cost–benefit studies, enforcement, monitoring, and planning. The index is computed using the expression

$$L_{dn} = 10 \log \{\tfrac{1}{24}[(15)(10^{L_d/10}) + (9)(10^{(L_n+10)/10})]\} \quad \text{dBA} \tag{3.8}$$

where $L_d = L_{eq}$ in the 15-hr daytime period 0700–2200 hr

$L_n = L_{eq}$ in the 9-hr nighttime period 2200–0700 hr

EXAMPLE 3.6 From noise measurements made in a particular community, it has been determined that the daytime L_{eq} is 78.2 dBA and the nighttime L_{eq} is 59.3 dBA. Using these data, determine the day–night A-weighted average sound level.

SOLUTION From (3.8),

$$L_{dn} = 10 \log_{10} \{\tfrac{1}{24}[(15)(10^{7.82}) + (9)(10^{6.93})]\} \quad \text{dB}$$
$$= 76.482 \text{ dBA}$$

3.5 MAJOR SOURCES OF NOISE

On May 28, 1975 [6], the EPA published a list of product types that are to be considered possible candidates for identification as major noise sources. The list is categorized into the following major subdivisions: Surface Transportation, Air Transportation, Construction/Industrial Equipment, Recreational Vehicles, Lawn Care, and Household Appliances. Some of these items have a direct impact on the relationship between industry and the community. For example, trucks are perhaps one of the most pervasive sources of noise and therefore present a very serious noise problem. Trucks are extensively employed by industry to facilitate the flow of raw materials and products. As such, these vehicles play a paramount role in the interrelationship between industry and the community.

On October 29 and 30, 1974, EPA issued two documents [7,8] which deal with truck noise. The first is the final noise emission standard for the motor carriers engaged in interstate commerce. In general, the sound level should not exceed 86 dBA at speeds less than 35 mph and 90 dBA at speeds in excess of 35 mph. The noise measurements [9] should be made on an open site with a fast meter response and at a distance of 50 ft perpendicular to the center line of travel. The second document recommends a schedule for the reduction of noise emissions for medium- and heavy-duty trucks as follows: 83 dBA in 1977, 80 dBA in 1981, and 75 dBA in 1983. Measurements are to be made in accordance with the Society of Automotive Engineers (SAE) Standard J366B. DOT is also engaged in programs to reduce truck noise. Their goal is to reduce truck noise to the range 75 to 78 dBA. For a detailed discussion of the noise sources associated with trucks (e.g., diesel engine noise and tire noise), the reader is referred to [10,11].

Building/construction equipment and home applicances, which are manufactured and/or employed by industry, are also potential major sources of noise. Representative A-weighted levels have been obtained for a wide variety of these devices and are contained in reference [12]. Another discussion of these noise sources, which contains their particular characteristics, an evaluation of their impact, and the industrial position relative to noise control for them, is contained in the article by Bender [13]. Also, a brief discussion of these sources is given in Chapter 5.

3.6 NOISE FROM INDUSTRIAL PLANTS

An industry in a community generates noise from a number of sources, both interior and exterior to the plant building. Some industrial plants generate even higher sound levels at nighttime than they do during the daytime hours. Whether the plant is a serious noise problem depends upon a number of factors, such as location—is it in an urban environment where transportation noise will mask plant noise, or is it in a rural community where the ambient noise level is generally low?

Other typical factors that affect community noise complaints are the magnitude of the noise, the background noise level, the character of the noise (i.e., broadband or tonal in nature), and whether the noise is steady or intermittent. There are also subjective factors, such as the community's history of prior exposure, the attitude of the local residents toward the owner of the noise source, and the like.

In general, industry is very sensitive to public relations and makes every effort to maintain a "good neighbor" image. They employ a number of devices, materials, and techniques, as outlined in the following chapters, for industrial noise abatement in the community. In fact, the plant design and layout are often done with noise control in mind. Nevertheless, except in some isolated instances, industiral noise problems are usually of much less importance than noise problems created by transportation systems. However, as the noise levels in the workplace are reduced through regulations, and the general public becomes better informed of the potential consequences of noise exposure, the importance of an industry's contributions to the overall noise level in the community will no doubt increase.

An assessment of the impact of an industry on the community can be obtained with modern measurement instrumentation. This instrumentation provides the capability to obtain some quantitative measure of the effects of noise through various rating systems. The rating schemes that are perhaps most commonly used employ L_N, L_{NP}, and L_{dn}.

3.7 TRENDS IN
STATE AND LOCAL GOVERNMENTS

State and local noise regulations will have an enormous impact on industry, and many states and municipalities within the United States have been active in promulgating these regulations for some time. Although noise regulations at the state and local level appear to be in a constant state of flux, the trend is clear. A number of states already have statewide noise restrictions [14], and every year more local governments enact legislation with quantitative noise emission limits. For example, the number of municipalities with noise regulations increased 21% from 1975 to 1976 [15].

The typical measurement parameters employed in state and local regulations have already been discussed. However, it is important to note that the regulations also generally contain restrictions on pure tones.

At present, noise regulations throughout the country are quite diverse in nature. Some regulations specify property-line noise limits while others dictate permissible noise limits at residential boundaries. Other regulations specify time-of-day maximum permissible noise levels in dBA for various zones. Typical zone classifications are residential, commercial, light industrial, and industrial. More restrictive ordinances classify land according to use and then specify the allowable octave-band sound pressure levels (dB) of sound emitted to any land class from another land class. Noise regulations are being incorporated into building codes in some cities. There are exceptions in some regulations for such things as safety valves and emergency vehicles.

In general, there is no accepted model for states to follow. However, the Model Community Noise Control Ordinance, developed by the National Institute of Municipal Law Enforcement Officers and the EPA Office of

Noise Abatement and Control, is a basic tool which both large and small communities can use to construct noise control ordinances suited to the local area.

Since the Noise Control Act of 1972 gives the state and local governments the power to establish and enforce controls on environmental noise, it would appear that the activity in this area in the future will increase.

PROBLEMS

3.1 A worker was exposed to noise according to the following schedule:

Exposure level (dBA)	Period of exposure (hr)
92	3
95	2
97	2
102	1

Does this daily noise dose exceed the OSHA standard?

3.2 The following data were collected at an industrial property line. Ten equal intervals from 75 dBA to 100 dBA were employed. Determine the equivalent sound level and its standard deviation.

Interval number	Sound level at center of interval (dBA)	Percentage of time spent in interval (%)
1	76.25	0.3
2	78.75	0.4
3	81.25	6.0
4	83.75	10.0
5	86.25	18.0
6	88.75	22.0
7	91.25	14.0
8	93.75	12.0
9	96.25	14.0
10	98.75	3.3

3.3 Determine the approximate values of L_{10}, L_{50}, and L_{90} for the data given in Problem 3.2.

3.4 Calculate the noise pollution level (L_{NP}) for the data given in Problem 3.2.

3.5 In a particular instance it was found that $L_{10} = 91$ dBA, $L_{50} = 85$ dBA, and $L_{90} = 77$ dBA. Furthermore, $L_{\text{eq}} = 89$ dBA and $\sigma = 3.2$. Determine the approximate expressions of equations (3.5) and (3.6).

3.6 Using the data of Problem 3.5, calculate L_{eq} by means of the approximate equation (3.7).

3.7 In a particular community, the daytime L_{eq} was 77 dBA and the nighttime L_{eq} was 58 dBA. Determine the day–night A-weighted average sound level.

REFERENCES

[1] LANG, W. W. "The Status of Noise Control Regulations in the U.S.A.," *Proceedings of the Inter-Noise Conference*, Sendai, Japan, Aug. 1975.

[2] "Public Health and Welfare Criteria for Noise," Environmental Protection Agency, Washington, D.C., July 1973.

[3] "Information on Levels of Environmental Noise Requisite to Protect Public Health and Welfare with an Adequate Margin of Safety," Environmental Protection Agency, Washington, D.C., Mar. 1974.

[4] MAGRAB, E. B. *Environmental Noise Control*, John Wiley & Sons, Inc., New York, 1975.

[5] DONOVAN, J. "A New Digital Data Analysis System for Community Noise Evaluation," *Sound and Vibration*, Sept. 1974, pp. 12–18.

[6] *Federal Register*, Vol. 40 (May 28, 1975), p. 23105.

[7] *Federal Register*, Vol. 39 (Oct. 29, 1974), p. 38208.

[8] *Federal Register*, Vol. 39 (Oct. 30, 1974), p. 38338.

[9] PETERSON, A. P. G. "Motor Vehicle Noise Measurements," *Sound and Vibration*, Apr. 1975, pp. 26–33.

[10] RENTZ, P. E. and L. D. POPE "Description and Control of Motor Vehicle Noise Sources," Vol. 2 of "Establishment of Standards for Highway Noise Levels," *Report No. NCHRP 3-7/3*, Highway Research Board, National Cooperative Highway Research Program, National Academy of Sciences, Washington, D.C., Feb. 1974. (Draft Report)

[11] WATERS, P. E. "Commercial Road Vehicle Noise," *J. Sound Vibration* Vol. 35, No. 2 (1974), pp. 155–222.

[12] "Noise from Construction Equipment and Operation, Building Equipment and Home Appliances," *Report No. NTID 300.1* Environmental Protection Agency, Washington, D.C., Dec. 1971.

[13] BENDER, E. K. "Noise from Construction, Home Appliances, and Building Equipment," *J. Environ. Sci.*, Sept./Oct. 1972, pp. 9–18.

[14] HAAG, F. G. "State Noise Restrictions—1976," *Sound and Vibration*, Dec. 1976, p. 27.

[15] BRAGDON, C. R. "Municipal Noise Ordinances—1976," *Sound and Vibration*, Dec. 1976, pp. 22–27.

4

INSTRUMENTATION

4.1 INTRODUCTION

The subject of instrumentation for noise measurements is an extremely broad one, hence we shall confine our discussion here to the most important principles and the most frequently used instrumentation for noise control.

Diagnostic instrumentation plays a fundamental role in noise control. With this instrumentation measurements are made to identify noise sources, determine the intensity of the noise at a particular distance from the source, measure the frequency characteristics of a noise source, and the like. These measurements then provide the data for comparison with specified standards and for the design and specification of control techniques.

The actual location of the measurements is very important. For example, measurements may be made at the location of the machine (i.e., *in situ* measurements) or in special rooms designed specifically for noise measurements (e.g., anechoic, reverberant, or semireverberant rooms). Because we are concerned in this text with industrial noise and vibration control, we will briefly consider the latter locations. The reader interested in measurements in these special locations is referred to references [1,2].

We shall represent the noise measurement system in the form shown in Figure 4.1. This figure illustrates the various components employed in the noise measurement system. We shall now examine each of these elements in some detail. Moreover, one must recognize that we cannot measure a noise field without disturbing the field; and that, furthermore, we cannot construct an instrument that will respond only to the particular parameter which we wish to measure. We shall now begin our discussion of instrumentation for noise measurements with this background information in mind.

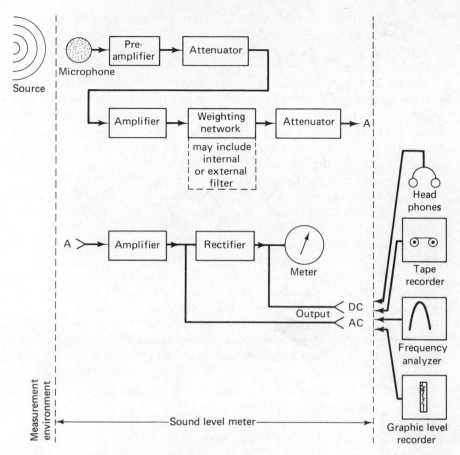

FIGURE 4.1 *Instrumentation system for noise measurements*

4.2 MEASUREMENT ENVIRONMENT

In industrial environments, noise sources exist both inside and outside the plant buildings. Various types of processing and production machinery may be found inside, and items such as motors, fans, cooling towers, and so on, may be located outside. Because we sometimes encounter different types of problems in these two environments, we shall briefly discuss some of the measurement considerations in each.

4.2.1 Indoor Measurements

To begin our discussion, we note that as we move away from a noise source, we expect the sound pressure level to decrease 6 dB for each doubling of distance. This relationship will be derived in Chapter 6. However, this relation requires that the noise source be in a free field, a situation that essentially never exists in an industrial environment. Thus, we must consider such things as the existence of a reverberant field, nearby reflecting surfaces, or background noise due to the presence of other noise sources.

Reverberation is a common occurrence in industrial environments, because machinery is normally placed in a plant with a cement floor and hard walls. Thus, multiple reflections from all the surfaces set up a reverberant field, which affects measurements. The presence of this field can be detected with a sound level meter, since in this case the sound field will be essentially constant in level throughout a considerable area.

If large reflecting surfaces exist near machinery, upon which noise measurements are being made, then reflections from these surfaces may contaminate the direct sound field. The magnitude of their effect will depend upon such things as the frequency of the noise, the size of the reflecting surface, and its distance from the machinery. Of particular interest is the problem that exists when the noise source emits relatively pure tones such as the hum of a transformer. In this case a standing-wave pattern may be set up between the noise source and the reflecting surface. This pattern yields a sequence of peaks and valleys in the sound pressure level between the source and reflecting surface as shown in Figure 4.2. If these spatial fluctuations in the noise level exist, then in general an average reading is taken and the fact that a standing wave was present is recorded with the data.

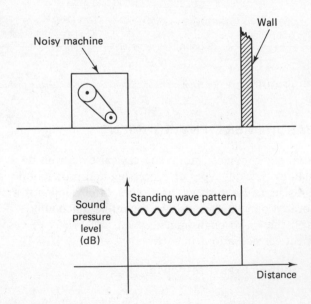

FIGURE 4.2 *Standing-wave pattern between a noise source and a reflecting surface*

The presence of background noise affects noise measurements in that it adds to the noise we wish to measure. In fact, if the background noise is very high, it may actually prevent us from measuring the noise level of a particular machine. In general, if the noise level of the machine under test exceeds the background noise level by more than 10 dB, the background noise level will have little or no effect on the measurements. If this difference is less than 10 dB, however, a correction factor must be applied to the total noise level to obtain the actual level of the source. For a difference in levels between 3 dB and 10 dB in any frequency band, the correction curve in Figure 1.3 can be applied.

EXAMPLE 4.1 In an industrial plant a sound pressure level measurement of 95 dB is obtained at a position with a particular machine under test, in operation. With the machine turned "off", the background noise measures 90 dB. What is the sound pressure level at the position in question due to the machine?

SOLUTION

$$\text{Difference} = 95 - 90 = 5 \text{ dB}$$

From Figure 1.3,

$$\text{correction factor} = 1.7 \text{ dB}$$
$$L_{P_{\text{machine}}} = 95 - 1.7 = 93.3 \text{ dB}$$

However, if the difference in levels is less than 3 dB, or if the source and background noise are known to be coherent, the actual noise level of the source cannot be accurately determined. Ideally, we would simply shut down the sources that create the background noise while measurements are being made. If this is not possible, a temporary barrier of the form shown in Figure 4.3 may be employed [3]. The hard barrier surface faces the machine that is providing the unwanted background noise and reflects this noise. The side of the barrier facing the machine under test is treated with sound-absorbing material to prevent the establishment of standing waves or a reverberant field between the barrier and the machine under test.

D4.1 With a certain machine "on," the sound pressure level in a plant is measured to be 86 dB and 80 dB with the machine "off." What is the sound pressure level due to the machine alone?

Ans. 84.7 dB.

EXAMPLE 4.2 Measurements are to be made on machine A as shown in Figure 4.3. Machine B ($h = 5$ ft or 1.52 m, and $l = 6$ ft or 1.83 m) introduces background noise which must be blocked by a barrier. Measurements indicate that the background noise is a low-frequency rumble in the range 100 to 200 Hz. Determine the barrier dimensions, H and L, indicated in Figure 4.3.

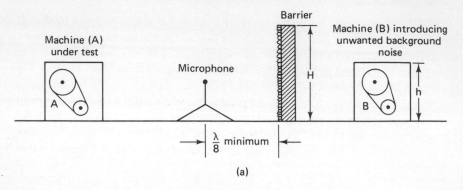

$$\frac{\lambda}{8} \text{ minimum}$$

(a)

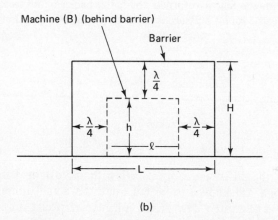

(b)

FIGURE 4.3 *Use of a portable barrier to reduce background noise in measurements:* (a) *side view of measurement configuration;* (b) *view of barrier from microphone*

SOLUTION From equation (1.2),

$$\lambda = \frac{c}{f} \quad \text{ft} \quad \text{or} \quad \text{m}$$

Assume that

$$c = 1128 \text{ ft/s} = 344 \text{ m/s}$$

and use $f = 100$ Hz. Then

$$\lambda = \frac{1128}{100} = 11.28 \text{ ft} = 3.44 \text{ m}$$

Therefore,

$$H = h + \left(\frac{\lambda}{4}\right) \quad \text{ft} \quad \text{or} \quad \text{m}$$

$$= 5 + \left(\frac{11.28}{4}\right) \text{ft} = 1.52 + \left(\frac{3.44}{4}\right) \text{m}$$

$$= 7.82 \text{ ft} \qquad = 2.38 \text{ m}$$

and

$$L = l + 2\left(\frac{\lambda}{4}\right) \quad \text{ft} \quad \text{or} \quad \text{m}$$

$$= 6 + 2\left(\frac{11.28}{4}\right) \text{ft} = 1.83 + 2\left(\frac{3.44}{4}\right) \text{m}$$

$$= 11.64 \text{ ft} \qquad = 3.55 \text{ m}$$

Finally, measurements of sound pressure level and/or sound level should, in general, be taken at least 2 ft or 60 cm from any major surface of the machine. However, in the process of determining the sound level at the worker's ear, one must obviously measure at the location of the ear regardless of the proximity to the machine.

D4.2 A certain machine must be examined for its contribution to the sound pressure level in a plant. The machine has dimensions of $h = 2$ m and $l = 4$ m, as shown in Figure 4.3. Because surrounding machines cannot be shut down for the measurements, a barrier must be constructed to reduce the background noise. The background noise has a frequency of 250 Hz. Find the barrier dimensions, H and L, as shown in Figure 4.3.

Ans. $H = 2.344$ m and $L = 4.688$ m.

4.2.2 Outdoor Measurements

The general considerations specified in the previous section on indoor measurements apply in this case also. Reverberation is generally not a problem; however, reflecting surfaces such as buildings may be present and therefore could affect the measurements. Two other factors that may play an important role are topography and weather conditions. The manner in which weather conditions affect measurements will be examined in a following section.

4.3 MICROPHONES

Microphones play a paramount and fundamental role in noise measurement instrumentation. The importance of these transducers stems from the fact that the measurement accuracy will be no better than that of the microphone. Because this transducer is the input device to our measurement system, as shown in Figure 4.1, we must understand its many ramifications so that it can be properly used to obtain accurate measurements.

4.3.1 Types of Microphones

The general types of microphones that are used in noise measurement systems are the condenser, ceramic, and dynamic microphones.

Condenser microphones employ a diaphragm as one side of a capacitor and when the diaphragm deflects due to a change in sound pressure, the resulting change in capacitance is converted to an electrical signal. These devices generally have relatively low capacitance, require a high dc polarizing voltage, employ a fragile diaphragm, and are adversely affected by high humidity conditions. However, they have good high-frequency response, long-term stability, are relatively insensitive to vibration excitation, and are viable transducers in extreme temperature and pressure environments.

The electret condenser microphone is constructed with a self-polarized plastic diaphragm and as such requires no external polarizing voltage. It is also relatively immune to high humidity. However, the frequency response of this device is not as good as the conventional condenser microphone, and its temperature limitations are more severe.

Ceramic microphones contain a piezoelectric crystal and therefore are often called piezoelectric microphones. In this microphone the crystal is placed behind the diaphragm, which acts as a force collector. The pressure field activates the diaphragm, which in turn strains the crystal material and thus generates an electrical signal. These microphones are rugged, reliable, have high capacitance and good dynamic range, and do not require a polarizing voltage. However, the high-frequency response and operational temperature range is somewhat limited.

Dynamic microphones produce an electrical signal by moving a coil, which is connected to a diaphragm, through a magnetic field. They have a relatively low impedance and therefore can be used in applications that require the use of long cables to the processing instruments and readout devices. However, they must not be used in the vicinity of devices that create magnetic fields (e.g., transformers).

Although the condenser and piezoelectric microphones are perhaps most often employed in the field, the type of microphone used in any given situation will depend upon the measurement conditions. A brief listing of typical microphone characteristics is given in Table 4.1. For detailed discussions of which microphone to use under various conditions, see references [4,5].

TABLE 4.1

Basic microphone types and characteristics

Microphone type	Sensing element	Typical frequency range (Hz)	Temperature and humidity stability
1. Crystal	Piezoelectric	10–10 k	Fair
2. Condenser	Capacitor	2–20 k	Fair
3. Ceramic	Piezoelectric	20–10 k	Good
4. Dynamic	Magnetic coil	25–15 k	Good

4.3.2 Sensitivity

A transducer's sensitivity is one of its most important characteristics. Another important characteristic is the frequency response, and in general, there is a trade-off between these two parameters (e.g., small microphones have low sensitivity but operate at both low and high frequencies, while large microphones have high sensitivity but are useful only at low frequencies). Typical microphones have diameters of $1, \frac{1}{2}$, and $\frac{1}{4}$ in., and generally the frequency response in a free field of a 1-in.-diameter microphone is virtually flat up to about 20 kHz, while that of a $\frac{1}{4}$-in.-diameter microphone is virtually flat up to about 100 kHz.

Microphone sensitivity (S) is defined by the expression

$$S = \frac{\text{electrical output}}{\text{mechanical input}} \tag{4.1}$$

and the microphone sensitivity level, L_s, often also termed simply "sensitivity", is defined by

$$L_s = 10 \log \left(\frac{E_{\text{out}}/p}{E_{\text{re}}} \right)^2 \quad \text{dBV}/\mu\text{bar} \tag{4.2}$$

where E_{out} = instrument output voltage (V)

E_{re} = reference voltage (1 V for an incident sound pressure of 1.0 μbar) (1.0 μbar equals 0.1 Pa)

p = rms pressure on the microphone (μbars)

Simplifying, one obtains the alternative expression,

$$L_s = 20 \log \left(\frac{E_{\text{out}}}{p} \right) \quad \text{dBV}/\mu\text{bar} \tag{4.3}$$

Typical microphone sensitivities are between 3 mV per μbar and 0.5 μV/μbar, or about between -50 dBV/μbar and -125 dBV/μbar. For example, microphones for general sound level measurements, 1 in. in diameter with a frequency range of 2 Hz to 18 kHz, have a sensitivity on the order of 3 mV/μbar (-50 dBV/μbar). On the other hand, a special-purpose 0.1-in.-diameter microphone, designed for measurement of frequencies beyond 100 kHz, may have a sensitivity of only 5 μV/μbar (-125 dBV/μbar). Upon rearranging equation (4.3) and solving for E_{out}, we obtain

$$E_{\text{out}} = p \left[\text{antilog} \left(\frac{L_s}{20} \right) \right] \quad \text{V} \tag{4.4}$$

79

Furthermore, the sound pressure level was defined by equation (1.14) as

$$L_p = 20 \log \left(\frac{p}{p_{re}}\right) \text{dB} \tag{4.5}$$

where p_{re} = reference pressure of 20×10^{-6} Pa or 0.0002 μbar

Also,

$$L_p = 20 \log \left(\frac{p}{0.0002}\right) \text{dB} \tag{4.6}$$

where we recall that here p is in μbars. We may now solve for the pressure (in μbars) associated with the sound pressure level, L_p. Rearranging, we obtain

$$p = 0.0002\left[\text{antilog} \left(\frac{L_p}{20}\right)\right] \quad \mu\text{bars} \tag{4.7}$$

which may be substituted into equation (4.4) to obtain

$$E_{out} = 0.0002\left[\text{antilog} \left(\frac{L_p}{20}\right)\right]\left[\text{antilog} \left(\frac{L_s}{20}\right)\right] \quad \text{V} \tag{4.8}$$

where E_{out} = actual voltage output of the instrument using a microphone of sensitivity level, L_s, and measuring a sound pressure level, L_p

EXAMPLE 4.3 Determine the output voltage of two microphones (a) with $L_{s_1} = -50$ dBV/μbar and (b) with $L_{s_2} = -125$ dBV/μbar, if the sound pressure level to which they are exposed is $L_p = 74$ dB, which corresponds to $p = 1$ μbar.

SOLUTION From equation (4.8), E_{out} can be expressed as

$$E_{out} = 2 \times 10^{-4}\left[\text{antilog} \left(\frac{L_p}{20}\right)\right]\left[\text{antilog} \left(\frac{L_s}{20}\right)\right] \quad \text{V}$$

(a) Then for $L_p = 74$ dB and $L_{s_1} = -50$ dBV/μbar,

$$E_{out_1} = 2 \times 10^{-4}\left[\text{antilog} \left(\frac{74}{20}\right)\right]\left[\text{antilog} \left(\frac{-50}{20}\right)\right] \quad \text{V}$$

$$= 3.1676 \text{ mV}$$

(b) And for $L_p = 74$ dB and $L_{s_2} = -125$ dB/μbar,

$$E_{out_2} = 2 \times 10^{-4}\left[\text{antilog}\left(\frac{74}{20}\right)\right]\left[\text{antilog}\left(\frac{-125}{20}\right)\right] \quad \text{V}$$

$$= 0.5636 \ \mu\text{V}$$

D4.3 Determine the microphone sensitivity level if the following parameters are noted: $E_{out} = 10$ V and $p = 1 \times 10^4$ μbar.

Ans. $L_s = -60$ dBV/μbar.

4.3.3 Special Application Considerations

Microphones may be calibrated for random, normal (pointing at the source), or grazing (pointing at an angle of 90° with the source) incidence. Therefore, to reduce measurement errors, care must be exercised to ensure that the directional orientation specified by the manufacturer is used. This phenomenon is particularly important at high frequencies.

The placement of a microphone in a sound field distorts the field. The importance of this phenomenon depends upon such things as the frequency of the sound, the direction of propagation, and the size and shape of the microphone. For example, at high frequencies where the wavelength of the sound is small in comparison to the dimensions of the microphone, reflections from the microphone cause the pressure acting on the microphone diaphragm to be quite different from that of the actual free-field sound pressure that is being measured. Therefore, a 1-in.-diameter microphone will not yield accurate free-field sound pressure measurements of noise in the frequency range above approximately 13 kHz, since a wavelength of 1 in. (the diameter of the microphone) corresponds to a frequency of 13,536 Hz.

If the microphone is used in a windy environment, special precautions must be taken to obtain representative data. The wind passing across a microphone produces turbulence, causing pressure fluctuations that generate low-frequency noise. If the wind speed is less than about 5 mph, no precautions are necessary for A-scale measurements. This is due to the large attenuation at low frequencies inserted by the A-weighting network. However, for linear or C-scale measurements, a wind screen should be employed if there is any wind at all, and of course for A-scale measurements as well if the velocity exceeds 5 mph. These devices are typically open-cell polyurethane foam spheres which cover the microphone. If the wind speed exceeds approximately 20 mph, a careful check should be made to determine if even the wind screen is effective.

Corrections may also be made for changes in atmospheric conditions such as temperature and pressure. For example, corrections for temperature and pressure referenced to 68°F and 30 in. Hg are given in Figure 4.4. These

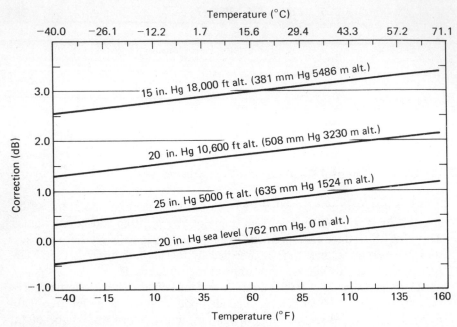

FIGURE 4.4 *Correction to 20°C, 760 mm Hg (68°F, 30 in. Hg) as a function of temperature and pressure*

curves are plotted according to the expression

$$C_{TP} = 10 \log\left[\left(\frac{460 + F}{528}\right)^{1/2}\left(\frac{30}{B}\right)\right] \quad \text{dB} \qquad (4.9)$$

where C_{TP} = correction to be added to the sound pressure level (dB)

   F = temperature (°F)

   B = barometric pressure (in. Hg)

In most industrial applications, changes in temperature and pressure are rather inconsequential and are rarely made.

EXAMPLE 4.4   At a cement plant located virtually at sea level, an outdoor operation produces a sound pressure level of 89 dB at a worker position when the temperature is 68°F. Determine L_p during the winter low of 20°F and the summer high of 100°F.

SOLUTION   Referring to Figure 4.4, we obtain

$$-0.2 \text{ dB} \quad \text{at } 20°F$$

and

$$+0.2 \text{ dB at } 100°\text{F}$$

Therefore,

$$L_p = 89 - 0.2 = 88.8 \text{ dB, winter}$$

and

$$L_p = 89 + 0.2 = 89.2 \text{ dB, summer}$$

Obviously, in such an application the correction is nonsense.

Other important aspects of measurements that require consideration are the presence of reflecting surfaces or, in fact, the presence of an individual making measurements. An individual will disturb the sound field and can cause significant measurement errors. To avoid this problem, the microphone may be placed on a tripod and remotely located from the other instrumentation.

Finally, when a microphone is remotely located via cable from the remainder of the measurement instrumentation, under certain harsh conditions, cable noise may develop. This noise is generated by the relative motion of different parts of the cable, and thus the cable should be restrained from moving and isolated from vibrations as much as possible. This problem is normally alleviated through the use of a preamplifier located at the microphone.

D4.4   A machine that produces a sound pressure level of 85 dB at the referenced temperature and pressure of 68°F and 30 in. Hg will produce what sound pressure level at 29 in. Hg and 100°F?

Ans.   85.3 dB.

4.4 SOUND LEVEL METER

One of the most useful and therefore important instruments in the analysis of noise is the sound level meter. This device amplifies the very small output signal from the microphone and makes it available for processing and for visual display by a readout meter contained within the unit.

These instruments must conform to the ANSI Standard S1.4-1971, "Specification for Sound Level Meters." Sound level meters are considered according to accuracy and functional requirements as either type 1, precision instruments, or type 2, general-purpose instruments. In general, the type 2 instruments are useful only for quickly surveying the sound field to determine the approximate sound level at various locations. The type 1 precision instrument must be employed to obtain accurate data for noise control purposes. Both types of instruments typically contain what are called A, B, and C weighting networks. The A and C networks can be used together to determine if low-frequency energy is present. From Figure 2.8 it can be seen that if measurements are taken on a source with both the A network and the C network, and if the dB reading obtained is much larger than the dBA reading, there is a considerable amount of energy present at low frequencies.

To ensure the accuracy of the measurements, an acoustic calibrator must be employed. A typical calibrator for field use is a single tone device that fits over the microphone, is attached to the sound level meter, and produces an accurate reference sound level to calibrate the entire instrument. One should also take care to record sufficient information to be able to identify all measuring and calibrating instruments at a later date if the need arises. As is usually the case in field work, the greater the care taken in recording the measured data, the more reliable and easier to evaluate the results are. In Figure 4.5 is shown a typical noise survey data sheet with spaces and tables provided for the usual information and data required.

Finally, precision sound level meters are also normally equipped to measure impulse-type noises. These instruments contain the necessary electronics to provide a very fast rise time and thus an accurate response to the pressure pulse. Application is found where machines such as punch presses or drop hammers are utilized.

4.5 SPECTRUM ANALYZERS

The readings obtained from the sound level meter as described above, although valuable, tell only part of the story. For example, the two noise signals shown in Figure 4.6 have quite different values across the frequency range illustrated, but they have the same dBA value. Since engineering control techniques are critically dependent upon the frequency spectrum of the noise, the manner in which the noise varies with frequency is not only useful; it is absolutely necessary.

A spectrum analyzer is a device that provides the capability for analysis of a noise signal in the frequency domain by electronically separating the signal energy into various frequency bands. This separation is performed by means of a set of filters.

In general, a filter is simply a two-port electrical network with a pair of terminals at each port. Simple filters can be constructed with as few as two passive electrical elements (e.g., a resistor and a capacitor). More complicated designs involve the use of many passive circuit elements, which may also be used in conjunction with active elements such as operational amplifiers (op amps).

Suppose that a sine-wave signal of the form $A \sin \omega t$ is applied at the input terminals of the filter. An examination of the filter's output will generally show that two things have happened to the input signal: a change in amplitude and a shift in phase have occurred. In other words, the filter characteristics have shaped the input signal. The amplitude and phase charac-

Noise survey data

Client: _____

Plant location: _____ Date: _____

Survey personnel: _____

Noise site layout:

Primary noise source: _____ | Other sources: _____

Equip: _____ | Equip: _____

Mod./serial No. _____ | Mod./serial No. _____

Oper. cond. _____ | Oper. cond. _____

Instrumentation:	Serial No.				
SLM: _____	_____	Time:			
Calibrator: _____	_____	Temp:			
Cable: _____	_____	% RH:			
Analyzer: _____	_____	mm Hg:			
Transducer: _____	_____	Wind dir.:			
Other: _____	_____	Wind mph.:			
		Calibration:			

| | | | | Sound pressure level (dB re 20 μN/M^2 rms) | | | | | | | | | | |

Test No.	Time	Posi-tion	Conditions	A scale level	Overall level	Octave-band center frequency (Hz)								
						31.5	63	125	250	500	1 k	2 k	4 k	8 k

Recommendations: _____
_____

FIGURE 4.5 *Typical noise survey data sheet*

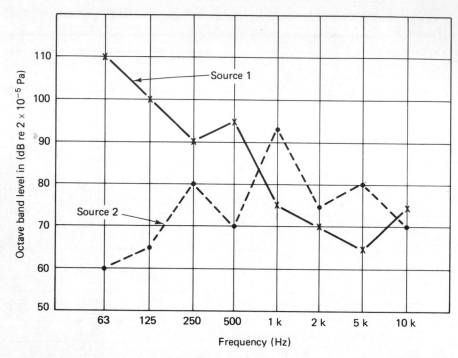

FIGURE 4.6 *Octave-band sound pressure levels for two noise sources with the same dBA reading, 93.4 dBA*

teristics of the filter can be derived by computing the ratio of the output of the filter to the input of the filter for all possible values of frequency. The ratio of output to input is commonly called the *transfer function*. The amplitude transfer function, or frequency response, of three typical filters is shown in Figure 4.7. Note that both the ideal and physically realizable characteristics are given.

Ideally, the low-pass filter passes all signals from dc up to a frequency of f_1 and totally rejects all signals with frequencies above f_1. The ideal high-pass filter passes all signals above frequency f_1 and totally rejects all signals with frequencies below f_1. In a similar manner, the ideal bandpass filter passes all signal frequencies between f_1 and f_2 and rejects all others. The actual filters that can be constructed with physical hardware have characteristics of the form shown on the right side of the figure. Consider, for example, the simple low-pass filter shown in Figure 4.8(a). The amplitude transfer function for this circuit is

$$|H(\omega)| = \left|\frac{E_0}{E_i}\right| = \frac{1}{\sqrt{1 + (\omega RC)^2}} \quad \text{(unitless)} \quad (4.10)$$

If we plot the amplitude of this function in dB versus frequency on semilog

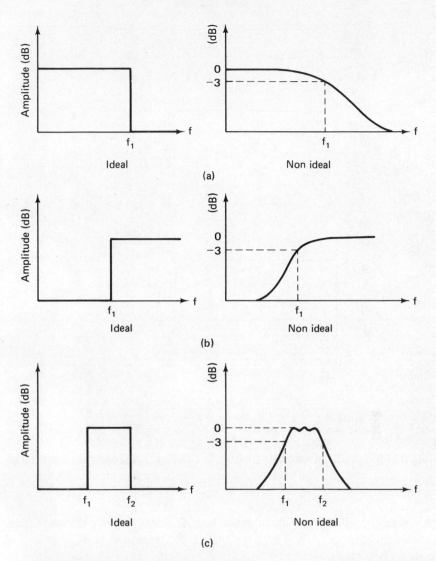

FIGURE 4.7 *Frequency response of ideal and physically realizable filters:* (a) *low-pass filter;* (b) *high-pass filter;* (c) *bandpass filter*

paper, we obtain a characteristic of the form shown in Figure 4.8(b). This characteristic is obtained as follows:

$$|H(\omega)| = 20 \log \left[\frac{1}{\sqrt{1 + (\omega RC)^2}} \right] \quad \text{dB}$$

$$= -20 \log [1 + (\omega RC)^2]^{1/2}$$

$$= -20 \log [1 + (2\pi f RC)^2]^{1/2} \quad \text{dB} \qquad (4.11)$$

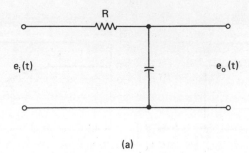

(a)

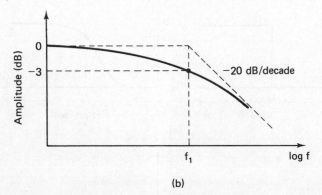

(b)

FIGURE 4.8 (a) *Single low-pass filter circuit;* (b) *frequency response of the low-pass filter*

Note that for small values of frequency, $(\omega RC)^2 \ll 1$, the equation reduces to

$$|H(\omega)| \simeq -20 \log [1] = 0 \quad \text{dB}$$

Therefore, for relatively small values of frequency, the value of the amplitude transfer function is 0 dB. For large values of frequency, $(\omega RC)^2 \gg 1$, the equation becomes

$$|H(\omega)| \simeq -20 \log (2\pi f RC) \quad \text{dB}$$

and therefore the amplitude characteristic has a negative slope of 20 dB/ decade. This asymptote is sometimes called the filter "skirt." The point f_1 where the two asymptotes meet is called the cutoff frequency and is given by

$$f_1 = \frac{1}{2\pi RC} \quad \text{Hz} \tag{4.12}$$

Note from equation (4.11) that at this cutoff frequency the amplitude has decreased by $1/\sqrt{2}$, or 3 dB. Therefore, the actual amplitude characteristic is shown solid in Figure 4.8(b). Since power is proportional to the square of the amplitude, the frequency f_1 is also called the *half-power point*. The amplitude response of complex filters can be derived in a similar manner.

To provide a vivid picture of the filter action, a number of simple filters were constructed and analyzed. In Figure 4.9(a) are shown the amplitude characteristics for both a low-pass and a high-pass filter. The spikes mark the frequency, and from left to right they are dc, 2 kHz, and 4 kHz. The input signal for the filters is shown in Figure 4.9(b). This signal is the sum of two sine waves, one whose frequency is 2 kHz and the other whose frequency is 4 kHz. The output of the low-pass filter is shown in Figure 4.9(c). Note that the 4-kHz frequency has been rejected and the 2-kHz signal has been passed.

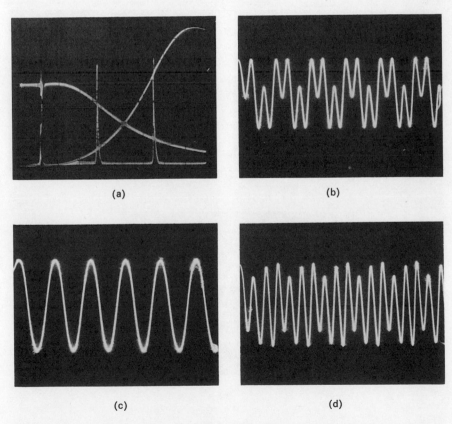

(a)   (b)

(c)   (d)

FIGURE 4.9 *Low-pass and high-pass filter characteristics:* (a) *amplitude response;* (b) *input signal;* (c) *low-pass filter output;* (d) *high-pass filter output*

The output of the high-pass filter is shown in Figure 4.9(d). The output is primarily a 4-kHz signal; however, some of the 2-kHz signal is present in the output, as illustrated by the uneven peaks. This result is not unexpected, since the skirt of the high-pass filter characteristic extends down past the 2 kHz range, as shown in Figure 4.9(a).

Figure 4.10(a) shows the amplitude characteristic for a bandpass filter. The spikes mark the frequency and from left to right they are 4 kHz, 5 kHz, and 8 kHz. The input signal, shown in Figure 4.10(b), is the sum of three sine waves of frequencies 4 kHz, 5 kHz, and 8 kHz. The output of the filter is the 5-kHz component shown in Figure 4.10(c).

Analyzers are typically classified according to the type of filters that are employed. For example, if an octave-band filter set is used, the device is called an octave-band analyzer.

With this information as background, let us now examine some of the

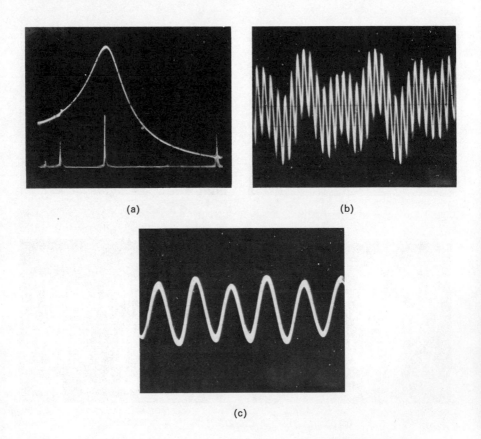

(a)   (b)

(c)

FIGURE 4.10 *Bandpass filter characteristics: (a) amplitude response; (b) input signal; (c) output signal*

typical spectrum analyzers that are employed to obtain the data required for the design and specification of control techniques.

D4.5 Calculate the sound level, dBA, of source 1 in Figure 4.6.

Ans.   93.4 dBA.

4.5.1 Octave-Band Analyzers

The *octave-band analyzer* is perhaps the most commonly used analyzer for industrial noise measurements. The device receives its name from the fact that it is capable of separating the noise signal spectrum into frequency bands that are 1 octave in width. An octave has a center frequency that is $\sqrt{2}$ times the lower cutoff frequency and an upper cutoff frequency that is twice the lower cutoff frequency. Therefore, if we define

$$f_1 = \text{lower cutoff frequency}$$
$$f_2 = \text{upper cutoff frequency}$$
$$f_0 = \text{center frequency}$$
$$\text{bw} = \text{bandwidth}$$
$$f_1 = \frac{f_0}{\sqrt{2}}$$
$$f_2 = \sqrt{2}\, f_0$$
$$f_2 = 2f_1$$
$$\text{bw} = f_2 - f_1$$

(4.13)

The octave bands in the audible range are identified by their center frequencies, which are 31.5, 63, 125, 250, 500, 1000, 2000, 4000, 8000, and 16,000 Hz. A typical frequency response of an octave-band filter set is shown in Figure 4.11.

EXAMPLE 4.5 Determine the lower and upper cutoff frequencies, f_1 and f_2, and the bandwidth, bw, of the octave band centered at $f_0 = 1000$ Hz.

SOLUTION According to equations (4.13), we obtain, for $f_0 = 1000$ Hz,

$$f_1 = \frac{f_0}{\sqrt{2}} = \frac{1000}{\sqrt{2}}$$
$$= 707 \text{ Hz}$$

and

$$f_2 = \sqrt{2}\, f_0 = \sqrt{2}(1000)$$
$$= 1414 \text{ Hz}$$

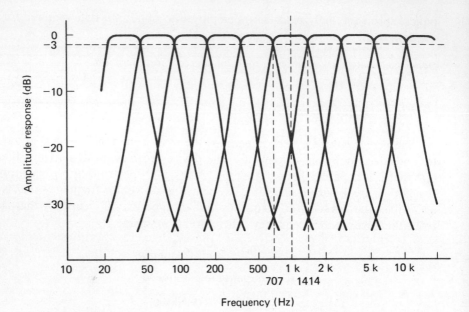

FIGURE 4.11 *Typical frequency response of an octave-band filter set*

Also,

$$\text{bandwidth} = f_2 - f_1$$
$$= 1414 - 707$$
$$= 707 \text{ Hz}$$

It is often desirable to perform a frequency analysis and obtain the spectrogram of the noise under investigation. In Figure 4.12(a) is shown the spectrogram of an air compressor, measured at a distance of 5 ft, using an octave-bandwidth analyzer, a paper speed of 3 mm/s, and a writing speed of 200 mm/s.

D4.6 Determine f_1, f_2, and bw for octave bands with the following center frequencies: (a) $f_0 = 2000$ Hz and (b) $f_0 = 5000$ Hz.
Ans. (a) $f_1 = 1414$ Hz, $f_2 = 2828$ Hz, bw = 1414 Hz   (b) $f_1 = 3536$ Hz, $f_2 = 7071$ Hz, bw = 3535 Hz.

4.5.2   Narrow-Band Analyzers

If the octave-band analyzer does not provide sufficient detail of the noise spectrum, then a filter set with a narrower bandwidth is used. In general, fractional-octave-band analyzers are excellent diagnostic devices for use in the field, especially in industrial environments, where the noise contains narrow-band signals such as pure tones. They are, however, tedious to use unless employed in conjunction with a display device which provides a simultaneous reading of all bands.

Two typical *narrow-band analyzers* are $\frac{1}{3}$-octave and $\frac{1}{10}$-octave analyzers. The $\frac{1}{3}$-*octave analyzer* is perhaps the most widely used. As the name implies, this device splits the octave into three parts. In order to compare a $\frac{1}{3}$-octave analysis with an octave analysis, the three $\frac{1}{3}$-octave levels within an octave are combined to obtain the octave level. Table 4.2 contains a comparison of

TABLE 4.2

Comparison of 1-octave and $\frac{1}{3}$-octave bands

1 OCTAVE			$\frac{1}{3}$ OCTAVE		
Lower cutoff frequency (Hz)	Center frequency (Hz)	Upper cutoff frequency (Hz)	Lower cutoff frequency (Hz)	Center frequency (Hz)	Upper cutoff frequency (Hz)
11	16	22	14.1	16	17.8
			17.8	20	22.4
			22.4	25	28.2
22	31.5	44	28.2	31.5	35.5
			35.5	40	44.7
			44.7	50	56.2
44	63	88	56.2	63	70.8
			70.8	80	89.1
			89.1	100	112
88	125	177	112	125	141
			141	160	178
			178	200	224
177	250	355	224	250	282
			282	315	355
			355	400	447
355	500	710	447	500	562
			562	630	708
			708	800	891
710	1,000	1,420	891	1,000	1,122
			1,122	1,250	1,413
			1,413	1,600	1,778
1,420	2,000	2,840	1,778	2,000	2,239
			2,239	2,500	2,818
			2,818	3,150	3,548
2,840	4,000	5,680	3,548	4,000	4,467
			4,467	5,000	5,623
			5,623	6,300	7,079
5,680	8,000	11,360	7,079	8,000	8,913
			8,913	10,000	11,220
			11,220	12,220	14,130
11,360	16,000	22,720	14,130	16,000	17,780
			17,780	20,000	22,390

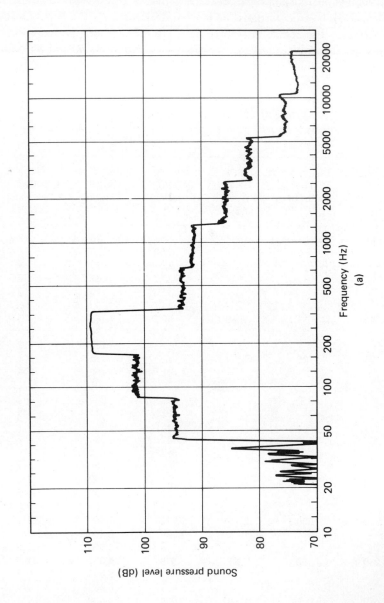

Sound pressure level (dB)

Frequency (Hz)

(a)

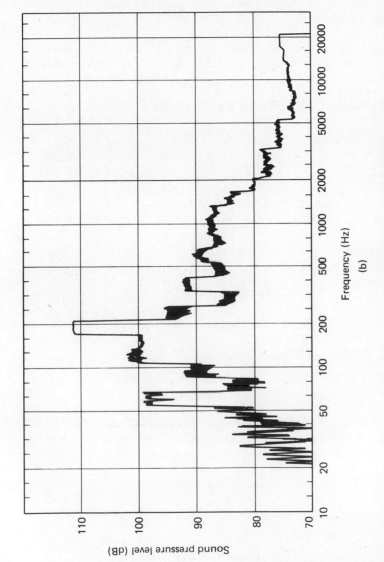

FIGURE 4.12 *Spectrograms of an air compressor (a) using an octave-band analyzer and (b) using a $\frac{1}{3}$ octave-band analyzer (both measured at a distance of 5 ft, paper speed of 3 mm/s, and writing speed of 200 mm/s) (courtesy C. E. Hickman, Southern Company Services, Inc., Birmingham, Ala.)*

the 1-octave and $\frac{1}{3}$-octave bands with respect to center, upper, and lower cutoff frequencies. This table is convenient for determining the three $\frac{1}{3}$-octave bands that fall within a given octave. Also, see Figure 4.12 for a comparison of the spectrograms of an air compressor obtained by means of a 1-octave and a $\frac{1}{3}$-octave bandwidth analyzer.

EXAMPLE 4.6   The sound pressure levels, L_p, in the 800-, 1000-, and 1250-Hz $\frac{1}{3}$-octave bands were found to be 87, 92, and 94 dB, respectively. Determine the sound pressure level in the equivalent octave band.

SOLUTION   In Table 4.2 we find that for the 800-Hz $\frac{1}{3}$-octave band,

$$f_1 = 708 \text{ Hz}$$

and that for the 1250-Hz $\frac{1}{3}$-octave band,

$$f_2 = 1413 \text{ Hz}$$

We note that the 1000-Hz octave band with $f_1 = 707$ Hz and $f_2 = 1414$ Hz is equivalent to the three $\frac{1}{3}$-octave bands in question. We now determine L_p of the equivalent 1000-Hz octave band by adding the three L_p's according to equation (1.19):

$$L_p = 10 \log \left(\sum_{i=1}^{n} 10^{L_{p_i}/10} \right) \quad \text{dB}$$

Substituting and expanding, we obtain

$$L_{p_{1000}} = 10 \log \left(10^{87/10} + 10^{92/10} + 10^{94/10} \right) \quad \text{dB}$$

and finally,

$$L_{p_{1000}} = 96.6 \text{ dB}$$

Note that while the upper cutoff frequency is equal to twice the lower cutoff frequency for the octave-band analyzer, the upper cutoff frequency for the $\frac{1}{3}$-octave analyzer is the lower cutoff frequency times the cube root of 2, that is,

$$f_2 = 2^{1/3} f_1 \quad \text{Hz} \tag{4.14}$$

In general, for an m-octave analyzer, the relationship between the upper and lower cutoff frequencies is

$$f_2 = 2^m f_1 \quad \text{Hz} \tag{4.15}$$

where $m = 1, \frac{1}{2}, \frac{1}{3}, \frac{1}{4}, \ldots$ octave

The center frequency, f_0, is in general the geometric mean of the product of the upper and lower cutoff frequencies, that is,

$$f_0 = (f_1 f_2)^{1/2} \quad \text{Hz} \tag{4.16}$$

Upon combining equations (4.15) and (4.16) and solving for/ the lower and upper cutoff frequencies in terms of the center frequency, we obtain

$$f_1 = 2^{-(m/2)} f_0 \quad \text{Hz} \tag{4.17}$$

and

$$f_2 = 2^{m/2} f_0 \quad \text{Hz} \tag{4.18}$$

respectively. The bandwidth, bw, defined by

$$\text{bw} = f_2 - f_1 \quad \text{Hz} \tag{4.19}$$

becomes

$$\text{bw} = [2^{m/2} - 2^{-(m/2)}] f_0 \quad \text{Hz} \tag{4.20}$$

upon substituting in equations (4.17) and (4.18).

EXAMPLE 4.7 Determine the lower and upper cutoff frequencies and the bandwidth of the 1000-Hz, $\frac{1}{10}$-octave band.

SOLUTION Taking advantage of equations (4.17) and (4.18), we obtain

$$f_1 = 2^{-(m/2)} f_0 \quad \text{Hz}$$
$$= 2^{-(0.1/2)}(1000)$$
$$= 966 \text{ Hz}$$

and

$$f_2 = 2^{m/2} f_0 \quad \text{Hz}$$
$$= 2^{0.1/2}(1000)$$
$$= 1035 \text{ Hz}$$

From equation (4.19), the bandwidth is

$$\text{bw} = f_2 - f_1 \quad \text{Hz}$$
$$= 1035 - 966$$

which yields

$$\text{bw} = 69 \text{ Hz}$$

Furthermore, the bandwidth can also be represented as a constant percentage of the center frequency of the band in question. For this reason, these analyzers are called *constant percentage bandwidth analyzers*. The bandwidth as a percent of the center frequency is expressed as

$$\text{percent} = \left(\frac{\text{bw}}{f_0}\right) 100\% \qquad (4.21)$$

and upon expanding by means of equation (4.20), one obtains

$$\boxed{\text{percent} = (2^{m/2} - 2^{-m/2})\, 100\%} \qquad (4.22)$$

where $m = 1\text{-}, \frac{1}{2}\text{-}, \frac{1}{3}\text{-}, \ldots, 1/n$-octave band

The bandwidth is therefore always a constant percentage of the center frequency to which the analyzer is tuned (i.e., as f_0 increases, the bandwidth increases).

D4.7 Determine the lower and upper cutoff frequencies and the bandwidth of the 2000-Hz fifth octave.

Ans. $f_1 = 1866\,\text{Hz}; f_2 = 2144\,\text{Hz}; \text{bw} = 278\,\text{Hz}.$

EXAMPLE 4.8 Determine the percentage bandwidth of a $\frac{1}{3}$-octave analyzer. What are the bandwidths in Hz of the 16-Hz, 1000-Hz, and 8000-Hz $\frac{1}{3}$-octave bands?

SOLUTION Using equation (4.22),

$$\text{percent} = (2^{m/2} - 2^{-m/2})100\%$$

we obtain

$$\text{percent} = [2^{(1/3)/2} - 2^{-(1/3)/2}]100\%$$
$$= 23\%$$

The desired bandwidths are:

$$16\,\text{Hz}: \quad \text{bw} = 0.23 \times 16 = 3.7\,\text{Hz}$$
$$1000\,\text{Hz}: \quad \text{bw} = 0.23 \times 1000 = 230\,\text{Hz}$$
$$8000\,\text{Hz}: \quad \text{bw} = 0.23 \times 8000 = 1840\,\text{Hz}$$

In contrast, a constant-bandwidth analyzer has a bandwidth that is fixed and independent of the center frequency to which it is tuned; if the bandwidth is 2 Hz, then it is 2 Hz at any center frequency. However, these analyzers are in general not as useful in an industrial environment as are the fractional-octave-band devices discussed above. In general, the smaller the bandwidth, the more detailed the analysis, the more costly will be the equipment, and the more time will be required to perform the analysis. This is one reason why analyzers which employ bands that are narrower than the $\frac{1}{3}$-octave bands are normally designed to be continuously tunable over the frequency range of the instrument and are usually confined to use in the laboratory. A comparison of the constant-percentage-bandwidth analyzer and the constant-bandwidth analyzer is shown in Figure 4.13.

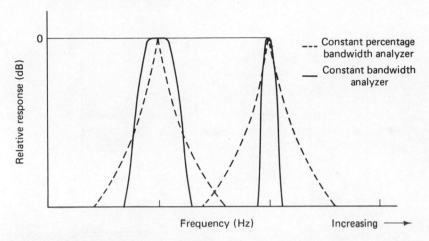

FIGURE 4.13 *Comparison of the frequency characteristics of constant-percentage-bandwidth and constant-bandwidth analyzers*

The same air compressor data used to obtain the spectrograms of Figure 4.12 were also analyzed with two constant-bandwidth analyzers. These spectrograms, shown in Figures 4.14(a) and (b), were obtained with 100-Hz and 31.6-Hz constant-bandwidth analyzers, respectively. Note the considerable difference as the bandwidth becomes more narrow. This type of analysis is particularly important where strong components of narrow-band noise are of concern.

4.6 MAGNETIC TAPE RECORDERS

These devices are extremely important in the analysis and control of industrial noise because they provide a permanent record of the noise data, which are then available for subsequent processing in a laboratory that may contain a wide variety of very sophisticated data analysis equipment. In addition, the tape recorder provides the capability for capturing transient or intermittent noise signals which it would otherwise not be possible to analyze.

Tape recorders are typically classified according to the type of modulation scheme employed for recording. The two types are *amplitude modulation* (AM) and *frequency modulation* (FM). Both types of instruments are avail-

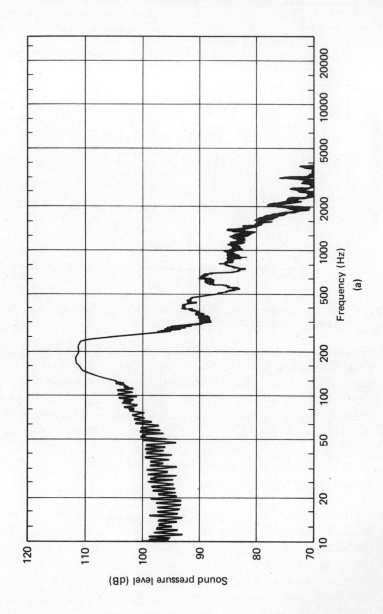

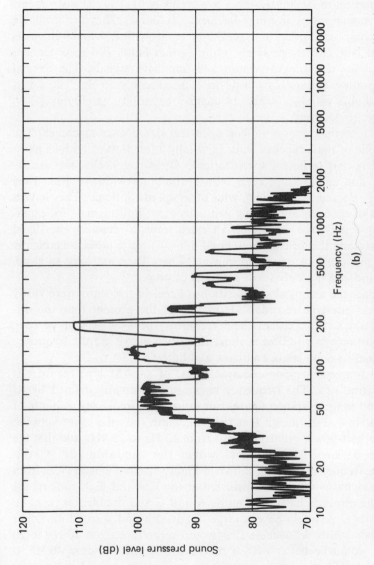

FIGURE 4.14 *Spectrograms of an air compressor (a) using a 100-Hz constant-bandwith analyzer and (b) using a 31.6-Hz constant-bandwidth analyzer (both measured at a distance of 5 ft, paper speed of 3 mm/s, and writing speed of 200 mm/s) (courtesy C. E. Hickman, Southern Company Services, Inc., Birmingham, Ala.)*

able as portable units for field use or as fixed units which are used in a laboratory environment.

Generally, the AM recorders, which are commonly referred to as *direct recorders*, are normally used for acoustical measurements in the approximate frequency range 30 to 20,000 Hz. They are typically lightweight instruments with good dynamic range and high-frequency response. They are available as 2-channel cassette recorders or as reel-to-reel units with 2, 4, and 8 channels on $\frac{1}{4}$-, $\frac{1}{2}$-, and 1-in. tape, respectively. On the other hand, FM recorders are typically used for recording vibration and impulsive signals. The normal range for vibration signals is from dc to approximately 50 Hz. These FM units are available with 4-, 7-, and 14-channel capability employing $\frac{1}{4}$-, $\frac{1}{2}$-, and 1-in. tape, respectively.

The tape recorders may be either open-reel or cassette machines. The selection of one of these types of units is usually a function of both its physical parameters and technical specifications. Open-reel decks are usually large, heavy, and employ tape reels with a minimum width of $\frac{1}{4}$ in. They operate at $3\frac{3}{4}$, $7\frac{1}{2}$, and 15 in./s (ips), with other speeds optional. They have a large number of channels and editing is usually easy. Editing may, for example, involve placing voice signals on an extra track to describe the signal being measured and the measurement conditions. Cassette tapes are typically small and light, use $\frac{1}{8}$-in. tape and run at $1\frac{7}{8}$ ips. They are easy to store, will not unwind and use only small amounts of power.

At one time, the technical specifications of reel-to-reel units were vastly superior to cassette-type recorders. However, at the present time the differences are almost nonexistent. The technical specifications of perhaps primary importance in addition to those described above are the frequency response, signal-to-noise ratio, and wow and flutter.

A typical frequency-response characteristic of an AM tape recorder is shown in Figure 4.15. The frequency response is normally defined by its bandwidth and range deviation figure. For example, a frequency response of 40 Hz to 15 kHz $\pm$ 3 dB means that the bandwidth (i.e., the range between the -3 dB or half-power points) extends from 40 Hz to 15 kHz and that the characteristic deviates from nominal within the bandwidth by ± 3 dB. Therefore, the frequency response shown in Figure 4.15 has a bandwidth from A to C. The signal is somewhat distorted at the low and high ends of the band since the power is halved at points A and C and doubled at point B.

The best tape recorders have a large bandwidth and a small deviation figure. In other words, a machine that would reproduce from 20 Hz to 20 kHz $\pm$ 2 dB would be better than a machine with the response 40 Hz to 15 kHz $\pm$ 3 dB. The response figure 20 Hz to 20 kHz $\pm$ 2 dB is typically the rating at $7\frac{1}{2}$ ips. Normally, this figure should improve at a speed of 15 ips and degrade at $3\frac{3}{4}$ ips and lower.

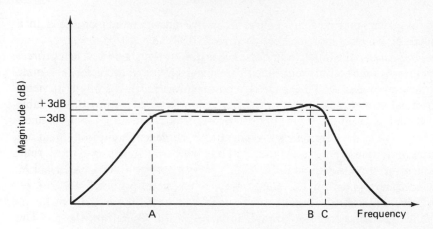

FIGURE 4.15 *Typical AM tape recorder frequency response*

The signal-to-noise ratio is a quantity that describes the ratio of the maximum undistorted signal-to-background noise. A figure of 40 to 50 dB is normal and a figure in the range 60 to 70 dB is very good. An important factor here, however, is that one must have a large enough input signal to ensure the full use of the recorder's signal-to-noise ratio.

Speed variations in the tape-drive system are described by two factors: wow and flutter. These figures are measured in percent of the signal value. For audio use, a good figure is 0.06%. The smaller the figure, the better. Wow is slow speed variation, usually due to drift in the drive capstan speed, and flutter is a faster speed variation caused by out-of-round components, mechanical noise, excessive friction in the drive path, and other factors. In noise analyses, any excessive wow and flutter will tend to mask low-frequency recordings.

In addition to the specifications above, it is important to note that multiple operating speeds provide the capability for frequency and time transformations which facilitate certain data-analysis procedures. For example, the exact shape of an impulse can be obtained by playing the tape back at a slower speed.

4.7 REAL-TIME ANALYZERS [6,7,8]

In general, frequency-response signal analysis performed by a spectrum analyzer indicates the manner in which signal energy is distributed as a function of frequency. This technique is extremely important because frequency characteristics of noise and vibration signals are generally more revealing than the time-domain characteristics.

Spectrum analyzers are concerned with measurements of the signal's spectrum, which is more correctly referred to as the *power spectrum* or *power spectral density* (PSD). The PSD is a measure of the signal's distribution of power as a function of frequency. The signal's average total power is equal to the area under the PSD curve. A power measurement is one that is proportional to the square of a signal (e.g., voltage squared, pressure squared, or similar quantity).

In noise and vibration measurements, three general types of signals are of interest: transient, broadband, and narrow band. In the case of transient signals the quantity of interest is the energy change during the transient period. The dimensions of energy are, for example, proportional to (volts squared) seconds. For broadband or random signals, the measurements are a function of the analyzer's resolution bandwidth. Therefore, the measurements are divided by this factor in order to obtain data that are calibrated on a per-Hz basis (i.e., power per unit frequency). This procedure is not necessary for narrow-band signals (e.g., sinusoids) since all the energy is contained in a very narrow band (i.e., the power spectrum of the signal $\sin 2\pi f_0 t$ is simply a vertical line in the frequency domain at f_0).

Many times the signal of interest is embedded in noise. Therefore, some type of scheme must be employed to extract the signal. Since the noise is generally statistically independent and has a mean value of zero, it can be eliminated through ensemble averaging in either the frequency or time domain.

A conventional technique for performing a spectrum analysis and one that illustrates the general concepts involved in the frequency investigation of a signal is the scanning-filter or swept-filter approach. A scanning-filter spectrum analyzer employs a single bandpass filter, the center frequency of which can be shifted throughout the analyzer's frequency range. This *tracking filter*, as it is often called, scans through the data, analyzing only one frequency band at a time. Obviously, the time required to analyze the spectrum of a signal is a function of the tracking filter's bandwidth. Therefore, this approach may require several hours to analyze a signal, because of the serial nature of the processing. However, a new era in spectrum analysis was born with the advent of the real-time analyzer. These highly sophisticated systems have enormous processing capability and are several hundred times faster than a scanning-filter spectrum analyzer.

In general, real-time analysis means that the data-processing system keeps up with the incoming data in such a fashion that all data are used, none are lost, and the results are displayed essentially instantly. One of the earlier attempts to develop a real-time analyzer resulted in the development of a constant-bandwidth multifilter system. This unit, which employed as many as 500 filters, used parallel processing. However, problems associated with the filter characteristics were encountered, and therefore the most viable

units of this type are the $\frac{1}{3}$- and full-octave analyzers which are used extensively in noise analysis. The more modern and rather complicated techniques
are time compression and fast Fourier transform analyses. Time compression is a hybrid technique using analog filtering and digital sampling; this
has been the most widely used approach.

The newer fast Fourier transform (FFT) approach employs both digital
sampling and digital filtering. Unlike the other techniques, which all use a
bandpass filter to measure analog amplitudes which represent the signal's
spectrum, the FFT system performs an efficient transformation on the signal
from the time domain to the frequency domain. It is capable of performing
essentially any analysis function on the signal at high speed and therefore is
an extremely powerful technique.

There are a number of factors which affect the performance of these
sophisticated systems. Since both of these systems employ digital sampling,
the system sampling rate must satisfy the criterion

$$f_s > 2w \qquad\qquad (4.23)$$

where f_s = sampling frequency (Hz)

w = highest frequency present in the signal

The sampling rate $f_s = 2w$ is called the *Nyquist rate*.

In order to illustrate the importance of the sampling rate, consider
Figure 4.16. Figure 4.16(a) illustrates a possible result of sampling a signal
below the Nyquist rate and Figure 4.16(b) shows the result of sampling the
same signal above the Nyquist rate. One might conclude from Figure 4.16(a)
that the signal was just a dc level, whereas Figure 4.16(b) shows clearly that
the signal was actually a sine wave.

Suppose that the sampling rate which is perhaps selected for a particular
frequency range of interest is not fast enough to satisfy the Nyquist criteria
(i.e., the frequency content of the signal exceeds the range specified by the
sampling theorem). Under this condition the spectrum is distorted because
it contains erroneous frequency information caused by the high-frequency
components of the signal above w. This distortion is called *aliasing* and is
eliminated through the use of a low-pass or what is commonly called an
anti-aliasing filter, which essentially band-limits the signal. From a purely
pragmatic standpoint, a spectrum analysis must be performed in a finite
amount time. However, the mean spectral characteristics of a stationary
random signal (i.e., one whose statistical characteristics do not change with
time) can only be determined exactly by measuring the signal for an infinite
period of time. Therefore, a spectrum analysis yields only an estimate of
the spectral parameters. In addition, many signals exist for only a short
time (i.e., they are duration-limited).

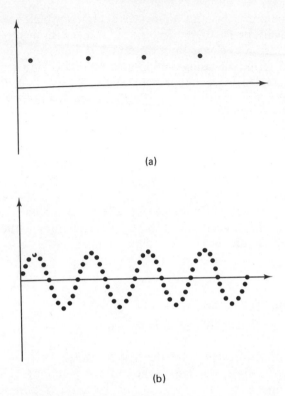

FIGURE 4.16 *Signal sampled above and below the Nyquist rate:* (a) $f_s < 2w$; (b) $f_s > 2w$

A problem of leakage, or what is known as *window error*, often occurs when the signal observation is duration-limited. There are two reasons for window error: the signal is not completely contained within the observation window, or it is not periodic within the window. This problem can often be corrected by means of a weighting or window function which modifies the shape of the observation window by tapering both leading and trailing edges of the data. Since the weighting function is applied to overcome duration limiting, it essentially increases the effective bandwidth. This increase is expressed by what is called the *noise bandwidth factor* (NBF). This factor, which ideally would be unity, is given by the expression

$$\text{NBF} = \frac{\text{effective bandwidth with window function}}{\text{effective bandwidth without window function}} \quad (4.24)$$

Another factor of importance in the use of weighting functions is the *side-lobe ratio* (SLR) or the highest side-lobe level. This selective characteristic is expressed as

$$\text{SLR} = \frac{\text{most sensitive out-of-band response}}{\text{center of bandwidth response}} \quad \text{dB} \qquad (4.25)$$

The best SLR is expressed by the largest negative number of decibels.

Table 4.3 shows the characteristics for three typical weighting functions. Note that a compromise must be achieved between the NBF and SLR.

TABLE 4.3

Typical weighting functions

Function	Shape	NBF (Hz)	SLR (dB)
Uniform	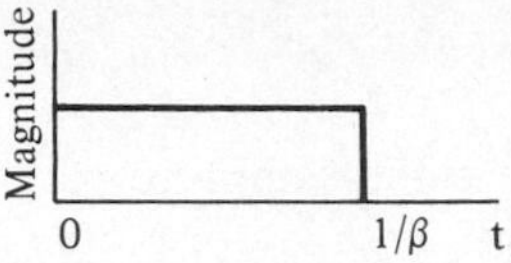	$1/\beta$	−13
Hanning $(\cos^2)$	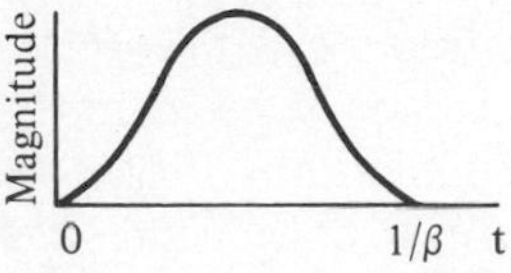	1.5β	−32
Hamming $(\cos^2 + 8\%$ pedestal$)$	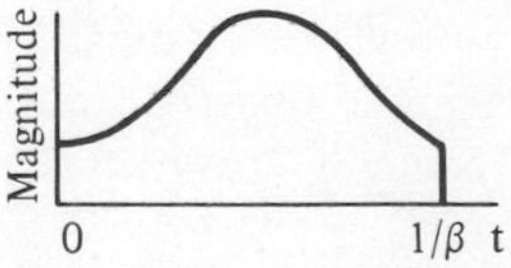	1.36β	−42

A parameter of primary importance which governs the ability of the spectrum analyzer to make accurate measurements is the frequency resolution. It has already been shown that if the sampling rate is not fast enough, erroneous conclusions can be drawn from the analysis (e.g., a signal that is actually a sine wave can be interpreted to be a dc level). An analogous situation exists with frequency resolution. If the resolution is not fine enough, the frequency spectrum of a signal may be completely misinterpreted and certain narrowband phenomena may go totally undetected.

In general, the resolution obtainable in the frequency domain is determined by the time length of measurement of the signal in the time domain.

Therefore, the frequency-domain resolution is given by the expression

$$\Delta f = \frac{1}{T} \quad \text{Hz} \tag{4.26}$$

where Δf = frequency resolution (Hz)

T = time length of measurement (s)

For example, in FFT analyzers the frequency range over which the Fourier transform is computed typically extends from dc to $f_s/2$. In this case the frequency resolution is given by the expression

$$\Delta f = \frac{1}{NT_s} \quad \text{Hz} \tag{4.27}$$

where N = number of samples in a block or frame which describes the function

$T_s = 1/f_s$ is the sampling period

The sampling frequency, and therefore the sampling period, are generally fixed by such things as the frequency range of interest and aliasing problems. Therefore, equation (4.27) indicates that to increase the resolution, the number of samples used to describe the signal must be increased. This approach is not necessarily viable either because the processing time increases with the block size or the block size is limited by memory-size constraints. Modern systems have overcome these limitations with a "zoom capability." This feature allows the spectrum analyzer to concentrate its entire resolution, whether it be 200 or 800 lines, in a small frequency interval which can be selected by the user. Therefore, with the zoom technique, arbitrarily fine frequency resolution is obtainable. However, one must keep in mind the trade-off between resolution and the required observation time. From equation (4.26) it can be seen that as the frequency resolution becomes finer, the observation time becomes longer.

> **EXAMPLE 4.9** If the frequency resolution desired for a signal is 4 Hz, then from equation (4.26) the observation time is $\frac{1}{4}$ s. However, if the frequency resolution is changed to 0.1 Hz, the observation time required is 10 s.

The system's memory-size requirements are also an important consideration. They are a function of both the sampling frequency and the frequency resolution. Typically, the memory is limited to a fixed number of words (e.g., 256 to 8K words). The following example indicates the manner in which memory size is dependent upon the sampling rate and the resolution.

> **EXAMPLE 4.10** Suppose that an analyzer is to process inputs over a 4-kHz range with a frequency resolution of 4 Hz. Since the frequency range of the

input signal is 4 kHz, the minimum sampling frequency according to equation (4.23) would be at least twice this rate. Therefore, suppose that a sampling frequency of 12 kHz is selected. If the frequency resolution is 4 Hz, then according to equation (4.26), the observation interval would be 0.25 s. In this observation interval a segment of data 0.25 s long must be stored and processed. Since the memory must store all the samples collected within a 0.25-s interval, the memory must be capable of storing 3000 samples or words.

In addition to the items mentioned above, there are other quantities of interest which affect system performance: for example, the number of bits employed in the A/D converter, the word size used by the processor, truncation or roundoff, and the like. The reader interested in exploring in detail the many performance characteristics of real-time analyzers is referred to the references given at the beginning of this section and the recent data sheets from the manufacturers of this equipment.

4.8 ENVIRONMENTAL NOISE MEASUREMENTS

Industry must not only control the noise within its plants, but it must also be concerned with the noise that it imparts to the environment. The sources of environmental noise pollution generated by industry include transportation and construction equipment as well as the normal industrial plant operations. The actual annoyance caused by industry in the environment is dependent upon many factors. For example, is the noise source broadband or a pure tone, and is it impulsive or continuous in nature? Some of the environmental factors that affect annoyance are the time of day at which the noise source is active, the land use of the affected area (e.g., recreational, residential, or commercial), the presence and history of other noise sources in the area, and the economic dependence of the affected area upon the noise source.

Environmental noise standards set forth by the government affect not only existing industrial plants but new and additional construction as well. The many facets and ramifications of environmental noise and the standards that regulate it are discussed in detail elsewhere [9]. However, one must remember that in order to reduce the environmental noise problem, in-plant noise sources must be controlled, using the techniques discussed in other chapters. Our purpose here is to describe some of the instrumentation that industry employs to evaluate this type of noise.

A system that can be used for an environmental noise analysis such as a plant boundary noise study is shown schematically in Figure 4.17. In general, the microphone and data-acquisition system are portable and environmentally protected for outdoor use. Either ac line or battery power may be employed. The microphone is typically a condenser or ceramic type of device. The data-acquisition system is in essence a sound level meter designed to

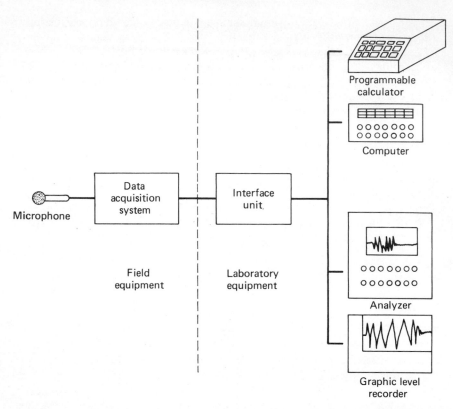

FIGURE 4.17 *Basic instrumentation system for the measurement and analysis of environmental noise*

meet ANSI type 1 or type 2 standards (depending upon the microphone employed) which operates in conjunction with a data sampling and storage system. Typical sampling rates are 1 to 10 s, the storage medium is, in general, an incremental magnetic tape cassette, and the data are recorded using *binary coded decimal* (BCD) digits.

The magnetic tape cassette generated by the field equipment can now be brought to the laboratory for processing. The interface unit reads data from the tape cassette and produces both analog and digital outputs, suitable as inputs, to the data processing and display equipment shown in Figure 4.17. In general, the tape playback is accelerated so that the data can be analyzed in only a fraction of the recording time.

The laboratory computational equipment can be used to process the data in order to determine such noise criteria as L_N, noise levels exceeded $N\%$ of the time; L_{eq}, the equivalent continuous sound level, in dBA; and L_{dn}, the day–night average sound level, in dBA. These values are used in federal, state, and local regulations and ordinances as compliance criteria. One should consult the regulations that pertain to a particular plant location prior to making measurements and calculations.

4.9 LABORATORY INSTRUMENTATION

When data taken in the field with a tape recorder are returned to the laboratory for processing, sophisticated, powerful, and expensive instrumentation is available for their detailed analysis. Equipment such as real-time analyzers, fast Fourier transform (FFT) processors, graphic level recorders, oscilloscopes equipped with cameras, mini and micro computers with their ancillary input–output equipment such as analog-to-digital and digital-to-analog converters, graphics terminals, and the like can be brought to bear to perform every conceivable analysis. The employment of the computer in the instrumentation system provides a myriad of possibilities for massaging the data to obtain the information vital for the design of control strategies.

PROBLEMS

4.1 A particular machine in a plant is under consideration. The sound pressure level of the machine is needed before more studies can be made. With the machine "on," the sound pressure level is measured to be 89 dB. With the machine "off," the sound pressure level is found to be 83 dB. What is the sound pressure level due to the machine?

4.2 A new machine will be placed in a large industrial plant. The plant area presently has a sound pressure level of 86 dB. When the machine is placed and turned "on," the sound pressure level of the area is measured and found to be 93 dB. Using Figure 1.3, determine the approximate sound pressure level associated with the new machine.

4.3 A particular machine in an area is under investigation for its contributions to the sound pressure level in that area. Because of the nature of the processes, shutdown is not practical. Therefore, a barrier must be constructed to reduce the background noise. Using Figure 4.3 as a model, given that $h = 6$ ft or 1.83 m and $l = 8$ ft or 2.44 m, find the dimensions of the required barrier. The disturbing frequency of the background noise can be taken to be 250 Hz.

4.4 Find E_{out} of a microphone with a sensitivity level of -100 dBV/μbar if it is exposed to a sound pressure level of 80 dB.

4.5 A microphone is exposed to a sound pressure level of 85 dB. The output voltage of the microphone is found to be 3×10^{-4} V. Determine the sensitivity level of the microphone.

4.6 A machine used for a cutting process is measured under the conditions of 68°F and 30 in. Hg at the plant and is found to produce a sound pressure level of 90 dB. The machine is to be used on a job site in a low-lying area in cold weather. The typical barometric pressure is 30.75 in. Hg and the typical temperature is 38°F. What sound pressure level will the machine produce at this job site?

4.7 Determine the lower cutoff frequency, f_1; the upper cutoff frequency, f_2; and the bandwidth, bw, for the octave bands, where:

    (a) $f_0 = 31.5$ Hz

    (b) $f_0 = 250$ Hz

4.8 Determine the lower and upper cutoff frequencies of the 4000-Hz, $\frac{1}{10}$-octave band, and find the bandwidth as a percentage of the center frequency.

REFERENCES

[1] CROCKER, M. J. "Use of Anechoic and Reverberant Rooms for Measurement of Noise from Machines," in *Reduction of Machinery Noise*, edited by M. J. Crocker, Purdue University, 1974, pp. 61–68.

[2] FAULKNER, L. L. "Fundamentals of Laboratory Facilities for Noise Rating and Product Development," *ASHRAE Trans.*, Vol. 80, Part 2 (1974), pp. 484–492.

[3] ANGEVINE, O. L. and M. V. BARSOTTELLI "Noise Measurements of Installed Machines" *ASHRAE Trans.*, Vol. 80, Part 2 (1974), pp. 505–515.

[4] BROCH, J. T. *Acoustic Noise Measurements*, B&K Instruments, Jan. 1973.

[5] PETERSON, A. P. G. and E. E. GROSS, JR. *Handbook of Noise Measurement*, 7th ed., General Radio, Concord, Mass., 1972.

[6] "Instrumentation for Dynamic Measurement," *Sound and Vibration*, Vol. 11, No. 3 (Mar. 1977).

[7] BICKEL, H. J. "Real-Time Spectrum Analysis," *Sound and Vibration*, Vol. 5, No. 3 (Mar. 1971).

[8] "How to Compare Real-Time and Fourier Analyzers," *Sound and Vibration*, Vol. 8, No. 3 (Mar. 1974).

[9] MAGRAB, E. B. *Environmental Noise Control*, John Wiley & Sons, Inc., New York, 1975.

BIBLIOGRAPHY

BERANEK, L. L. *Acoustics*, McGraw-Hill Book Company, New York, 1954.

BERANEK, L. L., ed. *Noise and Vibration Control*, McGraw-Hill Book Company, New York, 1971.

DIEHL, G. M. *Machinery Acoustics*, John Wiley & Sons, Inc., New York, 1973.

EBBING, C. E., T. S. KATRA and D. E. ROBERTSON "In-Place Machinery Noise Measurements," *ASHRAE J.*, June 1975, pp. 48–54.

KAMPERMAN, G. W. "Instrumentation for Noise Measurement," in *Proceedings of the Inter-Noise 72 Conference: Tutorial Papers on Noise Control*, Oct. 1972, Washington, D.C., pp. 84–97.

KEAST, D. N. *Measurements in Mechanical Dynamics*, McGraw-Hill Book Company, New York, 1967.

"Methods for Measurements of Sound Pressure Levels," *ANSI S1.13-1971*, American National Standards Institute, New York, 1971.

"Proposed USA Standard for Sound Level Meters," a revision of *ANSI S1.4-1961*, American National Standards Institute, New York, 1961.

TREE, D. R. "Instrumentation and Noise Measurements," in *Reduction of Machinery Noise*, edited by M. J. Crocker, Purdue University, 1974, pp. 51–61.

5

SOURCES OF NOISE

5.1 INTRODUCTION

A careful analysis of an industrial environment reveals that there exist essentially a myriad of noise sources. Many sources are peculiar to a particular industry, while others are common to a large number of widely diverse plants. Some typical noise sources commonly encountered are listed in Table 5.1. Other sources include such things as air hammers, conveyors, cranes, lathes, saws, transformers, welding machines, and the like, not to mention sources found in construction equipment and transportation elements.

Because of the enormous number of potential noise sources, we will

TABLE 5.1

Typical noise sources in industrial environs

Noise source category	Typical example
Combustion process	Furnace
Impact process	Punch, hammer
Electromechanical device	Motor, generator
Gas stream	Air intake, jet, vent
Metal contacting metal	Gear trains
Moving fluids in confined metal spaces	Ducts, pipe, valve
Moving metal surfaces contacting fluids	Compressor, fan, pump
Unbalanced rotating part	Shaft

simply examine some of the common ones typically found in industrial environments. Specifically, we will consider in some detail fans, electric motors, pumps, and compressors. Construction and building equipment, as well as certain household appliances, are treated briefly in several tables.

In general, one must always keep in mind that all methods of sound power estimation are exactly that—estimations. It is not uncommon for the estimation to be in error by more than 6 dB. This represents a power error greater than a factor of 4. Clearly, if possible, one should obtain measured data from the machine manufacturer or if such data are not available, it is preferred to determine the sound power level experimentally by the methods presented in Chapter 6.

Furthermore, all sound power and sound-power-level estimates presented in this chapter are total values for the four octave bands 500, 1000, 2000, and 4000 Hz. It is often reasonable and convenient to estimate the A-weighted sound level to be approximately equal to the sound pressure level calculated using the sound power level of these four bands.

5.2 ESTIMATION OF NOISE-SOURCE SOUND POWER

It is often necessary to estimate the expected sound power that a particular machine might introduce into an environment. One way in which such an estimate may be approached for a particular class of machine (e.g., electric motors, fans, etc.) is by means of the sound power conversion factor, F_n. This factor is defined as

$$F_n = \frac{P}{P_m} \quad \text{unitless} \tag{5.1}$$

where P = sound power of the machine (W)

P_m = power of the machine (W)

This relationship is valid for both mechanical and electrical machinery. The conversion factors for certain common noise sources are given in Table 5.2. (Note the large range of F_n for any given machine.)

This method of sound power estimation is probably the most simple and the least accurate; nevertheless, it is useful in making estimates when very little is known about the machine in question. In other sections of this chapter, more complex relations are presented for determining the sound power of certain common noise sources when a few of the machine's critical characteristics are known.

TABLE 5.2

Estimated sound power conversion factors for
certain common noise sources [1]
(Total sound power for the four octave bands from 500 to 4000 Hz)

	CONVERSION FACTOR (F_n)		
Noise source	*Low*	*Midrange*	*High*
1. Compressors, air (1–100 hp)	3×10^{-7}	5.3×10^{-7}	1×10^{-6}
2. Gear trains	1.5×10^{-8}	5×10^{-7}	1.5×10^{-6}
3. Loudspeakers	3×10^{-2}	5×10^{-2}	1×10^{-1}
4. Motors, diesel	2×10^{-7}	5×10^{-7}	2.5×10^{-6}
5. Motors, electric (1200 rpm)	1×10^{-8}	1×10^{-7}	3×10^{-7}
6. Pumps, over 1600 rpm	3.5×10^{-6}	1.4×10^{-5}	5×10^{-5}
7. Pumps, under 1600 rpm	1.1×10^{-6}	4.4×10^{-6}	1.6×10^{-5}
8. Turbines, gas	2×10^{-6}	5×10^{-6}	5×10^{-5}

EXAMPLE 5.1 Estimate the sound power level of an "average" 100-hp electric motor that operates at 1200 rpm.

SOLUTION From Table 5.2, we find that for average electric motors, $F_n = 1 \times 10^{-7}$. Thus, using equation (5.1) and 1 hp = 746 W,

$$P = F_n P_m = (1 \times 10^{-7})(100 \times 746) \text{ W}$$
$$= 7.46 \times 10^{-3} \text{ W}$$

is the total sound power of the motor. Then using equation (1.10),

$$L_w = 10 \log \left(\frac{P}{10^{-12}} \right) \text{ dB}$$

$$= 10 \log \left(\frac{7.46 \times 10^{-3}}{10^{-3}} \right)$$

$$= 99 \text{ dB}$$

which is the expected total sound power level of the motor.

D5.1 Using Table 5.2, estimate the sound power level for all three ranges in the four octave bands from 500 to 4000 Hz for a gas turbine rated at 20,000 hp.

Ans. 135 dB, 139 dB, 149 dB.

5.3 FAN OR BLOWER NOISE

In the absence of a manufacturer's actual test data obtained under controlled conditions, the sound power generated by a fan or blower can be estimated via the procedures outlined below. Techniques for predicting the sound power produced by centrifugal, axial, and induced draft fans will be described.

A technique for estimating the sound power from centrifugal and axial fans has been specified by Graham [2]. An estimate of the total sound power level in the four octave bands from 500 to 4000 Hz, at actual operating conditions, can be obtained from the expression

$$L_w = 10 \log F_r + 20 \log p_s + K_f \quad \text{dB} \tag{5.2}$$

where F_r = volume flow rate (ft³/min or m³/s)

   P_s = static pressure (in. H_2O or cm H_2O)

   K_f = sound-power-level constant, which is dependent on fan type and system of units (see Table 5.3)

Equation (5.2) assumes that the additional noise generated in the narrow band about the blade passage frequency falls outside the four octave bands from 500 to 4000 Hz. The fan constant, K_f, is given in Table 5.3 for a number of fan types, for both English and metric units.

The blade passage frequency, B_f, is calculated by

$$\boxed{B_f = \frac{N(\text{rpm})}{60} \quad \text{Hz}} \tag{5.3}$$

where N = number of blades

TABLE 5.3

Sound-power-level constants for centrifugal fans

Type of fan	K_f (English) (dB)	K_f (metric) (dB)
1. Axial, tube, or vane and centrifugal, radial	47	72
2. Centrifugal, airfoil blade, or forward or backward curved blade	34	59
3. Centrifugal, tubular	42	67
4. Propeller	52	77

Fans generate a narrow band of additional noise about this blade passage frequency which is rather troublesome in some instances. If B_f does fall within the four octave bands from 500 to 4000 Hz, then 3 dB should be added to the estimate for L_w obtained by equation (5.2). In the event an estimate of each of the individual octave bands is made, then an appropriate increment should be added to the band in which B_f occurs. If no better information is available, then 5 dB for the increment [3] is a reasonable assumption.

Clearly, these estimates are not intended to be highly accurate and should in no case be assumed to be such. More detailed estimation techniques are given in references [1,3,4] which may be useful where sufficient information about the fan and its installation are available.

Equation (5.2) for the total sound power level for each of the fans listed in Table 5.3 is plotted in Figures 5.1 through 5.8.

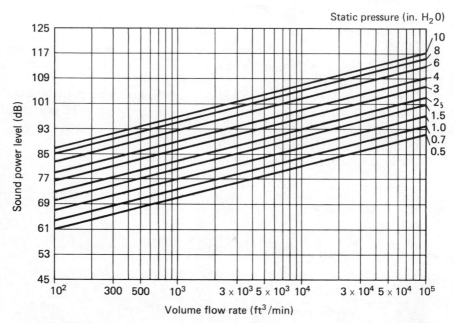

FIGURE 5.1 *Sound-power-level estimate for axial fans (tube or vane) and centrifugal fans (radial) in English units (total sound power level for the four octave bands from 500 to 4000 Hz)*

EXAMPLE 5.2 A certain 3.7-hp centrifugal fan, with backward-curved blades, produces a volume flow rate of 6000 ft³/min at a static pressure of 1.5 in. H₂O. The fan has 50 blades and operates at 1200 rpm.

(a) Calculate the blade passage frequency.

(b) Estimate the total sound power level in the four octave bands from 500 to 4000 Hz. Use equation (5.2).

SOLUTION The blade passage frequency is, from equation (5.3),

$$B_f = \frac{N(\text{rpm})}{60} \quad \text{Hz}$$

$$= \frac{(50)(1200)}{60} = 1000 \text{ Hz}$$

$$= 1000 \text{ Hz}$$

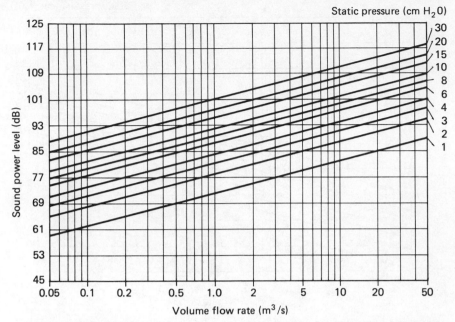

FIGURE 5.2 *Sound-power-level estimate for axial fans (tube or vane) and centrifugal fans (radial) in metric units (total sound power level for the four octave bands from 500 to 4000 Hz)*

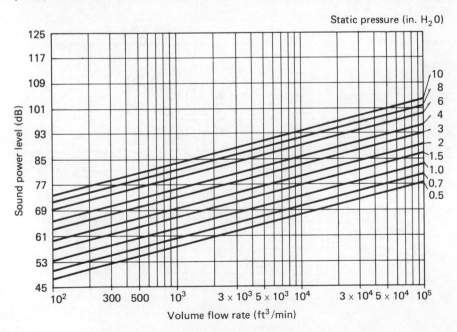

FIGURE 5.3 *Sound-power-level estimate for centrifugal fans (airfoil blade or forward or backward curved blade) in English units (total sound power level for the four octave bands from 500 to 4000 Hz)*

119

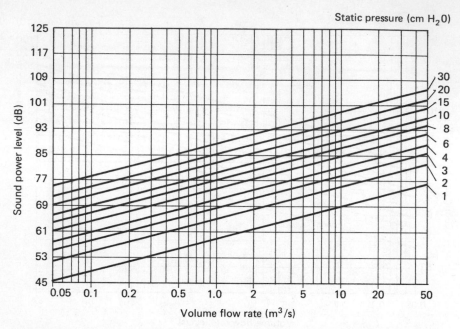

FIGURE 5.4 *Sound-power-level estimate for centrifugal fans (airfoil blade or forward or backward curved blade) in metric units (total sound power level for the four octave bands from 500 to 4000 Hz)*

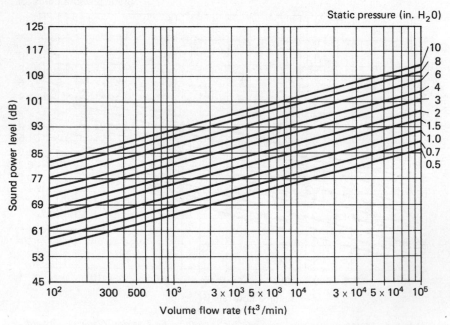

FIGURE 5.5 *Sound-power-level estimate for centrifugal fans (tubular) in English units (total sound power level in the four octave bands from 500 to 4000 Hz)*

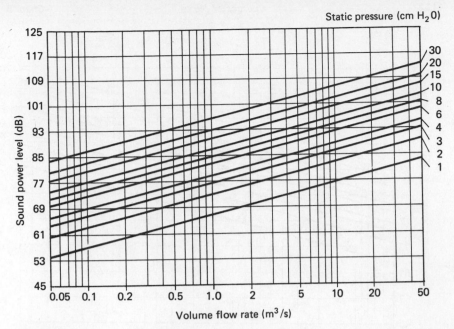

FIGURE 5.6 *Sound-power-level estimate for centrifugal fans (tubular) in metric units (total sound power level in the four octave bands from 500 to 4000 Hz)*

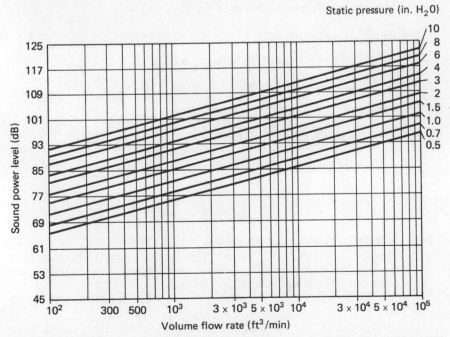

FIGURE 5.7 *Sound-power-level estimate for propeller fans in English units (total sound power level in the four octave bands from 500 to 4000 Hz)*

121

Static pressure (cm H$_2$0)

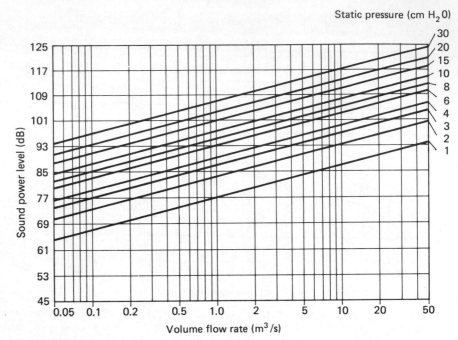

FIGURE 5.8   *Sound-power-level estimate for propeller fans in metric units (total sound power level in the four octave bands from 500 to 4000 Hz)*

Estimate L_w by equation (5.2):

$$L_w = 10 \log F_r + 20 \log P_s + K_f \quad dB$$
$$= 10 \log (6000) + 20 \log (1.5) + 34 \ dB$$
$$= 75 \ dB$$

Note that we obtain this same result, $L_w = 75$ dB, using Figure 5.3. Since $B_f = 1000$ Hz, it falls within the four octave bands in question. Therefore, we add 3 dB to the value of L_w obtained by equation (5.2). Thus,

$$L_w = 75 + 3 \ dB$$
$$= 78 \ dB$$

D5.2   Assume that a tube axial fan is rated at 3 hp and delivers 4 m³/s at a static pressure of 3 cm H$_2$O. If it has 40 blades and operates at 1200 rpm, determine: (a) the blade passage frequency   (b) L_w in the four octave bands from 500 to 4000 Hz by equation (5.2).

Ans.   (a) 800 Hz   (b) 91 dB.

An expression [5] is available for estimating the maximum sound power level at the discharge of induced draft fans at the blade passage frequency.

The maximum sound power level can be estimated by

$$L_w = 10 \log \text{hp} + 10 \log p_s + K_{fd} \quad \text{dB} \qquad (5.4)$$

where hp = rated horsepower (750–7500 hp)

P_s = static pressure in. H_2O (50–80 in.) or cm H_2O (125–200 cm)

K_{fd} = induced draft fan constant, 90 dB for English units and 86 dB for metric units

EXAMPLE 5.3 An induced draft fan has 12 blades and is rated at 1250 hp at 1200 rpm. It delivers 2540 ft³/s at a static pressure of 65 in. H_2O. Calculate B_f and determine L_w about this frequency.

SOLUTION The blade passage frequency is, from equation (5.3),

$$B_f = \frac{(12)(1200)}{60} \text{ Hz}$$

$$= 240 \text{ Hz}$$

Therefore, at this frequency the sound power level is, from equation (5.4), estimated to be

$$L_w = 90 + 10 \log (1250) + 10 \log (65) \text{ dB}$$

$$= 139 \text{ dB}$$

Finally, it is important to note that the noise produced by fans can be significantly increased if the fans are not of the proper size, do not operate at maximum efficiency, or do not have inlet and outlet duct connections that provide for straight and uniform air flow.

D5.3 Assume that an induced draft fan is rated at 1500 hp and has nine blades. It operates at 1200 rpm. If it delivers 50 m³/s with a static pressure of 120 cm H_2O, determine B_f and L_w about this frequency.

Ans. 180 Hz, 139 dB.

5.4 ELECTRIC MOTORS AS A SOURCE OF NOISE

Noise from electric motors originates from a number of sources. Some fundamental noises associated with the motor's operation are caused by such phenomena as rotational unbalance, rotor/stator interaction, and slot harmonics. Noise is also produced by excitation of the natural vibration frequency of the motor structure, air movement, and air resonant chambers. The amplitude of the noise is typically a function of speed and size, and the frequency distribution is normally a function of speed and construction.

National and international standards have been proposed for characterizing the noise emission of rotating electric machinery. For example, ISO Recommendation R1680, International Electrotechnical Commission Publication IEC 34-9 and BS4999:Part 51:1973, and NEMA MG1-12.49 and MG1-12.77. The noise generated by the machine should be measured in accordance with IEEE Standard 85-1973, IEEE Test Procedure for Airborne Sound Measurements on Rotating Electric Machinery. The standards provide a data base from which the evaluation of the noise performance of a particular motor can be determined relative to other motors which have the same rating and construction.

For an electric motor, the total sound power level in the four octave bands 500, 1000, 2000, and 4000 Hz may be estimated by the expression

$$L_w = 20 \log \text{hp} + 15 \log(\text{rpm}) + K_m \quad \text{dB} \qquad (5.5)$$

where   hp = rated horsepower (1–300 hp)

rpm = rated speed (revolutions per minute)

K_m = motor constant (13 dB)

More involved techniques [1,3,6] are available which suggest different motor constants for each of the octave bands of interest.

EXAMPLE 5.4   An electric motor rated at 100 hp operates at 1200 rpm. Estimate the sound power level in the four octave bands from 500 to 4000 Hz.

SOLUTION   Using equation (5.5),

$$L_w = 20 \log \text{hp} + 15 \log(\text{rpm}) + K_m \quad \text{dB}$$
$$= 20 \log(100) + 15 \log(1200) + 13 \text{ dB}$$
$$= 99 \text{ dB}$$

which is the total sound power in the four bands.

D5.4   Estimate the sound power level in the four octave bands from 500 to 4000 Hz for an electric motor with a rated horsepower and speed of 300 hp and 2400 rpm, respectively.

Ans.   113 dB.

5.5 PUMP NOISE

In general, pump noise originates from both hydraulic and mechanical sources, such as cavitation, pressure fluctuations in the fluid, the impact of mechanical parts, imbalance, resonance, misalignment, and the like. Hydraulic sources are normally the predominant noise generators. Pumps are prone to generate even more noise if they are not operated at rated speed and discharge pressure, if the rate of compression is high, if the inlet pressure is below atmospheric, or if the temperature is above 49°C. In addition, it is unfortunate that the noise can be transmitted via the fluid or structure to other system components, such as piping.

Pumps generate noise in two forms: discrete tones and broadband noise. The pump's fundamental pumping frequency is given by the product of the speed in revolutions per second and the number of pump chamber pressure cycles per revolution. In large pumps the noise emission is greatest at this fundamental frequency. As the pump size decreases, the frequency at which the maximum noise emission occurs increases—typically to a frequency that is several harmonics of the fundamental. Above approximately 3 kHz, the noise is broadband, with a spectrum that is essentially flat. This type of noise is caused by such things as cavitation and high-velocity flow.

The total sound power level of pumps in the four octave bands 500, 1000, 2000, and 4000 Hz can be estimated by the following expression:

$$\boxed{L_w = 10 \log \text{hp} + K_p \quad \text{dB}} \qquad (5.6)$$

where K_P = pump constant: 95 dB for centrifugal, 100 dB for screw, and 105 dB for reciprocating pumps (below rated speeds of 1600 rpm, subtract 5 dB)

One may also estimate the power in each of the four octave bands to be 6 dB less than the L_w calculated in equation (5.6). More detailed conditions are given in references [1,3,7,8,17] for estimating the sound power level in the various octave bands of interest.

Sound power level versus horsepower is plotted for centrifugal, screw, and reciprocating pumps in Figure 5.9 according to equation (5.6).

EXAMPLE 5.5 A particular 100-hp screw pump operates at 2400 rpm. Estimate: (a) the sound power level in the four octave bands from 500 to 4000 Hz; (b) the sound power level in each of the octave bands 500, 1000, 2000, and 4000 Hz.

SOLUTION Using equation (5.6),

$$L_w = 10 \log \text{hp} + K_P \quad \text{dB}$$
$$= 10 \log (100) + 100$$
$$= 120 \text{ dB} \quad \text{(same in Figure 5.9)}$$

which is the total for the four bands.

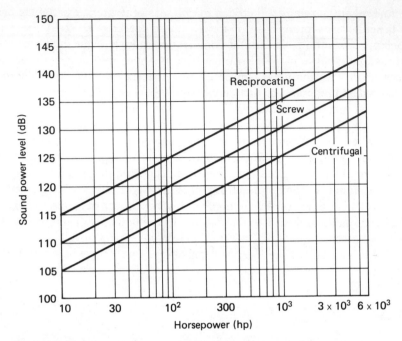

FIGURE 5.9 *Sound-power-level estimates for centrifugal, screw, and reciprocating pumps with a rated speed of 1600 rpm or more; subtract 5 dB if rated
under 1600 rpm (total sound power level for the four octave bands from 500 to
4000 Hz)*

In each of the four bands we obtain

$$L_{w_0} = L_w - 6 = 120 - 6$$
$$= 114 \text{ dB}$$

in each of the bands 500, 1000, 2000, and 4000 Hz.

D5.5 Estimate the sound power level in the four octave bands from 500 to 4000
Hz for a reciprocating pump rated at 150 hp and operating at 1200 rpm.
Also estimate the sound power level in each of the four bands individually.
Ans. 122 dB, 116 dB.

5.6 AIR COMPRESSOR NOISE

Compressors, which have been identified as one of the major sources of
noise, find extensive use in many types of major industries. They are used to
raise the pressure of a gas and are usually driven by a motor or turbine. The
noise emission characteristics of a compressor are a function of the type of

unit. Diehl [9] points out that in portable air compressors the driving engine is the major source of noise, not the compressor. The second largest noise source is the cooling fan. A reciprocating compressor typically generates a strong low-frequency pulsating noise, the characteristics of which are dependent upon the rotational speed and the number of cylinders. Diehl [10] indicates that the noise generated by centrifugal compressors is a function of a number of parameters such as the interaction of rotating and stationary vanes, the radial distance between impeller blades and diffuser vanes, the rotational speed, the number of stages, the inlet design, the horsepower input, turbulence, the gas molecular weight, and the mass flow.

As for fans, the blade passage frequency is an important frequency component in certain types of compressors. However, in diffuser-type machines, what is called the *blade-rate component* is of primary importance. The frequency of this component is determined by the equation

$$\boxed{f_{\text{BRC}} = \frac{N_r \times N_s}{K_{\text{BRC}}} \times \frac{\text{rpm}}{60} \quad \text{Hz}} \qquad (5.7)$$

where f_{BRC} = blade-rate component frequency (Hz)

N_r = number of rotating blades

N_s = number of stationary vanes

K_{BRC} = greatest common factor of N_r and N_s

EXAMPLE 5.6 The frequency of the blade-rate component of a diffuser-type compressor with $N_r = 8$ and $N_s = 12$ which operates at a speed of 6000 rpm is

$$f_{\text{BRC}} = \frac{8 \times 12}{4} \times \frac{6000}{60} \text{ Hz}$$

$$= 2400 \text{ Hz}$$

Note that if $N_r = 7$ and $N_s = 12$, then

$$f_{\text{BRC}} = \frac{7 \times 12}{1} \times \frac{6000}{60} \text{ Hz}$$

$$= 8400 \text{ Hz}$$

In this example, we note that f_{BRC} may fall well outside the frequency range of interest. If f_{BRC} falls within the range of interest, one should expect an increment of several decibels in the octave-band sound power level in which it occurs.

The total sound power level in the four octave bands 500, 1000, 2000, and 4000 Hz may be estimated for centrifugal and reciprocating compressors by the expression

$$\boxed{L_w = 10 \log \mathrm{hp} + K_c \quad \mathrm{dB}} \tag{5.8}$$

where K_c = air compressor constant; 86 dB (rated 1–100 hp)

This equation is plotted in Figure 5.10.

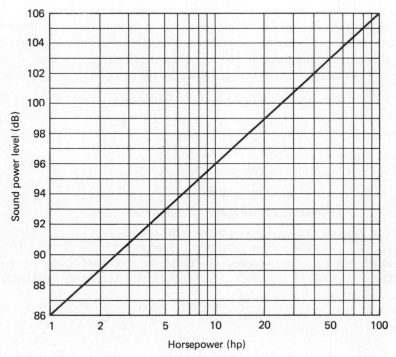

FIGURE 5.10 *Sound-power-level estimate for both centrifugal and reciprocating air compressors (total sound power level for the four octave bands from 500 to 4000 Hz)*

Again, it is also reasonable to estimate that the sound power level is equally divided among the four octave bands. Thus, each band level is 6 dB below the total determined by equation (5.8).

Much more detailed information is to be found in references [9–17].

EXAMPLE 5.7 A reciprocating air compressor is rated at 55 hp. Estimate the sound power level in the four octave bands from 500 to 4000 Hz, and in each of the four bands individually.

SOLUTION   Using equation (5.8), we obtain

$$L_w = 10 \log hp + K_c \quad dB$$
$$= 10 \log (55) + 86 \, dB$$
$$= 103 \, dB$$

for the four octave bands.

Then in each of the four bands, one obtains

$$L_{w_0} = L_w - 6 = 103 - 6 \, dB$$
$$= 97 \, dB$$

D5.6   Estimate the sound power level in the four octave bands from 500 to 4000 Hz, for a centrifugal air compressor rated at 40 hp.

Ans.   102 dB.

5.7 NOISE PRODUCED BY TYPICAL BUILDING AND CONSTRUCTION EQUIPMENT [3,18]

It is often desirable to estimate the sound power level of building equipment prior to actual purchase or prior to determining the specific equipment to be purchased. Table 5.4 contains an estimate listing of certain typical pieces of building equipment. Similarly, before construction is begun at a particular site, it is well to estimate the noise that will be produced. This may be done

TABLE 5.4

Sound-power-level estimates for typical building equipment
(Total sound power level in the four octave bands from 500 to 4000 Hz)

	SOUND POWER LEVEL (dB)		
Equipment	*Low*	*Midrange*	*High*
1. Air compressor	85	100	115
2. Air-cooled condenser	90	100	105
3. Boiler	65	80	100
4. Cooling tower	95	110	120
5. Pneumatic transport system	70	90	110
6. Pump	55	80	105
7. Rooftop air conditioner	80	90	100
8. Steam valve	70	85	105
9. Transformer	80	85	90
10. Unit heater	55	70	90

by first estimating the sound power level of the pieces of equipment that will be involved. A listing of certain typical construction equipment is given in Table 5.5.

In both cases, Tables 5.4 and 5.5, the total expected sound power level in the low, middle, and high ranges are given for the four octave bands 500, 1000, 2000, and 4000 Hz. The sound power level produced by these four octave bands is particularly useful, since, as is readily seen in Figure 2.8, the A-weighted sound level and the sound pressure level of the composite of these four bands are, in general, rather close together.

We may estimate the total sound pressure level in the four bands using the sound power levels given in Tables 5.4 and 5.5, if we can characterize the environment in which the equipment will be operating. In the case of construction equipment, one may often assume a simple outdoor or free field. Under this condition, and with a directivity of unity, as will be derived in Chapter 6, we have in English units,

$$\boxed{L_p = L_w - 10 \log (4\pi r^2) + 10 \quad \text{dB}}$$
(5.9)

TABLE 5.5

Sound-power-level estimates for typical construction equipment
(Total sound power level in the four octave bands from 500 to 4000 Hz)

Equipment	Sound Power Level (dB)		
	Low	Midrange	High
1. Backhoe	105	120	130
2. Compressor, stationary	110	115	120
3. Concrete mixer	110	115	125
4. Concrete pump	110	115	120
5. Crane, movable	110	115	120
6. Crane, stationary	115	120	125
7. Front loader	105	115	120
8. Generator, stationary	105	110	120
9. Jack hammer	115	125	135
10. Paver	115	120	125
11. Pile driver (peak value)	130	135	140
12. Pneumatic wrench	115	120	125
13. Pump, stationary	100	105	110
14. Rock drill	115	125	135
15. Scraper or grader	115	120	130
16. Tractor	110	120	130
17. Truck	115	120	130

or, in metric units,

$$\boxed{L_p = L_w - 10 \log (4\pi r^2) \quad \text{dB}}$$ (5.10)

where r = distance from the source (ft or m)

In Table 5.6 are calculated values of the sound level in dBA for a distance $r = 50$ ft for the construction equipment listed in Table 5.5. Note that we have, under this special condition, assumed that the sound levels, L, in dBA are equal to the calculated values of L_p. Furthermore, observe that if the distance, r, is doubled in equation (5.9), the value of L_p is reduced by 6 dB. Thus, in Table 5.6, the sound level at 100 ft may be obtained by subtracting 6 dBA from the value that one would expect at 50 ft.

EXAMPLE 5.8 Estimate the midrange sound pressure level in the four octave bands from 500 to 4000 Hz at a distance of 75 ft from a paver. Assume a directivity of unity.

TABLE 5.6

*Sound level estimates for the construction
equipment listed in table 5.5 (outdoors)*

	Sound Level (dBA) ($r = 50$ ft)		
Equipment	*Low*	*Midrange*	*High*
1. Backhoe	70	85	95
2. Compressor, stationary	75	80	85
3. Concrete mixer	75	80	90
4. Concrete pump	75	80	85
5. Crane, movable	75	80	85
6. Crane, stationary	80	85	90
7. Front loader	70	80	85
8. Generator, stationary	70	75	85
9. Jack hammer	80	90	100
10. Paver	80	85	90
11. Pile driver (peak value)	95	100	105
12. Pneumatic wrench	80	85	90
13. Pump, stationary	65	70	75
14. Rock drill	80	90	100
15. Scraper or grader	80	85	95
16. Tractor	75	85	95
17. Truck	80	85	95

SOLUTION   In Table 5.5 we find

$$L_w \text{ (paver, midrange)} = 120 \text{ dB}$$

Using equation (5.9),

$$L_p = L_w - 10 \log (4\pi r^2) + 10 \text{ dB}$$
$$= 120 - 10 \log [4\pi(75)^2] + 10 \text{ dB}$$

Thus,

$$L_p = 82 \text{ dB}$$

Note that if we double the distance and make $r = 150$ ft, we obtain

$$L_p = 76 \text{ dB}$$

which is 6 dB less than that obtained for 75 ft.

D5.7   Estimate the high-range sound level in dBA in the four octave bands from 500 to 4000 Hz, at a distance of 200 ft from a concrete mixer. Assume unity directivity.

Ans.   78 dBA.

5.8 HOME APPLIANCE NOISE [18]

A number of typical home appliances with estimates of their sound power levels are listed in Table 5.7 [3,18]. The sound power levels are estimates

TABLE 5.7

Sound-power-level estimates for typical home appliances
(Total sound power level for the four octave bands from 500 to 4000 Hz)

	SOUND POWER LEVEL (dB)		
Appliance	*Low*	*Midrange*	*High*
1. Air conditioner	55	70	80
2. Clothes dryer	55	70	75
3. Clothes washer	55	70	85
4. Dishwasher	60	75	85
5. Electric can opener	60	75	85
6. Electric shaver	55	70	80
7. Fan	45	65	80
8. Food blender	70	85	100
9. Food mixer	55	80	90
10. Food waste disposer	75	90	105
11. Hair dryer	65	70	75
12. Home shop tools	85	97	110
13. Refrigerator	40	50	65
14. Vacuum cleaner	70	80	95
15. Water closet	55	75	85

of the total sound power level in the four octave bands from 500 to 4000 Hz.

It is also reasonable, in many cases, to estimate the sound level, L, in dBA to be approximately equal to the sound pressure level, L_p, calculated using the L_w estimates of Table 5.7. Equations are derived in Chapter 6 with which to calculate L_p in various environments and positions once L_w is determined.

PROBLEMS

5.1 Estimate the sound power level in the four octave bands from 500 to 4000 Hz of a gear train that transfers 300 hp. Make estimates for expected levels in the low, middle, and high ranges.

5.2 Estimate the midrange sound power level in the four octave bands from 500 to 4000 Hz of an electric motor rated at 200 hp and operating at 1200 rpm. Compare this result with that calculated using equation (5.5).

5.3 A certain 10-hp centrifugal fan with airfoil blades produces a volume flow of 12,000 ft³/min at a static pressure of 2 in. H_2O. The fan has 45 blades and operates at 1200 rpm. Calculate the blade passage frequency and estimate the total sound power level in the four octave bands from 500 to 4000 Hz. [Estimate by means of equation (5.2).]

5.4 An induced draft fan has 16 blades and is rated at 1400 hp at 1200 rpm. The fan delivers 3000 ft³/s at a static pressure of 50 in. H_2O. Calculate B_f and estimate the sound power level generated about this frequency.

5.5 Estimate the sound power level in the four octave bands from 500 to 4000 Hz for an electric motor rated at 250 hp and operating at 2400 rpm. Use equation (5.5). Compare your result with the high-range value using Table 5.2.

5.6 A certain reciprocating pump rated at 165 hp operates at 2400 rpm. Estimate the sound power level in the four octave bands from 500 to 4000 Hz and the sound power level in each of the four individual octave bands. Compare your result with that obtained using a midrange value from Table 5.2.

5.7 Estimate the sound power level in the four octave bands from 500 to 4000 Hz for a centrifugal air compressor rated at 75 hp. Compare your result with that determined using a midrange value from Table 5.2.

5.8 Estimate the midrange sound pressure level in the four octave bands from 500 to 4000 Hz at a distance of 80 ft from a jack hammer. Assume outdoors and unity directivity. (Use Table 5.5)

REFERENCES

[1] HECKL, M. and H. A. MÜLLER, eds. *Taschenbuch der technischen Akustik*, Springer-Verlag, Berlin, 1975.

[2] GRAHAM, J. B. "How to Estimate Fan Noise," *Sound and Vibration*, May 1972, pp. 24–27.

[3] MAGRAB, E. B. *Environmental Noise Control*, John Wiley & Sons, Inc., New York, 1975.

[4] GROFF, G. C., J. R. SCHREINER, and C. E. BULLOCK "Centrifugal Fan Sound Power Level Prediction," *ASHRAE Trans.*, Vol. 73, Part II (1967), pp. V.4.1–V.4.18.

[5] HOOVER, R. M. and C. O. WOOD "Noise Control for Induced Draft Fans," *Sound and Vibration*, Apr. 1970, pp. 20–24.

[6] WEBB, J. D., ed. *Noise Control in Industry*, Sound Research Laboratories Limited, Holbrook Hall, Sudbury, Suffolk, Great Britain, 1976 (distributed by Halsted Press).

[7] MEYERSON, N. "Sources of Noise in Power Plant Centrifugal Pumps with Considerations for Noise Reduction," *Noise Control Eng.*, Vol. 2, No. 2, pp. 74–80.

[8] BECKER, R. J. "Noise of Hydraulic Equipment," Seventh Institute on Noise Control Engineering, Dearborn, Mich., Aug. 1973, Chap. 18.

[9] DIEHL, G. M. "Stationary and Portable Air Compressors," in *Proceedings of the Inter-Noise 72 Conference: Tutorial Papers on Noise Control*, Oct. 1972, Washington, D.C., pp. 154–158.

[10] DIEHL, G. M. "Centrifugal Compressor Noise Reduction," in *Reduction of Machinery Noise*, edited by M. J. Crocker, Purdue University, 1974, pp. 265–270.

[11] OSTERGAARD, P. B. "Industrial Noise Sources and Control," in *Proceedings of the Inter-Noise 72 Conference: Tutorial Papers on Noise Control*, Oct. 1972, Washington, D.C., pp. 65–68.

[12] BAADE, P. K. "Identification of Noise Sources," in *Proceedings of the Inter-Noise 72 Conference: Tutorial Papers on Noise Control*, Oct. 1972, Washington, D.C., pp. 98–114.

[13] "Bibliography on Noise Control," in *Proceedings of the Inter-Noise 72 Conference: Tutorial Papers on Noise Control*, Oct. 1972, Washington, D.C., pp. 519–548.

[14] HAMILTON, J. F. and D. R. TREE "Noise Reduction of Small Compressors," in *Proceedings of the 1974 Inter-Noise Conference*, Sept. 1974, Washington, D.C., pp. 337–342.

[15] PATTERSON, W. N.   "Portable Air Compressor Noise Diagnosis and Control," in *Proceedings of the 1974 Inter-Noise Conference*, Sept. 1974, Washington, D.C., pp. 519–524.

[16] BLAZIER, W. E., JR.   "Noise from Large Centrifugal Compressors," in *Proceedings of the 1971 Purdue Noise Control Conference*, July 1971, Purdue University, pp. 84–89.

[17] HEITNER, I.   "How to Estimate Plant Noises," *Hydrocarbon Processing*, Vol. 47, No. 12 (1968), pp. 67–74.

[18] "Noise from Construction Equipment and Operation, Building Equipment and Home Appliances," *Report No. NTID 300.1*, Environmental Protection Agency, Washington, D.C., Dec. 31, 1971.

6

ROOM ACOUSTICS

6.1 INTRODUCTION

In an industrial plant, or for that matter in an office or home, the sound behavior can be closely approximated from the knowledge of a few very fundamental noise source and room characteristics. Developments and derivations for sound pressure level, sound power level, reverberation time, and certain other less prominent associated equations are presented in this chapter.

As a starting point for the derivations, the solutions of the wave equations in velocity and pressure were chosen. Detailed solutions of these equations are found in Beranek [1]. Energy-density relationships for the direct and reverberant fields are then derived which provide the basis for the development of the desired, readily applicable equations. Several references [1–4] are provided which are recommended for those interested in further study.

6.2 SOUND FIELD DESIGNATION

Two sound fields, a direct and a multiple-reflected or reverberant field, exist in general in the total space of a room. As depicted in Figure 6.1, one notes that the direct field is only source- and distance-dependent. That is, the direct field is not affected by the size and reflective characteristics of the room. In contrast, in Figure 6.2 we find that the reverberant field is strongly dependent upon the size and reflectivity of the surfaces of the room. Finally, as shown in Figure 6.3, both sound fields exist simultaneously, with the magnitude of the reverberant field building up to a level determined by the acoustical losses of

136

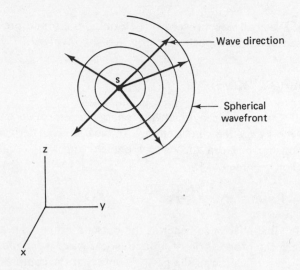

FIGURE 6.1 *Propagation of a spherical wavefront from an acoustic point source, s*

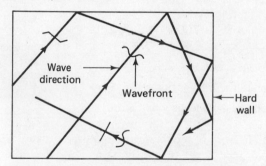

FIGURE 6.2 *Generation of an acoustical reverberant field by multiple reflections from the room walls*

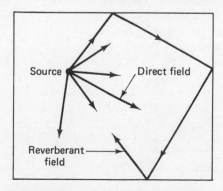

FIGURE 6.3 *Composite of both direct and indirect acoustical waves in a room*

137

the room. We shall now proceed to determine the sound pressure level of a room based upon the acoustical energy density involved.

6.3 ENERGY DENSITY

As is well known, we can break the energy up into two components, the potential energy and the kinetic energy. In the following derivation we require the relationship between the energy density and the sound source in its acoustical environment.

6.3.1 Potential Energy

For simplicity, let us consider at the outset a progressive sinusoidal wave in air. Since the progression of an acoustical wave is manifested by a compression and rarification of the gas, we note that the compression is, in fact, storing potential energy. The energy stored, or work done, is, from elementary thermodynamics,

$$E_p = -\int_v p_t \, dV' \quad \text{J} \tag{6.1}$$

where E_p = potential energy (J)

p_t = total instantaneous pressure (Pa)

V' = total instantaneous volume (m³)

The negative sign in front of the integral is required because when the volume becomes smaller (goes negative) by compression, the potential energy is positive.

We now require the bulk modulus which characterizes the medium in which the wave is propagating. The *bulk modulus* is defined as the ratio of the excess pressure to the expansion of a constant mass of the medium. That is, the adiabatic bulk modulus is given by

$$B = -V \frac{dp_t}{dV'} \quad \text{Pa} \tag{6.2}$$

where B = adiabatic bulk modulus (Pa)

V = static volume (m³)

V' = instantaneous volume (m³)

p_t = instantaneous pressure (Pa)

Rearranging equation (6.2), We obtain

$$dV' = -\frac{V}{B}\,dp_t \quad \text{m}^3 \tag{6.3}$$

which, substituted into equation (6.1), yields

$$E_p = \frac{V}{B}\int_{p_0}^{p_t} p_t\,dp_t = \frac{V}{B}\left[\frac{1}{2}p_t^2\right]_{p_0}^{p_t} \frac{V}{2B}(p_t^2 - p_0^2) \quad \text{J} \tag{6.4}$$

where p_0 = static pressure (Pa)

The instantaneous pressure, p_t, in its component parts, is

$$p_t = p' + p_0 \quad \text{Pa} \tag{6.5}$$

where p' = instantaneous time-varying component of pressure (Pa)

Upon substituting equation (6.5) into (6.4), we obtain

$$E_p = \frac{V}{2B}(p_t^2 - p_0^2) = \frac{V}{2B}[(p')^2 + 2p'p_0] \quad \text{J} \tag{6.6}$$

The adiabatic bulk modulus, B, however, can also be written as

$$B = \rho_0 c^2 \quad \text{Pa} \tag{6.7}$$

where ρ_0 = static gas density (kg/m^3)

   c = velocity of propagation (m/s)

Substituting equation (6.7) into (6.6) yields

$$E_p = \frac{V}{2\rho_0 c^2}[(p')^2 + 2p'p_0] \quad \text{J} \tag{6.8}$$

Moreover, it is convenient to take advantage of the relation

$$p' = \rho_0 c v \quad \text{Pa} \tag{6.9}$$

where v = instantaneous particle velocity (m/s)

Upon substitution of equation (6.9) into (6.8), we obtain

$$E_p = \frac{V}{2}\left[\frac{\rho_0^2 c^2 v^2 + 2\rho_0 c v p_0}{\rho_0 c^2}\right] \quad \text{J} \tag{6.10}$$

which expresses the potential energy, E_p, of the arbitrary volume, V, in terms of the medium characteristics and the particle velocity, v. In the case of air, the particle velocity is that of the gas molecules.

6.3.2 Kinetic Energy

Let us now leave the potential energy for the moment and turn our attention to a determination of the kinetic energy. The kinetic energy of the total volume of gas is given by

$$E_k = (\tfrac{1}{2}\rho_0 v^2)V \quad \text{J} \tag{6.11}$$

where E_k = kinetic energy (J)

We note that the expression for the kinetic energy of the volume of gas is an adaptation of the familiar expression from mechanics,

$$E_k = \tfrac{1}{2}mv^2 \quad \text{J} \tag{6.12}$$

where m = mass of the body (kg)

6.3.3 Total Energy Density

The total instantaneous energy in the volume in question is the summation of the potential and kinetic energies:

$$E = E_p + E_k \quad \text{J} \tag{6.13}$$

where E = total instantaneous energy (J)

We now expand equation (6.13) by substituting in E_p and E_k from equations (6.10) and (6.11), respectively. Thus,

$$E = \frac{V}{2}\left[\frac{\rho_0^2 c^2 v^2 + 2\rho_0 cvp_0}{\rho_0 c^2} + \rho_0 v^2\right] \quad \text{J} \tag{6.14}$$

Upon simplification, we obtain

$$E = V\left(\rho_0 v^2 + \frac{p_0}{c}v\right) \quad \text{J} \tag{6.15}$$

which is an expression for the total instantaneous energy within the volume. For our purposes in room acoustics it is advantageous to obtain the space average of the instantaneous energy density throughout the volume. The space-average energy density is readily found to be

$$\delta' = \frac{E}{V} = \rho_0 v^2 + \frac{p_0}{c}v \quad \text{J/m}^3 \tag{6.16}$$

where δ' = instantaneous average energy density throughout the volume (J/m³)

140

We are, however, interested in the time average of δ'. Measurements of room acoustics will be performed with integrating instrumentation rather than that which yields instantaneous values. Following this line of thought, we express the time average of the energy density by

$$\delta = \frac{1}{T} \int_0^T \delta' \, dt \quad \text{J/m}^3 \tag{6.17}$$

for an acoustical wave described by the particle velocity, v. The particle velocity, for the sinusoidal case, is

$$v = v_0 \cos(\omega t - kz) \quad \text{m/s} \tag{6.18}$$

where the wave is propagating in the positive z direction

$v =$ instantaneous particle velocity (m/s)

$v_0 =$ peak particle velocity (m/s)

$\omega =$ angular frequency, $2\pi f$ (rad/s)

$k =$ phase constant, $2\pi/\lambda$ (rad/m)

Upon substitution of equations (6.16) and (6.18) into equation (6.17), we obtain

$$\delta = \frac{1}{T} \int_0^T \left[\rho_0 v_0^2 \cos^2(\omega t - kz) + \frac{p_0 v_0}{c} \cos(\omega t - kz) \right] dt \quad \text{J/m}^3 \tag{6.19}$$

which can be further reduced to two integrals,

$$\delta = \frac{\rho_0 v_0^2}{T} \int_0^T \cos^2(\omega t - kz) \, dt + \frac{p_0 v_0}{cT} \int_0^T \cos(\omega t - kz) \, dt \quad \text{J/m}^3 \tag{6.20}$$

Integration yields

$$\delta = \tfrac{1}{2}\rho_0 v_0^2 \quad \text{J/m}^3 \tag{6.21}$$

which is an expression for the time and space average of the energy density within the volume, V, in question. One should note that this expression has been derived on the basis of a disturbance throughout the volume in the form of a uniform sinusoidal acoustic wave propagating in the z direction. We also note that in equation (6.21) the direction of propagation is of no consequence with regard to the final expression for the time- and space-average energy density, δ.

Since most acoustic instruments designed to measure pressure respond to the root-mean-square (rms) value of the pressure, it is expedient to express δ

in terms of the rms pressure, p. We accomplish this step by taking advantage of the plane-wave relationship,

$$v_0 = \frac{p\sqrt{2}}{\rho_0 c} \quad \text{m/s} \tag{6.22}$$

where $\rho_0 c$ = acoustic characteristic impedance of the gas in which the sound is propagating

This follows from the definition of the acoustic impedance, where the impedance is defined as the ratio of the dynamic pressure to the particle velocity in the medium. Upon substituting equation (6.22) into (6.21), we obtain

$$\delta = \frac{p^2}{\rho_0 c^2} \quad \text{J/m}^3 \tag{6.23}$$

or, upon rearranging,

$$p^2 = \rho_0 c^2 \delta \quad \text{Pa}^2 \tag{6.24}$$

which are two most useful equations relating the time and space average energy density to the rms pressure and bulk modulus, $\rho_0 c^2$.

Furthermore, the sound pressure level, L_p, may now be expressed in terms of the energy density. The sound pressure level, as defined by equation (1.13), is

$$L_p = 10 \log \left(\frac{p^2}{p_{re}^2}\right) \quad \text{dB} \tag{6.25}$$

which becomes

$$L_p = 10 \log \left(\frac{\rho_0 c^2 \delta}{p_{re}^2}\right) \quad \text{dB} \tag{6.26}$$

upon substitution of equation (6.24) for p^2.

In order to determine the sound pressure level in a room based upon the room characteristics, we now require an expression for the energy density as a function of these specific characteristics and the power of the noise source.

EXAMPLE 6.1 Determine the average energy density if the rms pressure is 0.1 Pa, the characteristic impedance $\rho_0 c = 407$ mks rayls, and $c = 344$ m/s.

SOLUTION

$$\delta = \frac{p^2}{\rho_0 c^2} \quad \text{J/m}^3$$
$$= \frac{(0.1)^2}{(407)(344)}$$
$$= 7.14 \times 10^{-8} \text{ J/m}^3$$

D6.1 If the peak particle velocity is 5.58×10^{-3} m/s at room temperature in air and the static gas density and sound velocity are 1.18 kg/m³ and 344 m/s, respectively, what is the rms pressure and the sound pressure level?

$$Ans. \quad p = 1.6 \text{ Pa}, L_p = 98 \text{ dB}.$$

6.3.4 Energy Density as a Function of the Room

Let us begin by asking ourselves the following question: What determines the sound pressure in a room with a continuous sound source? The answer to this fundamental question is deceptively simple. The sound pressure is determined by how much sound energy escapes or is absorbed by the room surfaces or contents. Following this thought to its logical conclusion, we note that in a room with no leaks and no absorption, the sound pressure would continue to increase as long as the sound source operates.

On the other hand, in a practical room with leaks and absorption, the sound pressure can only build up to a finite equilibrium value. Such a value is ordinarily attained very quickly. We will gain insight into the actual times involved in the following sections.

This development is based upon Sabine's theory of the growth and decay of acoustic energy in a room. The assumptions of homogeneous energy (completely diffuse sound in the room), and continuous, uniform absorption of energy by the boundary surfaces of the room are intrinsic to the theory. Under these assumptions, the rate of increase of energy within a room is equal to the rate of radiation by the sound source minus the rate of absorption by the bounding surfaces of the room. In equation form, this statement is expressed by

$$\boxed{V \frac{d\delta'}{dt} = w - w_s \quad \text{W}} \tag{6.27}$$

where V = volume of the room (m³)

         δ' = instantaneous space-average energy density (J/m³)

         w = power of the sound source (W)

         w_s = power absorbed by the room (W)

and the same energy density is assumed to exist throughout the room. Obviously, such uniformity exists only under idealized conditions.

Before we can undertake a solution of equation (6.27), we must determine the dependence of the absorption rate, w_s, on the energy density in the room. Consider the simple room of Figure 6.4. Our derivation of w_s will be based upon the differential volume, dV, as a source of sound energy and upon the differential surface of the room, ds, as an absorber of sound energy. Let us assume that dV radiates sound equally in all directions. Then that portion of energy subtended by the surface, ds, is given by

$$dE_s = \delta' \, dV \left(\frac{ds \cos \theta}{4\pi r^2} \right) \quad \text{J} \tag{6.28}$$

where E_s = energy striking the surface, ds (J)

r = distance from the volume element to the surface element (m)

θ = angle measured between the normal to the surface element (degree) and the straight line connecting the surface and volume elements

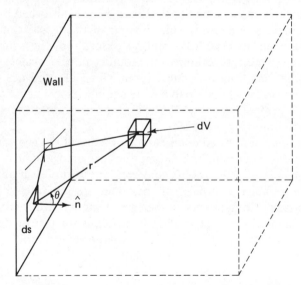

FIGURE 6.4 *Geometrical relationship between the differential volume, dV, and the differential surface, ds, for the derivation of w_s, the power absorbed by the room surfaces*

Furthermore, we can simplify the derivation by assuming dV to be a differential volume in a spherical shell centered on the differential surface, ds, as shown in Figure 6.5. This allows us to express dV as

$$dV = r^2 \sin \theta \, d\theta \, d\phi \, dr \quad \text{m}^3 \tag{6.29}$$

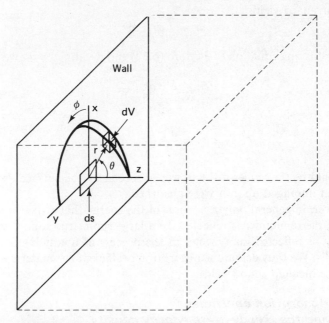

FIGURE 6.5 *Geometrical relationship of a radiating hemispherical shell to an element of wall surface*

which when substituted into equation (6.28) yields

$$dE_s = \frac{\delta' r^2 \sin\theta \cos\theta \, d\theta \, d\phi \, dr \, ds}{4\pi r^2} \quad \text{J} \tag{6.30}$$

The total energy arriving at the surface *ds* from the hemispherical shell of volume is

$$dE_s = \frac{\delta' \, ds \, dr}{4\pi} \int_{\phi=0}^{2\pi} \int_{\theta=0}^{\pi/2} \sin\theta \cos\theta \, d\theta \, d\phi \quad \text{J} \tag{6.31}$$

which upon integration yields

$$dE_s = \frac{\delta' \, ds \, dr}{4} \quad \text{J} \tag{6.32}$$

Recall that the velocity of propagation of sound is

$$c = \frac{dr}{dt} \quad \text{m/s} \tag{6.33}$$

which rearranged yields

$$dr = c \, dt \quad \text{m} \qquad (6.34)$$

Upon substituting equation (6.34) into (6.32) and dividing by dt, we obtain

$$\frac{dE_s}{dt} = \frac{\delta' c \, ds}{4} \quad \text{W} \qquad (6.35)$$

or

$$dw_s = \frac{\delta' c}{4} ds \quad \text{W} \qquad (6.36)$$

which would be the value of the power absorbed by the surface ds if all the energy that impinged upon it were absorbed.

However, in general only a portion of the energy that strikes a surface is absorbed; the remainder is reflected. In a large room, the sound waves will commonly be reflected many times in the process of finally being absorbed by the walls. We thus define the absorption coefficient for materials in terms of a diffuse incident sound field.

6.3.4.1 Absorption coefficient and the steady-state energy density

The *statistical absorption coefficient*, sometimes referred to as the *random-incidence sound-absorption coefficient*, α, which is a function of frequency, is defined as the ratio of the acoustical energy absorbed by the surface to the acoustical energy incident upon the surface when the incident sound field is perfectly diffused. In most applied situations, we note the materials of the room surfaces in question and look up the absorption coefficients in tables. A limited listing of certain common building materials is given in Table 6.1.

One should note in passing that most of the published absorption coefficients were obtained by a technique based upon the rate of decay of the acoustical energy in a reverberant chamber containing a sample of the material under investigation. This coefficient is more properly termed the *sabine absorption coefficient*. In certain instances the statistical and the sabine absorption coefficient may differ by 20 to 30%. However, in the context of industrial noise problems discussed here, we will simply refer to the "absorption coefficient" and employ equations derived from the viewpoint of a statistical absorption coefficient. No special regard is given to the one coefficient or the other as taken from the published literature; for industrial problems this should cause little or no difficulty. Those interested in a more thorough treatment of this problem and the associated problem with regard to the determination of the room constant, R, are referred to references [1,4,5].

Since different surfaces of a room will in general have different absorption coefficients, we require the average sound absorption coefficient, $\bar{\alpha}$, of the

TABLE 6.1

Absorption coefficients of general building materials
(Collected from the published literature and manufacturers' data)

	OCTAVE-BAND CENTER FREQUENCY (Hz)					
	125	*250*	*500*	*1000*	*2000*	*4000*
Brick, unglazed	0.03	0.03	0.03	0.04	0.05	0.07
Brick, unglazed, painted	0.01	0.01	0.02	0.02	0.02	0.03
Carpet on foam rubber	•0.08	0.24	0.57	0.69	0.71	0.73
Carpet on concrete	0.02	0.06	0.14	0.37	0.60	0.65
Concrete block, coarse	0.36	0.44	0.31	0.29	0.39	0.25
Concrete block, painted	0.10	0.05	0.06	0.07	0.09	0.08
Floors, concrete or terrazzo	0.01	0.01	0.015	0.02	0.02	0.02
Floors, resilient flooring on concrete	0.02	0.03	0.03	0.03	0.03	0.02
Floors, hardwood	0.15	0.11	0.10	0.07	0.06	0.07
Glass, heavy plate	0.18	0.06	0.04	0.03	0.02	0.02
Glass, standard window	0.35	0.25	0.18	0.12	0.07	0.04
Gypsum, board, $\frac{1}{2}$ in.	0.29	0.10	0.05	0.04	0.07	0.09
Panels, fiberglass, $1\frac{1}{2}$ in. thick	0.86	0.91	0.80	0.89	0.62	0.47
Panels, perforated metal, 4 in. thick	0.70	0.99	0.99	0.99	0.94	0.83
Panels, perforated metal with fiberglass insulation, 2 in. thick	0.21	0.87	1.52	1.37	1.34	1.22
Panels perforated metal with mineral fiber insulation, 4 in. thick	0.89	1.20	1.16	1.09	1.01	1.03
Panels, plywood, $\frac{3}{8}$ in.	0.28	0.22	0.17	0.09	0.10	0.11
Plaster, gypsum or lime, rough finish on lath	0.02	0.03	0.04	0.05	0.04	0.03
Plaster, gypsum or lime, smooth finish on lath	0.02	0.02	0.03	0.04	0.04	0.03
Polyurethane foam, 1 in. thick	0.16	0.25	0.45	0.84	0.97	0.87
Tile, ceiling, mineral fiber	0.18	0.45	0.81	0.97	0.93	0.82
Tile, marble or glazed	0.01	0.01	0.01	0.01	0.02	0.02
Wood, solid, 2 in. thick	0.01	0.05	0.05	0.04	0.04	0.04

room as a whole. This is calculated by

$$\bar{\alpha} = \frac{\sum_{i=1}^{n} S_i \alpha_i}{\sum_{i=1}^{n} S_i} \quad \text{dimensionless} \qquad (6.37a)$$

where S_i = area of the ith surface of the room (m² or ft²)

            α_i = absorption coefficient of the ith surface

It is important to remember that the absorption coefficient, α, as well as the average absorption coefficient, $\bar{\alpha}$, are both, in general, frequency-dependent. This is readily observable for the materials listed in Table 6.1.

Furthermore, if the room in question is large, it may be desirable to take atmospheric absorption into account at the higher frequencies, say 2000 Hz and above. Large room size enters into consideration because the larger the room, the larger the propagation path between reflections. Therefore, we will assume that the atmospheric absorption component is directly proportional to the mean free path of the sound wave in the room. The mean free path is derived later in this chapter and is found to be

$$r = \frac{4V}{S} \quad \text{m or ft}$$

as expressed by equation (6.54). The average excess absorption coefficient, $\bar{\alpha}_{ex}$, due to air absorption in the room is

$$\bar{\alpha}_{ex} = k\frac{4V}{S} \tag{6.37b}$$

where k = experimentally determined constant

A brief table of k as a function of humidity, temperature, and frequency is given in Table 6.2. A much more complete treatment appears in reference [6].

TABLE 6.2

Approximate values for the coefficient k in English and metric units [6]

			FREQUENCY (Hz)					
	TEMPERATURE		2000		4000		8000	
Relative humidity (%)	°F	°C	k (ft^{-1})	k (m^{-1})	k (ft^{-1})	k (m^{-1})	k (ft^{-1})	k (m^{-1})
30	68	20	0.0009	0.0030	0.0029	0.0095	0.0103	0.0340
30	77	25	0.0009	0.0029	0.0024	0.0078		
30	86	30	0.0009	0.0028	0.0022	0.0070		
50	68	20	0.0007	0.0024	0.0019	0.0061	0.0065	0.0215
50	77	25	0.0007	0.0024	0.0018	0.0059		
50	86	30	0.0007	0.0023	0.0018	0.0058		
70	68	20	0.0007	0.0021	0.0016	0.0053	0.0046	0.0150
70	77	25	0.0007	0.0021	0.0016	0.0053		
70	86	30	0.0006	0.0021	0.0016	0.0052		

In those instances in which it may be desirable to consider this effect, we obtain

$$\bar{\alpha}' = \bar{\alpha} + \bar{\alpha}_{ex} \tag{6.38a}$$

or

$$\bar{\alpha}' = \bar{\alpha} + k\frac{4V}{S} \tag{6.38b}$$

This adjusted value of $\bar{\alpha}'$ may then be used to determine the room constant, R, or the reverberation time, T. One must simply replace $\bar{\alpha}$ in the appropriate equations by $\bar{\alpha}'$ under the conditions that $k(4V/s)$ does not exceed 0.4 and that $\bar{\alpha}'$ does not exceed that allowed by the assumptions of the equation that is being used to calculate R or T.

In this text, only $\bar{\alpha}$ is used in the numerical exercises. Note that the units of k are ft^{-1} and m^{-1} in English and metric units, respectively.

EXAMPLE 6.2 Determine the average absorption coefficient at 1000 Hz for a room 40 ft × 70 ft with a 12-ft ceiling. The absorption coefficients for the floor, walls, and ceiling are 0.1, 0.2, and 0.7, respectively.

SOLUTION The areas of the various surfaces are:

floor: $S_1 = 40 \times 70 = 2800$ ft^2, $\alpha_1 = 0.1$

walls: $S_2 = 220 \times 12 = 2640$ ft^2, $\alpha_2 = 0.2$

ceiling: $S_3 = 40 \times 70 = 2800$ ft^2, $\alpha_3 = 0.7$

Taking advantage of equation (6.37a), we obtain

$$\bar{\alpha} = \frac{S_1\alpha_1 + S_2\alpha_2 + S_3\alpha_3}{S_1 + S_2 + S_3}$$

which becomes

$$\bar{\alpha} = \frac{(2800 \times 0.1) + (2640 \times 0.2) + (2800 \times 0.7)}{2800 + 2640 + 2800}$$

and finally,

$$\bar{\alpha} = 0.34$$

D6.2 In a room 5.18 × 6.10 m (17 ft × 20 ft) with a 3.66 m (12 ft) ceiling the absorption coefficients of the floor, walls and ceiling are (a) 0.2, 0.45, 0.6 (b) 0.07, 0.2, 0.4 (c) 0.5, 0.6, 0.8, respectively. What is the average absorption coefficient in each case?

Ans. (a) 0.428 (b) 0.215 (c) 0.622.

Our final step in determining w_s is to multiply equation (6.35) by the average absorption coefficient $\bar{\alpha}$ to allow for less than 100% absorption and to integrate over the total room surface. This yields

$$\bar{\alpha}\frac{dE_s}{dt} = dw_s = \frac{\delta'\bar{\alpha}c}{4}\,ds \quad \text{W}$$

which over the total room becomes

$$w_s = \int_s dw_s = \frac{\delta'\bar{\alpha}c}{4}\int_s ds \quad \text{W} \tag{6.39}$$

which yields

$$w_s = \frac{\bar{\alpha}cs\delta'}{4} \quad \text{W} \tag{6.40}$$

where s = total surface of the room (m² or ft²)

Upon having found the losses of the room due to the absorption of the room surfaces, we are now prepared to solve equation (6.27) for the time dependent space average energy density δ'. Beginning again with equation (6.27),

$$V\frac{d\delta'}{dt} = w - w_s \quad \text{W} \tag{6.41}$$

and substituting equation (6.40) for w_s, we obtain

$$V\frac{d\delta'}{dt} = w - \frac{\bar{\alpha}cs\delta'}{4} \quad \text{W} \tag{6.42}$$

Upon rearranging, one obtains

$$\frac{d\delta'}{dt} + \frac{\bar{\alpha}cs}{4V}\delta' = \frac{w}{V} \quad \text{W/m}^3 \tag{6.43}$$

Solving for δ', we obtain

$$\boxed{\delta' = \frac{4w}{\bar{\alpha}cs}[1 - e^{-(\bar{\alpha}cs/4V)t}] \quad \text{J/m}^3} \tag{6.44}$$

This is an expression for the growth of the space-average energy density in the room. Obviously, as it becomes very large, the exponential approaches zero. Thus, the steady-state time and space-average energy density δ, in the room is

$$\boxed{\delta = \frac{4w}{\bar{\alpha}cs} \quad \text{J/m}^3} \tag{6.45}$$

In later sections we shall return to equation (6.43) to derive an expression for the reverberation time of the room.

One can now take advantage of the two expressions for the time- and space-average energy density. We first obtained

$$\delta = \frac{p^2}{\rho_0 c^2} \quad \text{J/m}^3 \qquad (6.23)$$

which relates the average energy density to the mean-square pressure at the position in question. This time and space average is taken over a volume that may be quite small. That is, we may consider equation (6.23) as a "position function" in three-dimensional space.

On the other hand, the steady-state time- and space-average energy density throughout the entire volume of the room in question is expressed by

$$\delta = \frac{4w}{\overline{\alpha} c s} \quad \text{J/m}^3 \qquad (6.45)$$

The volume under consideration is therefore set by the volume of the room in contrast to the arbitrarily large or small volume allowed by equation (6.23).

D6.3 Consider a room that is $40\,\text{ft} \times 40\,\text{ft}$ $(12.19\,\text{m} \times 12.19\,\text{m})$ and has a 20-ft (6.08-m) ceiling; what is the energy density in the room due to a 2-W source 0.5 s after "start" if the average absorption coefficient is 0.05? What is the steady-state energy density in the room? Compare the two energy densities ($c = 1128$ ft/s or 344 m/s).
Ans. $\delta = 1.68 \times 10^{-5}$ J/ft³ or 5.93×10^{-4} J/m³ after 0.5 s; $\delta = 2.22 \times 10^{-5}$ J/ft³ or 7.83×10^{-4} J/m³ in steady-state.

D6.4 Determine the sound power of a source in a room 20 m by 20 m with a 7-m ceiling. The average absorption coefficient of the room is 0.06 and the reverberant sound pressure is 1.5 Pa. The acoustic characteristic impedance ($\rho_0 c$) is 407 mks rayls ($c = 344$ m/s).
Ans. 0.113 W.

6.3.4.2 Direct and reverberant energy

It is often advantageous in the determination of the sound pressure level in a room to be able to ascertain what portion is attributable to the direct field and what portion is due to the reverberant field. Clearly, only the reverberant portion of the total pressure can be reduced by placing absorbing materials on the surfaces of the room. Conversely, the direct field due to the sound source would exist even if the room were removed and the source were situated out-of-doors. Using equation (6.45) for the average energy density, we have for the total energy in the room

$$\delta V = \frac{4wV}{\bar{\alpha}cs} \quad \text{J} \tag{6.46}$$

We will now proceed to the determination of that portion of the total energy that is in the direct field. The energy in the reverberant field may then be found by a subtraction of the direct field energy from the total. Inasmuch as the direct field calculation will be based upon an assumed spherical radiator, we will proceed by representing a "reasonable room" by a sphere of equal volume.

6.3.4.2.1 ROOM-CONFIGURATION CONSIDERATION:

We shall assume that the direct energy is that portion of the total energy which has not struck a wall and which is contained in a sphere of equal volume to that of the room. It is also assumed that the sound source is located at the center of the sphere.

Assume a room whose dimensions are given by

$$l = \text{length (m)}$$
$$w = \text{width (m)}$$
$$h = \text{height (m)}$$

where $w = l$
          $h = 0.5l$

The volume of the room is

$$V = l \times l \times 0.5l = 0.5l^3 \quad \text{m}^3 \tag{6.47}$$

and its surface is

$$S = 2l^2 + l^2 + l^2 = 4l^2 \quad \text{m}^2 \tag{6.48}$$

We wish to determine the radius, r, of a sphere with the same volume, $0.5l^3$. The volume of a sphere is qiven by

$$V = \tfrac{4}{3}\pi r^3 \quad \text{m}^3 \tag{6.49}$$

It will prove to be more advantageous to relate the radius, r, of this sphere to the surface, S, and volume, V, of the room, rather than to the length, l, of the room. Equations (6.47) and (6.48) can be combined to yield

$$l = \frac{8V}{S} \quad \text{m} \tag{6.50}$$

Upon setting the volume of the sphere equal to that of the room, one obtains

$$\frac{4\pi r^3}{3} = \frac{l^3}{2} \quad m^3 \tag{6.51}$$

which yields

$$r = \left(\frac{3}{8\pi}\right)^{1/3} l \quad m \tag{6.52}$$

Upon substituting equation (6.50) into (6.52), we obtain

$$r = \left(\frac{3}{8\pi}\right)^{1/3} \frac{8V}{S} \quad m \tag{6.53}$$

which simplified yields the approximate value for the radius,

$$r = \frac{4V}{S} \quad m \tag{6.54}$$

This is the value for the radius of the sphere which we will use in the calculation of the direct energy. It is worth noting here that had one chosen a room of other proportions, a different value of the radius in terms of the volume and surface of the room would have been obtained. We find that the volume of the room and the sphere will be equal if the radius of the sphere is chosen to be

$$r = 0.62(V)^{1/3} \quad m \tag{6.55}$$

However, since an expression dependent upon both the volume and surface area of the room is required, equation (6.54) will be employed.

EXAMPLE 6.3   What is the radius of a sphere containing the same volume as a room 10 m × 10 m with a 5-m ceiling?

SOLUTION

$$r = \frac{4V}{s} \quad m$$

$$V = 10 \times 10 \times 5 = 500 \text{ m}^3$$

$$S = (2 \times 10 \times 10) + (4 \times 10 \times 5) = 400 \text{ m}^2$$

$$r = \frac{(4)(500)}{400} = 5 \text{ m}$$

6.3.4.2.2.   ENERGY DENSITY IN A PROGRESSIVE
          SPHERICAL WAVE:

It is necessary to obtain a measure of the sound pressure at any point a particular distance from the sound source. Toward this end we shall develop an expression for the direct energy density as a function of distance from the source. Furthermore, as has already been indicated, we require an expression for the total direct energy in the room. The direct energy is found by an integration of the radially dependent direct energy throughout a spherical volume equal to the volume of the room. It is convenient to first determine the average intensity as a function of pressure and particle velocity. In general, the intensity is

$$I = \text{Re}\, p_c^* v_c \cos \phi \quad \text{W/m}^2 \tag{6.56}$$

where I = average sound intensity (W/m²)

p_c^* = complex conjugate of the complex rms pressure (Pa)

v_c = complex rms particle velocity (m/s)

ϕ = angle between direction of travel and direction in which the intensity is determined (deg)

Re = real part of the complex expression

Since we are interested in the flow of energy outward through the imaginary sphere about the sound source, we shall always assume the direction of travel of energy and the direction in which the intensity is to be determined to be in the same radial direction. This allows the simplification of equation (6.56), since in this case $\phi = 0$. Therefore,

$$I = \text{Re}\, p_c^* v_c \quad \text{W/m}^2 \tag{6.57}$$

In those cases of interest here, assume that the point of interest is many wavelengths away from the center of the sound source. This assumption permits one to relate the complex rms particle velocity to the complex rms pressure through the expression for pressure given by equation (6.9). That is,

$$p_c = \rho_0 c v_c \quad \text{Pa} \tag{6.58}$$

Upon rearranging, one obtains

$$v_c = \frac{p_c}{\rho_0 c} \quad \text{m/s} \tag{6.59}$$

Substitution of this expression for v_c into equation (6.57) yields

$$I = \text{Re}\left(\frac{p_c^* p_c}{\rho_0 c}\right) \quad \text{W/m}^2 \tag{6.60}$$

which can be further simplified by noting that

$$\text{Re}\, p_c^* p_c = |p_c|^2 = p^2 \quad \text{Pa}^2 \tag{6.61}$$

where p = rms sound pressure (Pa)

We now combine equations (6.60) and (6.61) to obtain

$$\boxed{I = \frac{p^2}{\rho_0 c} \quad \text{W/m}^2} \tag{6.62}$$

The direct energy density, δ_d, was

$$\delta_d = \frac{p^2}{\rho_0 c^2} \quad \text{J/m}^3 \tag{6.23}$$

and therefore the intensity is related to the energy density by

$$I = c \delta_d \quad \text{W/m}^2 \tag{6.63}$$

Moreover, the intensity of an isotropic sound source is related to the power, w, of the source by

$$I = \frac{w}{4\pi r^2} \quad \text{W/m}^2 \tag{6.64}$$

Upon equating equations (6.63) and (6.64) and solving for δ_d, one obtains

$$\boxed{\delta_d = \frac{w}{4\pi r^2 c} \quad \text{J/m}^3} \tag{6.65}$$

6.3.4.2.3 SUM OF THE ENERGY DENSITY IN THE DIRECT AND REVERBERANT FIELDS:

We can readily determine the total direct energy, E_d, in the room by integrating the direct energy density, δ_d, within the equivalent spherical volume determined by the radius

$$r = \frac{4V}{s} \quad \text{m} \tag{6.54}$$

Thus, the total direct energy is

$$E_d = \int_V \delta_d \, dV \quad \text{J} \tag{6.66}$$

which expanded becomes

$$E_d = \int_0^{4V/S} \frac{w}{4\pi r^2 c} 4\pi r^2 \, dr \quad \text{J} \tag{6.67}$$

and upon integration yields

$$E_d = \frac{wr}{c} \Big|_0^{4V/S} \tag{6.68}$$

or

$$E_d = \frac{4wV}{cS} \quad \text{J} \tag{6.69}$$

This is the total energy in the room which has not been reflected from a wall.

We may now obtain the reverberant energy in the room by subtracting the direct energy from the total energy. Upon subtracting equation (6.69) from (6.46), we obtain

$$\delta_r V = \delta V - E_d \quad \text{J} \tag{6.70}$$

where δ_r = average energy density in the reverberant field (J/m³)

Upon expanding, one obtains

$$\delta_r V = \frac{4wV}{\bar{\alpha} cs} - \frac{4wV}{cs} \quad \text{J} \tag{6.71}$$

which simplified becomes

$$\delta_r V = \frac{4wV}{cs} \frac{1 - \bar{\alpha}}{\bar{\alpha}} \quad \text{J} \tag{6.72}$$

For purposes of repeated calculations, it is convenient to define a room constant, R, as

$$R \equiv \frac{\bar{\alpha} s}{1 - \bar{\alpha}} \quad \text{m}^2 \text{ or ft}^2 \tag{6.73}$$

The room constant, R, may then be substituted into equation (6.72) to obtain

$$\delta_r V = \frac{4wV}{cR} \quad \text{J} \tag{6.74}$$

which is a simplified expression of the total reverberant energy in the room. We also note that the reverberant energy density, δ_r, is

$$\boxed{\delta_r = \frac{4w}{cR} \quad \text{J/m}^3} \tag{6.75}$$

which can now be summed with δ_a to yield

$$\boxed{\delta = \frac{w}{4\pi r^2 c} + \frac{4w}{cR} \quad \text{J/m}^3}$$

(6.76)

where δ = the total energy density at any position within the room under the assumption that the source is an isotropic point radiator

EXAMPLE 6.4   Determine the energy density at a point 15 ft from a 2-W isotropic source in a room 25 ft × 20 ft × 12 ft which has an average absorption coefficient of 0.2.

SOLUTION

$$\delta = \frac{w}{4\pi r^2 c} + \frac{4w}{cR} \quad \text{J/ft}^3$$

$$R = \frac{\bar{\alpha} S}{1 - \bar{\alpha}}$$

$$S = (25 \times 20 \times 2) + (25 \times 12 \times 2) + (20 \times 12 \times 2) = 2080 \ \text{ft}^2$$

$$R = \frac{(0.2)(2080)}{1 - 0.2} = 520$$

$$\delta = \frac{2}{4\pi(15)^2(1128)} + \frac{(4)(2)}{(1128)(520)}$$

$$= 6.27 \times 10^{-7} + 1.36 \times 10^{-5}$$

$$= 1.43 \times 10^{-5} \ \text{J/ft}^3$$

Inasmuch as sources in general are not isotropic, we now require that δ_a be altered to allow for an arbitrary source. This is accomplished by means of the directivity factor, Q, which is discussed in detail in Chapter 1. It is expressed by

$$Q = \left(\frac{p_\theta^2}{\bar{p}^2} \right) \Bigg|_{r=r_1} \quad \text{unitless}$$

(1.31)

where p_θ = rms pressure in a particular direction from the source (Pa)

$\bar{p}$ = average rms pressure (Pa)

and both p_θ and $\bar{p}$ are measured at the same distance r, from the source. We shall now proceed to obtain an expression for Q in terms of energy density.

Upon taking advantage of equation (6.23), we have

$$\delta = \frac{p_\theta^2}{\rho_0 c^2} \Bigg|_{r=r_1} \quad \text{J/m}^3$$

(6.77)

and

$$\bar{\delta} = \frac{\bar{p}^2}{\rho_0 c^2}\bigg|_{r=r_1} \quad \text{J/m}^3 \tag{6.78}$$

where δ_θ and $\bar{\delta}$ = energy densities in a particular direction and due to an equivalent isotropic point source, respectively, both measured at a distance, r, from the source

Rearranging equations (6.77) and (6.78), one obtains

$$p_\theta^2 = \rho_0 c^2 \delta_\theta|_{r=r_1} \quad \text{Pa}^2 \tag{6.79}$$

and

$$\bar{p}^2 = \rho_0 c^2 \bar{\delta}|_{r=r_1} \quad \text{Pa}^2 \tag{6.80}$$

Substitution of equations (6.79) and (6.80) into (1.31) yields

$$Q = \frac{\delta_\theta}{\bar{\delta}}\bigg|_{r=r_1} \tag{6.81}$$

Furthermore, the direct energy density, δ_d, due to an isotropic source,

$$\delta_d = \frac{w}{4\pi r^2 c} \quad \text{J/m}^3 \tag{6.76}$$

may be multiplied by Q to obtain the direct energy density, δ_θ, of an arbitrary source:

$$\delta_\theta = \delta_d Q = \delta_d\left(\frac{\delta_\theta}{\bar{\delta}}\right) \quad \text{J/m}^3 \tag{6.82}$$

where, by definition, $\delta_d = \bar{\delta}$

Upon expansion, equation (6.82) becomes

$$\delta_\theta = \frac{wQ}{4\pi r^2 c} \quad \text{J/m}^3 \tag{6.83}$$

which may now be substituted for

$$\delta_d = \frac{w}{4\pi r^2 c} \quad \text{J/m}^3 \tag{6.65}$$

in (6.76) to obtain

$$\delta = \frac{wQ}{4\pi r^2 c} + \frac{4w}{cR} \quad \text{J/m}^3 \tag{6.84}$$

Equation (6.84) is the desired expression for the energy density, δ, at a general point a distance, r, from an arbitrary source located in a room whose characteristics are described by the room constant, R.

EXAMPLE 6.5   A nonisotropic radiator in free space has $Q = 2.5$ in the direction θ. If the sound power level of the source is 122 dB, determine the energy density in the direction θ at a point 25 m from the source ($c = 344$ m/s).

SOLUTION   The power, w, of the source is found from equation (1.12),

$$w = W_{\mathrm{re}} \times 10^{L_w/10}$$

where $W_{\mathrm{re}} = 10^{-12}$ W

This yields
$$w = 10^{-12} \times 10^{122/10} \text{ W}$$
$$= 1.59 \text{ W}$$

Substitution of the parameters into equation (6.83),

$$\delta_\theta = \frac{wQ}{4\pi r^2 c} \quad \text{J/m}^3$$

yields

$$\delta_\theta = \frac{(1.59)(2.5)}{4\pi (25)^2 (344)} \text{ J/m}^3$$

Thus,

$$\delta_\theta = 1.47 \times 10^{-6} \text{ J/m}^3$$

at the position in question.

6.4 PRESSURE AND POWER LEVELS

It is necessary to obtain an expression for the sound pressure level as a function of the distance, r, from the noise source in a particular room. This is important, since our noise measurements and criteria are commonly given in terms of dB levels. Furthermore, the sound power level, which is also in dB, is readily obtained by rearranging the expression for sound pressure level. The energy density, δ, as a function of the distance, r, serves as a base for obtaining both the pressure and power levels.

We continue by returning to equation (6.23) for the total energy density, δ,

$$\delta = \frac{p^2}{\rho_0 c^2} \quad \text{J/m}^3 \tag{6.23}$$

which in turn is substituted into equation (6.84) for δ to obtain

$$\frac{p^2}{\rho_0 c^2} = \frac{wQ}{4\pi r^2 c} + \frac{4w}{cR} \quad \text{J/m}^3 \tag{6.85}$$

Upon simplification, we obtain

$$\boxed{p^2 = w\rho_0 c \left(\frac{Q}{4\pi r^2} + \frac{4}{R}\right) \quad \text{Pa}^2} \tag{6.86}$$

where p^2 = mean-square pressure at a general point in the room

As already noted, however, we usually prefer to work with the sound pressure level, L_p, and the sound power level, L_w, in practical problems. The sound pressure level was defined as

$$L_p = 10 \log_{10} \left(\frac{p}{p_{re}}\right)^2 \quad \text{dB} \tag{1.13}$$

where p_{re} = reference pressure (20×10^{-6} Pa)

Furthermore, the sound power level was defined as

$$L_w = 10 \log \left(\frac{w}{w_{re}}\right) \quad \text{dB} \tag{1.10}$$

where w_{re} = reference power (10^{-12} W)

We may expand equation (1.13) to obtain

$$L_p = 10 \log p^2 - 10 \log (2 \times 10^{-5})^2 \quad \text{dB} \tag{6.87}$$

or

$$L_p = 10 \log p^2 - 20 \log (2 \times 10^{-5}) \quad \text{dB} \tag{6.88}$$

From equation (6.86), the expression for p^2 may be substituted into equation (6.88) to obtain

$$L_p = 10 \log w + 10 \log (\rho_0 c) + 10 \log \left(\frac{Q}{4\pi r^2} + \frac{4}{R}\right)$$
$$- 20 \log (2 \times 10^{-5}) \quad \text{dB} \tag{6.89}$$

Equation (1.10) may be expanded and rearranged to obtain

$$10 \log w = L_w - 120 \quad \text{dB} \tag{6.90}$$

which can now be substituted into equation (6.89) to yield

$$L_p = L_w + 10 \log \left(\frac{Q}{4\pi r^2} + \frac{4}{R} \right) + 10 \log \rho_0 C$$
$$- 20 \log (2 \times 10^{-5}) - 120 \quad \text{dB}$$

(6.91)

Let us now assume that we are operating near 22°C (71.6°F) and 751 mm Hg (29.6 in. Hg), which yields an acoustic characteristic impedance

$$\rho_0 c = 407 \text{ mks rayls (N-s/m}^3) \tag{6.92}$$

(The unit rayl is named in honor of Lord Rayleigh.) Upon simplification, equation (6.91) yields

$$L_p = L_w + 10 \log \left(\frac{Q}{4\pi r^2} + \frac{4}{R} \right) \quad \text{dB} \tag{6.93}$$

which is an expression for the sound pressure level in mks units.

It may be convenient to use r^2 and R in square feet rather than in square meters. We make this conversion by noting that

$$1 \text{ ft} = 0.3048 \text{ m} \tag{6.94}$$

Upon substituting equations (6.92) and (6.94) in (6.91), we obtain the expression for L_p in English units,

$$L_p = L_w + 10 \log \left(\frac{Q}{4\pi r^2} + \frac{4}{R} \right) + 10 \quad \text{dB} \tag{6.95}$$

where L_p = sound pressure level referenced to 20 μPa

L_w = sound power level referenced to 10^{-12} W

Q = directivity (unitless)

r = distance from source to point of measurement of L_p (ft)

R = room constant, $R = \bar{\alpha}s/(1 - \bar{\alpha})$, where s is the room surface and $\bar{\alpha}$ is the average absorption coefficient (ft²)

Obviously, equation (6.95) can be rearranged to obtain an expression for the sound power level, L_w, as

$$L_w = L_p - 10 \log \left(\frac{Q}{4\pi r^2} + \frac{4}{R} \right) - 10 \quad \text{dB} \tag{6.96}$$

Based upon this equation, one may calculate L_w with a singular measurement of L_p. In practice, however, usually a number of sound-pressure-level measurements are made in determining power levels. This is discussed in detail in Section 6.4.3.

Without doubt, equations (6.95) and (6.96), which yield the sound pressure level and the sound power level, respectively, are two of the most used equations in the abatement of industrial noise.

It is instructive to rearrange equation (6.95) to obtain

$$L_p - L_w = 10 \log \left(\frac{1}{4\pi r^2} + \frac{4}{R} \right) + 10 \quad \text{dB} \qquad (6.97)$$

where it is assumed that $Q = 1$

This equation is plotted in Figure 6.6. From the figure we see that in free space ($R = $ infinity), the difference $L_p - L_w$ is reduced by 6 dB for each doubling of the distance from the source. Furthermore, we note that as R becomes smaller and the room becomes more highly reverberant, the difference $L_p - L_w$ becomes more nearly a constant value, closer and closer to the sound source. Stated another way, the smaller the room constant, the more reverberant is the room.

For convenience in estimating the room constant, curves relating the room constant to the room surface area with the average absorption coefficient as a parameter are given in Figure 6.7. These curves are valid for both the metric and English system of units, dependent only upon the units in which the surface area of the room is measured.

Furthermore, in the analysis and design of lecture rooms, concert halls, theaters, and the like, it is also desirable to have an estimate of the range of the room constant required for effective speech communication. The range of the room constant for effective speech communication is plotted versus the room volume in Figure 6.8. It should also be noted that the room constant is based upon a 1000-Hz tone.

EXAMPLE 6.6 An isotropic sound source with a sound power level of 120 dB is in a room having a room constant, $R = 500$ ft². (a) Determine L_p a distance of 10 ft from the source. (b) What would L_p have been 10 ft from the source if the source were in free space?

SOLUTION (a) From Figure 6.6 we have for $R = 500$ ft² at 10 ft from the source,

$$L_p - L_w = -10.4 \text{ dB}$$

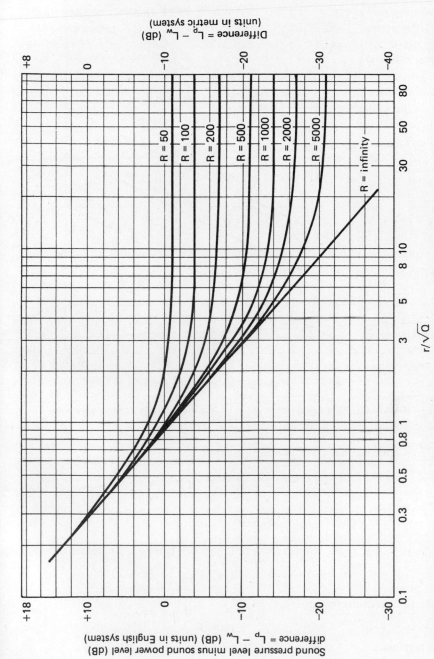

FIGURE 6.6 *Difference between the sound pressure level and the sound power level vs. the distance from the source with the room constant as a parameter*

163

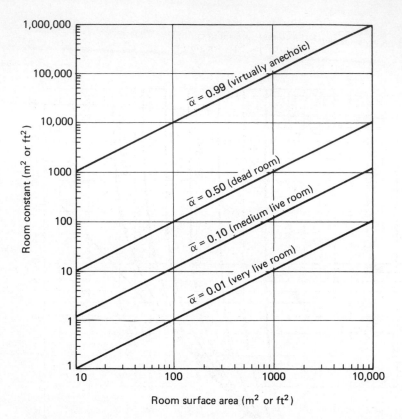

FIGURE 6.7 *Room constant vs. room surface area with the average absorption coefficient as a parameter*

Then, knowing that $L_w = 120$ dB, we obtain

$$L_p = 120 - 10.4 \text{ dB}$$

and

$$L_p = 109.6 \text{ dB} \quad \text{in the room}$$

(b) In free space, reading from the chart of Figure 6.6, we obtain

$$L_p - L_w = -21 \text{ dB}$$

At 10 ft from the source,

$$L_w = 120 \text{ dB}$$

Therefore,

$$L_p = 120 - 21 \text{ dB}$$
$$= 99 \text{ dB} \quad \text{in free space}$$

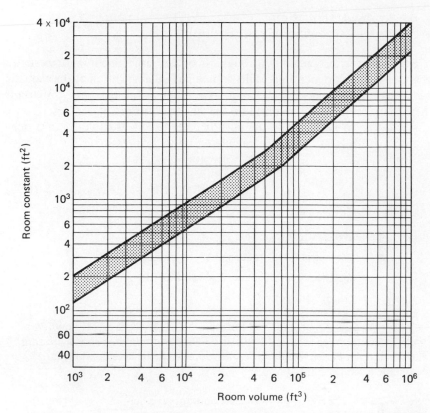

FIGURE 6.8 *Room constant, R, at 1 kHz as a function of room volume for effective speech communication* [1]

We note that the sound pressure level in the room at the position in question is approximately 11 dB more than at an equivalent position in free space.

D6.5   What is the room constant of a room with an average absorption coefficient of 0.3 if the dimensions are: (a) 40 ft $\times$ 40 ft $\times$ 20 ft   (b) 15 ft $\times$ 30 ft $\times$ 15 ft   (c) 10 ft $\times$ 50 ft $\times$ 20 ft ?

$\qquad\qquad\qquad\qquad$ *Ans.*   (a) 2743 ft^2 (b) 964 ft^2 (c) 1457 ft^2.

D6.6   What is the energy density 15 ft from a 1-W source in a room that has a room constant of 1025 ft^2 if the source radiates uniformly through a 2π steradian solid angle? ($c = 1128$ ft/s.)

$\qquad\qquad\qquad\qquad$ *Ans.* 4.1×10^{-6} J/ft^3.

D6.7   What is the sound pressure level 20 ft from a 2-W isotropic source if the source is in a room 20 ft $\times$ 20 ft $\times$ 15 ft with an average absorption coefficient of 0.2?

$\qquad\qquad\qquad\qquad$ *Ans.*   112 dB.

6.4.1 Air Attenuation by Atmospheric Absorption

Sound waves, propagating through air or any other medium, experience attenuation. This attenuation results from a partial absorption of the acoustic energy by the propagating medium. Air attenuation becomes important where long distances are involved, in particular outdoors.

According to Beranek [4], the approximate excess attenuation at a temperature of 20°C (68°F) may be calculated by the expression

$$A_{ex} = 7.4\left(\frac{f^2 r}{\phi}\right) 10^{-8} \quad \text{dB} \qquad (6.98)$$

where A_{ex} = excess attenuation (dB)

f = geometric mean frequency of the band (Hz)

r = distance between source and receiver (m)

ϕ = relative humidity (%)

EXAMPLE 6.7   Assume that an isotropic point source is located outdoors, with a sound power level of 120 dB at 2 kHz. Determine the sound pressure level at a point located 2500 ft from the source if the relative humidity is 65%. (Assume a temperature of 68°F.)

SOLUTION   The sound pressure level outdoors is a special case of equation (6.95); that is,

$$L_{p_0} = L_w + 10 \log \left(\frac{Q}{4\pi r^2}\right) + 10 \quad \text{dB}$$

This does not include atmospheric attenuation. We are given $Q = 1$, $r = 2500$ ft, and $L_w = 120$ dB. Thus,

$$L_{p_0} = 120 + 10 \log \left[\frac{1}{4\pi(2500)^2}\right] + 10 \quad \text{dB}$$

$$= 130 + 10 \log (1.273 \times 10^{-8}) \, \text{dB}$$

Without attenuation, $L_{p_0} = 51$ dB. The excess attenuation at 68°F is given by equation (6.98) to be

$$A_{ex} = 7.4\left(\frac{f^2 r}{\phi}\right) 10^{-8} \quad \text{dB}$$

where the variables are, in mks units, $f = 2000$ Hz, $r = 2500$ ft $= 762$ m, and $\phi = 65\%$

166

Thus,

$$A_{ex} = 7.4 \left[\frac{(2000)^2(762)}{65} \right] 10^{-8} \text{ dB}$$

$$= 3.47 \text{ dB}$$

Therefore,

$$L_p = L_{p_0} - A_{ex} \quad \text{dB}$$

$$= 51 - 3.47 \text{ dB}$$

and finally,

$$L_p = 47.53 \approx 48 \text{ dB}$$

with the excess attenuation included.

6.4.2 Sound Level

As already discussed, we are concerned with obtaining a sufficiently low A-weighted sound level in a given region to satisfy our abatement requirements. In those cases in which we have available the sound power levels and the characteristics of the room, for the various octave bands we can readily calculate the sound pressure level at a general position by means of equation (6.95),

$$L_p = L_w + 10 \log \left(\frac{Q}{4\pi r^2} + \frac{4}{R} \right) + 10 \quad \text{dB} \qquad (6.95)$$

Prior to determining the overall sound level at a particular position, it is only necessary to A-weight each of the sound pressure levels calculated from equation (6.95). See Table 2.2 for the conversion chart.

The following step-by-step procedure is convenient in determining the sound level from octave-band information (use the same procedure for $1/n$-octave bands also):

1. Obtain the sound power level, L_w, in the octave bands 31.5 Hz, 63 Hz, 125 Hz, 250 Hz, 500 Hz, 1000 Hz, 2000 Hz, 4000 Hz, and 8000 Hz. (If one must make this determination, a step-by-step procedure is given in Section 6.4.3.)
2. Apply equation (6.95) repeatedly for each of the octave bands.
3. A-weight each of the sound pressure levels for each octave band.
4. Sum the A-weighted sound levels for all bands for the total sound level at the position in question.

It is worth noting here that when making octave-band measurements, it is worthwhile to make a sound level measurement at each position as well. The procedure above can then be used to combine the measured sound pressure levels and check against the overall sound level. Such checks are most useful when making measurements under distracting industrial conditions.

> **EXAMPLE 6.8** A machine that has a sound power level of 120 dB in the 2-kHz band with a Q of 2 is in a room that has a room constant of 550 ft². What is the sound level in dBA 15 ft away in the band in question?

> **SOLUTION**

$$L_p = 120 + 10 \log\left(\frac{Q}{4\pi r^2} + \frac{4}{R}\right) + 10 \quad \text{dB}$$

$$= 120 + 10 \log\left[\frac{2}{4\pi(15)^2} + \frac{4}{550}\right] + 10$$

$$= 130 + 10 \log(0.00798)$$

$$= 109 \text{ dB}$$

Refer to Table 2.2. To convert to A weighting, add +1.2 dB. Therefore,

$$L = 109 + 1.2 \text{ dBA}$$

$$\approx 110 \text{ dBA} \quad \text{in the 2-kHz octave band}$$

> **D6.8** What is the total A-weighted sound level of a room that has a room constant of 625 ft² over all bands in question due to an isotropic source? The measurements are taken 20 ft away and the sound power levels in the octave bands of interest are as follows: 120 dB at 125 Hz, 122 dB at 250 Hz, 121 dB at 500 Hz, and 119 dB at 1000 Hz.
>
> *Ans.* 110 dBA.

6.4.3 Sound Power Level

Measurements in the direct field are not only useful in determining the directivity of a source, but are valuable in determining the sound power or sound power level as well. A knowledge of the sound power of a source, usually in octave bands, is required before one can determine those steps necessary to reduce the noise to the desired level.

Manufacturers are encouraged to determine the sound power level of their particular product in each of the octave bands of interest. In general, we are interested in the bands from 31.5 Hz through at least 8 kHz. Octave-band measurements are required because the sound power may vary considerably from band to band.

The sound power level is not measured directly, but rather the sound pressure level is determined and the sound power level is calculated by means of the equation

$$L_w = L_p - 10 \log \left(\frac{Q}{4\pi r^2} + \frac{4}{R} \right) - 10 \quad \text{dB} \qquad (6.96)$$

To obtain an accurate measure of L_w in each frequency band of interest, it is desirable to use the average sound pressure level, $\bar{L}_p$, for each of the corresponding bands, referred to a spherically radiating point source. Upon assuming that the measurements are either made in an anechoic chamber or outdoors, equation (6.96) becomes

$$L_w = \bar{L}_p - 10 \log \left(\frac{1}{4\pi r^2} \right) - 10 \quad \text{dB} \qquad (6.99)$$

Simplification yields

$$L_w = \bar{L}_p + 10 \log (4\pi r^2) - 10 \quad \text{dB} \qquad (6.100)$$

where $\bar{L}_p$ = average sound pressure level (dB)

  r = distance at which $\bar{L}_p$ was determined from the source (ft)

In the metric system of units, one obtains from equation (6.93) the expression

$$L_w = \bar{L}_p + 10 \log (4\pi r^2) \quad \text{dB} \qquad (6.101)$$

where r = distance (m) at which $\bar{L}_p$ was determined from the source

$\bar{L}_p$ is determined for this purpose by the same methods as for directivity calculations and is made for each frequency band of interest. Equations (1.28) and (1.39) are used when the measurements are taken over a sphere and over a hemisphere, respectively. In both instances, however, one obtains a value of $\bar{L}_p$ as though the radiator were a spherically radiating isotropic point source.

In many instances, however, one must determine the sound power level of a machine that is already installed in a plant. Under such circumstances it is usually virtually impossible to make precisely located sound pressure measurements about the machine in order to determine the average sound pressure level, $\bar{L}_p$. Given such conditions, one is well advised to make at least several calculations of L_w based upon individual measurements of L_p and then average the L_w's thus obtained.

A convenient step-by-step procedure for determining the sound power level in each of the octave bands of interest for an installed machine is as follows:

1. Measure the sound pressure level at a distance of at least twice the largest dimension (if practical) of the noise source in the octave bands 31.5 Hz, 63 Hz, 125 Hz, 250 Hz, 500 Hz, 1000 Hz, 2000 Hz, 4000 Hz, and 8000 Hz.

2. Calculate the room constant, R, from the dimensions of the room and the absorption coefficients or by means of the reverberation time in each of the octave bands.
3. Calculate the sound power level in each of the frequency bands using the expression

$$L_{w_i} = L_{p_i} - 10 \log \left(\frac{Q_i}{4\pi r_i^2} + \frac{4}{R} \right) \quad \text{dB}$$

where metric units are used, or

$$L_{w_i} = L_{p_i} - 10 \log \left(\frac{Q_i}{4\pi r_i^2} + \frac{4}{R} \right) - 10 \quad \text{dB}$$

in the event that English units are employed.
4. Repeat steps 1, 2, and 3 for several measurements of L_{p_i} at various positions about the machine to determine several independent measures of L_{w_i} in each octave band.
5. Calculate the average value of L_w for each of the octave bands:

$$L_w = 10 \log \left(\frac{1}{n} \sum_{i=1}^{n} 10^{L_{w_i}/10} \right) \quad \text{dB}$$

EXAMPLE 6.9 Determine the sound power level of a machine that has a sound pressure level of 110 dB measured 25 ft away and a directivity of 2.5 in the direction of the measurement in a room that has a room constant of 1025 ft^2.

SOLUTION

$$L_w = L_p - 10 \log \left(\frac{Q}{4\pi r^2} + \frac{4}{R} \right) - 10 \quad \text{dB}$$

$$= 110 - 10 \log \left[\frac{2.5}{4\pi(25)^2} + \frac{4}{1025} \right] - 10$$

$$= 100 - 10 \log (0.00422)$$

$$= 123.75 \approx 124 \text{ dB} \quad \text{based on a single measurement}$$

If it is desired to obtain the sound power rather than the sound power level,

$$L_w = 10 \log \left(\frac{W}{W_{\text{re}}} \right) \quad \text{dB} \tag{6.102}$$

may be solved for W, the acoustic power. Upon rearranging,

$$\log \left(\frac{W}{W_{\text{re}}} \right) = \frac{L_w}{10} \tag{6.103}$$

which becomes

$$W = W_{re} \text{ antilog} \left(\frac{L_w}{10}\right) \quad \text{W} \tag{6.104}$$

It is helpful in understanding the importance of the sound power to make an analogy with the power rating of an ordinary incandescent lamp. In the case of the lamp, only the power rating is invariable from application to application. For example, the same lamp will not equally light two rooms of different sizes or, for that matter, a white room and a black room of the same size. In the acoustical case we have a similar situation. A machine that radiates a specific acoustic power will not generate the same sound pressure level in two rooms of the same size if the overall absorption characteristics are different. This fact is made quite clear in equation (6.95), since a variation in the room constant, R, will cause a variation in the sound pressure level, L_p. However, once one has obtained the sound power and the room characteristics, the sound pressure level and thus the sound level can be determined with reasonable accuracy throughout the room.

D6.9 What is the sound power of an isotropic source in an anechoic chamber that produces a sound pressure level of 105 dB 10 ft away from the source?
Ans. 3.98 W.

6.4.3.1 Sound power level by means of a reference source

Quite often in an industrial environment, it is rather difficult to reasonably determine the effective value of the room constant, R, with respect to a particular machine. In such instances it is desirable to determine the approximate sound power level by means of a calibrated sound source. An adaptation and simplification of the techniques discussed in reference [7] for measurements in a reverberant field are presented here for general industrial applications.

The average sound pressure level of the machine, $\bar{L}_{p_x}$, is determined at a fixed radius just as for directivity measurements. Next, the reference noise source is placed in the location of the machine, if possible. An alternative location may be used if the environment is essentially the same as that about the machine. The average sound pressure level about the reference source, $\bar{L}_{p_s}$, is also determined at the same radius and in like manner as was used in determining $\bar{L}_{p_x}$ for the machine in question.

We may express the sound power level of the machine, L_{w_x}, and of the reference source, L_{w_s}, in metric units by, respectively,

$$L_{w_x} = \bar{L}_{p_x} - 10 \log \left(\frac{1}{4\pi r_x^2} + \frac{4}{R}\right) \quad \text{dB} \tag{6.105}$$

and

$$L_{w_s} = \bar{L}_{p_s} - 10 \log \left(\frac{1}{4\pi r_s^2} + \frac{4}{R} \right) \quad \text{dB} \qquad (6.106)$$

where $r_x = r_s = r_0$ (m)

   r_0 = convenient distance from the machine

Note also that the directivity factor, Q, of the machine and the reference source are both taken as unity. This is allowable because we are using the average sound pressure level in each case.

Upon subtracting equation (6.106) from (6.105) and rearranging, we obtain

$$\boxed{L_{w_x} = L_{w_s} + (\bar{L}_{p_x} - \bar{L}_{p_s}) \quad \text{dB}} \qquad (6.107)$$

where L_{w_x} = band power level of the machine (dB)

   L_{w_s} = band power level of the reference source (dB)

   $\bar{L}_{p_x}$ = average sound pressure level of the machine at a distance r_0 (dB)

   $\bar{L}_{p_s}$ = average sound pressure level of the reference source at a distance r_0 (dB)

A step-by-step procedure for determining the sound power level in each of the octave bands is as follows:

1. Measure the sound pressure level at a fixed distance, r_0, about the machine and determine $\bar{L}_{p_x}$ just as for a directivity determination.
2. Place the reference source in the position of the machine or in an equivalent environmental position and determine $\bar{L}_{p_s}$ at the same distance, r_0, as was done for the machine.
3. Obtain the value of L_{w_s} from the data on the reference source for the frequency band in question and determine the desired band power level by means of equation (6.107):

$$L_{w_x} = L_{w_s} + (\bar{L}_{p_x} - \bar{L}_{p_s}) \quad \text{dB}$$

Also note that this equation is valid in both the metric and English system of units.

EXAMPLE 6.10   A reference source with a sound power level of 90 dB in the 1-kHz octave band was used to determine the sound power level of a machine in this frequency band. At a distance of 6 m from the reference source, the average sound pressure level was found to be 71 dB; 6 m from the machine, it was found to be 80 dB. Find the sound power level of the machine in the 1-kHz octave band.

SOLUTION Referring to equation (6.107),

$$L_{w_x} = L_{w_s} + (\bar{L}_{p_x} - \bar{L}_{p_s}) \quad \text{dB}$$

We are given that $\bar{L}_{p_x} = 80$ dB, $\bar{L}_{p_s} = 71$ dB, and $L_{w_s} = 90$ dB. Upon substitution,

$$L_{w_x} = 90 + (80 - 71) \, \text{dB}$$
$$= 99 \, \text{dB}$$

which is the sound power of the source in the 1-kHz octave band.

6.5 REVERBERATION

Acoustical reverberation is encountered when the sound radiated by a source is reflected many times by the confining boundaries. For example, if a noise source is located in a room with hard walls, floor and ceiling, and with no leaks, then the sound energy will be reflected many times before it is absorbed. If we carry this a bit further and assume that the surfaces are perfectly reflecting, then the sound pressure will theoretically increase to a large value. In the practical situation the sound pressure level can and does grow to often surprisingly high levels in small hard rooms used to house noisy machines.

Let us return now to equation (6.93) in metric units,

$$L_p = L_w + 10 \log \left(\frac{Q}{4\pi r^2} + \frac{4}{R} \right) \quad \text{dB} \qquad (6.93)$$

which describes the sound pressure level in a room. Recalling that the room constant, R, is defined by

$$R = \frac{\bar{\alpha}S}{1 - \bar{\alpha}} \quad \text{m}^2 \qquad (6.73)$$

we note that as the average absorption coefficient, $\bar{\alpha}$, approaches zero, the room constant, R, approaches zero. That is,

$$R \longrightarrow 0 \quad \text{as} \quad \bar{\alpha} \longrightarrow 0 \qquad (6.108)$$

and therefore, from equation (6.93),

$$L_p = L_w + 10 \log (\infty) \quad \text{dB} \qquad (6.109)$$

or simply

$$L_p = \infty \, \text{dB} \qquad (6.110)$$

under the condition of zero absorption. This is, of course, unrealistic, but it does emphasize the fact that in a small room with hard surfaces, the sound pressure level can become surprisingly excessive.

In practical situations, where a pronounced reverberant field exists, it is characterized by a rather uniform or slowly varying sound pressure level throughout large regions of the reverberant field. Under reverberant field conditions,

$$\frac{4}{R} \gg \frac{Q}{4\pi r^2} \tag{6.111}$$

and equation (6.93) reduces to

$$L_p = L_w + 10 \log \left(\frac{4}{R}\right) \quad \text{dB} \tag{6.112}$$

or

$$L_p = L_w - 10 \log (R) + 6 \quad \text{dB} \tag{6.113}$$

in the metric system of units. In English units, equation (6.113) becomes

$$L_p = L_w - 10 \log(R) + 16 \quad \text{dB} \tag{6.114}$$

where R is in square feet

We note that L_p under reverberant conditions is not dependent upon the distance from the source, r. The sound power level, L_w, of the source and the room constant, R, determine a single value for L_p in the reverberant region in question.

In laboratories, reverberant rooms are often constructed for the purpose of carefully controlled conditions for accoustical measurements. Such reverberant rooms have highly reflective surfaces and usually a large, slowly rotating vane in the center of the room. The purpose of the vane is to break up standing waves that would otherwise disturb the measurements where narrow-band or discrete frequencies are concerned. However, we must note that measurements in such a room cannot be used to determine the directivity of a source since the direct field is not measurable.

EXAMPLE 6.11   A machine with a directivity of unity and a sound power level of 120 dB in the 2000-Hz octave band operates in a small room with a room constant of 9.29 m² (100 ft²) in this frequency range. (a) Determine the sound pressure level in the reverberant field. (b) How close must one be to the machine for this value to increase by 1 dB?

SOLUTION   (a) The sound pressure level due to the reverberant field is given by equation (6.113) to be

$$L_p = L_w - 10 \log R + 6 \text{ dB}$$

Upon substitution,

$$L_p = 120 - 10 \log (9.29) + 6 \quad \text{dB}$$
$$= 116 \text{ dB}$$

in the reverberant field.

(b) Begin with equation (6.93),

$$L_p = L_w + 10 \log \left(\frac{Q}{4\pi r^2} + \frac{4}{R} \right) \text{dB}$$

and solve—in this instance for r, the only unknown. We obtain, upon solving for r,

$$r = \left\{ \frac{Q}{4\pi \left[10^{(L_p - L_w)/10} - \dfrac{4}{R} \right]} \right\}^{1/2} \quad \text{m}$$

The value of L_p is found to be

$$L_p = 116 + 1 \text{ dB}$$
$$= 117 \text{ dB}$$

from the conditions of the problem and the results of part (a). Upon substitution,

$$r = \left\{ \frac{1}{4\pi \left[10^{(117-120)/10} - \dfrac{4}{9.29} \right]} \right\}^{1/2}$$

and finally,

$$r = 1.06 \text{ m}$$

or

$$r = 3.48 \text{ ft}$$

Therefore, if one is 1.06 m or 3.48 ft or more from the source, the sound pressure level will ideally vary by 1 dB or less.

6.5.1 Reverberation Time

We shall now define the reverberation time. Sabine defined reverberation time as the time required for the energy density in the acoustical field to reduce to a level 60 dB below its steady-state value. Equation (6.43) is an expression for the time-dependent energy density expressed by

$$\frac{d\delta'}{dt} + \frac{\bar{a}cS}{4V}\delta' = \frac{w}{V} \tag{6.43}$$

If we now assume that the noise source is "turned off" after the steady-state energy density is reached, we have

$$\frac{d\delta'}{dt} + \frac{\bar{a}cS}{4V}\delta' = 0 \tag{6.115}$$

where the value of δ' at $t = 0$ is $4w/\bar{a}cS$, the steady-state density

Upon solving for δ', one obtains

$$\delta' = \frac{4w}{\bar{a}cS}e^{-(\bar{a}cS/4V)t} \quad \text{J/m}^3 \tag{6.116}$$

which expresses the decay of the energy density with respect to time. Clearly, the rate of density decay in a given room is determined by the absorption of energy within the room.

The reverberation time, T, may now be obtained from equation (6.116) and the definition of reverberation time. One obtains

$$\frac{4w}{\bar{a}cS} \times 10^{-6} = \frac{4w}{\bar{a}cS}e^{-(\bar{a}cS/4V)T} \tag{6.117}$$

since in a time period, T, the steady-state density decays to $1/10^6$ of its original value. That is, it is reduced by 60 dB. Equation (6.117) may be simplified to yield

$$e^{-(\bar{a}cS/4V)T} = 10^{-6} \tag{6.118}$$

which becomes

$$\ln(10^6) = \frac{\bar{a}cS}{4V}T \tag{6.119}$$

or

$$T = \frac{4V}{\bar{a}cS}\ln(10^6) \quad \text{s} \tag{6.120}$$

In general, the absorption coefficient, $\bar{a}$, is not the same for all surfaces of the room. Thus, in general, we denote $\bar{a}S$ in equation (6.120) by

$$A \equiv \bar{a}S = \alpha_1 S_1 + \alpha_2 S_2 + \alpha_3 S_3 + \ldots + \alpha_n S_n \quad \text{ft}^2 \text{ or m}^2 \tag{6.121}$$

or

$$A \equiv \sum_{i=1}^{n} \alpha_i S_i \quad \text{ft}^2 \text{ (sabine) or m}^2 \text{ (metric sabine)} \tag{6.122}$$

where α_n = absorption coefficient

S_i = area of a component of the total surface

Substituting equation (6.121) into (6.120) and simplifying, one obtains

$$T = \frac{4 \ln (10^6)}{c} \frac{V}{A} \quad \text{s} \tag{6.123}$$

which may be reduced to the familiar form of the Sabine formulas for reverberation time

$$\boxed{T = 0.161 \left(\frac{V}{A}\right) \quad \text{s}} \tag{6.124}$$

where $c = 344$ m/s at 20°C

In English units, one obtains

$$\boxed{T = 0.049 \left(\frac{V}{A}\right) \quad \text{s}} \tag{6.125}$$

where $c = 1128$ ft/s at 20°C

Note that up to this point, we have not considered atmospheric absorption as a factor in the determination of reverberation time. In the case of a large room, air absorption can be important. It may be taken into consideration by replacing $\bar{\alpha}$ with $\bar{\alpha}'$ as given by equation (6.38b) in the definition of A in equation (6.121). Upon examination, one notes that these equations are in error for large values of absorption coefficient, say $\bar{\alpha} > 0.3$. For $\bar{\alpha} = 1$, the reverberation time, T, is in reality zero; however, equations (6.124) and (6.125) predict values of $0.161 V/S$ and $0.049 V/S$, respectively. In terms of a "reasonable room" in which the width and length are equal and the ceiling height is one-half the width, one obtains

$$T = 0.02l \quad \text{s}$$

where $l =$ width $=$ length of the room (m)
         $\bar{\alpha} = 1$

or

$$T = 0.006l \quad \text{s}$$

where $l =$ width $=$ length of the room (ft)
         $\bar{\alpha} = 1$

We therefore observe that the magnitude of the error, in this instance, is directly proportional to the dimensions of the room. For example, we find for a room 50 ft × 50 ft × 25 ft with $\bar{\alpha} = 1$ and using equation (6.125) a predicted reverberation time, $T = 0.3$ s, whereas in reality it should be zero.

Nonetheless, it should be pointed out that in the solution of industrial problems, equations (6.124) and (6.125) have been used most successfully; in this class of problem we are generally interested in rather acoustically "live" rooms.

In the event that it is desirable to work with a more accurate equation for large values of $\bar{\alpha}$, the Erying equation [2,5,8] is recommended:

$$T = 0.161 \frac{V}{S[-\ln(1 - \bar{\alpha})]} \quad \text{s} \qquad (6.126)$$

in metric units, or

$$T = 0.049 \frac{V}{S[-\ln(1 - \bar{\alpha})]} \quad \text{s} \qquad (6.127)$$

in English units.

This equation correctly predicts a reverberation time of zero for $\bar{\alpha} = 1$. In this text, however, only the Sabine formulas, equations (6.124) and (6.125), are employed.

Equations (6.124) and (6.125) are plotted in Figures 6.9 and 6.10, respec-

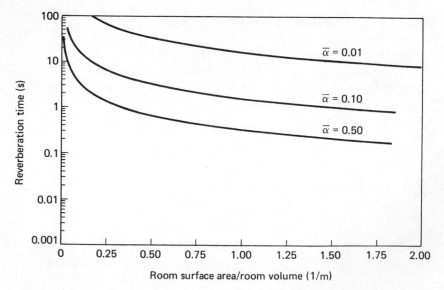

FIGURE 6.9 *Reverberation time vs. the ratio of the room surface to the room volume ; the average absorption coefficient is the parameter (metric units)*

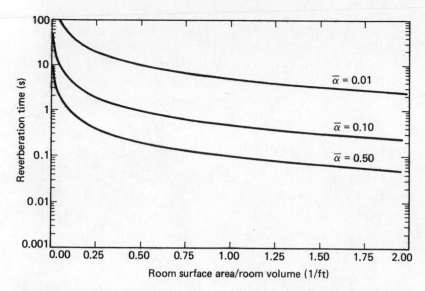

FIGURE 6.10 *Reverberation time vs. the ratio of the room surface to the room volume; the average absorption coefficient is the parameter (English units)*

tively. The average absorption coefficient has been taken as the parameter in both.

The range of reasonably acceptable reverberation times for most speech and music applications as a function of room volume is plotted in Figure 6.11. We note that the reverberation time is taken at 500 Hz. This is usually considered to be an adequate indicator. The interested reader is referred to reference [1] for a more detailed treatment.

EXAMPLE 6.12 What is the reverberation time of a room 20 ft × 15 ft × 12 ft if the average absorption coefficient is 0.2?

SOLUTION Employ equation (6.125) to obtain

$$T = 0.049(V/A) \quad \text{s}$$
$$V = 20 \text{ ft} \times 15 \text{ ft} \times 12 \text{ ft} = 3600 \text{ ft}^3$$
$$S = (20 \text{ ft} \times 15 \text{ ft} \times 2) + (20 \text{ ft} \times 12 \text{ ft} \times 2) + (15 \text{ ft} \times 12 \text{ ft} \times 2)$$
$$\quad = 1440 \text{ ft}^2$$
$$A = S\bar{\alpha} = 288 \text{ ft}^2$$
$$T = 0.049(3600/288) = 0.6 \text{ s} \quad \text{reverberation time}$$

These expressions for the reverberation time are adequate for most industrial noise calculations and design.

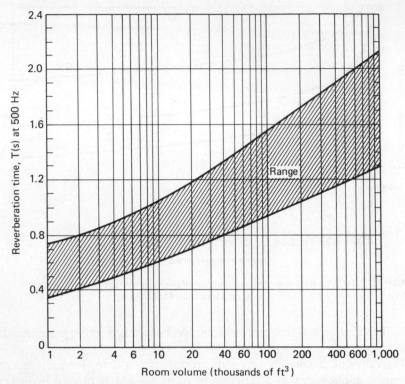

FIGURE 6.11 *Range of reasonably acceptable reverberation times for most speech and music applications* [1]

D6.10 How long will it take the steady-state energy density to decay by 40 dB in a room 20 ft × 20 ft × 12 ft with an average absorption coefficient of 0.3? ($c = 1128$ ft/s).

Ans. $T = 0.297$ s.

6.5.2 Room-Constant Determination from the Reverberation Time

Under some circumstances it is convenient to determine the room constant, R, from a measurement of the reverberation time, T. In equation (6.73) the room constant was defined as

$$R = \frac{\bar{\alpha}S}{1 - \bar{\alpha}} \quad \text{m}^2 \quad \text{or} \quad \text{ft}^2 \tag{6.73}$$

where

$$\bar{\alpha} = \frac{\alpha_1 S_1 + \alpha_2 S_2 + \alpha_3 S_3 + \ldots + \alpha_n S_n}{S_1 + S_2 + S_3 + \ldots + S_n} \tag{6.37}$$

which can also be related to A as defined by equation (6.121). Combining equations (6.121) and (6.73), we obtain

$$R = \frac{AS}{S - A} \quad \text{m}^2 \quad \text{or} \quad \text{ft}^2 \qquad\qquad (6.128)$$

Equations (6.124) and (6.125) may each be solved for A to yield

$$A = 0.161\left(\frac{V}{T}\right) \quad \text{m}^2 \qquad\qquad (6.129)$$

in metric units and

$$A = 0.049\left(\frac{V}{T}\right) \quad \text{ft}^2 \qquad\qquad (6.130)$$

in English units. One may now successively substitute equations (6.129) and (6.130) into (6.128) to obtain

$$R = \frac{S}{\dfrac{TS}{0.161V} - 1} \quad \text{m}^2 \qquad\qquad (6.131)$$

in metric units and

$$R = \frac{S}{\dfrac{TS}{0.049V} - 1} \quad \text{ft}^2 \qquad\qquad (6.132)$$

in English units, respectively. These relationships are most useful when a wide range and distribution of absorption coefficients characterize the room in question.

EXAMPLE 6.13   Determine the room constant of a room that is $15\,\text{m} \times 20\,\text{m}$ with a ceiling height of 3 m and a reverberation time of 2 s.

SOLUTION   Use equation (6.131),

$$R = \frac{S}{\left(\dfrac{TS}{0.161\,V}\right) - 1} \quad \text{m}^2$$

In this instance $S = 810\,\text{m}^2$, $V = 900\,\text{m}^3$, and $T = 2$ s. Upon substitution,

$$R = \frac{810}{\dfrac{(2)(810)}{(0.161)(900)} - 1} \quad \text{m}^2$$

which yields

$$R = 80 \text{ m}^2 \quad \text{the room constant}$$

D6.11 What is the average absorption coefficient of a room 25 ft × 25 ft × 15 ft if the reverberation time is 0.6 s?

Ans. $\bar{\alpha} = 0.28$.

PROBLEMS

6.1 A certain pneumatic machine is located in a room 7 m × 9 m with a ceiling 4 m high. The absorption coefficients in the 1000-Hz octave band for the floor, walls, and ceiling are 0.01, 0.3, and 0.8, respectively. The machine has a sound power level of 125 dB in this same band. What is the space average energy density in the room after the machine has been "on" for 10 ms? (Assume that $c = 344$ m/s.)

6.2 A room 12 ft × 15 ft with a 10-ft-high ceiling has an average absorption coefficient of 0.1 in the 2000-Hz octave band. A noise source located in the center of the room generates a 120-dB sound power level in the 2000-Hz band. What is the space average rms pressure in this band in the room after the source has been "on" for 1 ms? (Assume that $\rho_0 c = 407$ mks rayls and $c = 1128$ ft/s.)

6.3 A room 20 ft × 25 ft with a 10-ft ceiling has an average absorption coefficient of 0.15 in the 1-kHz octave band. There is a machine in the room that produces a sound power level of 120 dB in this band. What is the energy density in the 1-kHz band in the reverberant field after the source is "off" for 10 ms? (Assume that $c = 1128$ ft/s.)

6.4 A particular room 60 ft × 40 ft with a 20-ft ceiling was found to have an average absorption coefficient of 0.2 in the 4-kHz octave band. Two isotropic noise sources, separated by 39 ft, have been "on" for a long time period. Sources 1 and 2 produce 1 W and 4 W of acoustic power, respectively, in the 4-kHz band. If source 1 is turned "off" and then 1 s later source 2 is turned "off," what is the sound pressure level in the 4-kHz band at a position located 35 ft from source 1 and 25 ft from source 2 after source 2 is off for 1 s?

6.5 A room that is 60 ft × 100 ft with a 25-ft ceiling contains an isotropic noise source that produces a 125-dB sound power level in the 2-kHz octave band. The source is "on" and in the steady state. What will be the space average sound pressure level in this band in the room after the source is turned "off" for 2 s if the average absorption coefficient in this band is 0.2?

6.6 A machine that produces a sound power level of 128 dB in the 2-kHz octave band is located in a room 50 ft × 60 ft with a 30-ft ceiling. The floor and walls have an absorption coefficient of 0.1 and the ceiling 0.3 in the 2-kHz band. If the machine is in steady state prior to "turn-off," how long will it require for the average energy density in this band to decrease 40 dB below the steady-state value?

6.7 A noisy machine in a machine shop is under study. It produces 0.8 W of acoustic power in the 2-kHz octave band and has a directivity of 2 in the direction of position A. The shop is 20 ft × 30 ft with a 12-ft ceiling. The ceiling, walls, and floor have absorption coefficients of 0.7, 0.3, and 0.1, respectively, in the 2-kHz band. Determine the sound pressure level at position A located 15 ft from the machine.

6.8 A machine that radiates sound isotropically is operated in a room 20 ft × 40 ft with a 12-ft ceiling. The room has an average absorption coefficient of 0.5 in the 4-kHz octave band. At a point 16 ft from the machine the sound pressure level was found to be 94 dB. What is the acoustic power of the machine in the 4-kHz band?

6.9 A warehouse is 75 ft × 45 ft with a 25-ft ceiling. The walls, floor, and ceiling have absorption coefficients of 0.1, 0.1, and 0.4, respectively, in the 1000-Hz octave band. An acoustically isotropic pneumatic machine produces a sound pressure level of 96 dB at a distance of 30 ft from the machine. What is the sound pressure level in the 1-kHz band at a position 10 ft from the source?

6.10 A workbench is in a room 25 ft × 35 ft with a 15-ft ceiling. The average absorption coefficient in the 1000-Hz octave band is 0.01. A large machine in the room, which may be taken as isotropic, produces a sound pressure level of 95 dB in this band at a workbench that is 20 ft from the machine. What will be the sound pressure level in the 1-kHz band at the workbench if it is moved to within 5 ft of the machine?

6.11 A large room contains two noisy machines; the room is 30 ft × 50 ft with a 15-ft ceiling. The walls, ceiling, and floor have absorption coefficients of 0.1, 0.8, and 0.01 in the 2-kHz octave band. Machine 1 is 20 ft from position A and machine 2 is 15 ft from the same point. Machine one radiates 1 W and two radiates 0.5 W, both isotropically, in the 2-kHz band. Determine the sound pressure level at position A.

6.12 A machine with a sound power level of 112 dB and a directivity index of 4.8 dB in a certain direction in the 1-kHz octave band is located in a room 75 ft × 50 ft with a 25-ft ceiling. The walls, floor, and ceiling have absorption coefficients of 0.1, 0.06, and 0.75, respectively, in this band. What is the sound pressure level in the specified direction 35 ft from the machine?

6.13 A machine has a directivity factor of 6 in the 2-kHz octave band in a specified direction. The sound pressure level in this band and in this same direction is 98 dB at a distance of 20 ft from the source. Calculate the power of the source and the average sound pressure level in the 2-kHz band. (Assume an anechoic chamber.)

6.14 A machine was tested outdoors; the results yielded an average sound pressure level of 90 dB at a distance of 15 ft from the machine in the 1-kHz octave band. In a specified direction, also at 15 ft, it was found to be 96 dB. The machine is installed in a room which is 20 ft × 25 ft with a 12-ft ceiling. Furthermore, the absorption coefficients in this band for the walls, ceiling, and floor are 0.3, 0.3, and 0.01, respectively. What is the sound pressure level in this band at a distance of 12 ft in the specified direction with the machine installed in the room?

6.15 A machine with a sound power of 1 W in the 1-kHz band is located in a room that is 50 ft × 75 ft with a 20-ft ceiling. The reverberation time in this band was found to be 2 s. What is the sound pressure level in the 1-kHz band in the reverberant field of the room after the machine is "off" for 1 s?

6.16 A certain room is 20 ft × 30 ft with a 15-ft ceiling. The reverberation time in the 1-kHz octave band was found to be 0.7 s. An isotropic source with a sound power level of 120 dB in this band is placed in the room. Determine the sound pressure level in the 1-kHz band due to the source at a point 17 ft away.

6.17 An isotropic noise source with a sound power level of 120 dB in the 2-kHz octave band is located in a room which is 45 ft × 75 ft with a 20-ft ceiling. The reverberation time in the 2-kHz band was found to be 3 s. What is the sound pressure level in this band at a distance of 40 ft from the source?

REFERENCES

[1] BERANEK, L. L. *Acoustics*, McGraw-Hill Book Company, New York, 1954.

[2] HUNTER, J. L. *Acoustics*, Prentice-Hall, Inc., Englewood Cliffs, New Jersey, 1957.

[3] BERANEK, L. L. ed. *Noise Reduction*, McGraw-Hill Book Company, New York, 1960.

[4] BERANEK, L. L. ed. *Noise and Vibration Control*, McGraw-Hill Book Company, New York, 1971.

[5] MAGRAB, E. B. *Environmental Noise Control*, John Wiley & Sons, Inc., New York, 1975.

[6] HARRIS, C. M. "Absorption of Sound in Air," *J. Acoust. Soc. Amer.*, Vol. 40 (1966), pp. 148–159.

[7] "Methods for Determination of Sound Power Levels of Small Sources in Reverberation Rooms," *ANSI S1.21-1972*, American National Standards Institute, New York, 1972.

[8] RETTINGER, M. *Acoustic Design and Noise Control*, Chemical Publishing Co., New York, 1973.

7

ACOUSTICS OF WALLS, ENCLOSURES, AND BARRIERS

7.1 INTRODUCTION

In many practical applications an excessive sound level associated with a particular machine or group of machines is best handled by separating the offender from the worker by means of an acoustical partition. A wall separating two rooms for the purpose of containing the high sound level in the one is a fundamental configuration in understanding and applying this abatement technique. We will begin with such a configuration and proceed to enclosures and barriers.

7.2 TRANSMISSION LOSS AND TRANSMISSION COEFFICIENTS

In sharp contrast to sound-absorbing materials, which are light and porous, sound-isolation materials are massive and airtight. As such, these isolation materials form an effective sound-insulating structure between the noise source and receiver.

It is a common practice to specify the acoustical characteristics of an isolation wall by its transmission loss. This is most convenient and useful because the transmission loss is independent of the application and can therefore be obtained from tables for most common building materials.

The defining equation for the transmission loss is

$$TL = 10 \log \left(\frac{W_\alpha}{W_2} \right) \quad dB \tag{7.1}$$

where   TL = transmission loss   (dB)

W_α = acoustic power incident upon the wall, all of which is absorbed (W)

W_2 = acoustic power radiated by the wall into a perfectly absorbing second space (W)

The transmission coefficient is then defined by

$$\tau = \frac{p^2_{\text{transmitted}}}{p^2_{\text{incident}}} \tag{7.2}$$

or

$$\tau = \frac{W_2}{W_\alpha} \tag{7.3}$$

where τ = transmission coefficient

Clearly, τ is the ratio of the power that passed on through the wall to the power that was incident upon the wall. From equation (7.3) we observe that a desirably small transmission coefficient is obtained when

$$W_\alpha \gg W_2 \tag{7.4}$$

which means that only a small fragment of the power that was incident on the wall exits in the form of acoustical power in room 2.

It is desirable to combine equations (7.1) and (7.3) to obtain

$$\boxed{\text{TL} = 10 \log\left(\frac{1}{\tau}\right) \quad \text{dB}} \tag{7.5}$$

which relates the transmission loss to the transmission coefficient. For example, if $\tau = 0.01$, TL = 20 dB, and if $\tau = 0.001$, TL = 30 dB. It is not difficult to obtain a TL between 30 and 40 dB with ordinary high-mass building materials.

EXAMPLE 7.1   What is the transmission coefficient of a wall with a transmission loss of 32 dB?

SOLUTION

$$\text{TL} = 10 \log\left(\frac{1}{\tau}\right) \quad \text{dB}$$

$$32 = 10 \log\left(\frac{1}{\tau}\right)$$

$$3.2 = \log\left(\frac{1}{\tau}\right)$$

$$\text{antilog } 3.2 = \frac{1}{\tau}$$

$$\frac{1}{\text{antilog } 3.2} = \tau \quad \text{or} \quad \tau = \frac{1}{10^{3.2}}$$

$$\tau = 6.3 \times 10^{-4}$$

7.2.1 Transmission-Loss Frequency Dependence

The manner in which the transmission loss varies with frequency for a homogeneous wall is shown in Figure 7.1. At very low frequencies it is controlled primarily by the wall's stiffness. In general, at these low frequencies, the stiffer the wall, the better the transmission loss. As the frequency is increased, the TL curve enters a region where the transmission loss is controlled by the various resonant frequencies of the wall. Transmission loss in this region is limited by the damping of the wall.

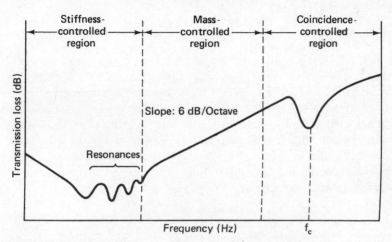

FIGURE 7.1 *Variation of the transmission loss, TL, with frequency for a homogenous wall with medium damping (f_c is called the critical frequency)*

At frequencies above the resonant frequencies, the transmission loss is controlled by the mass of the wall. In this region, the Mass Law, which is given by the approximate expression

$$\boxed{\text{TL} = (20 \log f) + (20 \log W) - C \quad \text{dB}} \tag{7.6}$$

where f = frequency (Hz)

W = surface density (lb/ft²/in. or kg/m²/cm)

$C = 33$ if W is in lb/ft²/in.

$C = 47$ if W is in kg/m²/cm

indicates that the TL increases 6 dB for each doubling of frequency or surface density. In practice, however, the actual increase in TL is somewhat less than that predicted by the Mass Law.

TABLE 7.1

Surface densities of certain common building materials

Material	SURFACE DENSITY	
	$lb/ft^2/(in.\ of\ thickness)$	$kg/m^2/(cm\ of\ thickness)$
Brick	10–12	19–23
Cinder concrete	8	15
Dense concrete	12	23
Wood	2–4	4–8
Common glass	15	29
Lead sheets	65	125
Gypsum	5	10

In Table 7.1 are given the surface densities for certain common building materials. Note that the surface densities are given per unit of thickness; as the thickness increases, the surface density increases accordingly.

Above the mass-controlled region lies the coincidence-controlled region. The model used to study the coincidence effect is that of an incident plane wave striking an infinite wall or panel as shown in Figure 7.2. One may question the validity of this model in studying real walls; however, this model does provide insight into the cause–effect relationships involving the transmission loss in applied situations.

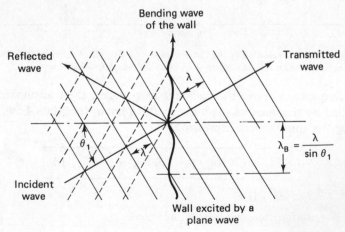

FIGURE 7.2 *Coincidence effect of an incident plane wave striking an infinite wall*

The coincidence effect arises from the fact that for all freqencies above the critical frequency, f_c, there is a certain angle of incidence, θ_1, at which a plane wave may excite the wall such that the sound wave will be transmitted through it with a reduced loss. This is possible, in that above the critical frequency, the wavelength of the bending wave, λ_B, in the wall may become equal to the projection of the wavelength in air, λ, upon the wall. This circumstance tightly couples the sound wave and the bending wave of the wall. In Figure 7.2 note that the projection of λ upon the wall is λ_B, that is,

$$\lambda_B = \frac{\lambda}{\sin \theta_1} \quad \text{m} \quad \text{or} \quad \text{ft} \qquad (7.7a)$$

and, upon rearranging, we obtain

$$\boxed{\sin \theta_1 = \frac{\lambda}{\lambda_B}} \qquad (7.7b)$$

This relation expresses the condition of wave coincidence. Furthermore, for every frequency above the critical frequency, if one fixes λ, there exists a θ_1 or, conversely, if one fixes θ_1, there exists a wavelength λ that satisfies equation (7.7b). If λ is fixed, θ_1 is called the *coincidence angle*, and if θ_1 is fixed, the frequency associated with λ is called the *coincidence frequency*.

The coincidence effect, which therefore occurs only when λ is less than or equal to λ_B, results in transmission losses lower than the mass law would indicate, as depicted in Figure 7.1. In order to reduce the coincidence effect, walls must be thick and very stiff, which decreases f_c, or heavy and limp, which greatly increases f_c.

In a homogeneous thin wall of infinite extent, the critical frequency is a function of the speed of sound in air, the thickness of the wall and its mass density, and the tensile modulus and Poisson's ratio of the wall. For example, the critical frequency, f_c, of common plywood, 1.27 cm or 0.5 in. in thickness, occurs at approximately 2000 Hz. In the case of glass with this same thickness, f_c is approximately 1300 Hz. A very useful set of curves given by Magrab [1] plot the thickness of several common building materials as a function of the critical frequency.

7.2.2 Sound Transmission Class

The Sound Transmission Class (STC) is a single-number classification used to rate the transmission loss of a wall. The higher the STC value, the more efficient the wall is in reducing sound transmission.

The general form of the standard reference STC contour used in the classification of walls is shown in Figure 7.3. The abscissa points are the same for

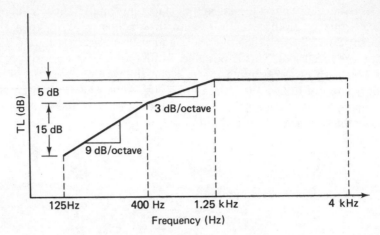

FIGURE 7.3 *Standard STC contour*

each contour; only the ordinate values change as the curve is shifted up or down.

If the TL for the wall at the 16 $\frac{1}{3}$-octave points from 125 Hz to 4 kHz are known and plotted, then the STC for the wall is determined by superimposing the contour shown in Figure 7.3 upon the TL curve such that (1) there is no more than a 8-dB deficiency between the TL and the STC contour at any $\frac{1}{3}$-octave frequency (i.e., no test point may be more than 8 dB below the STC contour), and (2) the total deficiency between the STC contour and the TL curve (i.e., the value of the STC contour minus the value of the TL curve summed at all $\frac{1}{3}$-octave frequencies from 125 Hz to 4 kHz) must be less than or equal to 32 dB. Once the curve has been adjusted to meet these two criteria, then the STC value of the wall is equal to the TL value of the contour at 500 Hz.

> **EXAMPLE 7.2** The TL curve for a wall is shown in Figure 7.4. The standard STC contour has been shifted into position to satisfy the two criteria above. It can be seen from the value of the contour at 500 Hz that the STC rating of the wall is approximately 43 dB.

7.3 NOISE REDUCTION OF A WALL

The noise reduction of a wall is of considerable interest. As depicted in Figure 7.5, acoustical walls are often used to separate excessively noisy machines from the majority of the workers in a plant situation. We will now obtain an expression for the noise reduction of a wall as defined by

$$\boxed{NR = L_{p_1} - L_{p_2} \quad dB}$$ (7.8)

where   NR = noise reduction (dB)

L_{p_1} and L_{p_2} = sound pressure level in rooms 1 and 2, respectively (dB) (both are measured near the wall, say within a wavelength or two)

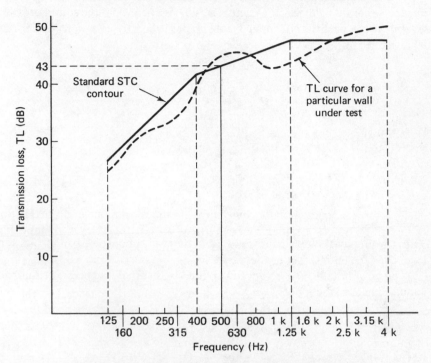

FIGURE 7.4 *Method of determining the STC value of a wall*

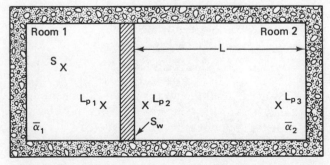

FIGURE 7.5 *Two rooms acoustically separated by a partition of area S_w*

Care must be taken in measuring L_{p_1} and L_{p_2} to ensure that the values obtained are representative. This is usually achieved by moving the microphone about in the region of interest and recording an average value. Note, however, that in designing a wall, both L_{p_1} and L_{p_2} are calculated, not measured, prior to wall construction.

We shall assume a noise source in room 1 which produces virtually a pure reverberant field near the partition. This is achieved if the sound pressure level near the partition in room 1 can be described by equation (6.92); that is, in metric units,

$$\boxed{L_{p_1} = L_{w_1} + 10 \log \left(\frac{4}{R_1}\right) \quad \text{dB}} \tag{7.9}$$

where

$$\frac{4}{R_1} \gg \frac{Q}{4\pi r^2} \tag{7.10}$$

R_1 = room constant of room 1

r = distance from the source in room 1 to the position of L_{p_1} in question

This requirement allows us to take advantage of a constant sound pressure level over the entire area of the partition.

The power in the reverberant field which is absorbed by the partition from room 1 is described by

$$W_\alpha = W_r \left(\frac{S_w \alpha_w}{S_1 \bar{\alpha}_1}\right) \quad \text{W} \tag{7.11}$$

where W_α = power absorbed by the wall (W)

W_r = power in the reverberant field (W)

S_w = area of the wall (m² or ft²)

α_w = absorption coefficient of the wall

S_1 = surface area of room 1 (m² or ft²)

$\bar{\alpha}_1$ = average absorption coefficient

We shall now assume that all the power which is incident upon the wall will be absorbed; that is, $\alpha_w = 1$. The power in the reverberant field may be expressed in terms of the power, W_1, of the source

$$W_r = W_1(1 - \bar{\alpha}_1) \quad \text{W} \tag{7.12}$$

which upon substitution into equation (7.11) yields

$$W_\alpha = W_1(1 - \bar{\alpha}_1)\frac{S_w}{S\bar{\alpha}_1} \quad \text{W} \tag{7.13}$$

We can now express the power, W_α, absorbed by the wall in terms of the room constant, R_1, of room 1. Equation (7.13) becomes

$$W_\alpha = W_1\frac{1 - \alpha_1}{S\bar{\alpha}_1}S_w = \frac{W_1 S_w}{R_1} \quad \text{W} \tag{7.14}$$

which may be combined with Eq. (7.3) to obtain

$$\boxed{W_2 = \frac{W_1 S_w \tau}{R_1} \quad \text{W}} \tag{7.15}$$

where $W_2 =$ power transmission into room 2 (W)

        $\tau =$ transmission coefficient of the wall

The energy density in both the direct and reverberant fields in room 2 can now be obtained with a knowledge of W_2 and the description of the room.

For the region near the wall in room 2, the direct field may be considered as a uniform plane wave progressing outward from the radiating wall. Thus, the portion of the energy in the direct field is equal to the product of the power transmitted into the room times the time required for the plane wave to reach the far end of the room. The time required is

$$t = \frac{L}{c} \quad \text{s} \tag{7.16}$$

where $L =$ length of the room (m or ft)

Applying this concept, the energy density in the direct field, δ_{d_2}, may be expressed as the direct field energy divided by the volume of the room:

$$\delta_{d_2} = W_2\left(\frac{L}{c}\right)V^{-1} \quad \text{J/m}^3 \tag{7.17}$$

where $V = S_w L =$ volume of room 2 (m³)

An expression for the reverberant energy density was previously obtained and is expressed by equation (6.75),

$$\delta_{r_2} = \frac{4W_2}{cR_2} \quad \text{J/m}^3 \tag{6.75}$$

Upon combining equations (7.17) and (6.75), we obtain for the total energy density near the wall,

$$\delta_2 = \frac{W_2}{c}\left(\frac{1}{S_w} + \frac{4}{R_2}\right) \quad \text{J/m}^3 \qquad (7.18)$$

We may now substitute in the expression for W_2 given in equation (7.15) to obtain

$$\delta_2 = \frac{W_1}{c}\frac{S_w\tau}{R_1}\left(\frac{1}{S_w} + \frac{4}{R_2}\right) \quad \text{J/m}^3 \qquad (7.19)$$

which upon rearranging becomes

$$\delta_2 = \frac{W_1}{c}\frac{4}{R_1}\left(\frac{1}{\tau}\right)^{-1}\left(\frac{1}{4} + \frac{S_w}{R_2}\right) \quad \text{J/m}^3 \qquad (7.20)$$

The mean-square pressure p_2^2, in room 2 according to equation (6.23) is

$$p_2^2 = \rho_0 c^2 \delta_2 \quad \text{Pa}^2 \qquad (7.21)$$

and upon expanding by means of equation (7.20), one obtains

$$p_2^2 = W_1\rho_0 c\frac{4}{R_1}\left(\frac{1}{\tau}\right)^{-1}\left(\frac{1}{4} + \frac{S_w}{R_2}\right) \quad \text{Pa}^2 \qquad (7.22)$$

We can now determine the sound presure level, L_{p_2}, from the defining equation

$$L_{p_2} = 10 \log \left(\frac{p_2}{p_{re}}\right)^2 \qquad (7.23)$$

where $p_{re} = 20 \times 10^{-6}$ Pa

Upon substitution of equation (7.22) into (7.23), one obtains

$$\begin{aligned} L_{p_2} = {} & 10 \log W_1 + 10 \log (\rho_0 c) - 10 \log (20 \times 10^{-6})^2 \\ & + 10 \log \left(\frac{4}{R_1}\right) - 10 \log \left(\frac{1}{\tau}\right) + 10 \log \left(\frac{1}{4} + \frac{S_w}{R_2}\right) \quad \text{dB} \end{aligned} \qquad (7.24)$$

Further simplification by use of the defining equation for power level,

$$L_{w_1} = 10 \log \left(\frac{W_1}{W_{re}}\right) \quad \text{dB} \qquad (7.25)$$

where $W_{re} = 1 \times 10^{-12}$ W

yields

$$L_{p_2} = L_{w_1} + 10 \log \left(\frac{4}{R_1}\right) - 10 \log \left(\frac{1}{\tau}\right) + 10 \log \left(\frac{1}{4} + \frac{S_w}{R_2}\right)$$
$$+ 10 \log (\rho_0 c) - 20 \log (2 \times 10^{-5}) - 120 \quad \text{dB} \tag{7.26}$$

Equation (7.26) may be reduced further to obtain

$$L_{p_2} = L_{w_1} + 10 \log \left(\frac{4}{R_1}\right) - 10 \log \left(\frac{1}{\tau}\right) + 10 \log \left(\frac{1}{4} + \frac{S_w}{R_2}\right) \quad \text{dB} \tag{7.27}$$

where we have assumed $\rho_0 c = 407$ mks rayls

Upon examining equation (7.27), we note that

$$L_{p_1} = L_{w_1} + 10 \log \left(\frac{4}{R_1}\right) \quad \text{dB} \tag{7.28}$$

under the conditions of a reverberant field near the wall in room 1. Furthermore, the transmission loss, TL, was defined as

$$\text{TL} = 10 \log \left(\frac{1}{\tau}\right) \quad \text{dB} \tag{7.29}$$

We shall now use equations (7.28) and (7.29) to simplify (7.27), to obtain

$$\boxed{L_{p_2} = L_{p_1} - \text{TL} + 10 \log \left(\frac{1}{4} + \frac{S_w}{R_2}\right) \quad \text{dB}} \tag{7.30}$$

where S_w = area of the wall (m^2 or ft^2)

R_2 = room constant of room 2 (m^2 or ft^2)

Using this expression, one can readily calculate the sound pressure level, L_{p_2}, near the wall in room 2, given that the characteristics of the wall and of room 2 are known.

Conversely, assuming that we know the desired value of L_{p_2}, which is often the case, we may rearrange equation (7.30) to determine the required transmission loss, TL, of the wall. We obtain

$$\boxed{\text{TL} = L_{p_1} - L_{p_2} + 10 \log \left(\frac{1}{4} + \frac{S_w}{R_2}\right) \quad \text{dB}} \tag{7.31}$$

which is also valid in both metric and English units. One may then go to tables of the transmission loss of various materials to choose the appropriate material and construction design for the situation in question.

A brief listing of representative transmission losses for certain common building materials is given in Table 7.2. An excellent tabulation is given in reference [2]. Typical noise reductions are given for several common wall types in Table 7.3.

TABLE 7.2

Transmission loss of general building materials

	TRANSMISSION LOSS (dB)					
Material	*125 Hz*	*250 Hz*	*500 Hz*	*1000 Hz*	*2000 Hz*	*4000 Hz*
Brick, 4 in.	30	36	37	37	37	43
Cinder block, $7\frac{5}{8}$ in., hollow	33	33	33	39	45	51
Concrete block, 6 in., lightweight, painted	38	36	40	45	50	56
Curtains, lead vinyl, $1\frac{1}{2}$ lb/ft^2	22	23	25	31	35	42
Door, hardwood, $2\frac{5}{8}$ in.	26	33	40	43	48	51
Fiber tile, filled mineral, $\frac{5}{8}$ in.	30	32	39	43	53	60
Glass plate, $\frac{1}{4}$ in.	25	29	33	36	26	35
Glass, laminated, $\frac{1}{2}$ in.	23	31	38	40	47	52
Panels, perforated metal with mineral fiber insulator, 4 in. thick	28	34	40	48	56	62
Plywood, $\frac{1}{4}$ in., 0.7 lb/ft^2	17	15	20	24	28	27
Plywood, $\frac{3}{4}$ in., 2 lb/ft^2	24	22	27	28	25	27
Steel, 18 gauge, 2 lb/ft^2	15	19	31	32	35	48
Steel, 16 gauge, 2.5 lb/ft^2	21	30	34	37	40	47
Sheet metal laminate, 2 lb/ft^2, visiolastic core	15	25	28	32	39	42

Moreover, upon rearranging, equation (7.31) becomes

$$L_{p_1} - L_{p_2} = \text{TL} - 10 \log \left(\frac{1}{4} + \frac{S_w}{R_2} \right) \quad \text{dB} \tag{7.32}$$

which in terms of the noise reduction, NR, is

$$\boxed{\text{NR} = \text{TL} - 10 \log \left(\frac{1}{4} + \frac{S_w}{R_2} \right) \quad \text{dB}} \tag{7.33}$$

TABLE 7.3

Typical noise reduction for certain common walls

	OCTAVE BAND (Hz)					
	125	*250*	*500*	*1000*	*2000*	*4000*
2 in. × 4 in. stud wall with 0.5-in. gypsum board both sides, 6 lb/ft²	22	30	35	40	41	40
2 in. × 4 in. staggered stud wall with 0.5-in. gypsum board both sides, 7 lb/ft²	36	37	40	47	52	45
Metal channel stud wall, 0.5-in. gypsum board both sides, 5 lb/ft²	25	30	38	47	48	44
0.5-in. plywood, one side of 2 in. × 4 in. studs, 2 lb/ft²	10	15	17	19	20	26

This expression allows us to determine the noise reduction (NR) afforded by a wall between two rooms. Attention is directed to the fact that this equation was derived under the condition of a reverberant field at the wall in room 1 with the sound pressure level determined near the wall in room 2.

EXAMPLE 7.3   Two rooms are separated by a common wall that has the dimensions 25 ft × 15 ft with a transmission loss of 30 dB. Room 1 contains a noise source that produces a reverberant field with a sound pressure level of 108 dB near the wall. Determine the sound pressure level near the wall in room 2 if room 2 has a room constant of 1500 ft².

SOLUTION   Use equation (7.30) to obtain L_{p_2}; that is,

$$L_{p_2} = L_{p_1} - \text{TL} + 10 \log \left(\frac{1}{4} + \frac{S_w}{R_2} \right) \quad \text{dB}$$

where $S_w = 25 \text{ ft} \times 15 \text{ ft} = 375 \text{ ft}^2$

        $\text{TL} = 30 \text{ dB}$

        $R_2 = 1500 \text{ ft}^2$

        $L_{p_1} = 108 \text{ dB}$

Upon substitution, one obtains

$$L_{p_2} = 108 - 30 + 10 \log \left(\frac{1}{4} + \frac{375}{1500} \right) \quad \text{dB}$$

which yields

$$L_{p_2} = 75 \text{ dB}$$

near the wall in room 2.

If room 2 supports virtually no reverberant field or if the wall is simply an outside wall such that it radiates outdoors, then R_2 becomes very large and equation (7.33) reduces to

$$\boxed{\text{NR} = \text{TL} + 6 \text{ dB}} \tag{7.34}$$

We note that in this special case NR exceeds TL by 6 dB.

7.3.1 Sound Pressure Level Away from the Wall

In many practical situations we will be interested in the sound pressure level at considerable distances from the wall. Under such conditions, if the room is reverberant, the reverberant field will be the determining factor. Under these circumstances the direct field is of little consequence. In Figure 7.5 is depicted such a situation in the event that we are interested in L_{p_3}, which is located a relatively large distance from the wall. Special attention should be given to the fact that we now require simply L_{p_3} based upon L_{p_1}. We cannot represent the difference $(L_{p_1} - L_{p_3})$ as noise reduction in the context of our definition.

We shall now derive an expression for L_{p_3} based upon the assumption that it will be determined by the energy density in the reverberant field. The reverberant energy density is given by equation (6.75) to be

$$\delta_{r_2} = \frac{4W_2}{cR_2} \quad \text{J/m}^3 \tag{6.75}$$

which is virtually the total energy density in the region of room 2 in question.

Inasmuch as the reverberant energy density will now be assumed to be the total energy density, δ_3, at a considerable distance from the wall, we have

$$\delta_3 = \delta_{r_2} = \frac{4W_2}{cR_2} \quad \text{J/m}^3 \tag{7.35}$$

Upon substituting equation (7.35) into (6.24), we obtain

$$p_3^2 = \rho_0 c^2 \delta_3 \quad \text{J/m}^3 \tag{7.36}$$

$$p_3^2 = W_2 \rho_0 c \frac{4}{R_2} \quad \text{Pa}^2 \tag{7.37}$$

where p_3^2 is the mean-square pressure in those regions of room 2 where the direct field emanating from the wall as a planar wave has dispersed to such a degree that the field is essentially a superposition of randomly reflected field components (i.e., a reverberant field).

Equation (7.15), which expresses W_2 in terms of W_1, may now be substituted into equation (7.37) to obtain

$$p_3^2 = W_1 \rho_0 c \frac{S_w \tau}{R_1} \frac{4}{R_2} \quad \text{Pa}^2 \qquad (7.38)$$

or, upon rearranging,

$$p_3^2 = W_1 \rho_0 c \frac{4}{R_1} \frac{1}{\tau} \frac{S_w}{R_2} \quad \text{Pa}^2 \qquad (7.39)$$

$$L_{p_3} = 10 \log \left(\frac{p_3}{20 \times 10^{-6}} \right)^2 \quad \text{dB} \qquad (7.40)$$

and upon assuming that $\rho_0 c = 407$ mks rayls, we obtain, upon expansion and simplification,

$$\boxed{L_{p_3} = L_{p_1} + 10 \log \left(\frac{S_w}{R_2} \right) - \text{TL} \quad \text{dB}} \qquad (7.41)$$

where $\quad L_{p_3} =$ sound pressure level a considerable distance from the wall under the assumption that the direct field is negligible compared to the reverberant field

D7.1 The common wall between two adjacent rooms is 20 ft $\times$ 12 ft and has a transmission loss of 28 dB. The room constants of rooms 1 and 2 are 2000 ft^2 and 3000 ft^2, respectively. Determine the sound pressure level near the wall in room 2 if the noise source in room 1 generates a sound power level of 123 dB and produces a reverberant field near the common wall. What is the noise reduction produced by the wall?

Ans. $L_{p_2} = 73$ dB; NR $= 33$ dB.

D7.2 In room 1 there is a noise source which produces a sound pressure level of 102 dB near a common wall. The transmission loss of the common wall is 20 dB. Room 2, adjacent to room 1, is 20 $\times$ 30 $\times$ 15 ft high with an average absorption coefficient of 0.15. The 20-ft wall is the common wall. What is the sound pressure level at a large distance from the common wall in room 2?

Ans. 80 dB.

7.4 COMPOSITION AND MULTIPLE CONSTRUCTION OF WALLS

The correct use of multiple wall construction can provide a higher transmission loss than can be achieved with a single wall, especially in the high-frequency range. The double-wall construction shown in Figure 7.6 will serve to illustrate some of the techniques used to obtain a high transmission loss. First, the transmission loss can be increased by increasing the weight of the wall. Furthermore, two walls with air space between them will increase the TL. However, it is very important to realize that there must not be any rigid connection between the two walls. Therefore, as shown in Figure 7.6, the studs are staggered and the electrical outlets are not placed back to back. In addition, the TL can be increased by placing an isolation blanket (e.g., lead-loaded vinyl) between the walls as shown in the figure.

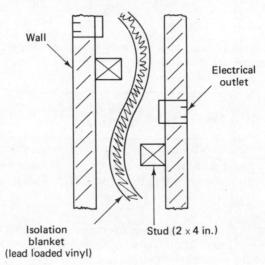

FIGURE 7.6 *High transmission loss, double-wall construction*

Analogous to the expression that a chain is only as strong as its weakest link, we find that for a composite wall which contains doors, windows, openings, or some of each, its transmission loss may be determined primarily by the element of the composite wall which itself has the smallest transmission loss. This is why any type of opening, such as a crack, can be so important.

A door may reduce the overall transmission loss of a wall because unless care is exercised, leaks will exist around the door which constitute a direct transmission path for noise, and the door may have a surface weight that is less than that of the wall. Windows also pose similar problems. Hence, for good sound isolation, doors should be heavy with a solid core and carefully sealed, and windows should be well sealed and made of thick or multiple panes. The importance of sound leaks cannot be overemphasized and so will be reiterated throughout this chapter.

The transmission loss for a composite wall consisting of n different elements (e.g., doors, windows, different types of materials) can be calculated from the expression

$$TL_{comp} = 10 \log \left(\frac{1}{\bar{\tau}}\right) \quad dB \qquad (7.42)$$

and

$$\bar{\tau} = \frac{S_1\tau_1 + S_2\tau_2 + \ldots + S_n\tau_n}{S_1 + S_2 + \ldots + S_n} \qquad (7.43)$$

$$\boxed{\bar{\tau} = \frac{1}{S} \sum_{i=1}^{n} \tau_i S_i} \qquad (7.44)$$

where $\bar{\tau}$ = average transmission coefficient of the composite wall

  τ_i = transmission coefficient of the ith element

  S_i = surface area of the ith element (m² or ft²)

  S = total surface area of the wall (m² or ft²)

The following example illustrates the use of this equation as well as some of the concepts discussed earlier.

EXAMPLE 7.4 A solid wall separating two areas in an industrial plant has a TL = 40 dB (i.e., $\tau = 10^{-4}$). In order to facilitate visual communication between the two areas, a large window is to be placed in the wall. The window will occupy 50% of the wall area and have a TL = 20 dB (i.e., $\tau = 10^{-2}$). The transmission loss of the composite structure is then calculated from equations (7.42) and (7.43). We have

$$TL_{comp} = 10 \log \left(\frac{1}{\bar{\tau}}\right) \quad dB$$

where

$$\bar{\tau} = \frac{S_1\tau_1 + S_2\tau_2}{S_1 + S_2}$$

In this case,

$$\bar{\tau} = \frac{S(0.5\tau_1 + 0.5\tau_2)}{S} = 0.5(\tau_1 + \tau_2)$$

Thus,

$$\bar{\tau} = 0.5(10^{-4} + 10^{-2}) = 0.005$$

Therefore,

$$\text{TL}_{\text{comp}} = 10 \log \left(\frac{1}{0.005}\right) \quad \text{dB}$$

$$= 23 \text{ dB}$$

which is the transmission loss for the composite wall.

If the window would occupy only 10% of the wall, we obtain

$$\bar{\tau} = \frac{S(0.9\tau_1 + 0.1\tau_2)}{S} = 0.9\tau_1 + 0.1\tau_2$$

$$= 0.9(10^{-4}) + 0.1(10^{-2})$$

$$= 0.0011$$

It then follows that

$$\text{TL}_{\text{comp}} = 10 \log \left(\frac{1}{0.0011}\right) \quad \text{dB}$$

$$= 30 \text{ dB}$$

D7.3 Find the transmission loss of a 40 ft $\times$ 12 ft composite wall consisting of a 3 ft $\times$ 7 ft door (TL = 25 dB) and 4 in. of plastered brick (TL = 47 dB).

Ans. TL = 38 dB.

D7.4 Two rooms have a common 12 ft $\times$ 8 ft concrete wall that has a transmission loss of 25 dB. Plans are to place a duct in this wall that will join the rooms; however, the transmission loss of the wall must not fall below 20 dB. What is the area of the largest allowable duct?

Ans. $S_{\text{vent}} = 0.659 \text{ ft}^2$.

7.5 INDIRECT SOUND PATHS

A number of indirect sound paths, or what might more appropriately be called sound leaks, are shown in Figure 7.7. These leaks must receive careful attention in order to insure that the integrity of an isolation system is not jeopardized.

The sound power through the wall can be reduced or minimized using the techniques described earlier in this chapter. The noise path through the air conditioning duct can be reduced using absorption material to line the duct as illustrated in Chapter 8. In addition, vibration isolation units, as described in Chapter 9, should be used to isolate the duct so that the duct structure does not transmit noise. The flanking path illustrated in Figure 7.8(a) can be reduced using the technique shown in Figures 7.8(b) and 7.8(c). [Note the use of vibration isolators in Fig. 7.8(c)].

Floors must be designed for protection against impact as well as airborne

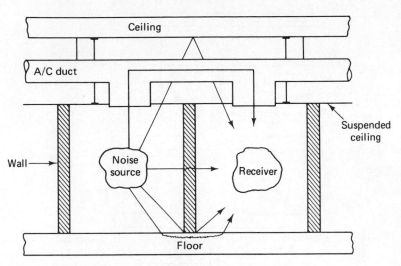

FIGURE 7.7 *Sound leaks*

noise. Typically, one can float the floor to reduce these types of noise. In addition impact noise can be reduced with carpet, rubber tile, or a cork blanket.

Careful consideration must also be given to a pipe or duct which must pass through a floor or ceiling. The pipe or duct must be well isolated in every situation of this type to prevent it from acting as a transmission medium for noise.

7.6 ENCLOSURES

Enclosures may be categorized in a number of ways. They may be full enclosures or partial ones, and if they are full, they may be large or small, depending upon whether the machinery occupies a large or small portion of the enclosed volume. In addition, an enclosure may be used to enclose people rather than machines. Before analyzing the various subdivisions, let us examine some of the factors that have significant influence on the performance of enclosures.

7.6.1 Factors that Affect the Performance of Enclosures

To begin with, we should never enclose more than is actually necessary. For example, if some component of a machine, such as a belt drive, gear box, or engine, is the primary noise source, then only this source should be enclosed. The enclosure walls should be constructed of a combination of

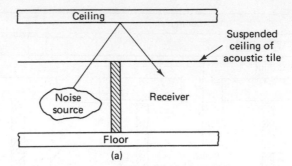

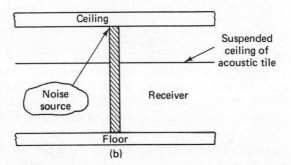

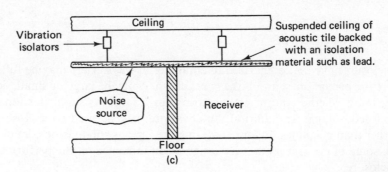

FIGURE 7.8 *Techniques for eliminating a flanking path: (a) noise flanking over the wall; (b) construction of the wall to the ceiling to block the flanking path; (c) suspended ceiling with backing to block the flanking path*

materials that will provide isolation, absorption, and damping, since all these functions are necessary, as will be shown later.

Any crack or leak in the structure, even though very small, can affect the noise reduction of the enclosure drastically. Mechanical, electrical, and utility connections must be carefully sealed. Openings that facilitate the flow of materials must be handled with care. Access doors for maintenance must be heavy and fit tightly. Windows used for inspection should be made from double panes and have no leaks.

The inside of the enclosure should be covered with appropriate absorbing material. This action will prevent the sound level from building up due to reflections, and thus minimize the wall vibration and the resultant radiation of the noise.

7.6.2 Full Enclosures or Hoods

Hoods are a special-purpose form of the acoustical enclosure designed for the express purpose of containing and absorbing the excessive acoustical energy of a machine. Conceptually, the design is quite straightforward; however, the major problem encountered is that due to acoustical leaks. Obviously, for a machine to be functional it must have both an input and output to the hood, as well as receive periodical maintenance. The required structural interruptions of the hood necessitate that great care be taken to keep all unavoidable leaks to a minimum.

We shall now proceed to describe the parameters and characteristics of hoods in a greatly simplified form. Excellent theoretical treatments have been presented by Jackson [3,4] and Junger [5] and are recommended for those interested in a more complete analysis of the problem. Two important concepts will be investigated, the noise reduction and the insertion loss of the hood. The insertion loss provides a means of determining the sound pressure level at any point removed from the hood.

7.6.2.1 *Noise reduction of a hood*

Most hoods of interest are designed such that the machine enclosed does not represent over about one-third of the total volume enclosed by the hood. Furthermore, we assume that the only absorber of acoustic energy inside the hood is its surface.

In the equilibrium condition, the total power, W_1, radiated by the source is absorbed by the walls of the hood. We also recall that the transmission coefficient is defined by

$$\tau = \frac{W_2}{W_1} \tag{7.45}$$

where W_1 = power incident upon the wall, all of which is absorbed (W)

W_2 = power transmitted through the wall (W)

We shall now proceed to determine the energy densities just inside and just outside the hood as designated in Figure 7.9 by δ_1 and δ_2, respectively.

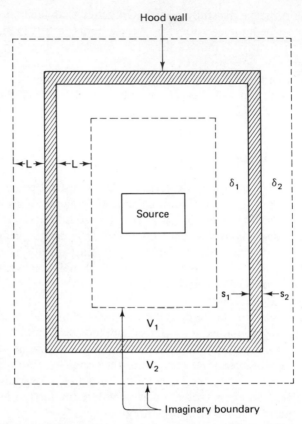

FIGURE 7.9 *Floor plan of a hood with energy densities designated by δ_1 and δ_2 just inside and just outside the hood*

Let us assume that

$$V = V_1 = V_2 \quad \mathrm{m^3} \tag{7.46}$$

This is a reasonable approximation for the region close to the hood.

The time required for the acoustic wave to travel the distance L is

$$t = L/c \quad \mathrm{s} \tag{7.47}$$

Therefore, the energy in the volume V_1 may be approximated by

$$E_1 = W_1(L/c) \quad \mathrm{J} \tag{7.48}$$

and the energy density by

$$\delta_1 = \frac{E_1}{V_1} = \frac{W_1(L/c)}{V} \quad \mathrm{J/m^3} \tag{7.49}$$

Similarly, the energy density, δ_2, in the volume V_2 is

$$\delta_2 = \frac{W_2(L/c)}{V} \quad \text{J/m}^3 \tag{7.50}$$

and upon substituting in equation (7.45), one obtains

$$\delta_2 = \frac{\tau W_1(L/c)}{V} \quad \text{J/m}^3 \tag{7.51}$$

Equation (7.49) may now be substituted into (7.51) to obtain

$$\delta_2 = \tau \delta_1 \quad \text{J/m}^3 \tag{7.52}$$

We can now readily obtain L_{p_1} and L_{p_2}, the sound pressure levels just inside and outside the hood, respectively. From equation (6.24), we have

$$p_1^2 = \rho_0 c^2 \delta_1 \quad \text{Pa}^2 \tag{6.24}$$

and

$$p_2^2 = \rho_0 c^2 \delta_1 \tau \quad \text{Pa}^2 \tag{7.53}$$

where p_1 = rms pressure just inside the hood (Pa)

p_2 = rms pressure just outside the hood (Pa)

Moreover, the sound pressure level was defined by

$$L_p = 10 \log \left(\frac{p}{p_{\text{re}}}\right)^2 \quad \text{dB} \tag{7.54}$$

into which we may substitute p_1^2 and p_2^2 to obtain L_{p_1} and L_{p_2}, respectively. Upon performing these substitutions, one obtains

$$L_{p_1} = 10 \log \left(\frac{\rho_0 c^2 \delta_1}{p_{\text{re}}^2}\right) \quad \text{dB} \tag{7.55}$$

and

$$L_{p_2} = 10 \log \left(\frac{\rho_0 c^2 \delta_1 \tau}{p_{\text{re}}^2}\right) \quad \text{dB} \tag{7.56}$$

We now need only substitute equations (7.55) and (7.56) into the defining equation for the noise reduction:

$$\text{NR} = L_{p_1} - L_{p_2} \quad \text{dB} \tag{7.8}$$

to obtain

$$\text{NR} = 10 \log \left(\frac{1}{\tau}\right) \quad \text{dB} \tag{7.57}$$

However, this is the defining equation for the transmission loss, TL; therefore,

$$NR = TL \quad dB \tag{7.58}$$

which states that for a hood in a virtual open space, the noise reduction from just inside the hood to a small distance external to the hood is equal to the transmission loss of the walls of the hood alone.

Care must be exercised in interpreting and applying equation (7.58). We now have that

$$\boxed{NR = L_{p_1} - L_{p_2} = TL \quad dB} \tag{7.59}$$

which is a reasonable estimate; however, one must remember that L_{p_1} is measured with the hood in place. Therefore, if when the hood is placed over a machine, the value of L_{p_1} is increased, then L_{p_2} will be correspondingly higher. If this is not considered and a hood is considered simply based on a tabulated value of TL, it is quite probable that an ineffective design will result. For this reason we are commonly interested in the insertion loss (IL) of the hood rather than simply the noise reduction.

EXAMPLE 7.5   The sound pressure level under the hood, with the hood in place, is 120 dB. The transmission loss of the walls of the hood is 22 dB. What is the sound pressure level just outside the hood? (Assume a free field.)

SOLUTION

$$TL = L_{p_1} - L_{p_2} \quad dB$$
$$L_{p_2} = L_{p_1} - TL$$
$$= 120 - 22$$
$$= 98 \text{ dB}$$

D7.5   A machine in a large room produces a sound power of 2 W. A hood that was $15 \times 15 \times 12$ ft with an internal average absorption coefficient of 0.65 was placed over the machine. Just outside the hood, the resultant sound pressure level was 87 dB. Measurements showed that the field just inside the hood was almost purely reverberant. What is the noise reduction of the hood? What is the transmission loss?

Ans.   NR = 19 dB, TL = 19 dB.

7.6.2.2   Hood insertion loss

In hood design and analysis we are more often interested in the insertion loss than in the noise reduction. This is the case because most hoods are designed after the fact to reduce the noise level at a particular location or in a specified region. The insertion loss pertains only to the sound pressure levels external to the hood, whereas the noise reduction refers to the sound pressure levels internal and external to the hood at the same time.

The insertion loss, IL, is defined by

$$\boxed{\text{IL} = L_{p_0} - L_{p_2} \quad \text{dB}} \tag{7.60}$$

where L_{p_0} = sound pressure level at the location in question without the hood (dB)

   L_{p_2} = sound pressure level at the location in question with the hood (dB)

We shall proceed to derive an approximate expression for the insertion loss with the hood in a room that is large compared to the dimensions of the hood. This requirement is easily met in most practical applications.

The sound pressure level, L_{p_0}, at any given point, A, without the hood installed is readily found by means of equation (6.93) to be

$$L_{p_0} = L_{w_0} + 10 \log \left(\frac{Q_0}{4\pi r^2} + \frac{4}{R}\right) \quad \text{dB} \tag{7.61}$$

Moreover, L_{p_2}, the sound pressure level at the same point A, with the hood installed is determined similarly by

$$L_{p_2} = L_{w_2} + 10 \log \left(\frac{Q_2}{4\pi r^2} + \frac{4}{R}\right) \quad \text{dB} \tag{7.62}$$

where L_{w_2} = sound power level of the machine and hood as a single unit

Our next task is to determine W_2, the total power radiated by the machine and hood as a single unit. This is readily done by taking advantage of the fact that the combination, machine and hood, is in an equilibrium condition. Therefore, the space average of the time-average energy density is a constant value under the hood. All the acoustic power being radiated by the machine is absorbed by the hood as losses or is radiated through the walls and outside the hood. That which reaches the outside of the hood is W_2.

The power W_2 can be approximated by means of equations (7.11) and (7.15). Based upon our assumption of equilibrium, we have

$$W_2 = W_0 \left(\frac{S_e \alpha_e}{S \bar{\alpha}}\right) \bar{\tau} \quad \text{W} \tag{7.63}$$

where W_2 = power radiated by the hood (W)

   W_0 = acoustic power of the source (W)

S_e = area of the hood walls and ceiling (m² or ft²)

α_e = absorption coefficient of the walls and ceiling

S = total area under the hood (m² or ft²)

$\bar{\alpha}$ = average absorption coefficient under the hood

$\bar{\tau}$ = average transmission coefficient of the hood, not including the floor

It is now desirable to further simplify equation (7.63) with regard to the hood surfaces. The total surface area under the hood is

$$S = S_e + S_f \quad \text{m}^2 \quad \text{or} \quad \text{ft}^2 \tag{7.64}$$

where S_f = area of the floor (m² or ft²)

We will assume that

$$S_e \gg S_f \tag{7.65}$$

and therefore we can make the approximation that

$$S = S_e \quad \text{m}^2 \quad \text{or} \quad \text{ft}^2 \tag{7.66}$$

Since the system is in equilibrium, it is reasonable to assume that

$$\alpha_e = 1 \tag{7.67}$$

which simply means that all the energy which impinges upon the surfaces of the hood walls is absorbed in one form or another.

Upon substituting equations (7.66) and (7.67) into (7.63), one obtains

$$W_2 = \bar{\tau} W_0 \left(\frac{1}{\bar{\alpha}} \right) \quad \text{W} \tag{7.68}$$

or

$$W_2 = \left(\frac{\bar{\tau}}{\bar{\alpha}} \right) W_0 \quad \text{W} \tag{7.69}$$

where $\bar{\tau} \leq \bar{\alpha} \leq 1$

These limits are set such that as $\bar{\alpha}$ approaches unity, the defining expression for τ is satisfied and, on the other hand, as $\bar{\alpha}$ approaches $\bar{\tau}$ virtually the total acoustic power of the source is radiated external to the hood. These limiting conditions are readily demonstrated experimentally.

We shall now assume that for our purposes in equations (7.61) and (7.62),

$$Q_0 = Q_2 = Q \tag{7.70}$$

which simply states that the Q of the hood–source combination is equal to that of the source alone. Now upon substituting equations (7.61) and (7.62) into (7.60), one obtains

$$IL = L_{w_0} - L_{w_2} \quad dB \tag{7.71}$$

or

$$IL = 10 \log \left(\frac{W_0}{W_2}\right) \quad dB \tag{7.72}$$

which may be further simplified by substituting in equation (7.69) to yield

$$\boxed{IL = 10 \log \left(\frac{\bar{\alpha}}{\bar{\tau}}\right) \quad dB} \tag{7.73}$$

where $\bar{\tau} \leq \bar{\alpha} \leq 1$

In practice, the average absorption coefficient, $\bar{\alpha}$, has a nonzero lower limit. This is due to three effects: (1) the air absorption inside the enclosure as given by equation (6.37b), (2) the viscous losses of waves near grazing incidence in the acoustical boundary layer on the inside of the enclosure, and (3) the change from adiabatic to isothermal compression as the inner walls are approached. The latter two of these effects may be taken into account by the expression [6]

$$\alpha_l = 1.8 \times 10^{-4}\sqrt{f} \tag{7.74}$$

where $f = $ frequency (Hz)

Upon adding the component due to air absorption, $\bar{\alpha}_{ex}$, as given by equation (6.37b) to that of (7.74), we obtain a minimum $\bar{\alpha}_{min}$, expressed by

$$\bar{\alpha}_{min} = \bar{\alpha}_{ex} + \alpha_l \tag{7.75}$$

or

$$\bar{\alpha}_{min} = K\left(\frac{4V}{S}\right) + 1.8 \times 10^{-4}\sqrt{f} \tag{7.76}$$

This minimum absorption coefficient for enclosures is also valid for reverberation rooms. In the final analysis, it determines the upper limit of the reverberation time that can be achieved in a reverberation chamber.

The two limiting cases for equation (7.73) are:

1. If $\bar{\alpha} = \bar{\tau}$, then

$$IL = 0 \text{ dB}$$

2. If $\bar{\alpha} = 1$, then

$$IL = 10 \log \left(\frac{1}{\tau}\right) \quad dB$$

or

$$IL = TL \quad dB$$

Case 1 is the worst case and case 2 is the best case considered for the insertion loss of the hood. In a practical situation we shall endeavor to keep $\bar{\alpha}$ as near unity as possible, and $\bar{\tau}$ must be much less than unity for an effective hood.

It should also be noted that in this analysis we have considered only high frequencies with a diffuse field inside the enclosure. A more complete treatment, including low-frequency effects, is given in reference [7]. Furthermore, owing to the necessary assumptions that were made in the derivation of the hood equation, one must always recognize that results are only approximations for engineering designs.

EXAMPLE 7.6   What is the insertion loss of a hood with an average absorption coefficient of 0.25 and a transmission coefficient of 0.001?

SOLUTION

$$IL = 10 \log \left(\frac{\bar{\alpha}}{\bar{\tau}}\right) \quad dB$$

$$= 10 \log \left(\frac{0.25}{0.001}\right)$$

$$= 10 \log (250)$$

$$= 23.98 \approx 24 \ dB$$

D7.6   The sound pressure level at a designated point due to a machine is 126 dB. A hood is placed over the machine, which reduces the sound pressure level at the same point to 106 dB. The transmission loss of the hood is 22 dB. What is the average absorption coefficient of the hood?

Ans. $\bar{\alpha} = 0.63$.

7.6.3   Small Enclosures

If the enclosure fits closely around the noise source, some interesting and detrimental phenomena may take place at certain frequencies. These frequencies are a function of the standing-wave resonances within the enclosure and the resonant frequencies of the enclosure walls.

If the machine has flat surfaces that are parallel to the enclosure walls, then standing-wave resonances at frequencies that are integer multiples of the half-wavelengths of the generated noise may render the enclosure practically useless. This condition can, however, be corrected through the use of sound-absorbing material which is at least one-quarter-wavelength thick, applied to the inside of the enclosure.

If the noise is generated at the resonant frequencies of the enclosure walls, the IL will not only be small, it may actually be negative! This situation may be corrected through structural damping or any other means which will ensure that the walls are not resonant at frequencies at which a high IL is necessary. In general, a reduction in noise levels of approximately 10 dBA is typical with small enclosures.

It is important to note that the phenomena which occur with small enclosures result from the fact that the machine and enclosure form a total system consisting of two parts (the machine and enclosure) which are coupled through the air medium.

The interested reader is referred to reference [7] for a more complete treatment of this topic.

7.6.4 Partial Enclosures

These elements are most useful in specialized cases, and in general provide protection in the form of a multicomponent barrier. They can be useful in a person–machine interaction situation in that some noise reduction can be provided while allowing personnel to interact with the machine. These devices can be of limited help but must always be used with caution.

In an industrial environment some pieces of equipment are inherently noisy, and when all other forms of control fail, the use of a partial or total enclosure must be considered. Although operator and maintenance personnel often dislike enclosures because of the lack of accessibility which they cause, nevertheless, these devices may be the only viable approach.

7.7 ACOUSTIC BARRIERS

The acoustic barrier provides a possible means to reduce the sound pressure level due to the direct field of the source at a given position. The barrier may therefore be regarded as a means of altering the effective directivity of the source in the direction of the position in question. In this process the diffraction of sound waves at the edges of the barrier with respect to the distances between the barrier, the source, and the receiver are most important. Furthermore, the reverberation characteristics of the environment must also be considered for a valid solution.

The problem of the diffraction at the edge of a semiinfinite barrier has been treated by Sommerfeld [8], McDonald [9], Redfearn [10], and others. In general, these more elegant analyses obtain solutions expressed in terms of Fresnel integrals. Maekawa [11] suggested that one might obtain a measure of the effectiveness of a finite-sized barrier by summing the diffracted fields of each of the edges. Furthermore, Moreland and Musa [12] went on to develop a well-defined and experimentally justified expression for the insertion loss of rectangular finite-sized barriers for industrial applications. The barrier analysis presented here is based upon their work.

We shall now develop expressions both for the sound pressure level and the insertion loss in the shadow region of a finite barrier. Emphasis will be placed on rectangular, planar barriers constructed from materials with a high transmission loss relative to the expected insertion loss of the barrier. The following development is based upon the concept that the sound pressure level at the receiver is the sum of that due to the diffracted field around the barrier and the composite reverberant field of the room.

7.7.1   Sound Pressure Level without the Barrier

Before the barrier is in position as shown in Figure 7.10, the mean square pressure at the receiver as derived in Chapter 6 is given by equation (6.86) as

$$p_0^2 = W \rho_0 c \left(\frac{Q}{4\pi r^2} + \frac{4}{R} \right) \quad \text{Pa}^2 \tag{6.86}$$

where p_0^2 = mean-square pressure without the barrier

Furthermore, the sound pressure level is given by equation (6.93) as

$$L_{p_0} = L_w + 10 \log \left(\frac{Q}{4\pi r^2} + \frac{4}{R} \right) \quad \text{dB} \tag{6.93}$$

where L_{p_0} = sound pressure level without the barrier (dB)

L_w = power level of the source (dB)

Q = directivity of the source

R = room constant without the barrier (m²)

r = distance from the source to the receiver (m)

7.7.2   Sound Pressure Level with the Barrier

The mean-square pressure, p_2^2, at the receiver as shown in Figure 7.10 with the barrier is expressed by

$$p_2^2 = p_{r_2}^2 + p_{b_2}^2 \quad \text{Pa}^2 \tag{7.77}$$

where $p_{r_2}^2$ = mean-square pressure at the receiver due to the reverberant field (Pa²)

$p_{b_2}^2$ = mean-square pressure at the receiver due to the diffracted field around the barrier edges (Pa²)

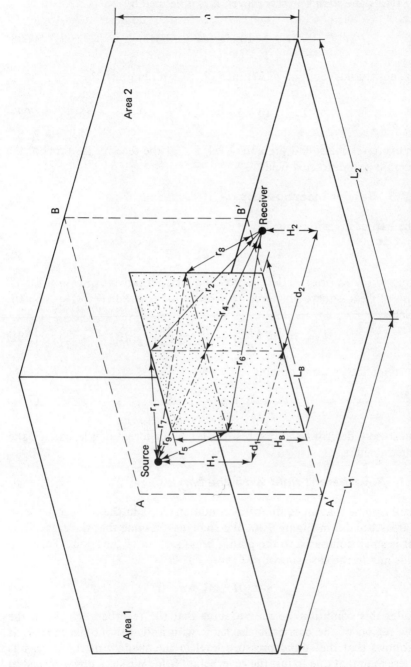

FIGURE 7.10 Schematic of a room with the barrier in position; the plane of AA'. BB' denotes the boundary between areas 1 and 2

215

As we recall, the sound pressure level, L_{p_2}, is defined by

$$L_{p_2} = 10 \log \left(\frac{p_2^2}{p_{\text{re}}^2} \right) \quad \text{dB} \tag{7.78}$$

Upon substituting equation (7.27) into (7.78), we obtain

$$L_{p_2} = 10 \log \left(\frac{p_{r_2}^2 + p_{b_2}^2}{p_{\text{re}}^2} \right) \quad \text{dB} \tag{7.79}$$

which expresses the sound pressure level, L_{p_2}, at the receiver in terms of the reverberant and diffracted fields.

7.7.3   Barrier Insertion Loss

The barrier insertion loss, IL, is expressed by

$$\text{IL} = L_{p_0} - L_{p_2} \quad \text{dB} \tag{7.80}$$

in terms of sound pressure levels. We find an alternative expression for IL in terms of mean-square pressures by substituting equation (7.78) into (7.80); this yields

$$\text{IL} = 10 \log \left(\frac{p_0^2}{p_{\text{re}}^2} \right) - 10 \log \left(\frac{p_2^2}{p_{\text{re}}^2} \right) \quad \text{dB} \tag{7.81}$$

Upon simplification, one obtains

$$\text{IL} = 10 \log \left(\frac{p_0^2}{p_2^2} \right) \quad \text{dB} \tag{7.82}$$

We must now determine an expression for p_2^2 in terms of the barrier and room parameters.

7.7.3.1   *Reverberant field with the barrier*

We shall assume that an equilibrium condition exists in the total room, that is, in areas 1 and 2 in Figure 7.10. We shall also assume that the area of the barrier is small compared to the planar cross section of the room. This condition is met in the case shown in Figure 7.10 if

$$L_2 h \gg L_B H \quad \text{m}^2 \quad \text{or} \quad \text{ft}^2 \tag{7.83}$$

Under this condition we may assume that the reverberant field in the shadow region of the barrier is the same with and without the barrier. It then follows that the sound pressure level in the shadow of the barrier is never less than that due to just the reverberant field. We shall now proceed to determine the assumed uniform reverberant energy density in the room.

The space average of the time-average reverberant energy density, δ_r, in the total room without the barrier is, from equations (6.75) and (6.23), given by

$$\delta_r = \frac{4W}{cR} = \frac{p_r^2}{\rho_0 c^2} \quad \text{J/m}^3 \tag{7.84}$$

Moreover, under the condition of equation (7.83),

$$\delta_r = \delta_{r_1} = \delta_{r_2} \quad \text{J/m}^3 \tag{7.85}$$

where δ_{r_1} = reverberant energy density in area 1 (J/m³)

   δ_{r_2} = reverberant energy density in area 2 (J/m³)

Therefore, equation (7.84) may be rearranged to obtain

$$p_{r_2}^2 = W\rho_0 c \frac{4}{R} \quad \text{Pa}^2 \tag{7.86}$$

where $p_{r_2}^2$ = mean-square pressure in the reverberant field of the shadow zone of the barrier (Pa²)

Equation (7.86) expresses the assumed uniform reverberant mean-square pressure in area 2 as a function of the source acoustic power, W, and the composite acoustic characteristics of areas 1 and 2. We now require the mean-square pressure in area 2 at the receiver position due to the diffracted field around the edges of the barrier.

7.7.3.2 Barrier diffracted field

The theory of diffraction due to the barrier has been well developed for the semiinfinite screen, that is, for the barrier which is very long compared to its height. For those interested in this theoretical development, Born and Wolf [13] and Redfearn [10] are excellent references. Furthermore, Maekawa [11] did extensive experimental work that yielded results in reasonable agreement with theory. Tatge [14] and Moreland and Musa [15] went on to suggest and reasonably verify a very much simplified expression for the diffracted mean-square pressure. The presentation given here is an extension of the work by Moreland and Musa.

The mean-square pressure, $p_{b_2}^2$, in the diffracted field, according to Moreland and Musa as reasonably verified by experiment, is given by

$$p_{b_2}^2 = p_{d_2}^2 \sum_{i=1}^{n} \frac{1}{3 + 10N_i} \quad \text{Pa}^2 \tag{7.87}$$

where $p_{d_2}^2$ = mean-square pressure due to the direct field before the barrier is inserted (Pa²) and the Fresnel number, N_i, is defined by

$$N_i \equiv \frac{2\delta_i}{\lambda} \qquad (7.88)$$

where δ_i = difference in direct path and diffracted path between the source and receiver (m or ft)

λ = wavelength (m or ft)

For example, in Figure 7.10, for the shadow zone, we have

$$\delta_1 = [(r_1 + r_2) - (r_3 + r_4)] \quad \text{m or ft}$$
$$\delta_2 = [(r_5 + r_6) - (r_3 + r_4)] \quad \text{m or ft} \qquad (7.89)$$
and
$$\delta_3 = [(r_7 + r_8) - (r_3 + r_4)] \quad \text{m or ft}$$

Equation (7.87) may be expanded to yield

$$p_{b_2}^2 = p_{d_2}^2\left(\frac{1}{3 + 10N_1} + \frac{1}{3 + 10N_2} + \frac{1}{3 + 10N_3}\right) \quad \text{Pa}^2 \qquad (7.90)$$

In the application of rectangular barriers, the three values of δ_i used in equation (7.90) are generally adequate for the shadow zone of the barrier.

D7.7 In a room are a source, barrier, and receiver. The source is isotropic and is 5 ft in front of the center of the barrier. The receiver is 4 ft behind the center of the barrier; both the source and the receiver are assumed to be 3 ft above the ground. The barrier is 6 ft wide and 8 ft high. Determine r_1 through r_8 as shown in Figure 7.10.

Ans. $r_1 = 7.1$ ft, $r_2 = 6.4$ ft, $r_3 = 5$ ft, $r_4 = 4$ ft, $r_5 = r_7 = 5.8$ ft, $r_6 = r_8 = 5$ ft.

The mean-squared pressure p_d^2 due to the direct field is easily derived from equations (6.23) and (6.83) to be

$$p_{d_2}^2 = W\rho_0 c\frac{Q}{4\pi r^2} \quad \text{Pa}^2 \qquad (7.91)$$

where r = direct distance from the source to the receiver (m or ft)

Equation (7.91) may be further simplified by noting that the diffraction coefficient, D, as defined by Moreland and Musa [15], is

$$D \equiv \sum_{i=1}^{n} \frac{1}{3 + 10N_i} \qquad (7.92)$$

D7.8 Determine the Fresnel numbers and the diffraction coefficient for the configuration in drill problem D7.7 in the 2000-Hz octave band. (Use $c = 1128$ ft/s.)

$$\text{Ans.} \quad N_1 = 16, \, N_2 = N_3 = 6.4, \, D = 0.04.$$

In the interest of completeness, it should be pointed out that an alternative definition [12] for the diffraction coefficient is

$$D \equiv \sum_{i=1}^{n} \frac{1}{3 + 20N_i} \tag{7.93}$$

This definition, which was proposed by Rathe, is also often used in the literature.

We may now substitute equations (7.91) and (7.92) into (7.87) to obtain

$$p_{b_2}^2 = W\rho_0 c \frac{QD}{4\pi r^2} \quad \text{Pa}^2 \tag{7.94}$$

It is convenient to define

$$\boxed{Q_B \equiv QD} \tag{7.95}$$

where Q_B = the effective directivity of the source in the direction of the shadow zone of the barrier

Therefore, one obtains

$$\boxed{Q_B = Q \sum_{i=1}^{n} \frac{1}{3 + 10N_i}} \tag{7.96}$$

or upon substituting equation (7.88) into (7.96), one obtains

$$\boxed{Q_B = Q \sum_{i=1}^{n} \frac{\lambda}{3\lambda + 20\delta_i}} \tag{7.97}$$

for the composite directivity of the source. The mean-square pressure $p_{b_2}^2$, due to the diffracted field as given by equation (7.94), may now be expressed as

$$p_{b_2}^2 = W\rho_0 c \frac{Q_B}{4\pi r^2} \quad \text{Pa}^2 \tag{7.98}$$

7.7.3.3 *Barrier-insertion-loss formulation*

We are now ready to substitute equations (7.86) and (7.98) into (7.77) to obtain the desired expression for the total mean-square pressure, Pa², at the receiver in the presence of the barrier; we obtain

$$p_2^2 = W\rho_0 c\left(\frac{Q_B}{4\pi r^2} + \frac{4}{R}\right) \quad \text{Pa}^2 \tag{7.99}$$

Upon comparing equation (7.98) to (6.86), one observes that the two are analogous equations. The sound pressure level, L_{p_2}, at the receiver in the shadow zone of the barrier in metric units is

$$L_{p_2} = L_w + 10 \log\left(\frac{Q_B}{4\pi r^2} + \frac{4}{R}\right) \quad \text{dB} \tag{7.100}$$

where

$$Q_B = Q \sum_{i=1}^{n} \frac{\lambda}{3\lambda + 20\delta_i} \tag{7.101}$$

In English units, we have

$$\boxed{L_{p_2} = L_w + 10 \log\left(\frac{Q_B}{4\pi r^2} + \frac{4}{R}\right) + 10 \quad \text{dB}} \tag{7.102}$$

By way of summary and clarification, we note that in determining L_{p_2} at the receiver, one requires

L_w = sound power level of the source (dB)

Q = directivity of the source

λ = wavelength of frequency in question (m or ft)

r = shortest distance from the source to the receiver without the barrier (m or ft)

δ_i = [(diffracted path length) − (direct path length)] (m or ft)

In terms of the notation of Figure 7.10, the three δ's required for a rectangular barrier are:

$$\delta_1 = [(r_1 + r_2) - (r_3 + r_4)] \quad \text{m or ft}$$
$$\delta_2 = [(r_5 + r_6) - (r_3 + r_4)] \quad \text{m or ft} \tag{7.103}$$
$$\delta_3 = [(r_7 + r_8) - (r_3 + r_4)] \quad \text{m or ft}$$

Note also that in equation (7.101) we allowed i to go from 1 through 3 as denoted in (7.103).

The insertion loss, IL, given in equation (7.80) as

$$\text{IL} = L_{p_0} - L_{p_2} \quad \text{dB}$$

now becomes, upon substituting in equations (6.95) and (7.100),

$$IL = 10 \log \left(\frac{\dfrac{Q}{4\pi r^2} + \dfrac{4}{R}}{\dfrac{Q_B}{4\pi r^2} + \dfrac{4}{R}} \right) \quad dB \qquad (7.104)$$

which is a general expression in the metric or English system of units for the insertion loss of the barrier for a receiver in the shadow zone, as depicted in Figure 7.10.

7.7.3.4 Special cases of the insertion loss

Two special cases of the insertion loss are of particular interest. They are (1) the barrier located in a free field only, and (2) the barrier located in a highly reverberant environment.

Case 1: Barrier in a Free Field In a free field, that is, if the barrier is outdoors or in a room, where the average absorption coefficient approaches unity, we obtain

$$R \approx \infty \qquad (7.105)$$

Equation (7.104) becomes

$$IL = 10 \log \left(\frac{Q}{Q_B} \right) \quad dB \qquad (7.106)$$

but since $Q_B = QD$, we obtain

$$IL = 10 \log (1/D) \quad dB \qquad (7.107)$$

where

$$D = \sum_{i=1}^{n} \frac{\lambda}{3\lambda + 20\delta_i} \qquad (7.108)$$

If we now expand equation (7.108) for the rectangular barrier and substitute into (7.107), we obtain

$$IL = -10 \log \left[\lambda \left(\frac{1}{3\lambda + 20\delta_1} + \frac{1}{3\lambda + 20\delta_2} + \frac{1}{3\lambda + 20\delta_3} \right) \right] \quad dB \qquad (7.109)$$

for the insertion loss of the barrier in a free field or a highly acoustically absorptive environment. Furthermore, we observe that if the barrier is of semiinfinite extent; that is, if in Figure 7.10, $L_B \approx \infty$, equation (7.109) becomes

$$\boxed{\text{IL} = -10 \log \left(\frac{\lambda}{3\lambda + 20\delta_1} \right) \quad \text{dB}} \qquad (7.110)$$

EXAMPLE 7.7   As an example of a free-field environment we shall use the anechoic room chosen by Moreland and Musa [15]. The room we shall use is about 19 ft × 16 ft with a ceiling of 11 ft. In the room is a source that is 0.5 ft above the floor and is in the center and 2 ft directly behind the 6-ft-wide × 4-ft-high rectangular barrier. The barrier used was a 1-in.-thick compressed-wood-chip Micarta-faced panel with a relatively high transmission loss. A sound-level meter 2 ft above the floor and directly in front of the barrier is used to measure the sound pressure level at various distances from the source. The sound source was located 7.9 ft from the shorter of the two walls. We shall now proceed to calculate the insertion loss for a separation between the source and receiver of 5 ft in the 1000-Hz octave band. We begin by calculating δ_1, the difference in the direct path and the diffracted path between the source and receiver, as given in equation (7.89):

$$\begin{aligned}
\delta_1 &= [(r_1 + r_2) - (r_3 + r_4)] \, \text{ft} \\
&= [(4 + 3.4) - (5)] \\
&= 2.4 \, \text{ft} \\
\delta_2 &= [(r_5 + r_6) - (r_3 + r_4)] \, \text{ft} \\
&= [(3.6 + 4) - (5)] \\
&= 2.6 \, \text{ft} \\
\delta_3 &= [(r_7 + r_8) - (r_3 + r_4)] \, \text{ft} \\
&= [(3.6 + 4) - (5)] \\
&= 2.6 \, \text{ft}
\end{aligned}$$

Having calculated the δ_i's, we now are ready to obtain the diffraction coefficient, D. The diffraction coefficient is defined by equation (7.108) as

$$D = \sum_{i=1}^{n} \frac{\lambda}{3\lambda + 20\delta_i}$$

The wavelength, λ, at 1000 Hz is

$$\begin{aligned}
\lambda &= c/f \quad \text{ft} \\
&= 1128/1000 \\
&= 1.128 \, \text{ft}
\end{aligned}$$

Therefore,

$$D = \lambda\left(\frac{1}{3\lambda + 20\delta_1} + \frac{1}{3\lambda + 20\delta_2} + \frac{1}{3\lambda + 20\delta_3}\right)$$

$$= 1.128\left[\frac{1}{3(1.128) + 20(2.4)} + \frac{1}{3(1.128) + 20(2.6)} + \frac{1}{3(1.128) + 20(2.6)}\right]$$

$$= 0.063$$

We can then calculate the insertion loss of equation (7.107):

$$IL = 10 \log (1/D) \quad dB$$
$$= 10 \log (1/0.063)$$
$$= 12 \ dB$$

Figures 7.11 and 7.12 are graphs of the insertion loss in the anechoic chamber as functions of frequency and distance between source and receiver. These graphs include the insertion loss as derived by the preceding equations and also measured data adapted from Moreland and Musa [15].

Case 2: Barrier in a Reverberant Environment As we shall proceed to show, barriers are ineffective in a highly reverberant environment. The acoustic field reaches the receiver by rebounding from the surfaces of the room to the extent that blockage of the direct field due to the barrier becomes inconsequential.

In a reverberant room we have

$$\frac{4}{R} \gg \frac{Q}{4\pi r^2} \quad \text{and} \quad \frac{4}{R} \gg \frac{Q_B}{4\pi r^2}$$

Therefore, IL as expressed by equation (7.104) becomes

$$IL = 10 \log (1) \quad dB$$

which yields

$$IL = 0 \ dB$$

EXAMPLE 7.8 As an example of a reverberant environment, we shall choose the source, barrier, and meter configuration used in case 1 and depicted in Figure 7.10. The high-transmission-loss barrier is 4 ft high × 6 ft wide, and the source and receiver are situated with respect to the barrier just as in case 1. The dimensions of the reverberant room in this case were about 15 × 20 ft with a 12-ft ceiling. The room constants, R, for the reverberant room are given in Figure 7.13 as a function of frequency in the appropriate octave

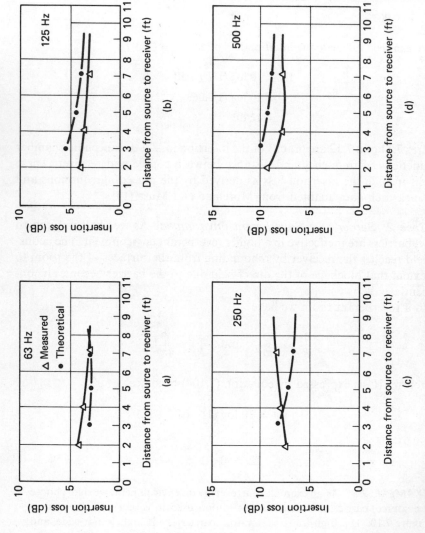

FIGURE 7.11 *Insertion loss in octave bands due to a barrier as described by Example 7.7: (anechoic environment) (measured data adapted from* [15])

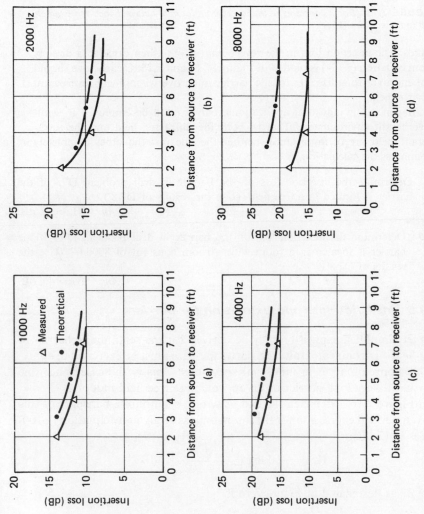

FIGURE 7.12 *Insertion loss in octave bands due to a barrier as described in Example 7.7:* (anechoic environment) (measured data adapted from [15])

225

Room constant (ft^2)	OCTAVE-BAND CENTER FREQUENCY (Hz)							
	63	*125*	*250*	*500*	*1000*	*2000*	*4000*	*8000*
	55	44	32	33	46	55	53	89

FIGURE 7.13 *Room constants of the reverberant room in octave bands for Example 7.8 (adapted from [15])*

bands. The insertion loss in the reverberant room when calculated directly from equation (7.104) is plotted in Figures 7.14 and 7.15. Clearly, one should not endeavor to utilize an acoustic barrier in an overwhelmingly reverberant environment.

In general, in industrial applications, barriers will be employed in semi-reverberant environments. That is, both the diffracted field and the reverberant field are of importance: neither the one nor the other completely dominates the calculation.

D7.9 Determine the insertion loss of the barrier in drill problem D7.7 if the barrier is located in a free field. (Use the results of D7.8.)

Ans. IL = 14 dB.

D7.10 Determine the insertion loss of the barrier in drill problem D7.7 if the barrier is located in a room with a room constant of 8000 ft² (Use the results of D7.8.)

Ans. IL = 4.4 dB.

7.7.3.5 Insertion-loss approximation for barriers

A rather simplified expression may be derived for the semiinfinite barrier in a free-field environment. However, particular care must be exercised in applying this approximation in industrial situations. This is the case, since one rarely has a very long barrier or an approximate free field indoors.

In Figure 7.16 is depicted the end view of a semiinfinite barrier. For this case, our general expression for the insertion loss as given by equation (7.104) reduces to (7.110):

$$\text{IL} = -10 \log \left(\frac{\lambda}{3\lambda + 20\delta_1} \right) \quad \text{dB}$$

where δ_1 as designated in Figure 7.16 is

$$\delta_1 = [(R^2 + H^2)^{1/2} + (D^2 + H^2)^{1/2} - (R + D)] \quad \text{m} \quad \text{or} \quad \text{ft} \quad (7.111)$$

Upon rearranging, we obtain

$$\delta_1 = \left\{ R\left[\left(1 + \frac{H^2}{R^2}\right)^{1/2} - 1 \right] + D\left[\left(1 + \frac{H^2}{D^2}\right)^{1/2} - 1 \right] \right\} \quad \text{m} \quad \text{or} \quad \text{ft} \quad (7.112)$$

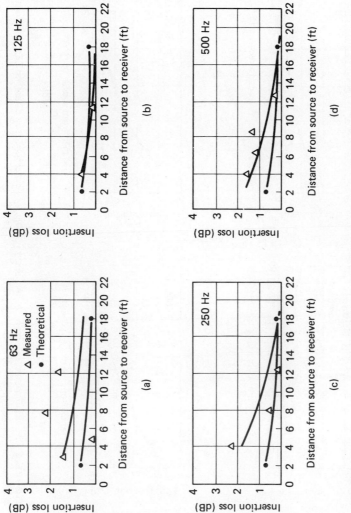

FIGURE 7.14 *Insertion loss in octave bands due to a barrier as described by Example 7.8: (reverberant environment) (measured data adapted from [15])*

227

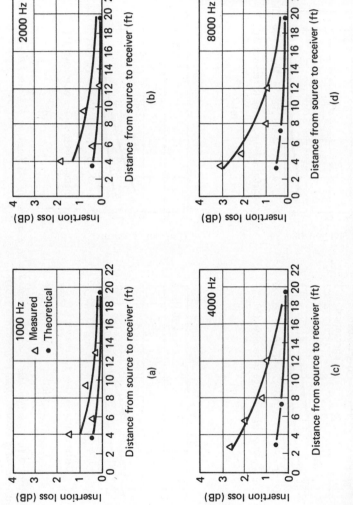

FIGURE 7.15 *Insertion loss in octave bands due to a barrier as described by Example 7.8; (reverberant environment) (measured data adapted from [15])*

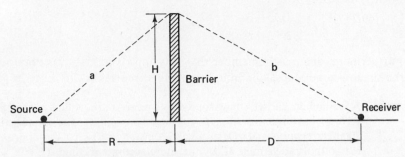

FIGURE 7.16 *Simplified end view of a semiinfinite barrier*

We shall now assume that

$$D \gg R \geq H$$

which allows us to reduce δ_1 to

$$\delta_1 \approx R\left[\left(1 + \frac{H^2}{R^2}\right)^{1/2} - 1\right] \quad \text{m} \quad \text{or} \quad \text{ft} \tag{7.113}$$

Applying the binomial theorem and keeping only the first two terms, one obtains

$$\left(1 + \frac{H^2}{R^2}\right)^{1/2} \approx \left(1 + \frac{H^2}{2R^2}\right) \tag{7.114}$$

which may be substituted into equation (7.113) to obtain

$$\delta_1 \approx \frac{H^2}{2R} \quad \text{m} \quad \text{or} \quad \text{ft} \tag{7.115}$$

We now substitute equation (7.115) into (7.110), which yields

$$\text{IL} \approx -10 \log\left[\frac{\lambda}{3\lambda + (10H^2/R)}\right] \quad \text{dB} \tag{7.116}$$

and now assuming that

$$\frac{10H^2}{R} \gg 3\lambda$$

one obtains

$$\boxed{\text{IL} \approx 10 \log\left(\frac{10H^2}{\lambda R}\right) \quad \text{dB}} \tag{7.117}$$

where $\quad D \gg R \geq H$

$$\frac{10H^2}{R} \gg 3\lambda$$

Furthermore, one must remember that this approximation was obtained for the shadow zone of a semiinfinite barrier in a free field.

D7.11　A semiinfinite barrier, sound source, and receiver are configured as in Figure 7.16. In this instance: $R = H = 6$ ft and $D = 18$ ft. The sound is in the 1000-Hz octave band. Determine the insertion loss by the approximate method; also determine IL by the more accurate method. (Use $c = 1128$ ft/s.)

Ans.　IL(approx.) $= 17$ dB, IL(accurate) $= 18$ dB.

PROBLEMS

7.1　A large room, room 1, shares a common wall 20 ft long with room 2, which is 20 ft $\times$ 30 ft with a 15-ft ceiling. The transmission loss of the common wall is 20 dB, and the average absorption coefficient of room 2 is 0.15, both in the 1-kHz octave band. If a noise source generates a uniform sound pressure level of 102 dB in this same band, near the common wall in room 1, what will be the sound pressure level 25 ft from the common wall in room 2?

7.2　Two rooms are joined by a common wall 20 ft long; rooms 1 and 2 have dimensions of 20 ft $\times$ 20 ft and 60 ft $\times$ 20 ft, respectively; both have 15-ft ceilings. The room constants of the rooms are 300 ft^2 and 500 ft^2 for 1 and 2, respectively. A machine with a sound power level of 125 dB in the 2-kHz octave band is located in room 1 such that the sound pressure level at the common wall is virtually uniform. What must be the value of the transmission coefficient in this band in order to obtain a 75-dB sound pressure level at a point in room 2—50 ft from the common wall?

7.3　A certain area has the floor plan shown. Rooms 1 and 2 have average absorption coefficients of 0.1 and 0.02, respectively, in the 1-kHz octave band. Furthermore, the wall and door (3 ft $\times$ 7 ft) that separate the two rooms have transmission losses of 20 dB and 25 dB, respectively, in the 1-kHz band. The door was installed with a 1-in. crack at the bottom. A machine, which generates a sound power level of 125 dB in this band, is located at point A in room 1 such that it produces a virtually uniform sound pressure level over the entire common wall. Determine the sound pressure level at position B in room 2.

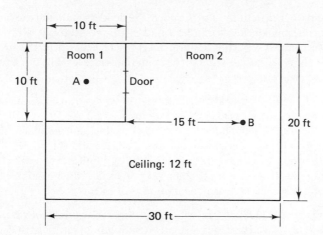

7.4 A certain room, 10 ft × 10 ft with a 10 ft-ceiling, has an average absorption coefficient of 0.1 in the 2-kHz octave band. This room has an outside wall with a transmission coefficient of 0.001 in this same band. A noise source with a sound power level of 120 dB in the 2-kHz band is located in the room. A uniform field was found to exist near the outside wall. Calculate the expected sound pressure level near the outside wall external to the building in this band.

7.5 A particular isotropic noise source produces a sound pressure level of 95 dB in the 2-kHz octave band, 10 ft from the source in a large room 20 ft × 50 ft with a 20-ft ceiling. The room presently has an average absorption coefficient of 0.1. The room is to be divided into two rooms with average absorption coefficients remaining 0.1. Rooms 1 and 2 are to be 20 ft × 20 ft and 20 ft × 30 ft, respectively. The noise source is to be placed in room 1 such that a uniform field is produced at the common wall. If the dividing wall has a transmission coefficient of 0.001 in the 2-kHz band, what will the sound pressure level be in room 2 near the common wall in this same band?

7.6 A certain area has the floor plan shown. A machine that generates 2 W of acoustical power in the 2-kHz octave band is located at point S in the small room. The average absorption coefficients of the small and large rooms are

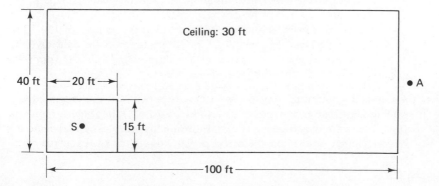

0.1 and 0.3, respectively; furthermore, the dividing wall and the outside walls have transmission losses of 25 dB and 30 dB, respectively, all in the 2-kHz octave band. A virtually uniform sound field exists near the dividing wall inside the small room. What is the sound pressure level in this band at point A near the outside wall?

7.7 Two adjoining rooms have a common wall with a transmission coefficient of 0.0005 in the 1-kHz octave band. Rooms 1 and 2 are 30 ft × 20 ft and 30 ft × 60 ft, respectively; both have a 15-ft ceiling. Furthermore, the average absorption coefficients of rooms 1 and 2 are 0.3 and 0.1, respectively, in the 1-kHz band. A machine in room 1 produces a uniform sound field of 98 dB sound pressure level near the common wall in this same band. Determine the sound pressure level near the common wall in room 2.

7.8 A large room, 30 ft × 80 ft, with an average absorption coefficient of 0.3 in the 1-kHz octave band, has a small room adjoining it on each end. Each of the small rooms is 30 ft × 30 ft; all three rooms have a 20-ft ceiling. The average absorption coefficient of the room and the transmission loss of the common wall are 0.1 and 20 dB, respectively, for small room 1 and 0.2 and 30 dB, respectively, for small room 2 in the band in question. Noise sources in rooms 1 and 2 produce 1 W and 2 W, respectively, of acoustical power in the 1-kHz band. The sound fields in both small rooms may be taken as diffuse for calculation purposes. Determine the sound pressure level in the center of the large room due to the two sources.

7.9 Two rooms, 1 and 2, are adjoined by means of a common wall with a trans-mission loss of 30 dB in the 1-kHz octave band. Room 1 is 20 ft × 40 ft and room 2 is 40 ft × 40 ft; both have a 20-ft ceiling. The average absorption coefficients in this band are 0.1 and 0.6 for rooms 1 and 2, respectively. A noise source located in room 1 produces 1 W of acoustical power in the 1-kHz band; the sound pressure level near the common wall is assumed uniform. Determine the sound pressure level near the common wall in room 2.

7.10 In a certain plant, a long narrow space, 6 ft × 120 ft with a 10-ft ceiling, is divided into three equal areas, each 6 ft × 40 ft in size. Each of the dividing walls has a transmission loss of 30 dB in the 2-kHz octave band; furthermore, each of the three new rooms has an average absorption coefficient of 0.02 in this same band. A machine that produces a sound power level of 130 dB in the 2-kHz band is located in one of the end rooms far away from the dividing wall. Determine the sound pressure level near the dividing wall in the opposite end room from that where the machine is located.

7.11 An upright cylindrical special purpose area is buried 30 ft in the ground and extends 20 ft above the ground for a total length of 50 ft; it has a radius of 10 ft. The top of the cylinder has a transmission loss of 25 dB, and the interior has an average absorption coefficient of 0.6, both in the 1-kHz octave band.

A machine generates 2 W of acoustical power at the floor in this same band. What is the sound pressure level just above the top of the cylinder?

7.12 A machine that radiates 4 W of acoustical power isotropically in the 1-kHz octave band is enclosed under a hood that is 10 ft × 10 ft × 10 ft with an average absorption coefficient of 0.5 in this same band. The machine is located in a room that is 100 ft × 75 ft with a 30-ft ceiling; it has an average absorption coefficient of 0.2 in the 1-kHz band. The sound pressure level 60 ft from the enclosed source was found to be 60 dB. Determine the transmission coefficient of the band.

7.13 A hood is to be designed with an insertion loss of 36 dB in the 2-kHz octave band; the construction material has a transmission coefficient of 0.0002 in this band. Determine the required average absorption coefficient inside the hood.

7.14 Two machines, A and B, are located in a room that is 50 ft × 75 ft with a 30-ft ceiling; the average absorption coefficient of the room is 0.2 in the 1-kHz octave band. Machine A, which generates 4 W of acoustical power in the 1-kHz band, is enclosed by a hood with an average absorption coefficient and transmission coefficient of 0.6 and 0.001, respectively, in this same band. Machine B, located 35 ft from machine A, produces 2 W of acoustical power in the 1-kHz band; both sources may be considered isotropic. Determine the sound pressure level at a position located 20 ft from machine A and 50 ft from machine B.

7.15 A certain hood which is 15 ft × 15 ft × 15 ft has an insertion loss of 30 dB in the 2-kHz octave band. The absorption coefficients of the top, bottom, and walls are 0.9, 0.1, and 0.5, respectively in this band. What is the transmission loss of the hood?

7.16 A room that is 12 ft × 25 ft with a 10-ft ceiling has an average absorption coefficient of 0.1 in the 1-kHz octave band. A machine, enclosed by a 6-ft × 6-ft × 6-ft hood, is located in the room. The machine, which radiates isotropically, produces a sound power level of 120 dB in the 1-kHz band. The hood has an average absorption coefficient and a transmission coefficient of 0.7 and 0.001, respectively, in this band. With the hood in place, what is the sound pressure level at a position 10 ft from the machine?

7.17 A particular barrier is used as shown. The directivity of the 2-W source, point S, is 2 in the direction of position A. The frequency of the source is 1000 Hz; the velocity of propagation is assumed to be 1128 ft/s. The room has an average absorption coefficient of 0.75 in the 1-kHz octave band. A barrier, with a high transmission loss, is used as depicted in the figure. Determine the sound pressure level at position A, with and without the barrier. Also determine the insertion loss of the barrier in this configuration and room.

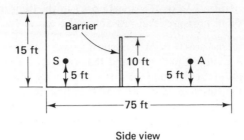

Side view

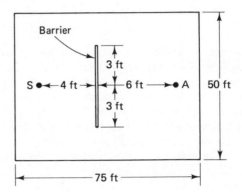

Top view

7.18 A semiinfinite barrier as depicted in Figure 7.16 is used in a configuration outdoors. The source is 25 ft from the barrier; the receiver is 80 ft from the barrier; and the barrier, which has a high transmission loss, is 15 ft high. The source radiates 3 W of acoustical power isotropically at 1 kHz; determine the sound pressure level at the receiver with an assumed velocity of propagation of 1128 ft/s.

REFERENCES

[1] MAGRAB, E. B. *Environmental Noise Control*, John Wiley & Sons, Inc., New York, 1975.

[2] *Compendium of Materials for Noise Control*, Department of Health, Education and Welfare, Washington, D.C., June 1975, prepared under Contract No. HSM-99-72-99.

[3] JACKSON, R. S.   "The Performance of Acoustic Hoods at Low Frequencies," *Acustica*, Vol. 12 (1962), pp. 139–152.

[4] JACKSON, R. S.   "Some Aspects of the Performance of Acoustic Hoods," *J. Sound Vibration*, Vol. e: 1 (1966), pp. 82–94.

[5] JUNGER, M. C.   "Sound Transmission Through an Elastic Enclosure Acoustically Closely Coupled to a Noise Source," *ASME Publication No. 70-WA/DE-12* (1970).

[6] CREMER, L.   *Statistische Raumakustik*, Hirzel Verlag, Stuttgart, 1961, Chap. 29.

[7] VER, I. L.   "Reduction of Noise by Acoustic Enclosure," in *Isolation of Mechanical Vibration, Impact, and Noise*, by J. C. Snowdon and E. E. Ungar, ASME Design Engineering Technical Conference, AMD-Vol. 1, 1973, pp. 192–220.

[8] SOMMERFELD, A.   *Math. Ann.*, Vol. 47 (1896), pp. 317–374.

[9] McDONALD, H. M.   *Proc. London Math. Soc.*, Vol. 14 (1915), pp. 410–427.

[10] REDFEARN, S. W.   *Phil. Mag.*, Vol. 30 (1940), pp. 223–236.

[11] MAEKAWA, F.   *Appl. Acoustics*, Vol. 1, American Elsevier Publishing Co., New York (1968), pp. 157–173.

[12] MORELAND, J. B. and R. S. MUSA   *International Conference on Noise Control Engineering, Oct. 4–6, 1972, Proceedings*, pp. 95–104.

[13] BORN, M. and E. WOLF   *Principles of Optics*, Pergamon Press, London, 1959.

[14] TATGE, R. B.   "Noise Reduction of Barrier Walls," presented at the 1972 Arden House Conference.

[15] MORELAND, J. B. and R. S. MUSA   "The Performance of Acoustic Barriers," *Noise Control Eng.* Vol. 1, No. 2 (1973), pp. 98–101.

BIBLIOGRAPHY

BELL, L. H.   *Fundamentals of Industrial Noise Control*, Harmony Publications, Trumbull, Conn., 1973.

BERANEK, L. L.   *Acoustics*, McGraw-Hill Book Company, New York, 1954.

BERANEK, L. L.   *Noise and Vibration Control*, McGraw-Hill Book Company, New York, 1971.

BERANEK, L. L.   *Noise Reduction*, McGraw-Hill Book Company, New York, 1960.

BISHOP, D. E.   "Partial Enclosures to Reduce Noise in Factories," *Noise Control*, Mar. 1951.

COPLEY, L. G.   "Control of Noise by Partitions and Enclosures," in *Proceedings of the Inter-Noise 72 Conference: Tutorial Papers on Noise Control*, pp. 22–23.

CROCKER, M. J.   "Noise Control by the Use of Enclosures and Barriers," in *Reduction of Machinery Noise*, edited by M. J. Crocker, Purdue University, 1974, pp. 102–112.

DIEHL, G. M.   *Machinery Acoustics*, John Wiley & Sons, Inc., New York, 1973.

FADAR, B.   "Practical Designs for Noise Barriers Based on Lead," *Amer. Ind. Hygiene Assoc. J.*, Vol. 27 (1966), pp. 520–525.

KURZE, U. J.   "Noise Reduction by Barriers," *J. Acoust. Soc. Amer.*, Vol. 55, No. 3 (1974), pp. 504–518.

MAGRAB, E. B.   *Environmental Noise Control*, John Wiley & Sons, Inc., New York, 1975.

YERGES, L. F.   "Cost/Effectiveness Approach to Machinery Noise Control," *J. Sound Vibration*, July 1974, pp. 30–32.

8

ACOUSTICAL
MATERIALS AND STRUCTURES

8.1 INTRODUCTION

In this chapter we shall form a dichotomy of acoustical materials and structures and treat them under the categories of absorption and isolation. Materials for these two applications are quite dissimilar, and hence they form a natural subdivision.

8.2 SOUND-ABSORBING MATERIALS

As shown in Figure 8.1(a), the sound energy from a source reaches the receiver from both direct and reflected paths. Sound-absorption material can be applied to the ceiling, floor, and walls to remove as much *reflected* sound as possible or desirable. Ideally, we would often like to remove all reflected sound energy, as shown in Figure 8.1(b), so that the room appears to be like free space in which the sound pressure level decays 6 dB for each doubling of the distance from the sound source. Total absorption is usually not possible or practical; nevertheless, the use of materials with high absorption coefficients can be effectively used to limit reflections and thus reduce the sound pressure level in the room, thereby creating a more comfortable acoustical environment.

We shall consider next the materials and structures used in sound-absorption systems—their physical properties, the mechanisms they employ for energy absorption, and the frequency range in which they are applicable.

237

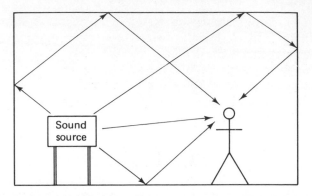

(a) Untreated hard room with a direct and diffuse field

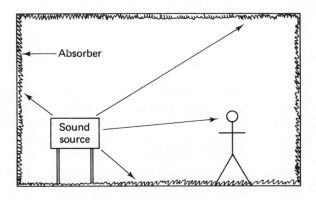

(b) Treated room with a direct field only

FIGURE 8.1 *Removal of the diffuse field by means of absorbing materials: (a) untreated hard room with direct and diffuse field; (b) treated room with direct field only*

8.2.1 Porous Materials

Porous acoustic materials are materials that possess a cellular structure of interlocking pores. It is within these interconnected open cells that the sound energy is converted into thermal energy. Thus, the sound-absorbing material is a dissipative structure which acts as a transducer, to convert acoustic energy into thermal energy. The actual loss mechanisms in the energy transfer are viscous flow losses caused by wave propagation in the material and internal frictional losses caused by motion of the material's fibers. The absorption characteristics of a material are dependent upon its thickness, density, porosity, flow resistance, fiber orientation, and the like.

Common porous absorption materials are made from vegetable and mineral fibers and elastomeric foams, and come in various forms. The materials may be prefabricated units, such as glass blankets, fiberboards, or lay-in tiles; the material may also be sprayed or troweled on the surface; or it may be a foam or open-cell plastic. Each type of material has its inherent advantages and disadvantages, and quite often the particular application dictates which form of absorbent material to use. For example, the aesthetics of the environment often prove to be the factor that governs the choice of material. In addition to the acoustical efficiency of the material, one must also consider its cost, installation, maintenance, and resistance to wear and environmental factors. For a complete list of architectural acoustical materials, their characteristics and performance data, the reader is referred to the bulletins published by the Acoustical and Insulating Materials Association (AIMA).

Suppose now that we have a piece of sound-absorbing material and that a sound wave is incident upon that material, as shown in Figure 8.2(a). If the frequency of the sound wave is very low, the whole material moves as shown in Figure 8.2(b), and therefore very little absorption occurs because there is very little relative motion of the internal material fibers and therefore very little viscous flow or frictional losses. However, if the incident sound wave is of high frequency, as shown in Figure 8.2(c), there is a lot of relative motion of the internal material fibers and therefore good absorption. This is a major factor in the variation of the absorption coefficient of the material with frequency, and also helps to explain why the wavelength of the incident sound becomes very important relative to the thickness of the absorbing material. This discussion also provides an indication of the difficulty of controlling low-frequency noise.

8.2.2 Noise-Reduction Coefficient

One of the indices used to describe a sound-absorbing material is the *noise-reduction coefficient* (NRC). The NRC is defined to be the arithmetic average of the material's sound-absorption coefficients at 250, 500, 1000, and 2000 Hz:

$$\text{NRC} = \frac{\alpha_{250} + \alpha_{500} + \alpha_{1000} + \alpha_{2000}}{4} \tag{8.1}$$

As such, the NRC is an index of the sound-absorbing efficiency of the material. Although in general this is a useful quantity, it must be used with caution in view of the fact that the index is an average value and therefore in a particular instance the best absorption capability may not be in the required frequency range.

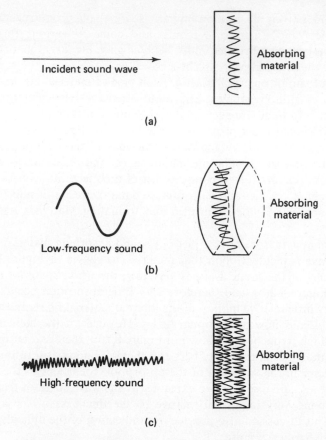

FIGURE 8.2 *Material performance as a function of incident sound wave frequency: (a) experimental configuration (material unexcited); (b) exaggerated motion of the material due to low-frequency excitation; (c) high-frequency excitation*

EXAMPLE 8.1 Consider sound absorption coefficients for the two typical materials listed below:

	OCTAVE BAND					
	125 Hz	*250 Hz*	*500 Hz*	*1000 Hz*	*2000 Hz*	*4000 Hz*
1. Fiberboard absorbing material	0.60	0.75	0.82	0.80	0.62	0.38
2. Thick porous absorbing material	0.32	0.39	0.78	0.99	0.82	0.88

SOLUTION   The NRC for material 1 is

$$NRC = \frac{0.75 + 0.82 + 0.80 + 0.62}{4} = 0.75$$

and the NRC for material 2 is

$$NRC = \frac{0.39 + 0.78 + 0.99 + 0.82}{4} = 0.75$$

Note that although the NRC for the two materials is the same, they have quite different absorption capabilities in the low-frequency (i.e., 250-Hz) range. Note also the difference that occurs at 4000 Hz.

D8.1   Use Table 6.1, Absorption Coefficients of General Building Materials, to determine the noise-reduction coefficient of the following materials: (a) painted concrete block   (b) fiberglass panels 1.5 in. thick   (c) plywood panels $\frac{3}{8}$ in. thick.

Ans.   (a) 0.0675   (b) 0.805   (c) 0.145.

Another factor that is related to the noisiness of an industrial environment is its reverberation. The reverberation time, T, is defined as the time required for the sound to decay 60 dB (i.e., one millionth of the energy density it had when the source was suddenly stopped). This quantity is a basic acoustic index of a room which depends only upon the room dimensions and the total absorption of the room. It was shown in Chapter 6 that this quantity in English units is expressed by

$$T = 0.049\left(\frac{V}{A}\right) \quad s \tag{6.125}$$

and in metric units is given by

$$T = 0.161\left(\frac{V}{A}\right) \quad s \tag{6.124}$$

where T = reverberation time (s)

V = volume of the room (ft^3 or m^3)

A = total absorption of the room (sabins or metric sabins)

Note that T varies directly with the volume of the room and inversely with the room absorption. This expression is a viable one if the average absorption coefficient of the room surfaces is about 0.2, or less. For higher values of the average absorption coefficient, the more accurate Erying equation (6.128) is recommended.

As we have indicated, the absorption properties of materials vary with frequency and, hence, the reverberation time does also. Quite often, however, the reverberation time is calculated for the single frequency of 500 Hz. This is due to the fact that historically the measurements made to obtain proper reverberation times were made at this single frequency.

EXAMPLE 8.2 A room in an industrial plant measures 80 ft × 40 ft × 15 ft. The absorption coefficients at 500 Hz of the walls, ceiling, and floor are 0.25, 0.05, and 0.15, respectively. We want to first calculate the reverberation time in the room with no acoustical treatment and then calculate the reverberation time if 80% of the wall surface is treated with absorption material having an $\alpha = 0.68$ at 500 Hz.

The reverberation-time analysis performed at 500 Hz is as follows:

1. Without acoustical treatment, the room volume in cubic feet = 80 × 40 × 15 = 48,000 ft^3. The room absorption in sabins is

$$
\begin{aligned}
\text{walls—} 2 \times 40 \times 15 \times 0.25 &= 300 \\
2 \times 80 \times 15 \times 0.25 &= 600 \\
\text{ceiling—} 80 \times 40 \times 0.05 \quad\; &= 160 \\
\text{floor—} 80 \times 40 \times 0.15 \quad\;\; &= 480 \\
\hline
\text{Total, A} &= 1540 \text{ sabins}
\end{aligned}
$$

The reverberation time without acoustical treatment is then

$$
T = 0.049 \left(\frac{V}{A} \right)
$$
$$
= \frac{(0.049)(48,000)}{1540}
$$
$$
= 1.5 \text{ s at 500 Hz}
$$

2. With acoustical treatment, the room absorption in sabins is

$$
\begin{aligned}
\text{walls—} 2 \times 40 \times 15 \times 0.68 \times 0.80 &= 653 \\
2 \times 80 \times 15 \times 0.68 \times 0.80 &= 1306 \\
900 \times 0.20 &= 180 \\
\text{ceiling—} &= 160 \\
\text{floor—} &= 480 \\
\hline
\text{Total, A} &= 2779 \text{ sabins}
\end{aligned}
$$

The new reverberation time is

$$
T = \frac{(0.049)(48,000)}{2779}
$$
$$
= 0.85 \text{ s at 500 Hz}
$$

Note that we have ignored the surfaces of any equipment contained in the room and we have assumed that the room is unoccupied. In a detailed analysis, both of these factors should be considered because they will affect the reverberation time.

D8.2 A particular room has dimensions 10 m $\times$ 8 m $\times$ 3 m for the length, width, and ceiling height, respectively. The average absorption coefficient, $\bar{\alpha}$, for the room at 1 kHz is 0.07 before acoustical treatment. Acoustic panels with an $\bar{\alpha} = 0.8$ at 1 kHz are to be mounted on 15% of the wall surfaces only. Determine the reverberation time before and after treatment.

Ans. $T_{\text{before}} = 2.06$ s, $T_{\text{after}} = 1.26$ s.

In a room or enclosure the sound energy will build up due to multiple reflections from the enclosing surfaces. This reflected sound energy is taken as essentially uniform throughout the enclosed area and is dependent upon two quantities—the source sound power output and the total absorption of the room. An important measure or performance index which is sometimes used to express the reduction in reflected sound energy achieved by the application of absorbing materials is the reverberant *noise reduction* (NR). In Chapter 6 we derived an expression for the sound pressure level due to a reverberant field only. We obtained

$$L_{p_{r_1}} = L_w + 10 \log \left(\frac{4}{R_1} \right) \quad \text{dB} \qquad (6.112)$$

in metric units

where $L_{p_{r_1}}$ = sound pressure level in the reverberant field before acoustical treatment (dB)

         L_w = sound power level of the source (dB)

         R_1 = room constant before acoustical treatment (m²)

We shall now assume that the average absorption coefficient, $\bar{\alpha}$, is about 0.3 or less after acoustical treatment. Using this assumption we may now approximate R_1 as

$$R_1 = A_1 = \bar{\alpha} S_1 \qquad (8.2)$$

where A_1 = total room absorption before treatment (metric sabins)

Upon taking advantage of this assumption and substituting A_1 into Equation (6.112), one obtains

$$L_{p_{r_1}} = L_w + 10 \log \left(\frac{4}{A_1} \right) \quad \text{dB} \qquad (8.3)$$

Similarly, after acoustical treatment,

$$L_{p_{r_2}} = L_w + 10 \log \left(\frac{4}{A_2} \right) \quad \text{dB} \qquad (8.4)$$

where $L_{p_{r_2}}$ = sound pressure level in the reverberant field after acoustical treatment (dB)

A_2 = total room absorption after treatment (metric sabins)

The reverberant noise reduction (NR) is then defined as

$$NR = L_{p_{r_1}} - L_{p_{r_2}} \quad dB \tag{8.5}$$

or upon substituting equations (8.3) and (8.4) into (8.5), we have

$$\boxed{NR = 10 \log \left(\frac{A_2}{A_1} \right) \quad dB} \tag{8.6}$$

which is valid for both metric and English units.

It is important to remember that use of this equation assumes that the average absorption coefficient of the room is less than about 0.3 and that the room itself is essentially regular in proportion (i.e., one room dimension is not much greater than the other). The equation for the reverberant noise reduction (NR) versus the absorption ratio (A_2/A_1) is plotted in Figure 8.3 for convenience.

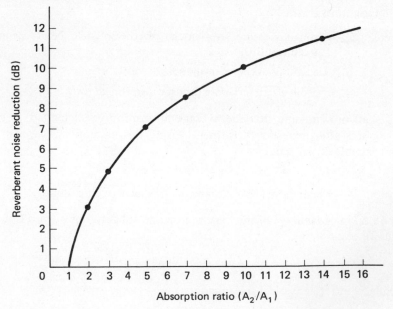

FIGURE 8.3 *Reverberation noise reduction (NR) versus the absorption ratio* (A_2/A_1)

EXAMPLE 8.3 A room measures 40 ft × 40 ft × 15 ft. The absorption coefficients for the materials of which the room is made are as follows:

	OCTAVE-BAND CENTER FREQUENCY (Hz)					
	125	*250*	*500*	*1000*	*2000*	*4000*
Walls—painted concrete block	0.1	0.05	0.06	0.07	0.09	0.08
Floor—hardwood	0.15	0.11	0.10	0.07	0.06	0.07
Ceiling—concrete	0.01	0.01	0.015	0.02	0.02	0.02

Now suppose that the room is refinished in order to create a better acoustical environment by installing a suspended acoustical tile ceiling and heavy carpet on the floor. These two items have the following absorption coefficients:

	OCTAVE BAND CENTER FREQUENCY (Hz)					
	125	*250*	*500*	*1000*	*2000*	*4000*
Suspended acoustical tile ceiling	0.76	0.93	0.83	0.99	0.99	0.94
Carpet on foam rubber	0.08	0.24	0.57	0.69	0.71	0.73

We shall now determine the reduction in reverberation noise which can be achieved through the installation of these sound-absorbing materials. The areas of the walls, ceiling, and floor are:

$$\text{wall area} = 4 \times 40 \times 15 = 2400 \text{ ft}^2$$
$$\text{ceiling area} = 1600 \text{ ft}^2$$
$$\text{floor area} = 1600 \text{ ft}^2$$

An example calculation for the total absorption at the 125-Hz frequency is as follows:

$$A_1 \text{ at } 125 \text{ Hz} = 2400(0.1) + 1600(0.15) + 1600(0.01)$$
$$= 496 \text{ sabins}$$
$$A_2 \text{ at } 125 \text{ Hz} = 2400(0.1) + 1600(0.08) + 1600(0.76)$$
$$= 1584 \text{ sabins}$$

Similar calculations yield the following table:

	OCTAVE-BAND CENTER FREQUENCY (Hz)					
	125	*250*	*500*	*1000*	*2000*	*4000*
A_1 (sabins)	496	252	288	276	332	312
A_2 (sabins)	1584	1930	2336	2822	2924	2838
NR (dB)	5.0	8.8	9.1	10.1	9.5	9.6

Note that the reverberant component of the sound pressure level is reduced from 5 to 10 dB, and thus the addition of the sound-absorbing materials may produce a significant reduction of the reverberant noise in a room. An increase of 10 times the original absorption, which yields a NR of 10 dB, is often considered a practical limit.

D8.3 A certain room is 6 m × 7 m with a 3-m-high ceiling. The absorption coefficients of the walls, ceiling, and floor are 0.06, 0.07, and 0.07, respectively, at 1 kHz. If an acoustical tile ceiling is installed which has an absorption coefficient of 0.8 at 1 kHz, what reverberant noise reduction (NR) can be achieved?

Ans. NR = 6 dB.

The mounting of acoustical absorbing materials is very important because the absorption coefficient varies widely with frequency for different mountings. For example, mineral wool and similar materials may be mounted directly on a wall, or may be mounted in such a way as to leave an air space between the material and the wall. The latter mounting technique will improve the low-frequency performance of the porous material. The same is true for acoustical tiles, which can be mounted directly on the ceiling or hung in a suspension system away from it. In general, the manufacturer of absorbing materials will specify the manner in which the absorption coefficient varies for different mounting methods.

In addition to mounting considerations, it is important to realize that many sound-absorbing materials are not sufficiently durable, abrasion-resistant, etc., to be applied without a sound transparent protective facing. These facings, which tend to degrade the high-frequency performance and improve the low-frequency performance, range from those which are very simple and plain to those which are highly decorative. In general, the open area of the facings ranges from 5 to 50%. However, even a sheet of light, flexible, thin material such as Mylar can be employed as a facing because it will efficiently transmit the sound pressure wave to the absorbing material.

Porous materials and their protective facings (when employed) must be cleaned and maintained with care to avoid destroying the absorption properties of the material. For example, if acoustical ceiling tile is painted, the holes in the tile must not be clogged. In general, the facing protecting a porous material should not be painted. However, if it is painted, it must be done with a very thin coat that will not reduce the ability of the facing to transmit a sound pressure wave. Other properties concerning a porous material which are important from an architectural standpoint are the material's flame-spread index, light reflectance, moisture resistance, appearance, and the like.

We have discussed a number of ways in which porous-type materials can be used. One other method in which porous materials can be applied effectively is in the form of a *space absorber*. This device is a free-hanging unit consisting of porous material and built as a hollow cone, cylinder, or other desired shape. They also come in panels that can be hung horizontally or vertically. Some typical units are shown in Figure 8.4. These units can be effectively used in industrial areas where such things as lighting fixtures and overhead cranes make difficult the use of other types of absorbers. They provide very good absorption but are also relatively expensive. Their efficiency obviously depends upon how densely they are placed in the room.

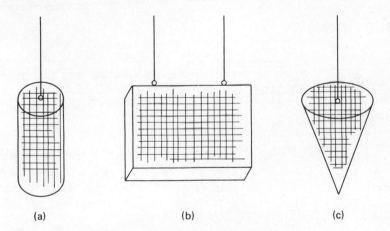

(a)   (b)   (c)

FIGURE 8.4 *Some typical space absorbers*

8.2.3 Panel Absorbers

A sound wave that strikes a panel causes the panel to vibrate at either the fundamental or a harmonic frequency of the incident wave. Owing to the inherent damping of the panel, an energy loss occurs and hence some sound absorption takes place. A low-frequency incident wave is much more effective in causing the panel to flex than a high-frequency one. This fact is intuitively obvious, since the panel will perform in a manner similar to that shown in Figure 8.2 for a sheet of absorbing material. The panel, being a relatively solid structure, may simply reflect the high-frequency sound waves with little or no energy loss unless special care is taken to provide for their absorption also.

The absorption coefficient as a function of frequency is dependent upon the mass and stiffness of the panel and therefore is determined by the specific design configuration. In general, the absorption coefficients for a basic panel are greater at low frequencies than they are at high frequencies. Typical values of α for 125 Hz are 0.2 to 0.5.

The panel absorber can be useful as a component in an absorption system by designing the panel so that it is tuned to a specified frequency and thus acts as a trap for a particular low-frequency sound. If a panel absorber is installed next to a wall with a space between the panel and the wall, and if the space is then filled with some type of porous absorber, the low-frequency performance will be increased and the region of good absorption will be extended to higher frequencies.

Panel absorbers can be constructed of such materials as metal, plywood, and glass and, like different types of porous absorbers, must be considered in view of their cost, resistance to wear, and similar practical parameters.

8.3 DUCT NOISE

A few basic principles and precautions should be applied in the design of ducting systems to prevent introducing excessive noise with the system. A brief summary is presented here.

8.3.1 Flow Noise in Ducts

One of the major sources of noise in a duct system is the fan. In general, fans are either axial or centrifugal, and the frequency spectrum of the noise generated is dependent upon the fan type. For example, fans will generate pure tones at the blade passage frequency, B_f, and its harmonics. This frequency is defined by the expression

$$\boxed{B_f = \frac{KN}{60} \ \text{Hz}} \tag{8.7}$$

where $K =$ number of fan blades

        $N =$ rpm of the fan

Thus, a 12-blade fan operating at 900 rpm has a B_f of 180 Hz and harmonics at 360 Hz, 540 Hz, 720 Hz, etc. The reader interested in pursuing further the subject of fan noise and its attenuation is referred to Chapter 5, the literature published by the Buffalo Forge Co., and the most recent issue of the *ASHRAE Guide and Data Book*.

Another important source of noise in a duct system is simply that generated by the air flow through the duct. This is particularly true at such elements as bends, branches, and dampers. The amount of noise depends upon the element, its size, and the turbulence and flow rate of the air. In fact, noise within a duct system will cause the duct walls to vibrate and thus radiate noise into the adjacent area.

8.3.2 Duct Design Considerations

Duct systems should be designed as closely as possible for smooth uniform air flow. Although important throughout the entire duct system, this is particularly important at the fan inlet. If possible, the duct system should be designed to have a long section of straight duct (5 to 10 duct diameters) just prior to the fan inlet so that the air is essentially uniform as it enters the fan. If the air entering the fan is turbulent, the fan performance will be degraded and the sound power level increased. This point is important regardless of fan application. All duct fittings should be designed to reduce turbulence and avoid possible sudden changes in flow direction or velocity. This is particularly important if the duct is near a "quiet area."

Excessive velocities in the duct system should be avoided. Of course, a smaller duct system requiring less material which is used in conjunction with duct silencers may represent a viable alternative. However, this trade-off must be evaluated from an economic standpoint.

Two elements (e.g., bends or branches) should be separated by 5 to 10 duct diameters so that turbulence generated by one element does not immediately enter another. If this separation does not exist, then the noise generated by the second element will be significantly larger than it would be if the elements were far apart.

8.4 MUFFLERS

Mufflers have many applications, which range from air-conditioner systems to jet engines, and are classified into two basic types—absorptive and reactive. *Absorptive mufflers* employ fibrous or porous materials as sound absorbers to reduce noise. *Reactive mufflers*, on the other hand, do not primarily depend upon absorptive materials for their effectiveness.

In practice, muffler performance is generally described by one or more of the following performance specifications:

1. *Insertion loss* is defined to be the difference in sound pressure levels measured at a specified point in space before and after a muffler is inserted between the noise source and the point of measurement.
2. *Dynamic insertion loss* is identical to the definition above with one exception—that the measurements are made under rated flow conditions.
3. *Attenuation* is the decrease in sound power between two points in the system.
4. *Noise reduction* has already been defined. Within the present context it is the difference in sound pressure levels measured at the muffler inlet and outlet.
5. *Muffler transmission loss* (TL) is 10 times the logarithm to the base 10 of the ratio of the incident sound power on the muffler to the transmitted sound power of the muffler.

8.4.1 Absorptive Mufflers

In general, the noise reduction in this type of muffler occurs through the use of a blanket of absorption material. The absorption coefficient of this material will increase with thickness. This type of muffler is very useful for broadband noise (or narrow-band noise, if the frequency shifts with operating speed) and provides good absorption at high frequencies. The absorption material may require a special protective facing if placed in a duct with a high-velocity or high-temperature air stream. For example, mineral wool with a corten or stainless steel facing may be required for use with a gas turbine. If the gas stream contains an oil mist, then something like Mylar can be employed to protect the sound-absorbing material. With these general comments as background, we shall now examine the most important types of absorptive mufflers.

8.4.1.1 Lined ducts

This device, as shown in Figure 8.5, is perhaps the simplest form of absorptive muffler. The sound-absorbing material attenuates the noise as it travels down the duct. The amount of attenuation is dependent upon the frequency of the noise present, length of the duct, the type and thickness of the absorbing lining, the presence or absence of facing material, and the cross-sectional area of the air stream. In addition, a discontinuity in cross section at the duct inlet and outlet causes reflection and scattering of the sound waves, and this process tends to increase the attenuation.

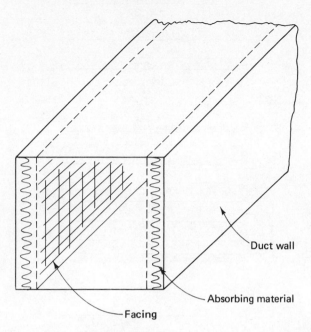

FIGURE 8.5 *Lined duct*

8.4.1.2 *Parallel and blocked-line-of-sight baffles* [1]

These devices are shown schematically in Figure 8.6 and are essentially extensions of the lined duct. Therefore, the performance of these baffles is determined by the same type of parameters that govern the performance of a lined duct. For example, above 250 Hz there is a significant improvement in attenuation as either the thickness of the absorbing material or the length of the baffle is increased. In addition, the baffle with a blocked line of sight provides much better attenuation than the baffle shown in Figure 8.6(a) in the frequency range from 1 kHz to 8 kHz. Therefore, the high-frequency performance of these devices exceeds their performance at low frequencies; however, their low-frequency performance can be improved by employing some of the techniques outlined earlier in the section on sound-absorbing materials.

8.4.1.3 *Lined bends*

When sound waves must change direction, they generally do so through surface reflections. If these surfaces which the sound waves strike are treated with sound-absorbing material, then a portion of the sound energy can be absorbed. Consider the lined bend shown in Figure 8.7. If we assume that

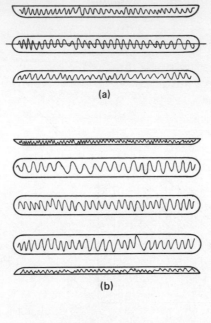

(a)

(b)

(c)

FIGURE 8.6 *Cross sections of certain parallel (a) and (b) and blocked-line-of-sight (c) baffles*

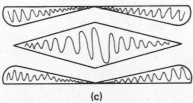

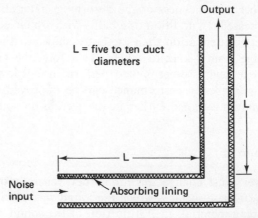

Output

L = five to ten duct diameters

L

L

Noise input

Absorbing lining

FIGURE 8.7 *Schematic of lined bend for noise attenuation*

the waves which enter the bend are essentially axial, then these waves are partially absorbed at the wall and partially reflected back toward the source. The waves that successfully pass through the bend are attenuated due to the multiple reflections downstream of the bend. Once again, the amount of attenuation will depend upon the factors outlined earlier in the discussion of lined ducts.

Lined bends, however, scatter the sound waves and therefore yield greater attenuation than an equivalent section of lined duct. Greatest attenuation occurs at higher frequencies. The attenuation characteristic of a typical lined bend varies linearly from very little attenuation ($\simeq 1$ dB) at 63 Hz to about 10 dB at 4 kHz and remains essentially constant from 4 kHz to 8 kHz. The low-frequency performance can be improved by employing a thicker lining.

Another ramification of the lined bend is to employ a 180° bend. This type of bend gives better performance than the 90° bend, but it also increases the pressure drop. The performance of these bends can be evaluated by measuring the insertion loss downstream of the bend.

8.4.1.4 Plenum chambers

These units are large volume chambers which interconnect two ducts, as shown in Figure 8.8. The interior of the chamber is lined with absorbing material, and thus part of the sound energy which enters the chamber is absorbed due to multiple reflections within the unit. Facing material may or may not be required, depending upon the temperature and velocity of the gas stream. Although classified here as an absorptive muffler, the plenum

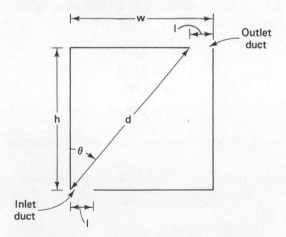

FIGURE 8.8 *Layout of a single chamber plenum (end view)*

chamber also acts as a reactive device, owing to the discontinuities that exist at both the inlet and outlet of the chamber.

Typically, the attenuation ranges from about 10 dB at low frequencies to approximately 20 dB in the frequency range above 1 kHz. This performance can be improved by increasing the thickness of the absorbing lining, blocking the direct line of sight from chamber inlet to outlet either through the use of a sound-absorbing panel or by using multiple chambers, and by increasing the cross-sectional area of the chamber for a given size of inlet and outlet duct.

References [2,3] suggest the following approximate expression for determining the attenuation of the plenum shown in Figure 8.8:

$$\text{attenuation} = -10 \log \left[S \left(\frac{\cos \theta}{2\pi d^2} + \frac{1 - \bar{\alpha}}{\bar{\alpha} S_w} \right) \right] \quad \text{dB} \qquad (8.8)$$

where $\bar{\alpha}$ = average absorption coefficient of the plenum lining

         S = plenum inlet or exit area (m² or ft²)

        S_w = plenum wall area (m² or ft²)

         d = slant distance from input to output (m or ft)

       $d^2 = (W - l)^2 + h^2$ (m² or ft²)

    $\cos \theta = h/d$

This equation yields results that are fairly accurate at high frequencies, conservative for low frequencies; it thus provides a reasonable, practical measure of the plenum's performance.

EXAMPLE 8.4   A plenum chamber with the following characteristics is used in an air-conditioning system:

$$S_w = 374 \text{ ft}^2$$
$$S = 24 \text{ ft}^2$$
$$\theta = 30°$$
$$d = 12 \text{ ft}$$

The absorptive coefficients of the acoustic lining of the plenum are as follows:

Frequency (Hz)	125	250	500	100	2000	4000
$\bar{\alpha}$	0.2	0.53	0.85	0.93	0.83	0.78

We may now find the frequency-dependent attenuation by means of equation (8.8):

$$\text{attenuation} = -10 \log \left[S \left(\frac{\cos \theta}{2\pi d^2} \right) + \frac{1 - \bar{\alpha}}{\bar{\alpha} S_w} \right) \right] \quad \text{dB}$$

SOLUTION Using the data above, the expression for the attenuation becomes

$$\text{attenuation} = -10 \log \left\{ 24 \left[0.000957 + \frac{1 - \bar{\alpha}}{\bar{\alpha}(374)} \right] \right\} \quad \text{dB}$$

Upon substituting in the absorption coefficients, one obtains

Frequency (Hz)	125	250	500	1000	2000	4000
Attenuation (dB)	3.9	6.8	7.9	8.1	7.9	7.7

These data can now be used to determine if the required attenuation in the octave bands is satisfactory for a particular system application.

D8.4 A certain plenum chamber, as depicted in Figure 8.8, is described by the following: $S_w = 54$ m², $S = 1.5$ m², $\theta = 45°$, $d = 4$ m, and $\bar{\alpha}$ of the lining is 0.8 at 1 kHz. Determine the attenuation of the chamber at 1 kHz.

Ans. Attenuation = 18 dB.

8.4.2 Reactive Mufflers

Reactive mufflers employ one or more chambers that serve to reflect and attenuate the incident sound energy. They are most useful when the noise contains discrete tones, because in general they are narrowband devices. These mufflers are viable for applications where the gas stream is dirty and would perhaps block or clog an absorptive-type muffler. The devices are economical and have little pressure drop; however, their performance deteriorates at high frequencies and when used with large-diameter ducts.

We shall now consider some of the important types of reactive mufflers.

8.4.2.1 Expansion chambers

An expansion chamber, as shown in Figure 8.9, has discontinuities in cross section at both the inlet and outlet. The design of the chamber is dependent upon the wavelength of the incident sound; that is, the chamber length is chosen so that the device is "tuned" to cancel a very narrow band of frequencies and thus acts as a notch filter. Because expansion chambers are inherently frequency-selective, they are ineffective in reducing broadband noise. However, a number of expansion chambers can be employed in series and their frequencies of maximum attenuation staggered to yield an overall system with somewhat broadband characteristics.

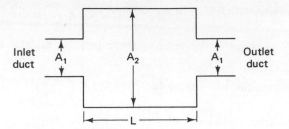

FIGURE 8.9 *Simple expansion chamber muffler*

The actual amount of attenuation that the silencer produces is dependent upon the parameter, m, defined as

$$m = \frac{\text{cross-sectional area of the expansion chamber}}{\text{cross-sectional area of the connecting ducts}} = \frac{A_2}{A_1} \qquad (8.9)$$

The attenuation increases as m increases. Another way to improve the attenuation for a fixed m is by placing two identical expansion chambers in series.

The transmission loss for the expansion chamber shown in Figure 8.9 has been shown to be [4,2]

$$\text{TL} = 10 \log \left[1 + \tfrac{1}{4}(m - 1/m)^2 \sin^2 kL \right] \quad \text{dB} \qquad (8.10)$$

where k = wave number = $\dfrac{2\pi}{\lambda} = \dfrac{2\pi f}{c}$ (m^{-1} or ft^{-1})

L = length of the muffler (m or ft)

This equation, which is valid as long as the greatest transverse dimension of the muffler cross section is less than about 0.8λ, defines the family of curves shown in Figure 8.10. Note that the equation is periodic in kL and that it repeats every π radians. Maximum transmission loss occurs when

$$kL = \frac{n\pi}{2}; \quad n = 1, 3, 5, 7, \ldots$$

Zero transmission loss occurs at

$$kL = n\pi; \quad n \times 1, 2, 3, 4, \ldots$$

Therefore, the chamber length of the silencer should be

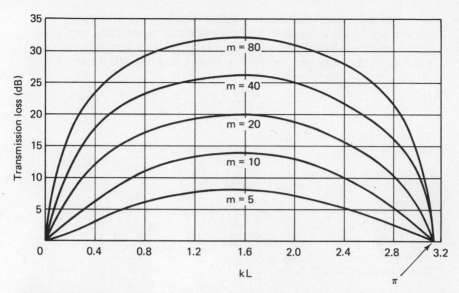

FIGURE 8.10 *Transmission loss, TL, of an expansion chamber as a function of kL with m as a parameter (if kL is between π and 2π, subtract π and use the curves; similarly, for values between 2π and 3π, subtract 2π, etc.)*

$$L = \frac{n\pi}{2k} = \frac{n\pi}{2(2\pi/\lambda)} = \frac{n\lambda}{4}; \quad n = 1, 3, 5, \ldots \quad (8.11)$$

where L is in meters or feet

i.e., an odd multiple of a quarter-wavelength of the sound that is to be attenuated. A maximum transmission loss is obtained when $L = \lambda/4$. Other maxima occur at $L = n\lambda/4$, $n = 3, 5, 7, \ldots$. The following example illustrates the use of the preceding material in the design and selection of an expansion chamber.

EXAMPLE 8.5 Suppose that the noise produced by a system occurs primarily at 125 Hz. The particular frequency of the noise that emanates from the system is controlled by the operating speed of the engine within the system. We want to determine the design parameters of an expansion chamber which when used with the system will provide 25 dB of transmission loss.

For a 25-dB TL, m may be taken as 40, as shown in Figure 8.10. The length of the chamber is then determined by the expression

$$kL = \frac{\pi}{2}$$

$$L = \frac{\pi}{2k} = \frac{\pi}{2(2\pi/\lambda)} = \frac{\lambda}{4} = \frac{c}{4f} \quad \text{m} \quad \text{or} \quad \text{ft}$$

In general, from equation (1.5), $c = 49.03\sqrt{R°}$ ft/s, where $R°$ is the absolute temperature of the air in degrees Rankine. If the temperature were given in centigrade, one would have used equation (1.7). If the temperature at the point where the expansion chamber is to be inserted is 200°F, then $R° = (459.7 + 200) = 659.7°$, and hence

$$c = 49.03\sqrt{659.7} = 1259 \text{ ft/s}$$

Therefore, the chamber length is

$$L = \frac{1259}{(4)(125)} = 2.52 \text{ ft}$$

From Figure 8.10, note that for $m = 40$, the required transmission loss spans the range

$$1.09 \leq kL \leq 2.0$$

Therefore, for $L = 2.52$ ft, we obtain

$$1.09 \leq \frac{2\pi f}{c}(2.52) \leq 2.0$$

or

$$87 \text{ Hz} \leq f \leq 159 \text{ Hz}$$

Note that the chamber not only provides a transmission loss of 25 dB at 125 Hz, but that it is effective in the narrow frequency range from 87 Hz to 159 Hz.

D8.5 An expansion chamber must provide a transmission loss of 20 dB about a center frequency of 600 Hz in a gas stream where the temperature is 250°F. If m is chosen as 30 and n as 1, over what band of frequencies will a minimum TL of 20 dB be obtained and what will be the TL at 500 Hz?
    Ans. 306 Hz $\leq f \leq$ 841 Hz, TL = 24 dB.

Expansion chambers can be obtained in many different forms other than that shown in Figure 8.9. Some of the various ramifications of the simple expansion chamber are shown in Figure 8.11. The transmission-loss characteristics of these chambers in general differ from one another, and hence each must be considered in view of the particular application.

8.4.2.2 Cavity resonators

The general form of a *cavity resonator* (also called a *Helmholtz resonator*) is shown in Figure 8.12. The air within the cavity acts as a spring which is forced in and out of the cavity, by way of the neck, by the periodic air flow

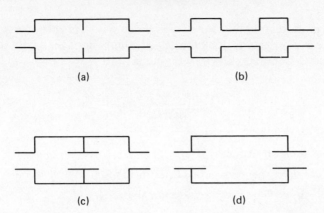

FIGURE 8.11 *Schematics of different types of expansion chambers*

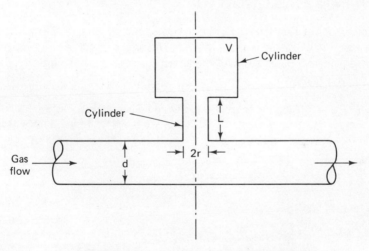

FIGURE 8.12 *General form of a cavity resonator*

(which acts as a mass). The cavity and neck are designed to be tuned or resonant to a particular low frequency, and thus this device is a very selective low-frequency narrow-band absorber. Because the problem of controlling low-frequency noise is very difficult, this type of absorber can be extremely useful.

Although this device has a definite absorption peak in the low-frequency range, this peak can be flattened so as to extend into the high-frequency range by filling the cavity with some form of porous material, such as fiberglass. One may also accomplish the attenuation of a wide range of frequencies by employing a number of these devices, each tuned to a different frequency within the range.

If we ignore the viscous losses associated with the movement of gases through the cavity opening, then the design equation for the cavity resonator, based on the dimensions shown in Figure 8.12, is

$$f_r = \frac{c}{2\pi}\sqrt{\frac{A}{lV}} \quad \text{Hz} \tag{8.12}$$

where f_r = resonant frequency (Hz)

c = velocity of sound in the field (m/s or ft/s)

A = m² or ft² = neck cross-sectional area

L = neck length (m or ft)

V = volume of the cavity (m³ or ft³)

$l = L + 0.8\sqrt{A}$ (m or ft)

At the resonant frequency, f_r, the acoustic energy that is transmitted to the cavity by way of the neck is returned to the pipe or duct out of phase and thus is transmitted back toward the source, causing cancellation of the incident wave.

EXAMPLE 8.6   The noise generated at one end of an air duct has a predominant component at approximately 200 Hz. We wish to design a spherical cavity resonator to eliminate the noise at this frequency. Using equation (8.7),

$$f_r = \frac{c}{2\pi}\sqrt{\frac{A}{lV}} \quad \text{Hz}$$

where f_r = 200 Hz

c = 1128 ft/s

we obtain

$$200 = \frac{1128}{2\pi}\sqrt{\frac{A}{lV}}$$

Note that this equation has three unknowns. Hence, we must assume two of them, solve for the third, and see if the result is reasonable. Rearranging and simplifying yields

$$1.241 = \frac{A}{lV}$$

We now assume that the neck (cylindrical) has a one-unit diameter and that it is one unit in length, i.e., $r = 0.5$ in. (0.042 ft) and $l = 1.0$ in. (0.083 ft).

Therefore,

$$1.241 = \frac{\pi(0.042)^2}{(0.083)V}$$

or

$$V = 0.054 \text{ ft}^3$$

The radius of the spherical chamber required is

$$r = \left(\frac{3V}{4}\right)^{1/2}$$

Upon substituting in our required value of V, one obtains

$$r = 0.11 \text{ ft} \quad \text{or} \quad 1.36 \text{ in.}$$

One notes that the chamber is small compared to the wavelength of the 200-Hz tone that we wish to attenuate.

We have only introduced this important reactive muffler component here. The reader interested in more detail is referred to reference [2].

D8.6　A cavity resonator as depicted in Figure 8.12 is to be designed which has an absorption peak at 250 Hz. It is desired to make $l = 5$ cm and $r = 2$ cm. Assume the sound velocity to be 344 m/s. Determine the volume required in the resonant cavity.

Ans.　$1.2 \times 10^{-3} \text{ m}^3$.

An important example of a cavity resonator in the construction industry is that formed by cutting slots in a standard concrete block as shown in Figure 8.13. When the blocks are closed at the top, they can be stacked to form a wall of cavity resonators. Their performance can be improved by filling the cavities with porous material. This type of resonator, in contrast to the soft porous materials, is extremely rugged and durable.

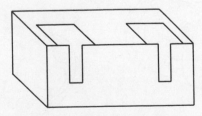

FIGURE 8.13 *Cavity resonators formed into a standard-sized concrete block*

It is important to note that if a porous absorbing material is placed behind a protective facing in which holes or slits are cut, then, in effect, a form of cavity resonator has been created. These types of devices usually have an open area of 15% or less and are subject to the same comments stated earlier concerning the maintenance of porous materials with protective facings. An excellent treatment of this topic for high-level sound pressure levels is found in reference [5].

8.4.3 Prefabricated Mufflers

Prefabricated mufflers may be absorptive, reactive, or a combination of these two basic types. Because they are carefully designed and engineered, they are compact, economical, durable, and can be specially fabricated for corrosive or high-temperature application. Perhaps the most important feature of a prefabricated muffler is that its performance is proven through extensive testing and actual use, and these data are documented by the manufacturer. Their applications, which span the spectrum, include gas turbines, engines, blowers, compressors, vacuum pumps, and vents and blowdowns. Some typical tubular silencers manufactured by the Industrial Acoustics Company are shown in Figure 8.14.

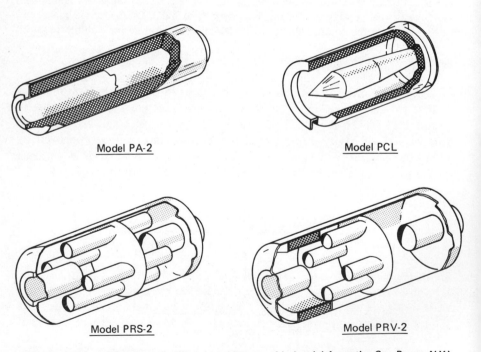

Model PA-2   Model PCL

Model PRS-2   Model PRV-2

FIGURE 8.14 *Prefabricated mufflers (compliments of Industrial Acoustics Co., Bronx, N.Y.)*

8.5 PIPE-WRAPPING MATERIALS

A great deal of noise can be generated by high-pressure gas flowing through pipes and valves. In addition, noise is produced by systems such as hydraulic and ventilating systems. This noise can be reduced or confined via the technique shown in Figure 8.15. In this approach, an impermeable membrane such as thin sheet metal, lead-loaded vinyl, or specially coated asbestos cloth is used to cover a porous blanket of material such as mineral wool or polyurethane foam insulation. Lead is a very popular material because it is heavy, limp, highly damped, flexible, resistant to fire and chemicals, and provides maximum attenuation with minimum thickness.

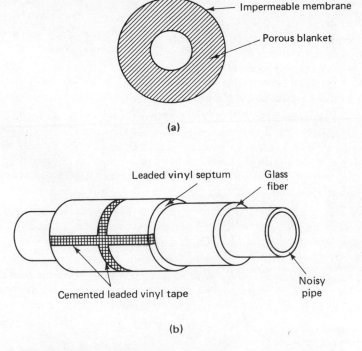

(a)

(b)

FIGURE 8.15 (a) *Fundamental approach to acoustic wrapping for pipes;* (b) *example of pipe wrapping using leaded vinyl*

In the construction of the noise-control scheme shown in Figure 8.15, the wrapping material for the pipes should be overlaid a few inches and sealed with a vinyl adhesive. In addition, no mechanical connection between the pipe and the impermeable membrane should exist; otherwise, this connection will flank the porous blanket and the noise will be transmitted directly to the outer surface, which, in turn, will radiate the sound.

263

Dear [6] studied 19 different pipe coverings consisting of various materials used in different configurations. His work indicates that good broadband noise reduction can be obtained using a 2-in.-thick molded fiberglass blanket (4 lb/ft^3) covered with a single layer of lead-impregnated vinyl (0.87 lb/ft^2). He also found that certain configurations of materials were best in specific limited frequency ranges, and that in general none of the pipe coverings tested were good for low-frequency noise control.

PROBLEMS

8.1 Determine the noise-reduction coefficients of the following materials based upon the absorption coefficient data presented in Table 6.1:
    (a) carpet on concrete
    (b) mineral fiber ceiling tile
    (c) 1-in.-thick polyurethane foam

8.2 A particular room is 10 m $\times$ 9 m $\times$ 3 m in length, width, and ceiling height, respectively. The floor is concrete, the walls are painted concrete block, and the ceiling is 0.5-in.-thick gypsum board. Determine the number of square meters of 1-in.-thick polyurethane foam which must be installed on the walls to reduce the reverberation time to 1.5 s at 500 Hz. (Use Table 6.1 for absorption coefficients.)

8.3 A certain machine shop, 40 ft $\times$ 50 ft $\times$ 12 ft, presently has an average absorption coefficient of 0.1 at 1 kHz. What must the new value of $\bar{\alpha}$ be after acoustical treatment in order to obtain a reverberant noise reduction (NR) of 5 dB?

8.4 In a particular application, two acoustical treatments are under consideration to reduce the reverberant noise. Treatments A and B yield absorption ratios of 7 and 14, respectively. What is the noise reduction (NR) in each case?

8.5 A particular 6-bladed fan is operated at 1800 rpm. Determine the blade passage frequency.

8.6 A plenum chamber has the basic design depicted in Figure 8.8. It is cubic, 5 ft on each edge, with a 1-ft-wide inlet and outlet duct the full 5-ft width of the plenum. The walls are lined with polyurethane foam with $\bar{\alpha} = 0.16, 0.45, 0.84,$ and 0.97 in the 125-Hz, 500-Hz, 1000-Hz, and 2000-Hz octave bands, respectively. Determine the attenuation in each of these octave bands.

8.7 A certain plenum chamber is to have a wall surface of 500 ft^2 and input and output ducts each 12 ft^2 in area; the lining is characterized by $\bar{\alpha} = 0.46$ in the 500-Hz octave band. The basic configuration is that shown in Figure 8.8.

The dimensions are adjusted such that $d = 12$ ft with the following changes in θ:

  (a) $\theta = 0°$
  (b) $\theta = 15°$
  (c) $\theta = 30°$
  (d) $\theta = 45°$

Determine the attenuation in the 500-Hz octave band for each value of θ.

8.8 An expansion chamber is designed as basically depicted in Figure 8.9. The cross-sectional area of the chamber is 0.2 m² and that of the inlet and outlet ducts is 0.0067 m²; the length is 0.5 m. The gas temperature through the chamber may be taken as 150°C. Determine the frequencies associated with $n = 1, 3, 5, 7$ and the band about each within which the transmission loss is a minimum of 20 dB.

8.9 Determine the length and m for an expansion chamber that must yield a transmission loss of 25 dB at a frequency of 125 Hz. Also determine the frequency band over which TL is at least 20 dB. (Assume that $c = 1128$ ft/s.)

8.10 A particular cavity resonator design, as depicted in Figure 8.12, has a neck 3 in. long with a radius of 0.75 in. Several cylindrical resonators with the following dimensions are available for attachment to the neck:

  (a) radius 2 in. and height 4 in.
  (b) radius 3 in. and height 5 in.
  (c) radius 4 in. and height 5 in.

Determine the resonant frequency in each case. (Use $c = 1128$ ft/s.)

8.11 A cavity resonator for use in a gas stream with a temperature of 180°C is required with a resonant frequency at 160 Hz. The neck must be 5 cm long and the volume of the resonator must be seven times the volume of the neck. Determine the cross-sectional area of the neck.

REFERENCES

[1] SANDERS, G. J. "Silencers: Their Design and Application," *Sound and Vibration*, Feb. 1968, pp. 6–13.

[2] BERANEK, L. L. ed. *Noise and Vibration Control*, McGraw-Hill Book Company, New York, 1971.

[3] *ASHRAE Guide and Data Book*, ASHRAE, New York, 1970, Chap. 33, Systems.

[4] HARRIS, C. M. ed. *Handbook of Noise Control*, McGraw-Hill Book Company, New York, 1957.

[5] INGARD, U. and H. ISING "Acoustic Nonlinearity of an Orifice," *J. Acoustic Soc. Amer.*, Vol. 42, No. 1 (July 1967), pp. 6–17.

[6] DEAR, T. A. "Noise Reduction Properties of Selected Pipe Covering Configurations," in *Proceedings of the Inter-Noise 72 Conference: Tutorial Papers on Noise Control*, Oct. 1972, Washington, D.C., pp. 138–144.

BIBLIOGRAPHY

BELL, L. H. *Fundamentals of Industrial Noise Control*, Harmony Publications, Trumbull, Conn., 1973.

CROCKER, M. J. "Mufflers," in *Proceedings of the Inter-Noise 72 Conference: Tutorial Papers on Noise Control*, Oct. 1972, Washington, D.C., pp. 40–44.

EGAN, M. D. *Concepts in Architectural Acoustics*, McGraw-Hill Book Company, New York, 1972.

DIEHL, G. M. *Machinery Acoustics*, John Wiley & Sons, Inc., New York, 1973.

DOELLE, L. L. *Environmental Acoustics*, McGraw-Hill Book Company, New York, 1972.

KINZLER, L. E. and A. R. FREY *Fundamentals of Acoustics*, John Wiley & Sons, Inc., New York, 1962.

THUMANN, A. and R. K. MILLER *Secrets of Noise Control*, The Fairmont Press, Atlanta, 1974.

YERGES, L. F. *Sound, Noise, and Vibration Control*, Van Nostrand Reinhold, New York, 1969.

9

VIBRATIONAL CONTROL SYSTEMS
FOR INDUSTRIAL APPLICATIONS

9.1 INTRODUCTION

In our study of noise control thus far, we have omitted one very common source of noise, noise generated through vibration. The numerous sources of vibration in an industrial environment span an extremely wide range, included are impact processes, such as blasting, pile driving, hammers, and presses; machinery, such as motors, fans, compressors, engines, and pumps; turbulence in fluid systems; and transportation vehicles, such as trucks, railroads, and aircraft. These are not only troublesome themselves, but may induce problems elsewhere. For example, it was shown in Chapters 7 and 8 that a noise source in one room will, in general, cause the walls to vibrate and thus produce noise in an adjacent room. In addition, floor vibration can cause serious problems if delicate operations are being performed, or if sensitive instruments are present nearby.

The following two simple examples illustrate not only the manner in which vibration noise is generated but also some of the techniques used to control it. Consider, for example, a small motor or pump mounted rigidly to a large concrete floor in an industrial plant. Under these conditions the device will generate very little noise due to vibration. However, if this motor or pump is mounted on something like a relatively small table, the device will set the table in vibration and generate a large amount of noise. Similarly, the standard washing machine found in the home will generate very little noise when properly balanced on the floor. However, if one of the mounting feet is shortened, the machine will generate a considerable amount of noise when the machine is in its spin cycle.

In general, vibrations that generate noise require control because they result in either one or a combination of the following undesirable factors: human discomfort or pain, structure or mechanism failure, more frequent and thus less economical maintenance.

9.2 VIBRATION SYSTEMS

9.2.1 The Model

The model that will be employed in our vibration analysis is the single-degree-of-freedom mechanical system. The characteristics and ramifications of this model will be studied in some detail. It is important to note that even though this is the simplest system to analyze, nevertheless the concepts and principles developed form the basis for understanding and solving more complex practical vibration problems.

The single-degree-of-freedom system is shown in Figure 9.1. The system has only one degree of freedom because it is constrained by guides to move

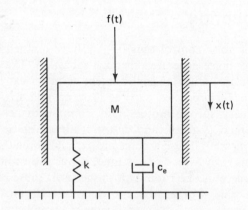

FIGURE 9.1 *Model for the single-degree-of-freedom mechanical system*

only in the vertical direction. The constant-coefficient linear differential equation derived from Newton's second law that describes the motion of this system is

$$m\ddot{x} + C_e\dot{x} + kx = f(t) \tag{9.1}$$

where m, C_e, and k represent the system mass, damping constant, and spring constant, respectively, and x, $\dot{x}$, and $\ddot{x}$ represent position, velocity, and acceleration, so that

$m\ddot{x}$ = inertia force

$C_e\dot{x}$ = viscous damping force

kx = linear elastic force

$f(t)$ = external forcing function

Note that of the three normal types of damping—coulomb or dry friction damping, hysteresis damping, and viscous damping—only the latter is considered here. The external forcing function may take many forms; for example, there may not be a forcing function, or the forcing function may be impulsive, sinusoidal, or random in nature.

9.2.2 Free Vibration

We begin our analysis of the unforced vibration system with the model defined by equation (9.1). Dividing this equation through by m, we obtain

$$\ddot{x} + \frac{C_e \dot{x}}{m} + \frac{k}{m} x = 0 \tag{9.2}$$

Because of some special features, Laplace transforms are employed in this analysis to solve the linear differential equation that describes the system. However, the reader may readily employ other techniques as well to obtain the same results.

Assuming for the moment that the initial conditions are now zero, the equation above can be written using the Laplace transform variable s as

$$s^2 X(s) + \frac{C_e}{m} s X(s) + \frac{k}{m} X(s) = s x(0) + \dot{x}(0) + \frac{C_e}{m} x(0)$$

or

$$\left(s^2 + \frac{C_e}{m} s + \frac{k}{m}\right) X(s) = \left(s + \frac{C_e}{m}\right) x(0) + \dot{x}(0) \tag{9.3}$$

Setting the first term in parentheses equal to zero yields

$$s^2 + \frac{C_e}{m} s + \frac{k}{m} = 0 \tag{9.4}$$

This equation can be put in the special form

$$s^2 + 2\xi\omega_n s + \omega_n^2 = 0 \tag{9.5}$$

where $\quad 2\xi\omega_n = \dfrac{C_e}{m}$

$\qquad\quad \omega_n^2 = \dfrac{k}{m}$

This equation is called the *characteristic equation* (in standard form) and is extremely important because the roots of this equation basically determine the response of the system [position of the mass as a function of time, i.e., $x(t)$], as will be illustrated in the discussion that follows.

Solving for the roots s_1 and s_2 of equation (9.5) yields

$$s_1, s_2 = \frac{-2\xi\omega_n \pm \sqrt{(2\xi\omega_n)^2 - 4\omega_n^2}}{2}$$

$$= -\xi\omega_n \pm \sqrt{\xi^2\omega_n^2 - \omega_n^2}$$

$$= -\xi\omega_n \pm j\omega_n\sqrt{1 - \xi^2} \tag{9.6}$$

The roots s_1 and s_2, when plotted in the s plane (where $s = \sigma + j\omega$ and $\omega = 2\pi f$), appear in general as shown in Figure 9.2(a). In plotting the roots as shown in the figure, we have assumed that the system is stable and that the value of ξ is $0 < \xi < 1$. The pertinent terms in the equation are

$$\xi = \text{damping ratio}$$

$$\omega_n = \text{undamped natural frequency}$$

and from Figure 9.2(a) we note that

$$\xi = \cos\theta \tag{9.7}$$

The damping ratio ξ and the natural frequency ω_n are the key factors that determine the roots of the characteristic equation and thus the response of the system. Basically, there are four cases of interest.

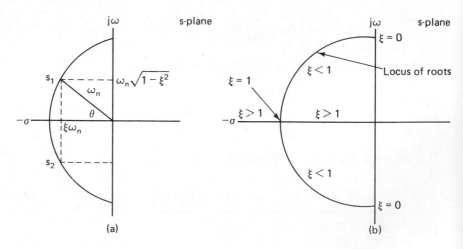

FIGURE 9.2 *Standard definitions and root locations for the second-order dynamic system*

Case 1: $\xi < 1$—*underdamped system* The roots are $s_1, s_2 = -\xi\omega_n \pm j\omega\sqrt{1 - \xi^2}$ and the system response is of the form

$$x(t) = Ae^{-\xi\omega_n t} \cos (\omega_n \sqrt{1 - \xi^2}\, t + \theta)$$

Case 2: $\xi > 1$—*overdamped system* The roots are $s_1, s_2 = -\xi\omega_n \pm \omega_n \sqrt{\xi^2 - 1}$, and the system response is of the form

$$x(t) = Ae^{-(\xi\omega_n + \omega_n\sqrt{\xi^2 - 1})t} + Be^{-(\xi\omega_n - \omega_n\sqrt{\xi^2 - 1})t}$$

Case 3: $\xi = 1$—*critically damped* The roots $s_1, s_2 = -\omega_n$, and the system response is of the form

$$x(t) = Ate^{-\omega_n t} + Be^{-\omega_n t}$$

Case 4: $\xi = 0$—*undamped system* The roots are $s_1, s_2 = \pm j\omega_n$ and the system response is of the form

$$x(t) = A \cos (\omega_n t + \theta)$$

The location of the roots in the s plane as a function of the damping ratio ξ for a fixed ω_n is shown in Figure 9.2(b). The heavy line indicates the locus of the roots.

To understand the importance of the system damping, suppose that the system in Figure 9.1 is subjected to an initial displacement of x_0 and that no external forcing function is present. Then the response of the system as a function of the root locations, which is determined by the damping ratio with ω_n held constant, is as shown in Figure 9.3.

A number of important features are demonstrated in Figure 9.3. For example, if we compare Figure 9.2 with Figure 9.3(a) and (b), we can determine what is meant by critical damping. In Figure 9.3(a), the system response is very sluggish and the system returns to the equilibrium position very slowly. In Figure 9.3(b), the system response is fast, but it overshoots the equilibrium position and eventually the oscillation is damped out. The dividing line between the overdamped system in Figure 9.3(a) and the underdamped system in Figure 9.3(b) is the critically damped system in which the system returns to the equilibrium position as fast as possible with no overshoot. Figure 9.2(b) and (a) illustrate, respectively, that the roots for the underdamped system (case 1) lie in the plane and appear in complex-conjugate pairs, and that the roots for the overdamped system (case 2) are real and unequal and lie on the negative real axis. For the critically damped system (case 3), the roots are real and equal and, as shown in Figure 9.2(c), represent the transition between the overdamped and underdamped systems. A comparison of Figure

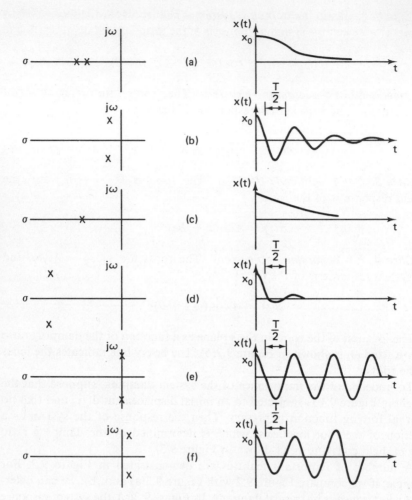

FIGURE 9.3 *System response as a function of root location*

9.3(b) and (d) indicates the difference in response between a lightly damped and heavily damped system. Note that the response of the heavily damped system in Figure 9.3(d) dies out very quickly. This is an extremely important point because it shows us that if we are dealing with a complex higher-order system and that some of the roots of the characteristic equation lie far out in the left half of the s plane, then it is often possible to approximate this higher-order system with a second-order system containing a pair of "dominant" roots located close to the $j\omega$ axis as shown in Figure 9.3(b). The word "dominant" is used because the response due to other roots with a large damping coefficient dies out so quickly that the total system response appears to depend only upon the roots that are close to the $j\omega$ axis.

If the roots lie on the $j\omega$ axis as shown in Figure 9.3(e), then no damping is present and hence the system as shown will simply oscillate *ad infinitum*. Finally, if any roots appear in the right half of the s plane, then the response will increase monotonically with time and the system is unstable. Hence, the $j\omega$ axis is the dividing line between stability and instability. The systems with which we shall be dealing will always be stable.

A comparison of equations (9.4) and (9.5) indicates that

$$\omega_n^2 = \frac{k}{m} \qquad (9.8)$$

and

$$2\xi\omega_n = \frac{C_e}{m} \qquad (9.9)$$

Therefore, the undamped natural frequency of the system is

$$\omega_n = \sqrt{\frac{k}{m}}$$

and the damping ratio is

$$\xi = \frac{C_e}{2\sqrt{km}} \qquad (9.10)$$

Since critical damping occurs when $\xi = 1$, we define $C_e = C_c$ for this value of $\xi = 1$, and thus

$$\xi = 1 = \frac{C_c}{2\sqrt{km}}$$

and hence from the expression above we obtain

$$C_c = 2\sqrt{km} \qquad (9.11)$$

as the critical amount of damping. Therefore, the damping ratio is often written as

$$\xi = \frac{C_e}{C_c} \qquad (9.12)$$

for the critically damped case.

EXAMPLE 9.1   Suppose that the system in Figure 9.1 has the following parameters in metric units: $W = 22.68$ N, $C_e = 0.0579$ N-s/cm, $k = 0.357$ N/cm, $g = 980.7$ cm/s². Find the undamped natural frequency of the system, its damping ratio, and the type of response the system would have if the mass were initially displaced and released.

SOLUTION   The mass is determined from the expression

$$m = \frac{W}{g}$$

$$= \frac{22.68}{9.807} = 2.31 \text{ kg}$$

Then, from equation (9.8),

$$\omega_n^2 = \frac{35.7}{2.31} = 15.43$$

$$\omega_n = 3.93 \text{ rad/s}$$

From equation (9.10),

$$\xi = \frac{C_e}{2\sqrt{km}}$$

$$= \frac{5.79}{2\sqrt{(35.7)(2.31)}}$$

$$= 0.318$$

The damping ratio is less than unity, and hence the response is underdamped. Therefore, the system response is of the form shown in Figure 9.3(b).

EXAMPLE 9.2   Determine the amount of additional damping required for the system in the previous example so that it will be critically damped.

SOLUTION   From equation (9.11), the system will be critically damped if the damping coefficient is

$$C_c = 2\sqrt{km}$$

$$= 2\sqrt{(35.7)(2.31)}$$

$$= 18.16 \text{ N-s/m} = 0.1816 \text{ N-s/cm}$$

Therefore, the additional damping ΔC required is

$$\Delta C = 0.1816 - 0.0579$$

$$= 0.1237 \text{ N/cm/s}$$

Note that with this additional damping the differential equation describing the motion of the system is

$$\ddot{x}(t) + \frac{C_e}{m}\dot{x}(t) + \frac{k}{m}x(t) = 0$$

and the characteristic equation in transform variable form is

$$s^2 + \frac{18.16}{2.31}s + \frac{35.7}{2.31} = 0$$

$$s^2 + 7.85s + 15.43 = 0$$

$$(s + 3.93)^2 = 0$$

The last equation indicates that case 3 exists, which is the critically damped condition.

D9.1 The system in Figure 9.1 has a $W = 34$ N and $k = 20.6$ N/cm. Determine the system's natural frequency.

Ans. 24.37 rad/s.

D9.2 If the viscous damping coefficent in problem D9.1 is $C_e = 0.1$ N/cm/s, what is the damping ratio for this system?

Ans. 0.059.

D9.3 What is the free vibration equation of motion for the system described in drill problems D9.1 and D9.2?

Ans. $\ddot{x}(t) + 2.88\dot{x}(t) + 593.66x(t) = 0.$

D9.4 If the system described by the previous drill problems is initially displaced and released, what is the form of the system response?

Ans. Underdamped.

In general, the response of the system shown in Figure 9.1 to an initial displacement can be determined as shown below. The differential equation in standard form is

$$\boxed{\ddot{x}(t) + 2\zeta\omega_n\dot{x}(t) + \omega_n^2 x(t) = 0} \qquad (9.13)$$

Using the Laplace transform, we obtain

$$s^2 x(s) - sx(0) - \dot{x}(0) + 2\zeta\omega_n[sx(s) - x(0)] + \omega_n^2 X(s) = 0$$

Assuming that $x(0) = x_0$ and that $\dot{x}(0) = 0$, the equation above reduces to

$$\boxed{X(s) = x_0\frac{s + 2\zeta\omega_n}{s^2 + 2\zeta\omega_n s + \omega_n^2}} \qquad (9.14)$$

Employing the graphical residue technique outlined and described in Appendix B (see Example B.7), we find that the inverse transform for equation (9.14), or equivalently the solution of equation (9.13) with the initial condition $x(t = 0) = x_0$, is

$$x(t) = \frac{x_0}{\sqrt{1 - \xi^2}} e^{-\xi \omega_n t} \cos (\omega_n \sqrt{1 - \xi^2}\, t - \pi/2 + \cos^{-1} \xi) \quad (9.15)$$

Using the fact that $\cos (x - \pi/2) = \sin x$, equation (9.15) reduces to

$$\boxed{x(t) = \frac{x_0}{\sqrt{1 - \xi^2}} e^{-\xi \omega_n t} \sin (\omega_n \sqrt{1 - \xi^2}\, t + \theta)} \quad (9.16)$$

whereas in equation (9.7),

$$\theta = \cos^{-1} \xi \quad (9.17)$$

EXAMPLE 9.3   The system shown in Figure 9.1 is initially displaced and released. The equation that describes the resultant motion is

$$x(t) = 2.0e^{-2.59t} \sin (9.66t + 75°)$$

Determine the damping ratio and natural frequency of the system, and its initial displacement.

SOLUTION   From equations (9.16) and (9.17),

$$\theta = 75° = \cos^{-1} \xi$$
$$\xi = 0.259$$

Also,

$$\xi \omega_n = 2.59$$
$$\omega_n = 10$$

The initial condition is determined from the relation

$$2.0 = \frac{x_0}{\sqrt{1 - \xi^2}}$$

$$= \frac{x_0}{\sqrt{1 - (0.259)^2}}$$

$$x_0 = 1.932$$

or, in fact, x_0 can be simply derived from

$$x_0 = 2e^{-0} \sin (0 + 75°)$$

The response described by equation (9.16) is shown in Figure 9.4. The condition $x_1 > x_0$ shown in the figure illustrated a general case (e.g., an

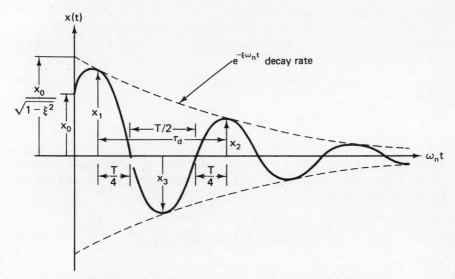

FIGURE 9.4 *Underdamped response of a second-order system*

initial velocity may be present). Note that the actual damping is $\xi\omega_n$. The damped period is defined as

$$\tau_d = \frac{2\pi}{\omega_n\sqrt{1-\xi^2}} \tag{9.18}$$

From Figure 9.4 it can be seen that the decay rate of the free oscillation is related to the system damping (i.e., the more damping, the faster the decay rate).

In order to determine a relationship between these two quantities, we must first determine the time response at two distinct points, each of which is one-quarter period from the crossover points in the first two lobes. These points are the points at which the sine function is equal to unity, and are not the peak points of the damped response as shown in Figure 9.4. From Figure 9.4 and equation (9.16), we note that

$$x_1 = \frac{x_0}{\sqrt{1-\xi^2}}\, e^{-\xi\omega_n t_1} \sin\left(\omega_n\sqrt{1-\xi^2}\, t_1 + \theta\right)$$

and

$$x_2 = \frac{x_0}{\sqrt{1-\xi^2}}\, e^{-\xi\omega_n(t_1+\tau_d)} \sin\left(\omega_n\sqrt{1-\xi^2}(t_1 + \tau_d) + \theta\right)$$

Taking the ratio of the two amplitudes and noting that in this case the sine functions are equal to unity yields

$$\frac{x_1}{x_2} = \frac{e^{-\xi\omega_n t_1}}{e^{-\xi\omega_n(t_1+\tau_d)}}$$ (9-19)

Using the properties of the exponential function, equation (9.19) can be written as

$$\frac{x_1}{x_2} = e^{-\xi\omega_n(t_1-t_1-\tau_d)} = e^{\xi\omega_n\tau_d}$$

The logarithmic decrement δ is defined as the natural logarithm of the ratio of two points such as x_1 and x_2, and thus from the previous expression we can derive the equation

$$\boxed{\delta = \ln\left(\frac{x_1}{x_2}\right) = \frac{2\pi\xi}{\sqrt{1-\xi^2}}}$$ (9.20)

Note that the logarithmic decrement δ is directly related to the damping ratio of the system.

It is important to note that the negative of the value of the function at t_3 can also be used to determine the logarithmic decrement, but in this case τ_d should be replaced by $\tau_d/2$, since a half-period is used in the evaluation. In this situation

$$\delta = 2\ln\left(\frac{x_1}{-x_3}\right)$$

Quite often this method yields better resolution than the approach indicated above. However, one should realize that, in general, the values of the function at any two times can be used to determine the two unknowns ξ and ω_n.

It is important to note that the ratio of the points in the response curve (i.e., x_1/x_2, x_2/x_3, etc.) will be identical only when viscous damping is present.

EXAMPLE 9.4 Using an oscilloscope, the free response of a system such as that shown in Figure 9.1 is measured and it is found that the ratio of the response at the t_1 and t_2 points is 1.72. Determine the approximate damping ratio using equation (9.20).

SOLUTION

$$\delta = \ln 1.72 = \frac{2\pi\xi}{\sqrt{1-\xi^2}}$$

$$0.294 = \frac{4\pi^2\xi^2}{1-\xi^2}$$

$$\xi = 0.086$$

D9.5   A system such as that shown in Figure 9.1 is initially displaced and released. If the equation that describes the resultant motion is

$$x(t) = 4.0e^{-at} \sin (10.4t + 60°)$$

determine the actual damping a.

Ans. $a = 6.0$.

9.2.3   Forced Vibration

9.2.3.1   *Harmonic excitation*

Harmonic excitation is an important forcing function because it arises naturally in many mechanical systems. It results from unbalance in rotating machinery, such as motors, generators, and fans. Let us suppose that this forcing function is of the form $F_0 \sin \omega t$. Then the differential equation describing the motion of the system shown in Figure 9.1 is

$$m\ddot{x} + C_e\dot{x} + kx = F_0 \sin \omega t \qquad (9.21)$$

Taking the Laplace transform of this equation and assuming that the initial conditions are zero yields

$$(ms^2 + C_e s + k)X(s) = F_0\left(\frac{\omega}{s^2 + \omega^2}\right) \qquad (9.22)$$

and hence

$$X(s) = \frac{F_0}{ms^2 + C_e s + k}\left(\frac{\omega}{s^2 + \omega^2}\right) \qquad (9.23)$$

In general, the solution $x(t)$ will consist of two parts—a complementary solution and a particular solution. These two parts of the solution correspond to the transient part and steady-state part, respectively. The transient portion of the solution is of the form of equation (9.16) and is determined by the residues at the complex poles, where the complex poles are the solution of

$$ms^2 + C_e s + k = 0$$

The steady-state solution of equation (9.21) will be a sinusoidal oscillation of the form

$$x(t) = x(\omega) \sin (\omega t - \phi) \qquad (9.24)$$

and is determined by the residues at the complex poles $s = \pm j\omega$. The exact solution can be derived by following the procedure illustrated at the end of the appendix on Laplace transforms. The solution of equation (9.21) can also be easily derived using standard differential-equation methods. However, we employ the Laplace transform method here, not only because it converts the differential equation to an algebraic equation, but also because of the ease with which it provides frequency and stability information.

The magnitude of the steady-state oscillation can be obtained from

$$X(s) = \frac{F_0}{ms^2 + C_e s + k}$$

which was derived from equations (9.23) and (9.24). From this equation one can determine the magnitude of the oscillation as a function of frequency. Since $s = \sigma + j\omega$ and $\omega = 2\pi f$, substituting $s = j\omega$ in the expression above yields

$$x(j\omega) = \frac{F_0}{-m\omega^2 + j\omega C_e + k}$$

The denominator is a complex number and the magnitude of this number is $[(k - m\omega^2)^2 + (\omega C_e)^2]^{1/2}$. Therefore, the oscillation magnitude as a function of frequency is

$$x(\omega) = \frac{F_0}{\sqrt{(k - m\omega^2)^2 + (C_e\omega)^2}}$$

$$= \frac{F_0/k}{\sqrt{[1 - (m/k)\omega^2]^2 + (C_e\omega/k)^2}} \qquad (9.25)$$

Using the definitions in equations (9.8) through (9.12), the equation above can be written as

$$\boxed{\frac{X(\omega)}{F_0/k} = \frac{1}{\sqrt{[1 - (\omega/\omega_n)^2]^2 + [2\xi(\omega/\omega_n)]^2}}} \qquad (9.26)$$

Therefore, the magnitude is a function of only two quantities, the ratio ω/ω_n and the damping ratio ξ. The phase angle ϕ is also a function of these quantities and is given by the expression

$$\boxed{\tan \phi = \frac{2\xi(\omega/\omega_n)}{1 - (\omega/\omega_n)^2}} \qquad (9.27)$$

The system is in resonance when the excitation frequency $f = \omega/2\pi$ is equal to the natural frequency of the dynamic system $f_n = \omega_n/2\pi$. The magnitude of the oscillation peak at resonance, and the sharpness of this resonant peak, are related to the damping ratio ξ.

TABLE 9.1

Frequency	Response	Controlling parameter
$\omega^2 \ll \omega_n^2$	$x(\omega) = \dfrac{F_0}{k}$	Stiffness controlled
$\omega^2 \gg \omega_n^2$	$x(\omega) = \dfrac{F_0}{m\omega^2}$	Mass controlled
$\omega^2 = \omega_n^2$	$x(\omega) = \dfrac{F_0}{C_e\omega}$	Damping controlled

Consider now the system response as a function of frequency. Notice that the response varies with frequency, as shown in Table 9.1. The relationships shown in the table are very important because they demonstrate that each of the parameters—stiffness, mass, and damping—effectively control the response only within a limited region (i.e., damping is primarily effective only at resonance). The importance of these relationships stems from the fact that a vibration correction scheme that is to be chosen for any given situation will depend upon whether the excitation frequency is less than, greater than, or equal to the resonant frequency.

The magnification factor is defined as

$$\text{M.F.} = \frac{x(\omega)}{F_0/k} = \frac{1}{\sqrt{[1 - (\omega/\omega_n)^2]^2 + [2\xi(\omega/\omega_n)]^2}} \qquad (9.28)$$

At resonance $\omega = \omega_n$ and hence

$$\text{M.F.}_{\text{res}} = \frac{1}{2\xi} \qquad (9.29)$$

The bandwidth of the resonant peak is the range between what is called the *half-power points*. The half-power points are the two points each side of the resonant peak which have a magnitude equal to $1/\sqrt{2}$ of the value of the peak (power is proportional to the square of the magnitude). These quantities are illustrated in Figure 9.5.

Letting $r = \omega/\omega_n$, then at the half-power points,

$$\frac{1}{2\sqrt{2}\,\xi} = \frac{1}{\sqrt{(1 - r^2)^2 + (2\xi r)^2}} \qquad (9.30)$$

Solving this equation for r^2 by means of simple algebra yields

$$r^2 = 1 - 2\xi \pm 2\xi\sqrt{1 + \xi^2} \qquad (9.31)$$

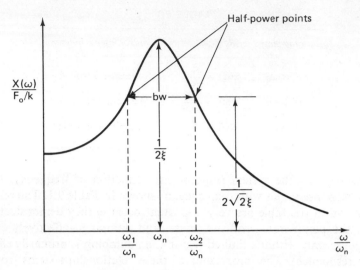

FIGURE 9.5 *Definitions for a resonant peak*

Assuming small values of damping (i.e., $\xi \ll 1$) and neglecting second-order terms in the expression above yields

$$r^2 = 1 \pm 2\xi \qquad (9.32)$$

Then

$$r_1 = \left(\frac{\omega_1}{\omega_n}\right)^2 = 1 - 2\xi$$

and

$$r_2 = \left(\frac{\omega_2}{\omega_n}\right)^2 = 1 + 2\xi$$

and therefore

$$r_2 - r_1 = \frac{\omega_2^2 - \omega_1^2}{\omega_n^2} = 4\xi$$

To obtain an expression for the bandwidth, we approximate

$$\frac{\omega_2^2 - \omega_1^2}{\omega_n^2} \approx \frac{2(\omega_2 - \omega_1)}{\omega_n} \qquad (9.33)$$

Then

$$\text{bw} = \frac{\omega_2 - \omega_1}{\omega_n} = \frac{\Delta\omega}{\omega_n} = 2\xi \qquad (9.34)$$

The quality factor Q is defined as the reciprocal of the bandwidth and is

$$Q = \frac{\omega_n}{\Delta\omega} = \frac{1}{2\xi} \qquad (9.35)$$

Therefore, if a frequency response on a system is made to determine a curve such as that shown in Figure 9.5, then this figure, together with equation (9.35), provide a technique for determining the equivalent viscous damping of the system.

9.2.3.2 Impulse excitation

At times in an industrial environment we encounter a force, such as that produced by a hammer, which acts for a very short period of time. This type of force can be represented by what is called an *impulse* or *delta function*. The delta function is defined as follows. Given a pulse starting at time $t = \epsilon$ of width a and height $1/a$, the function that results from taking the limit of this pulse as $a \longrightarrow 0$ is one with essentially zero width and infinite height such that

$$\int_0^\alpha \delta(t - \epsilon)\, dt = 1, \quad 0 < \epsilon < \alpha$$

where $\delta(t - \epsilon) = $ impulse function at $t = \epsilon$

This equation states that the area or strength of the impulse is unity. If we begin with a pulse of width a and height A/a, then we say that the impulse function $A\delta(t - \epsilon)$ has strength A at $t = \epsilon$.

The response of the system in Figure 9.1 to an impulse force of strength A is determined by the equation

$$m\ddot{x} + C_e\dot{x} + kx = A\delta(t)$$

Since $\mathcal{L}[A\delta(t)] = A$, and if the initial conditions are zero, then

$$X(s) = \frac{A}{ms^2 + C_e s + k}$$

If the system is lightly damped, then the response, which can be easily calculated using the techniques in Appendix B, is

$$x(t) = \frac{A/k}{\omega_n\sqrt{1 - \xi^2}} e^{-\xi\omega_n t} \sin \omega_n\sqrt{1 - \xi^2}\, t \qquad (9.36)$$

where ω_n and ξ have been previously defined

EXAMPLE 9.5 Suppose that the constants for the system shown in Figure 9.1 are $W = 9.09$ N, $k = 8.93$ N/cm, and $C_e = 0.089$ N/cm/s. If the maximum value of the harmonic force that excites the system is 13.61 N (i.e., $F = 13.61 \sin \omega t$), then determine the amplitude at resonance, which for small damping ratios is the maximum amplitude of the steady-state motion of the mass.

SOLUTION From equation (9.10),

$$\xi = \frac{C_e}{2\sqrt{km}}$$

$$= \frac{8.9}{2\sqrt{(893)(0.927)}}$$

$$= 0.15$$

Then, from equation (9.29),

$$\frac{x}{F_0/k} = \frac{1}{2\xi}$$

Hence, x at resonance is

$$x_{res} = \frac{13.61/893}{0.31} = 0.0492 \text{ m} = 4.92 \text{ cm}$$

D9.6 A sinusoidal forcing function, $2 \sin 10t$, is applied to the system shown in Figure 9.1 with parameters $W = 50$ N, $C_e = 0.08$ N/cm/s, and $k = 0.56$ N/cm. What parameter essentially controls the system response?

Ans. Mass is the controlling parameter.

D9.7 Determine the approximate bandwidth of the system in drill problem D9.6.

Ans. Bandwidth = 0.472.

9.2.3.3 Static deflection

The static deflection of the system shown in Figure 9.1 is the deflection that occurs under the static or deadweight load of the supported mass and is given by the expression

$$\delta_{st} = \frac{W}{k} = \frac{mg}{k}$$

where g = acceleration of gravity

The static deflection is related to the natural frequency of the system since

$$f_n = \frac{1}{2\pi} \sqrt{\frac{k}{m}}$$

$$= \frac{1}{2\pi} \sqrt{\frac{k}{W/g}}$$

$$= \frac{1}{2\pi} \sqrt{\frac{g}{W/k}}$$

$$\boxed{f_n = \frac{1}{2\pi} \sqrt{\frac{g}{\delta_{st}}}} \qquad (9.37)$$

EXAMPLE 9.6 A set of springs is used to support a large machine weighing 453.6 N. If the static deflection of the springs is 0.51 cm, determine the undamped natural frequency.

SOLUTION

$$f_n = \frac{1}{2\pi} \sqrt{\frac{980.7}{0.51}}$$

$$= 7 \text{ Hz}$$

D9.8 The second-order system shown in Figure 9.1 has a mass weight $W = 200$ N. If the static deflection of the spring is 6 cm, determine the spring constant k.

Ans. $k = 33.3$ N/cm.

9.3 VIBRATION CONTROL

9.3.1 Transmissibility

Assuming that the forcing function is harmonic in nature, we shall consider two cases of vibration transmission—one in which force is transmitted to the supporting structure, and one in which the motion of the supporting structure is transmitted to the machine.

9.3.1.1 Force excitation

Consider the system shown in Figure 9.6, where $f(t)$ is the harmonic force acting on the system and $f_T(t)$ is the force transmitted to the supporting structure or base. The force transmitted through the spring and damper to the supporting structure is

$$f_T(t) = kx + C_e \dot{x}$$

The magnitude of this force as a function of frequency is

$$F_T = \sqrt{[kx(\omega)]^2 + [C_e \omega X(\omega)]^2} \qquad (9.38)$$

$$= kx(\omega)\sqrt{1 + (C_e \omega/k)^2}$$

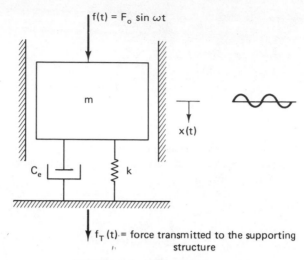

$f(t) = F_o \sin \omega t$

m

$x(t)$

C_e k

$f_T(t) =$ force transmitted to the supporting
structure

FIGURE 9.6 *Force excitation model*

Substituting equation (9.25) into the equation above yields

$$F_T = \frac{F_0\sqrt{1 + (C_e\omega/k)^2}}{\sqrt{\left(1 - \dfrac{m\omega^2}{k}\right)^2 + \left(\dfrac{C_e\omega}{k}\right)^2}}$$

Employing our standard definitions for ω_n and ξ, we obtain

$$T = \frac{F_T}{F_0} = \frac{\sqrt{1 + \left(2\xi\dfrac{\omega}{\omega_n}\right)^2}}{\sqrt{\left[1 - \left(\dfrac{\omega}{\omega_n}\right)^2\right]^2 + \left(2\xi\dfrac{\omega}{\omega_n}\right)^2}} \qquad (9.39)$$

T is defined as the *transmissibility* and represents the ratio of the amplitude of the force transmitted to the supporting structure to that of the exciting force.

9.3.1.2 Motion excitation

The system that illustrates motion excitation is shown in Figure 9.7. The motion of the dynamic system is represented by the variable x and the harmonic displacement of the supporting base is represented by the variable y. The equation that describes the dynamics of the system is

$$m\ddot{x} + C_e(\dot{x} - \dot{y}) + k(x - y) = 0 \qquad (9.40)$$

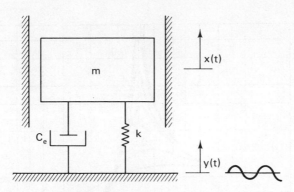

FIGURE 9.7 *Motion excitation model*

or

$$m\ddot{x} + C_e\dot{x} + kx = C_e\dot{y} + ky$$

Then the ratio of the magnitudes of the displacements as a function of frequency, which is the transmissibility, is given by the expression

$$T = \frac{x}{y} = \frac{\sqrt{k^2 + (C_e\omega)^2}}{\sqrt{(k - m\omega^2)^2 + (C_e\omega)^2}} \qquad (9.41)$$

This expression, when written in terms of ω_n and ξ, is

$$T = \frac{x}{y} = \frac{\sqrt{1 + [2\xi(\omega/\omega_n)]^2}}{\sqrt{[1 - (\omega/\omega_n)^2]^2 + [2\xi(\omega/\omega_n)]^2}} \qquad (9.42)$$

Note that the transmissibility expressions for both force and motion excitation are identical. Therefore, it would appear that the engineering principles employed to protect the supporting structure under force excitation are the same as those used to protect the dynamic system from motion excitation.

9.3.2 Design Curves

9.3.2.1 Transmissibility vs. damping ratio

These curves, which are derived from equations (9.39) and (9.42), are shown in Figure 9.8. Note that if the ratio of the disturbing frequency to the natural frequency $\omega/\omega_n < \sqrt{2}$, then the transmissibility $T > 1$ and the input disturbance is amplified. Note also that in the region $\omega/\omega_n > \sqrt{2}$, T decreases

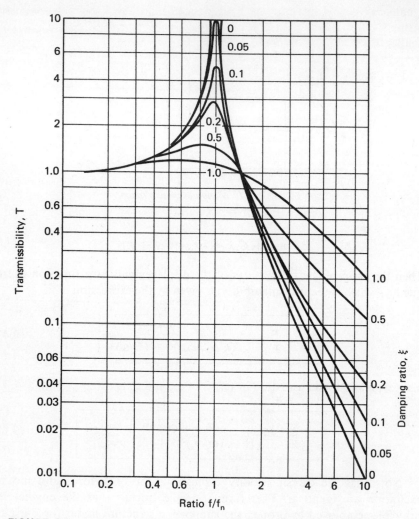

FIGURE 9.8 *Design curves for the transmissibility vs. the frequency ratio f/f_n as a function of the damping ratio ξ for a linear single-degree-of-freedom system*

with increasing ξ. If $\omega/\omega_n = 1$, a resonance condition exists and the transmissibility can be quite large. If $\omega/\omega_n > \sqrt{2}$, then $T < 1$, and hence this is the only region where isolation is possible. In this region T decreases with decreasing ξ, and hence better isolation is achieved with very little or no damping.

The curve in Figure 9.8 demonstrates the effectiveness of an isolator to reduce vibration. Figure 9.8 also indicates a number of important concepts: (a) isolators should be chosen so as not to excite the natural frequencies of the system; (b) damping is important in the range of resonance whether the dynamic system is operating near resonance or must pass through resonance during start-up; (c) in the isolation region, the larger the ratio ω/ω_n (i.e., the smaller the value of ω_n), the smaller the transmissibility will be. Since $\omega_n = \sqrt{k/m}$, ω_n can be made small by selecting soft springs. Soft springs used in conjunction with light damping provides good isolation. Note also that ω_n can be reduced by increasing the mass. However, remember that increasing the mass also increases the dead weight of the dynamic system which the mounting structure must support.

9.3.2.2 *Isolation efficiency vs. ω and ω_n*

Another graphical method of illustrating the regions of isolation and amplification as a function of the disturbing frequency and the natural frequency of the system is shown in Figure 9.9. In using this figure we must note that *percent isolation* is defined by the expression

$$\text{percent isolation} = (1 - \text{transmissibility})100\%$$
$$\%\text{I} = (1 - T)100\%, \quad T \leq 1 \tag{9.43}$$

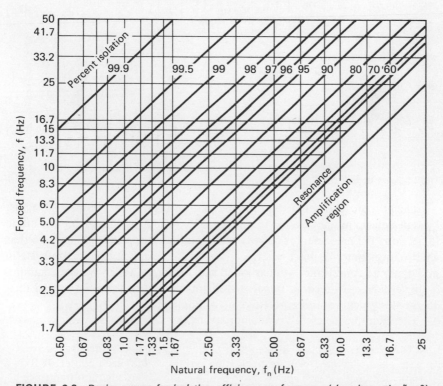

FIGURE 9.9 *Design curves for isolation efficiency vs. frequency (damping ratio, $\xi = 0$)*

The forcing frequency on the ordinate and the percent isolation lines in the graph locate a point, the abscissa of which is the natural frequency of the system necessary to achieve the required isolation. The system parameters may then be selected or adjusted to obtain this desired natural frequency.

9.3.2.3   Static deflection vs. natural frequency

As shown in an earlier example, the static deflection is the deflection of an isolator that occurs due to the dead weight load of the mounted equipment. Since the static deflection is given by the expression $\delta_{st} = W/k$, and since the undamped natural frequency of a single-degree-of-freedom system is determined by the equation

$$f_n = \frac{1}{2\pi}\sqrt{\frac{kg}{W}}$$

Then

$$f_n = \frac{1}{2\pi}\sqrt{\frac{980.7}{\delta_{st}}}$$

$$\boxed{f_n = 4.98\sqrt{\frac{1}{\delta_{st}}} \ \ \text{Hz}} \qquad\qquad (9.44a)$$

where δ_{st} is in centimeters

and f_n is given by

$$\boxed{f_n = 3.13\sqrt{\frac{1}{\delta_{st}}} \ \ \text{Hz}} \qquad\qquad (9.44b)$$

where δ_{st} is in inches

The graphical presentation of this equation is given in Figure 9.10. Thus, we can determine the natural frequency of a system by measuring the static deflection. This statement is correct provided that the spring is linear and that the isolator material possesses the same type of elasticity under both static and dynamic conditions. As mentioned previously, however, we are assuming a single-degree-of-freedom linear system throughout our analysis, and thus all the design curves presented above are applicable.

The examples that follow demonstrate the use of these design curves.

EXAMPLE 9.7   A pump in an industrial plant is mounted rigidly to a massive base plate. The base plate rests on four springs, one at each corner. If the static deflection of each spring is two centimeters, then the natural frequency

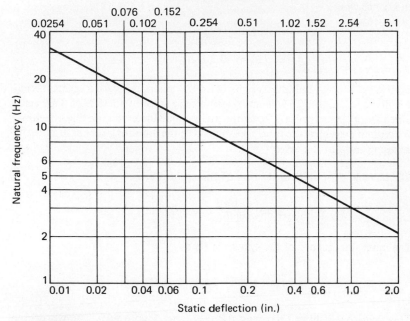

FIGURE 9.10 *Design curve for static deflection vs. natural frequency for a linear single-degree-of-freedom system*

of the system is given by equation (9.44a) as

$$f_n = 4.98\sqrt{\frac{1}{\delta_{st}}}$$

$$= 4.98\sqrt{\frac{1}{2}}$$

$$= 3.52 \text{ Hz}$$

Note that this could also have been obtained from the static deflection chart in Figure 9.10.

EXAMPLE 9.8 A small fluidic system, which is mounted at four points on a bench, operates at a disturbing frequency of 25 Hz. It is estimated that the noise it produces will be reduced to a value much lower than the ambient noise if 80% of the disturbing vibration is attenuated. Hence, the natural frequency of the isolators obtained from Figure 9.9 is approximately 10.4 Hz. This value can be checked by using equation (9.39) or (9.42). For typical values of T and no damping, these equations reduce to the following for $f/f_n > \sqrt{2}$.

$$T = \frac{1}{(f/f_n)^2 - 1}$$

In this case, an isolation of 80% corresponds to a transmissibility of 0.2, and hence

$$0.2 = \frac{1}{(25/f_n)^2 - 1}$$

$$f_n = 10.2 \text{ Hz}$$

EXAMPLE 9.9   A unit that weighs 181.43 N is supported on a spring having a stiffness value of 892.91 N/cm. A disturbing force of 45.36 N at 3600 rpm, which is produced from a rotating mass unbalance, acts on the machine. We wish to determine the force transmitted to the mounting base assuming the damping ratio of the system is 0.1.

SOLUTION   The static deflection of the springs is

$$\delta_{st} = \frac{181.43}{892.91} = 0.2032 \text{ cm}$$

The natural frequency of the system is then

$$f_n = 4.98 \sqrt{\frac{1}{.2032}} = 11 \text{ Hz}$$

This value could have been read directly from Figure 9.10. The disturbing frequency caused by the mass unbalance is

$$\text{disturbing frequency} = \frac{3600}{60} = 60 \text{ Hz}$$

Then

$$\frac{\omega}{\omega_n} = \frac{f}{f_n} = \frac{60}{11} = 5.45$$

The transmissibility for $\omega/\omega_n = 5.45$ and a $\zeta = 0.1$ is obtained from Figure 9.8 as

$$T = \frac{F_T}{F_0} = 0.052$$

Therefore, the force transmitted to the mounting structure is

$$F_T = (0.052)(45.36)$$

$$= 2.36 \text{ N}$$

EXAMPLE 9.10   A vibrating single-degree-of-freedom dynamic system is to be isolated from its supporting base. What is the required damping ratio that must be achieved by the isolator to limit the transmissibility at resonance to $T = 5$?

SOLUTION From Figure 9.8 we can see that a damping ratio of not less than $\xi = 0.1$ will be necessary to limit the transmissibility. This value could also be obtained from equation (9.29). Upon setting $\omega = \omega_n$, one obtains

$$\xi = \frac{1}{2\sqrt{T^2 - 1}}$$

$$= \frac{1}{2\sqrt{25 - 1}} = 0.1$$

Thus, $\xi = 0.1$, which checks.

D9.9   It is assumed that a small machine mounted on a table can be modeled as shown in Figure 9.1. If the forcing frequency and damping ratio of the system are 10 Hz and 0.2, respectively, what is the desired, natural frequency of the system in order to obtain 90% isolation?

Ans. $f_n = 2 \text{ Hz}.$

D9.10   A mechanical system is mounted on four springs. If the forcing frequency and natural frequency of the system are 15 Hz and 1.5 Hz, respectively, what percent isolation do the springs provide?

Ans. $I \cong 98.7\%.$

D9.11   A vibrating system is mounted on a set of springs. A measurement of the springs under load indicates that they have deflected 1 cm. What is the approximate natural frequency of the system?

Ans. $f_n = 5 \text{ Hz}.$

9.3.3   Control Techniques

In the control of noise we basically considered three areas: the source, the path, and the receiver. The reader is encouraged to watch for analogies to these areas in the discussion that follows.

Vibration control may involve one or a combination of the following techniques.

9.3.3.1   Source alteration

In the control of vibration it is important to first check and see if the noise or vibration level can be reduced by altering the source. This may be accomplished by making the source more rigid from a structural standpoint, changing certain parts, balancing, or improving dimensional tolerances. The system mass and stiffness may be adjusted in such a way so that resonant frequencies of the system do not coincide with the forcing frequency. This process is called *detuning*. Sometimes it is also possible to reduce the number of coupled resonators that exist between the vibration source and the receiver of interest. This technique is called *decoupling*. Although these techniques can be applied during design or construction, they are perhaps more often used as a correction scheme. However, it is also important to ensure that the application of these schemes does not produce other problems elsewhere.

9.3.3.2 Isolation

In general, vibration isolators can be broken down into three categories: (1) metal springs, (2) elastomeric mounts, and (3) resilient pads. Before examining each of these areas, a few general comments can be made which are pertinent to all categories. We must always remember that we are assuming a single-degree-of-freedom system, and therefore our analysis will not be exact in every case. However, practical systems are normally reduced to this model because it is the only one that we understand thoroughly.

When building or correcting a design, always ensure that the machine under investigation and the element that drives it both rest on a common base. Always design the isolators to protect against the lowest frequency that can be generated by the machine. Design the system so that its natural frequency will be less than one-third of the lowest forcing frequency present. The isolation device should also reduce the transmissibility at every frequency contained in the Fourier spectrum of the forcing function.

The reasoning behind these guidelines will become clear as we progress through this section.

9.3.3.2.1 METAL SPRINGS

Metal springs are widely used in industry for vibration isolation. Their use spans the spectrum from light, delicate instruments to very heavy industrial machinery. The advantages of metal springs are: (1) they are resistant to environmental factors such as temperature, corrosion, solvents, and the like; (2) they do not drift or creep; (3) they permit maximum deflection; and (4) they are good for low-frequency isolation. The disadvantages of springs are: (1) they possess almost no damping and hence the transmissibility at resonance can be very high; (2) springs act like a short circuit for high-frequency vibration; and (3) care must be taken to ensure that a rocking motion does not exist.

Careful engineering design will minimize the effect of some of these disadvantages. For example, the damping lacked by springs can be obtained by placing dampers in parallel with the springs. Rocking motions can be minimized by selecting springs in such a way that each spring used will deflect the same amount. In addition, the use of an inertia block that weighs from one to two times the amount of the supported machinery minimizes rocking, lowers the center of gravity of the system, and helps to uniformly distribute the load. High-frequency transmission through springs caused by the low damping ratio can be blocked by using rubber pads in series with the springs. A typical damping ratio for steel springs is 0.005.

The *design procedure* for selecting springs for vibration isolation is outlined below:

1. Determine the weight of the machinery to be isolated, the lowest expected forcing frequency, the degree of isolation, and the number of mounting points.
2. The percent isolation yields the transmissibility T, and the lowest forcing frequency is the quantity f. Hence, if the damping ratio is essentially zero, the transmissibility equation reduces to

$$T = \frac{1}{(f/f_n)^2 - 1}$$

where $\quad \xi \approx 0$

Combining this equation with the equation for the natural frequency

$$f_n = 4.98\sqrt{\frac{1}{\delta_{st}}}$$

We obtain

$$\delta_{st} = \frac{(1 + T)24.8}{Tf^2} \qquad (9.45)$$

where $\quad \xi \approx 0$

3. The total weight of the machinery in newtons and the number of mounting points yields W_{mp}, which is the weight per mounting point. The k coefficient for the spring is then calculated from the expression

$$k = \frac{W_{mp}}{\delta_{st}} \quad \text{N/cm} \qquad (9.46)$$

This value and/or the static deflection can be used to select a spring from the manufacturer's catalog.

EXAMPLE 9.11 A machine set operating at 2400 rpm is mounted on an inertia block. The total system weighs 907 N. The weight is essentially evenly distributed. We want to select four steel springs upon which to mount the machine. The isolation required is 90%.

SOLUTION
1. Weight $= 907$ N
   Lowest forcing frequency $= 2400$ rpm
   Percent isolation $= 90\%$
   Number of mounting points $= 4$
2. 90% isolation yields $T = 0.1$

$$f = \frac{2400}{60} = 40 \text{ Hz}$$

Therefore, from equation (9.45),

$$\delta_{st} = \frac{(1.1)(24.8)}{(0.1)(40)^2}$$

$$= 0.171 \text{ cm}$$

3. $W_{mp} = \dfrac{907}{4} = 226.75$ N

Therefore,

$$k = \frac{226.75}{0.171} = 1330 \text{ N/cm}$$

At this point the reader should note from Figure 9.10 that as the static deflection increases, the natural frequency decreases. Figure 9.8 indicates that as the ratio f/f_n increases (i.e., f_n decreases), the percent vibration isolation increases. Hence, we would select a set of springs with a stiffness coefficient of about 1300 N/cm and a $\delta_{st} \geq 0.17$ cm.

EXAMPLE 9.12 A motor operating at 1200 rpm drives a fan operating at 600 rpm. The two elements are mounted on a common base. The load is distributed such that the weight per mounting point for the two points under the motor is 181.4 N, and the weight per mounting point for the two points under the fan is 90.7 N. Select the set of springs that will produce 85% vibration isolation.

SOLUTION
1. Lowest forcing frequency $= 600$ rpm produced by the fan
   Percent isolation $= 85\%$
   Number of mounting points $= 4$
2. 85% isolation yields $T = 0.15$

$$f = \frac{600}{60} = 10 \text{ Hz}$$

and, from equation (9.45),

$$\delta_{st} = \frac{(1.15)(24.8)}{(0.15)(100)}$$

$$= 1.9 \text{ cm}$$

3. The k for the springs beneath the motor is

$$k = \frac{181.4}{1.9} = 95.5 \text{ N/cm}$$

The k for the springs beneath the fan is

$$k = \frac{90.7}{1.9} = 47.7 \text{ N/cm}$$

Therefore, two springs with a k of 90 N/cm and two springs with a k of 40 N/cm would be satisfactory. Recall that the static deflection of all the springs should be the same, to minimize rocking motions.

Recall that we stated earlier that the natural frequency f_n should be about one-third the lowest forcing frequency. From Figure 9.8 we see that if f/f_n is 3, then the degree of isolation is about 90%, which is a good general criterion. In addition, note that if f/f_n is 3 for the lowest forcing frequency, then even better isolation is achieved at higher forcing frequencies.

D9.12 A mass of 400 kg is evenly distributed over four mounting points. Springs with a static deflection of 1 cm are used to isolate the system. What is the lowest forcing frequency for which the system will provide 90% isolation?
Ans. 16.5 Hz.

9.3.3.2.2 ELASTOMERIC MOUNTS

Elastomeric mounts consist primarily of natural rubber and synthetic rubber materials such as neoprene. In general, elastomeric mounts are used to isolate small electrical and mechanical devices from relatively high forcing frequencies. They are also useful in the protection of delicate electronic equipment. In a controlled environment, natural rubber is perhaps the best and most economical isolator. Natural rubber contains inherent damping, which is very useful if the machine operates near resonance or passes through resonance during "startup" or "shutdown." Synthetic rubber is more desirable when the environment is somewhat hazardous.

Rubber can be used in either tension, compression or shear; however, it is normally used in compression or shear and rarely used in tension. In compression it possesses the capacity for high-energy storage; however, its useful life is longer when used in shear. Rubber is classified by a durometer number.

Rubber employed in isolation mounts normally ranges from 30-durometer rubber, which is soft, to 80-durometer rubber, which is hard. The typical damping ratio for natural rubber and neoprene is $\xi = 0.05$.

One word of caution when dealing with rubber: it possesses different characteristics depending upon whether the material is used in strips or bulk, and whether it is used under static or dynamic conditions.

The steps for selecting an elastomeric mount are essentially those enumerated in the previous section on metal springs. The following examples will illustrate the procedure.

EXAMPLE 9.13   A drum weighing 120 N and operating at 3600 rpm induces vibration in adjacent equipment. Four vertical mounting points support the drum. Choose one of the isolators shown in Figure 9.11 so as to achieve 90% vibration isolation.

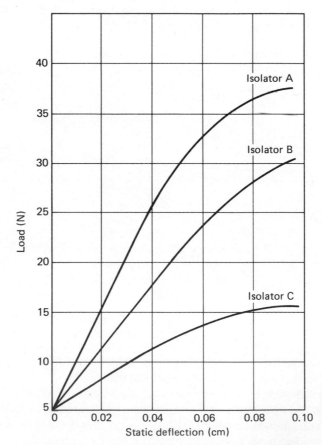

FIGURE 9.11   *Typical load vs. deflection curve for an elastomeric mount*

SOLUTION

(1) $W = 120\text{ N}$
Speed $= 3600$ rpm
$\%I = 90\%$
Number of mounting points $= 4$
(2) 90% isolation yields $T = 0.10$

$$f = \frac{3600}{60} = 60\text{ Hz}$$

$$\delta_{st} = \frac{(1.1)(24.8)}{(0.1)(60)^2} \quad \text{from equation (9.45)}$$

$$= 0.076\text{ cm}$$

The equation above for the static deflection can be employed because the damping ratio $\xi \approx 0$. The same answer could have been obtained by entering Figure 9.8 at $T = 0.1$ and reading $f/f_n = 3.3$ for $\xi = 0.05$. Since $f = 60$ Hz, then $f_n = 18.2$ Hz. Then, as shown in Figure 9.10, this f_n corresponds to a static deflection of 0.076.

(3) $W_{mp} = \dfrac{120}{4} = 30\text{ N}$

Hence, Figure 9.11 indicates that for a static deflection of 0.076 cm and a weight per mounting point of 30 N, isolator B is the best choice. Note that isolator B is chosen because at a load of 30 N it has a $\delta_{st} > 0.076$ cm. Recall that as shown in Figure 9.10, as δ_{st} increases, f_n decreases. A decrease in f_n increases the ratio f/f_n, which, as shown in Figure 9.8, increases the percent isolation.

EXAMPLE 9.14 Some electronic gear used in conjunction with a system that samples air pollution is mounted on four points in a mobile trailer. The equipment weighs 60 N, and the weight is essentially evenly distributed. From a careful analysis it has been estimated from the motion of the vehicle that an isolator with a natural frequency of 18 Hz would provide proper isolation of the equipment. From the manufacturer's data on isolators A, B, C, and D shown in Figure 9.12, choose the proper isolator for this application.

SOLUTION With a weight per mounting point of 15 N, we can see from Figure 9.12 that isolator B will provide even a lower natural frequency for the load. The argument of selecting isolator B is the same as that given at the end of the previous example, and thus will not be repeated here.

In order to achieve specific and sometimes complicated characteristics, multiple isolators are employed in various configurations. For example, high and low damping rubbers can be used in parallel in an attempt to take advantage of the desirable properties of each material.

One word of caution must be expressed in the use of rubber isolators. At

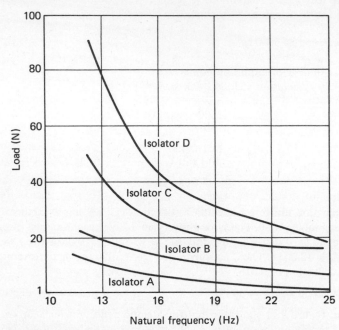

FIGURE 9.12 *Typical natural frequency vs. load curves for elastomeric mounts*

high frequencies, where the dimensions of the isolator become equal to an integer number of multiples of the half-wavelength of the elastic wave through the isolator, wave resonances occur. These wave effects, which in general do not occur in heavily damped isolators, cause the high-frequency trailing edge of the transmissibility curve to have resonant peaks, and thus the isolation actually obtained may not be the same as the model predicts.

D9.13 In a system it is desired to use one of the isolators in Figure 9.11 to isolate a vibrating part from the housing. An analysis indicates that the weight per mounting point is 10 N and the δ_{st} is 0.04 cm. Which isolator in Figure 9.11 should be used?

Ans. None of the isolators will work!

D9.14 If the weight per mounting point in drill problem D9.13 were changed to 35 N, which isolator should be chosen?

Ans. Isolator A.

D9.15 Equipment that requires vibration isolation has a weight per mounting point of 5 N. It is estimated that the natural frequency required is 16 Hz. Which isolator in Figure 9.12 should be chosen?

Ans. None of the isolators will work!

D9.16 If the weight per mounting point in drill problem D9.15 were changed to 50 N, which isolator should be chosen?

Ans. Isolator D.

9.3.3.2.3 ISOLATION PADS

The materials in this particular classification include such things as cork, felt, and fiberglass. In general, these items are easy to use and install. They are purchased in sheets and cut to fit the particular application, and can be stacked to produce varying degrees of isolation. Cork, for example, can be obtained in squares (like floor tile) 1 to 2.5 cm in thickness or in slabs up to 15 cm thick for large deflection applications. Cork is very resistive to corrosion and solvents and is relatively insensitive to a wide range of temperatures. Some of the felt pads are constructed of organic material and hence should not be employed in an industrial environment where solvents are used. Fiberglass pads, on the other hand, are very resistant to industrial solvents. A typical damping ratio for felt and cork is $\xi = 0.05$ to 0.06.

EXAMPLE 9.15 The inertia base of a machine is placed on a slab of cork. The cork, which has a density of B, is described by the manufacturer by the curve in Figure 9.13. The cork slab is loaded at 4 N/cm², and the forcing frequency of the vibration is 45 Hz. If the damping ratio of the cork is 0.05, calculate the transmissibility.

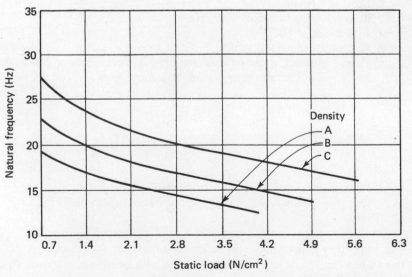

FIGURE 9.13 *Typical natural frequency vs. static load curves for cork*

SOLUTION From Figure 9.13 we can see that the cork of density *B* when loaded at 4 N/cm² has a natural frequency of 15 Hz. Hence, the ratio of the forcing frequency to the natural frequency f/f_n is $\frac{45}{15} = 3$. Therefore, from Fig. 9.8 the transmissibility is 0.17 for $\xi = 0.05$.

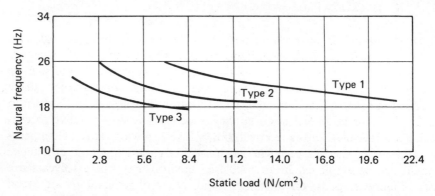

FIGURE 9.14 *Typical natural frequency vs. static load curves for fiberglass*

EXAMPLE 9.16 A large machine is mounted on a concrete slab. The lowest expected forcing frequency is 60 Hz. If the isolator will be loaded at 7 N/cm², choose the proper fiberglass isolator from the manufacturer's data shown in Figure 9.14 to produce 80% isolation. Assume that the damping ratio of the material is $\xi = 0.05$.

SOLUTION Following the design procedure illustrated earlier:

1. Load = 7 N/cm²
   Lowest forcing frequency = 60 Hz
   % Isolation = 80%
2. 80% isolation yields $T = 0.2$
   From Figure 9.8, it can be seen that for a damping ratio of $\xi = 0.05$ and $T = 0.2$, the ratio $\omega/\omega_n = 2.6$. Therefore, since

$$\frac{f}{f_n} = 2.6$$

we find f_n to be

$$f_n = 23 \text{ Hz}$$

3. For the given static load and natural frequency requirement, it can be seen from Figure 9.14 that the type 2 fiberglass isolator will be satisfactory.

An important point which can be gleaned from the curves in Figure 9.14 is this: since the lowest natural frequency of the fiberglass is about 18 Hz and since isolation can be achieved only for frequencies greater than $\sqrt{2}$

times the natural frequency, these fiberglass isolators are only useful for forcing frequencies above $18\sqrt{2}$ Hz, or about 25 Hz.

D9.17 The desired natural frequency for a system with a cork isolator loaded at 2 N/cm² is 20 Hz. Which isolator in Figure 9.13 should be employed?

Ans. Isolator B.

D9.18 If the load on the isolator in drill problem D9.17 is doubled, which isolator would be chosen?

Ans. Isolator C.

9.3.3.3 Inertia blocks

Isolated concrete inertia blocks play an important part in the control of vibration transmission. Large-inertia forces at low frequencies caused by equipment such as reciprocating compressors may cause motion that is unacceptable for proper machine operation and transmit large forces to the supporting structure. One method of limiting motion is to mount the equipment on an inertia base. This heavy concrete or steel mass limits motion by overcoming the inertia forces generated by the mounted equipment.

Low natural frequency isolation requires a large deflection isolator such as a soft spring. However, the use of soft springs to control vibration can lead to rocking motions which are unacceptable. Hence, an inertia block mounted on the proper isolators can be effectively used to limit the motion and provide the needed isolation.

Inertia blocks are also useful in applications where a system composed of a number of pieces of equipment must be continuously supported. An example of such equipment is a system employing calibrated optics.

Thus, inertia blocks are important because they lower the center of gravity and thus offer an added degree of stability; they increase the mass and thus decrease vibration amplitudes and minimize rocking; they minimize alignment errors because of the inherent stiffness of the base; and they act as a noise barrier between the floor on which they are mounted and the equipment that is mounted on them. One must always keep in mind, however, that to be effective, inertia blocks must be mounted on isolators.

9.3.3.4 Absorption

Consider the system shown in Figure 9.15. The components m_2 and k_2 represent a vibration absorber. The equations of motion that describe the system are

$$m_1\ddot{x}_1 + k_1x_1 + k_2(x_1 - x_2) = f$$
$$m_2\ddot{x}_2 + k_2(x_2 - x_1) = 0$$

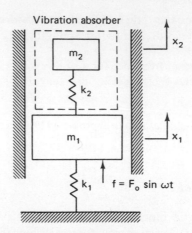

FIGURE 9.15 *Model for the analysis of a vibration absorber*

The magnitude of the frequency response is obtained from the following equations:

$$-m_1\omega^2 X_1 + (k_1 + k_2)X_1 - k_2 X_2 = F_0$$
$$-k_2 X_1 - \omega^2 m_2 X_2 + k_2 X_2 = 0$$

or

$$\boxed{\begin{aligned} (-m_1\omega^2 + k_1 + k_2)X_1 - k_2 X_2 &= F_0 \\ -k_2 X_1 + (-m_2\omega^2 + k_2)X_2 &= 0 \end{aligned}} \tag{9.47}$$

The two simultaneous equations can be written in matrix form as

$$\begin{bmatrix} -m_1\omega^2 + k_1 + k_2 & -k_2 \\ -k_2 & -m_2\omega^2 + k_2 \end{bmatrix} \begin{bmatrix} X_1 \\ X_2 \end{bmatrix} = \begin{bmatrix} F_0 \\ 0 \end{bmatrix} \tag{9.48}$$

Now note what happens to the equations above if the forcing frequency ω is equal to the natural frequency of the vibration absorber (i.e., $\omega_2 = k_2/m_2$). Under this condition

$$\boxed{\begin{aligned} x_1 &= 0 \\ X_2 &= -\frac{F_0}{k_2} \end{aligned}} \tag{9.49}$$

Therefore, we can see that if the natural frequency of the vibration absorber is tuned to the vibration forcing frequency, the motion of the main mass is ideally zero and the spring force of the absorber is at all times equal and opposite to the applied force, F_0. Hence, no force is transmitted to the supporting structure. Because of the problem of tuning the natural frequency of the absorber to that of the vibration forcing frequency, the absorber is normally used only with constant speed (synchronous) machinery. Thus, although isolation is useful over a broad range of frequencies, absorption is primarily useful only for very narrow band or single-frequency control.

If damping is present in the system, the analysis becomes more complicated; however, proper tuning of the absorber will still limit the motion of the main mass. The vibration absorber attached to a machine can be used to limit the force transmitted by the machine to the supporting structure, or it can be used to limit the motion of the machine if the foundation is in vibration. Vibration absorbers can also be used to limit resonant vibration response in lightly damped systems. Figure 9.16 indicates schematically the manner in which this is done.

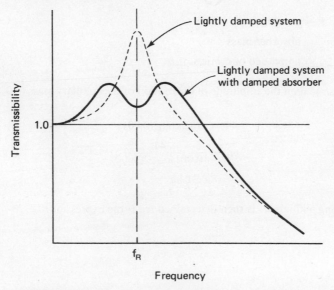

FIGURE 9.16 *Effect of a damped absorber on a lightly damped system*

Finally, one must always keep in mind that if the absorber is not properly designed, the vibration may actually increase rather than decrease. However, when properly applied, the absorber can be useful for both force and motion excitation. The dynamic absorber or auxiliary mass damper, as it is called, can be effectively applied to control the amplitude of a vibrating system. If the device has little or no damping and is applied in a constant vibration frequency application, it is called a *dynamic absorber*. If the auxiliary mass system employs some damping which otherwise cannot be incorporated, the term *auxiliary mass damper* is used.

> **EXAMPLE 9.17** A large two-pole synchronous motor weighing 2400 N operates at a synchronous speed of 3600 rpm. The vibration frequency of the machine is given by the expression $f = pn/120$, where p is the number of poles and n is the machine speed in rpm. Thus, the vibration frequency is 60 Hz. It has been found that this unit induces vibration into the surrounding area through its pedestal mount. The magnitude of the exciting force is approximately 44 N. Determine the parameters of the dynamic absorber which when mounted on the pedestal will reduce the vibration.
>
> **SOLUTION** Since the objective is to have the pedestal remain essentially motionless, the amplitude of the motion of the dynamic absorber should be equal and opposite to that of the exciting force. Therefore,
>
> $$F = m_2 \omega^2 x_2$$
>
> where $F =$ exciting force
>
> $m_2 =$ absorber mass
>
> $x_2 =$ amplitude of motion of m_2
>
> If we wish to limit the amplitude of the motion of the auxiliary mass to 0.1 cm, then
>
> $$44 = m_2 (2\pi 60)^2 (0.001)$$
>
> $$m_2 = \frac{44}{(0.001)(4\pi^2)(3600)}$$
>
> $$= 0.31 \text{ kg}$$
>
> The spring stiffness k_2 is then determined from the expression
>
> $$\omega^2 = \frac{k_2}{m_2}$$
>
> Hence,
>
> $$k_2 = (4\pi^2)(3600)(0.31)$$
>
> $$= 44{,}060 \text{ N/m} = 440.6 \text{ N/cm}$$

9.3.3.5 Active systems

In contrast to the passive isolators described earlier, active isolation and absorption systems normally consist of a feedback control system. The control system elements may be electromagnetic, electronic, fluidic, mechanical, pneumatic, etc., or a combination of these. The applications of these systems include precision instruments such as electron microscopes, lasers, stabilized platforms, and the like where a very high degree of isolation is required. A high degree of isolation may also be required when the air-conditioning system is mounted on the top floor of a building directly over offices.

In our previous analyses we have seen that the natural frequency is inversely proportional to the degree of isolation (i.e., the lower the natural frequency of the mounting system, for a given forcing frequency, the higher the degree of isolation). The usefulness of active systems stems from the fact that they are capable of isolating very low frequencies (below 1 Hz) with zero static deflection. In addition, they have the flexibility to provide tracking capability for vibration absorbers when the natural frequency to which the absorber is tuned may change.

In general, active vibration isolation or absorption systems are not off-the-shelf items but rather systems constructed for a particular application. They should be used only when passive isolators or absorbers cannot do the job because they are more expensive, and require power and space.

9.3.3.6 Damping

Our model for vibration studies is defined by three parameters—mass, damping, and stiffness. Of these three, damping is the most difficult parameter to deal with. When properly employed, damping performs a number of important functions: (1) it reduces transmission of vibration energy along a structure, (2) it reduces amplitudes of vibration at resonances; and (3) it reduces free vibrations or vibrations due to impact. And, of course, since vibrations generate noise, a reduction of vibration energy yields a corresponding reduction in noise. In fact, quite often the dominant component of noise in a system is generated by resonating structural surfaces. In these vibrations, controlling the noise is equivalent to reducing structural vibrations.

Basically, structural damping employs the use of high-energy dissipating mechanisms to reduce vibration levels at frequencies of resonance and has little or no effect on the response away from resonance. For instance, consider the example transmissibility curves shown in Figure 9.17. The transmissibility here is the ratio of the amplitude of the harmonic vibration at some point on the structure to the amplitude of the harmonic forcing function. The undamped structure has resonant peaks that occur at a number of frequencies in the spectrum. Damping materials can significantly reduce the vibration amplitude at these frequencies of structural resonance.

The following discussion will illustrate some of the important measures of damping and the mechanisms from which damping is derived.

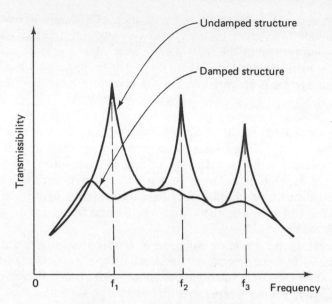

FIGURE 9.17 *Effect of structural damping*

9.3.3.6.1 Measures of Damping

One of the most important measures of damping is what is called the *system loss factor*. This term, η_s, is defined as the ratio of the damping energy loss per radian to the peak potential energy. Assuming a sinusoidal oscillation of the form

$$x(t) = X \sin (\omega t + \phi) \qquad (9.50)$$

Then

$$\dot{x}(t) = \omega X \cos (\omega t + \phi) \qquad (9.51)$$

The damping force $C_e \dot{x}(t)$ dissipates an amount of energy $C_e \dot{x}(t)\,dx$ as it moves through the position increment dx. Employing the relationship

$$C_e \dot{x}(t)\,dx = C_e \dot{x}^2(t)\,dt = \frac{1}{\omega} C_e \dot{x}^2(t)\,d(\omega t) \qquad (9.52)$$

the energy dissipated in a cycle is

$$\frac{C_e}{\omega} \int_0^{2\pi} \dot{x}^2(t)\,d(\omega t) = C_e \omega X^2 \int_0^{2\pi} \cos^2 (\omega t + \phi)\,d(\omega t) = \pi C_e \omega X^2 \qquad (9.53)$$

and hence the energy dissipated per radian is $(C_e \omega / 2) X^2$. Since the peak potential energy is $\frac{1}{2} k X^2$, the system loss factor is

$$\eta_s = \frac{(C_e/2)\omega X^2}{\frac{1}{2}kX^2}$$

$$= \frac{C_e\omega}{k} \tag{9.54}$$

Now employing equations (9.8) and (9.9), the system loss factor becomes

$$\boxed{\eta_s = 2\xi\frac{\omega}{\omega_n}} \tag{9.55}$$

At resonance the expression above reduces to

$$\boxed{\eta_s = 2\xi} \tag{9.56}$$

Note that systems with a large loss factor are highly damped.

It is interesting to note that for small values of damping,

$$\boxed{\eta_s = 2\xi = \frac{\delta}{\pi} = \frac{1}{MF_{\text{res}}} = \text{bw}} \tag{9.57}$$

where the log decrement, the magnification factor at resonance, and the bandwidth of the resonant peak are given by equations (9.20), (9.29), and (9.34), respectively.

Other terms which are important in the analysis of structural damping are complex stiffness and complex moduli. From our previous analyses we know that the steady-state magnitude of the harmonic oscillation of our vibration system model is given by the expression

$$(-\omega^2 m + j\omega C_e + k)X = F_0 \tag{9.58}$$

However, this expression can be written in the form

$$(-\omega^2 m + k^*)X = F_0 \tag{9.59}$$

where

$$k^* = k + j\omega C_e \tag{9.60}$$

Note that the imaginary stiffness is the damping.

Using equation (9.54), we can write this complex stiffness in the form

$$k^* = k(1 + j\eta_s) \tag{9.61}$$

Hence, from a vibration standpoint, a system with a mass and complex stiffness as defined by equation (9.61) is completely equivalent to the original system defined by equation (9.58).

In a manner similar to that outlined above, it can be shown that the complex Young's modulus for damping material can be written in the form

$$E^* = E(1 + j\eta_M) \tag{9.62}$$

where E = real part of the Young's modulus of the damping material

      η_M = loss factor of the damping material (η_s is the loss factor of the system structure plus damping material)

For good damping, η_M and E must be large. Other complex moduli can be defined similarly.

The loss factors and complex moduli vary with temperature and frequency. It is important to understand the manner in which these quantities vary with temperature and frequency for a particular damping material so that the proper damping treatment can be selected for a given application.

It must be remembered that in our analyses we have assumed that the damping unit was viscous damping, and thus the damping force is proportional to velocity and independent of amplitude. In many cases, however, the damping of structural material is primarily dependent upon vibration amplitude, and hence the assumption of viscous damping must always be used with caution.

9.3.3.6.2 Damping Mechanisms

Damping is the conversion of mechanical energy into thermal energy. Every technique for performing this conversion is a damping mechanism. Although there are numerous damping mechanisms, we shall confine ourselves here to those which appear to be the most practical and useful.

In addition to reducing vibration amplitudes, damping materials also reduce sound transmission. For example, if a sound wave strikes a structure as shown in Figure 9.18(a), causing the surface to vibrate, the vibrating surface then produces a reflected wave and a transmitted wave. The transmission loss through the structure at a given temperature varies with frequency in the manner illustrated in Figure 9.18(b). Note that the damping control region lies between the low-frequency region, where stiffness is the most important parameter, and the higher-frequency region, where mass is the controlling parameter. Between these two regions is one in which many natural vibration modes of the structure exist. Hence, it is only in this region, where transmission loss is very dependent upon resonance conditions, that structural damping is the controlling parameter.

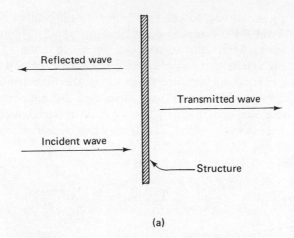

(a)

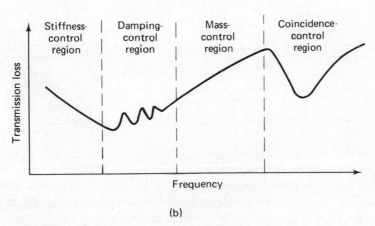

(b)

FIGURE 9.18 *Characteristics of sound transmission: (a) illustration of sound transmission; (b) transmission loss through a structure as a function of frequency*

It is extremely important that the reader pause here to correlate these facts not only with those given in Table 9.1 but also with the material on damping discussed in Chapter 8 on sound-isolation materials.

The damping property of materials also varies with temperature. The temperature range of interest is very wide and ranges from a low of approximately $-130°C$ to a high of approximately $1200°C$. In general, elastomers are most useful in the low-temperature range, plastics in the middle-temperature range, and enamels in the high-temperature range. For materials that can be represented by a complex modulus in the form of equation (9.62), the parameters E and η_M are strong functions of temperature and exhibit the behavior shown in Figure 9.19. Typically η_M varies from about 10^{-4} to about 1.0 or 2.0 and E varies from approximately $2\,\mathrm{N/cm^2}$ to about 2×10^6 $\mathrm{N/cm^2}$. It is important to note that damping materials exhibit good damping properties only in a narrow temperature range called the *transition region*. However, the transition region itself will vary from very low temperature to very high temperature, depending upon the type of material used. The mechanisms in which these various types of materials can be used are outlined below.

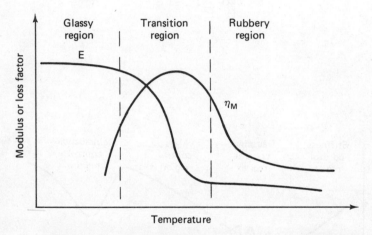

FIGURE 9.19   *Variation of E and n_M as a function of temperature*

Internal damping: To eliminate the noise emanating from a structure, we might completely replace the given structure with one made from materials that exhibit very high internal damping properties. Examples of such materials are ferromagnetic materials and certain alloys of magnesium and cobalt. However, in general, these materials are not very strong and are very costly.

Tuned dampers: A typical tuned damper is shown in Figure 9.20(a). It consists of a mass attached to a point of vibration through a spring and dashpot or a viscoelastic spring. In general, this device is not very useful because it is not economical and it is efficient only at a single frequency or in a very narrow frequency band. In addition, the device must be operated in the rubbery region of Figure 9.19, because any change in temperature will change the tuning frequency of the damper.

The following two mechanisms are based upon adding viscoelastic layers to an undamped structure. If such a structure, which consists of different layers of materials, is bent, the layers will extend or deform in shear. This

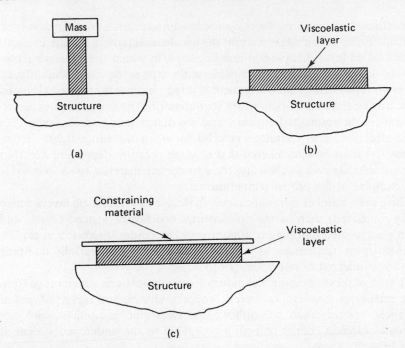

FIGURE 9.20 *Damping mechanisms: (a) tuned damper; (b) free viscoelastic layer; (c) constrained viscoelastic layer*

deformation causes energy dissipation, and this energy dissipation forms the basis for the two following mechanisms.

Free viscoelastic layer—Extensional damping: This method involves the addition of a free (i.e., uncovered) viscoelastic layer to a structure as shown in Figure 9.20(b). The ratio of the damping material thickness to the structure or wall thickness should be about 1 to 3; and, in general, one thick layer on one side of the structure is better than two small ones, one on each side of the structure. The application of this technique results in an easy-to-use, very economical, highly damped structure.

Constrained viscoelastic layer—Shear damping: This technique involves the addition of a covered (i.e., constrained) viscoelastic layer to a structure as shown in Figure 9.20(c). This arrangement, which resembles a sandwich, yields a highly damped structure. However, the resulting structure is not easy to construct and is therefore quite expensive.

To be effective, the damping must store a significant amount of the energy present in the entire system. Damping should be applied at points where stretching or bending is the greatest, because these are the points where the maximum energy storage occurs. One can achieve fairly good results through constrained layer damping with very little trouble; however, excellent results require very careful design. Incidentally, a simple example of constrained layer damping is damping tape, in which the adhesive represents the damping layer and the outside of the tape is the constraining layer.

The η_s for systems employing these layered treatments can be calculated by measuring the system bandwidth [equation (9.57)]. E can then be determined from the resonant frequency and the dimensions and characteristics of the structure. This technique is valid for η_s in the range 0.001 to 0.6. However, it must be remembered that η_s is temperature-dependent and thus one must exercise care in choosing the appropriate material so as to provide peak damping at the proper temperature.

There are a number of ramifications in the use of viscoelastic layers which can be employed, such as multiple constrained layers, spaced layers, and Lazan's corrugated damping configuration. The reader interested in exploring these other techniques as well as the properties of various damping materials is referred to references [1-6].

The use of these damping techniques spans applications which range from walls, enclosures, barriers, conveyers, hoppers, chutes, and racks to types of specialized systems such as sophisticated electronic equipment, and can provide a resonant energy reduction compared to the undamped system of up to 90 + %.

In summary, we have presented a number of control techniques for the reduction of vibrations. We must always remember, however, that our simple model does not tell the whole story. For example, we consider only the single-degree-of-freedom system. We ignore the loading effect upon the source. We assume that a foundation is absolutely rigid, and this may not always be the case at high frequencies. We ignore other things, too, and yet individuals working in the area of vibration control have found that the concepts and techniques studied through this simple model will either lead directly to a solution or at least give one a good idea about the solution of many practical problems.

9.4 STRUCTURAL SUPPORT CONSIDERATIONS [7]

In the previous sections of this chapter we have designed vibration isolation systems based upon the standard efficiency curves. The use of these design curves tacitly assumes that the foundation or structure upon which the system rests does not move. This assumption, and it is a very important one, can lead to problems. In practical installations, the structure is rarely absolutely rigid. Soil can move also, and therefore it cannot be assumed to be absolutely rigid.

To be more explicit, consider the case of a modern building with a long beam span. Mechanical equipment that vibrates at ω_0 is placed on these beams. Because the beam has a long span, it will also have a large deflection. Therefore, this structure will have a low natural frequency ω_{ns}. However, to isolate the mechanical equipment, isolators must be chosen so that the entire mechanical equipment system will have a low natural frequency ω_{ne} and, of course, the problem which invariably arises is that ω_{ne} may be very close to ω_{ns}. Therefore, one can easily see that the natural frequency of the structure upon which the vibrating equipment is placed is an important parameter in any vibration-isolation problem.

It is interesting to note that since $\omega_n = \sqrt{k/m}$, the addition of material to make the structure stiffer also adds weight, which increases the mass. An attempt to increase the stiffness of the structure is an attempt to raise its natural frequency above the operating frequency of the machinery so that the natural vibration modes of the structure will not be excited. However, there exist situations where it may not be economically feasible to accomplish this. Under these circumstances the machinery is brought up to speed as quickly as possible in an attempt to pass through the natural frequency of the structure fast so as not to excite it.

A good rule of thumb for isolator design is to design the isolation system whenever possible so that the natural frequency of the support structure is three to five times that of the isolation system. Under these circumstances the support structure appears to be a stiff member in the system and thus is not excited.

EXAMPLE 9.18 Assume that a piece of rotating machinery that rotates at 600 rpm is supported by a 6-m beam. It is often assumed that the allowable beam deflection is $\frac{1}{240}$ of the beam span and that the probable deflection is $\frac{1}{5}$ of the allowable deflection. The static deflection and natural frequency for both the allowable and probable beam deflections are as follows:

Allowable deflection	*Probable deflection*
$\delta_{st} = \dfrac{600 \text{ cm}}{240}$	$\delta_{st} = \dfrac{600 \text{ cm}}{5 \times 240}$
$\delta_{st} = 2.5 \text{ cm}$	$\delta_{st} = 0.5 \text{ cm}$
$f_n = \dfrac{4.98}{\sqrt{\delta_{st}}}$	$f_n = \dfrac{4.98}{\sqrt{\delta_{st}}}$
$f_n = 3.15 \text{ Hz}$	$f_n = 7.04 \text{ Hz}$

Since the forcing frequency is 10 Hz, if it is assumed that the natural frequency of the beam is 7.04 Hz, then an isolator can be chosen so as to produce an isolation system with a natural frequency of approximately 2.4 Hz. Note that the natural frequency of the beam is then approximately three times that of the isolator, and the ratio of the forcing frequency to the natural frequency of the isolator has been chosen to provide good isolation. However, why wasn't the natural frequency of the beam selected as 3.15 Hz?

The answer to this question is simply one of economics. In general, it is extremely expensive to base the design upon the allowable deflection. Since there is normally a large safety factor employed in building design, practicing engineers in the field have found that designs based upon probable deflections yield good results. Note, however, that this problem becomes even more acute as the beam span increases.

The performance of machinery may also provide an indication of structural support problems. In this regard consider the following example.

EXAMPLE 9.19 A large industrial fan driven by a 200-hp motor was mounted approximately 20 m in the air on a steel support structure. The fan was used in conjunction with an electrostatic precipitator at a paper mill installation as shown in Figure 9.21(a). The fan was balanced by the

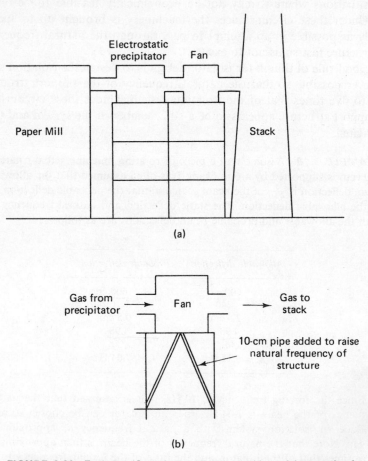

(a)

(b)

FIGURE 9.21 *Technique for reducing vibration problems: (a) example of a fan used at a paper mill; (b) method of adding stiffness to the support to reduce vibration problem*

manufacturers at their plant but the fan was not and could not be balanced in its permanant location on the support structure. A very detailed structural analysis indicated that the operating frequency of the fan was close to a resonant frequency of the structure. The problem was solved by adding two 10-cm pipes to the structure, as shown in Figure 9.21(b). These two support pipes added additional stiffness to the structure, which in turn increased its natural frequency so that the structure was not excited by the fan. Under these conditions the fan operated in a balanced mode.

9.5 CRITICAL SHAFT SPEEDS [7]

Although this subject is covered in more detail in the following chapter, it is informative at this point to briefly illustrate that the relationship between the static deflection and the vibration frequency is an important one which is very useful in the design of rotating machinery. For example, fans that have shafts mounted between two bearings should be designed so that the shaft deflection will yield a critical vibration frequency that does not coincide with the operating frequency of the system. The shaft will actually "jump rope" at this critical vibration frequency, and hence the system should always be designed to operate at a point removed from this frequency. In addition, one must consider the natural frequency of the supporting structure.

Typical design rules often applied are to design the fan shaft so that its natural frequency is 30–40% above the maximum forcing frequency and to design the support steel to have a natural frequency of $\frac{1}{2}$ to $\frac{1}{3}$ of the operating frequency. The 30% figure is used when the fan is motor driven and the 40% figure is used if the fan is turbine driven, because a turbine can overspeed. It is desirable to pass through the natural frequency of the support steel as quickly as possible. It is simply not economical to design the support steel with a natural frequency above the operating speed. On the other hand, the natural frequency of the support steel should be far away from the operating speed but not so small that the beam deflections would be very large. Thus, the range of $\frac{1}{2}$ to $\frac{1}{3}$ of the operating frequency represents this compromise.

EXAMPLE 9.20 A large industrial fan is to be turbine-driven and operate at 900 rpm. Determine the critical resonant frequency of the shaft, its desired static deflection, and the required resonant frequency of the support steel.

SOLUTION The critical resonant frequency of the shaft should be

$$f_n = \frac{(1.4)(900)}{60}$$

$$= 21 \text{ Hz}$$

Thus, the static deflection should be [from equation (9.44a)]

$$\delta_{st} = \frac{(4.98)^2}{(21)^2}$$

$$= 0.056 \text{ cm}$$

The resonant frequency of the support steel should be in the range 7 to 10 Hz.

EXAMPLE 9.21   A fan driven by a 200-hp motor is to operate at 880 rpm. The motor and fan weigh 22,240 N. This system is to be mounted overhead on a 20-ft or 6.1-m steel beam.

Suppose that a design engineer who wants to be sure that he has more than enough support decides to select from the *Manual of Steel Construction* a W1049 (10-in.-wide flange, 49 lb/ft) steel beam with a static deflection of only 0.05 in. (0.127 cm). This is indeed a strong beam for this application. Is the design a good one?

SOLUTION   The beam certainly appears to be strong enough; however, let us see if there might be problems due to vibration. The resonant frequency of the beam is

$$f_n = \frac{4.98}{\sqrt{0.127}}$$

$$= 13.97 \text{ Hz}$$

A quick calculation shows that the forcing frequency is 14.67 Hz. This indicates that the operating frequency of the system is almost exactly the same as the natural frequency of the supporting beam, and thus the design is a poor one. A better choice of beam would be a W820 (8 in.-wide flange, 20 lb/ft) with a static deflection of 0.2 in. (0.508 cm). The natural frequency of this beam is 6.99 Hz, which is between $\frac{1}{3}$ and $\frac{1}{2}$ of the forcing frequency.

This example illustrates that although from a structural standpoint one might be permitted to overestimate loads for safety considerations, when dealing with vibrations, additional calculations are required.

Chapter 10 is devoted to vibration analysis and dynamic balancing in rotating machinery and so we shall not pursue this topic further here.

9.6 VIBRATION IN DUCTS [7]

Another area in industrial environments where vibration problems may be encountered is in duct work. Generally speaking, vibration problems in ducts can be solved by one or more of the following techniques:

1. Employ isolators and expansion joints between the vibrating machinery and the duct.

2. Reinforce (i.e., stiffen) the duct to raise its natural frequency above that of the vibration forcing frequency.
3. Alter the flow within the duct to eliminate flow-induced vibrations.

The following example illustrates one of the problems encountered when employing ducts.

EXAMPLE 9.22 At a particular industrial plant, the fan and duct system shown in Figure 9.22 was employed. The fan was driven by a 200-hp motor and the duct dimensions were approximately 1.5 m × 2.5 m. The plant manager became concerned when he noticed that near the top of the duct, the vibrations were so severe that they were causing cracks in the duct welding. The problem was compounded by the fact that the location of the problem was approximately 60 m in the air. The consulting engineer who examined the system suspected that the vibration problems were caused by high-velocity turbulence of gases in the duct. Hence, an experiment was designed to test this theory. The model shown in Figure 9.23 was built to simulate the actual system.

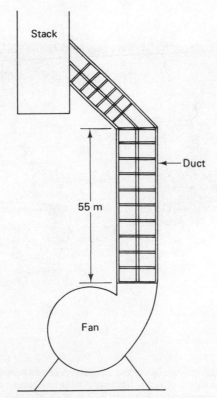

FIGURE 9.22 *Schematic illustration for Example 9.22*

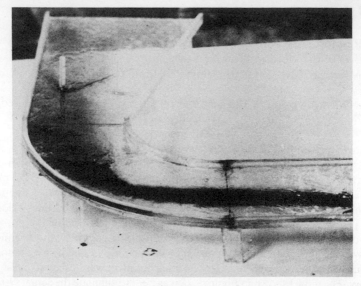

(a)

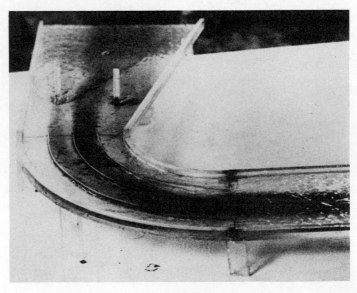

(b)

FIGURE 9.23 *Simulation of the vibration problem described in Example 9.22: (a) simulation of actual flow conditions in Example 9.22; (b) alteration of flow conditions in Example 9.22 with turning vanes (compliments of Robert E. Perry, Barron/A.S.E., Inc., Leeds, Alabama)*

Comparable dimensions and Reynolds numbers were chosen for the experiment. Figure 9.23(a) models the actual duct system shown in Figure 9.22. Note the turbulence and unsymmetric velocity profile in the duct. In an attempt to correct the situation, the use of turning vanes was simulated, as shown in Figure 9.23(b). Note the symmetric steady flow that results from the use of the vanes. As a result of the experiment, turning vanes were installed in the actual system to correct the vibration problem.

The previous example illustrates that significant vibration problems can be caused by such things as vortex shedding in the duct work due to unsymmetric velocity profiles in the cross section of the duct, or exceptional high-velocity turbulence, which acts like a hammer and pounds against the duct.

9.7 VIBRATION MEASUREMENTS

Measurements should be made to produce the data needed to draw meaningful conclusions from the system under test. These data can be used to minimize or eliminate the vibration and thus the resultant noise. There are also examples where the noise is not the controlling parameter, but rather the quality of the product produced by the system. For example, in process control equipment, excessive vibration can damage the product, limit processing speeds, or even cause catastrophic machine failure.

9.7.1 Measurement Quantities

In general, vibrations may be either deterministic or random. If they are deterministic, they may be periodic or nonperiodic, and if they are random, they may be stationary or nonstationary. For simplicity we shall only briefly discuss some of the important aspects of these signals.

The most elementary form of vibration is that described by simple harmonic motion. It has already been shown that this deterministic, periodic motion can be represented mathematically by a sine wave as shown in Figure 9.24(a). The period of the wave, T, is shown in the figure, and since the period is the reciprocal of the frequency (i.e., $T = 1/f$), it is obvious that the vibration described by the wave in Figure 9.24(b) is of higher frequency than the vibration described by Figure 9.24(a).

The displacement of the vibration in Figure 9.24 is described by the equation

$$x(t) = X \sin \omega t \qquad (9.63)$$

where X = amplitude

ω = frequency of the signal $x(t)$

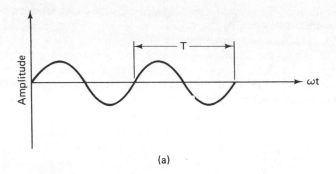

(a)

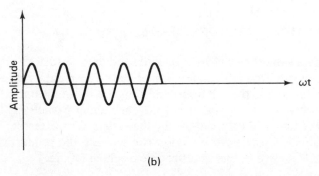

(b)

FIGURE 9.24 *Simple harmonic vibration motion: (a) periodic motion represented by a sine wave; (b) periodic motion represented by a sine wave of higher frequency than in (a)*

Then the velocity is

$$\dot{x}(t) = \omega X \cos \omega t$$
$$= \omega X \sin (\omega t + \pi/2) \tag{9.64}$$

and the acceleration is

$$\ddot{x}(t) = -\omega^2 X \sin \omega t$$
$$= \omega^2 X \sin (\omega t + \pi) \tag{9.65}$$

The importance of these relationships stems from the fact that if the vibration pickup in the measurement system is an accelerometer, the device produces a signal proportional to the vibration acceleration. From this measurement of acceleration the velocity and displacement can be obtained by integrating the acceleration signal, and this integration can be performed electronically.

The magnitude of harmonic vibration signals, such as those shown in Figure 9.24, can be described by a number of quantities, such as the peak, rms (root mean square), and average absolute value. Since $x(t)$ is a sine wave, all three of these quantities are directly related mathematically. However, the quantity of most importance is the rms value. This value is obtained from the expression

$$X_{rms} = \left[\frac{1}{T} \int_0^T x^2(t)\, dt \right]^{1/2} \tag{9.66}$$

Substituting equation (9.63) into (9.66) yields

$$X_{rms} = \frac{X}{\sqrt{2}} \tag{9.67}$$

This rms value is a very useful quantity because it is a useful description of a function, but, more important, because of its direct relationship to vibration energy. For example, note that in the vibration system model, both the energy dissipated per radian and the peak potential energy are directly related to $X_{rms}^2 = X^2/2$, as shown in equation (9.54). In fact, it can be shown that the overall rms value of a vibration signal is a meaningful quantity whether the signal is periodic or random in nature [8].

If the vibrations are periodic but nonharmonic and satisfy the Dirichlet conditions [9], then we can represent the signal by a Fourier series expansion of the form

$$x(t) = X_0 + \sum_{i=1}^{\infty} X_i \sin (i\omega t + \phi_i) \tag{9.68}$$

For example, suppose that a frequency analysis of a vibration signal measured by an accelerometer can be represented by the function

$$x(t) = X_1 \sin \omega_1 t + X_3 \sin 3\omega_1 t + X_5 \sin 5\omega_1 t$$

Then the frequency spectrum of the vibration is shown in Figure 9.25 and can be obtained by taking the Fourier transform of the function $x(t)$. Note that the frequency spectra of the signal consists of discrete lines. This result, which is true in general for the nonharmonic periodic functions, is in sharp contrast to the random vibrations, which have a continuous frequency spectra.

If the vibration is random in nature, the waveform that describes the vibration might look something like that shown in Figure 9.26(a). If the statistical characteristics (i.e., mean value, variance, etc.) do not change with time, the random vibration is said to be stationary. If the statistical properties of the vibration are time-dependent, the vibration is nonstationary.

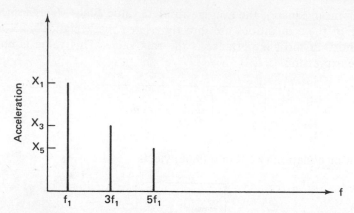

FIGURE 9.25 *Frequency spectrum of a periodic nonharmonic vibration signal*

Stationary vibration of an ergodic process can be described by the time function, called the *time-average autocorrelation function*, which is defined as

$$\psi(\tau) = \lim_{T \to \infty} \frac{1}{T} \int_0^T f(t) f(t + \tau)\, dt \tag{9.69}$$

where $f(t)$ = magnitude of the vibration signal at time t

$f(t + \tau)$ = magnitude of the signal at time $t + \tau$

Thus, $\psi(\tau)$ is obtained by displacing the vibration signal by τ to obtain $f(t + \tau)$, multiplying the signal $f(t + \tau)$ by $f(t)$, and then taking the average in the limit as the interval T approaches infinity. This quantity tells us how the instantaneous value of the signal depends upon values nearby in time. In a manner analogous to that for periodic vibrations, we can calculate the power spectral density (power spectrum) of this random vibration by taking the Fourier transform of the time-average autocorrelation function. Suppose that the power spectrum for this function in Figure 9.26(a) is shown in Figure 9.26(b). Then this figure indicates that the vibration energy all lies in the frequency range from $f = 0$ to $f = f_0$. This information is very important for the design and specification of control measures.

9.7.2   Measurement Systems

The basic measurement system used for diagnostic analyses of vibrations consists of the three system components shown in Figure 9.27. Since many of the components employed in this system are also used in noise measurements described in Chapter 4, the discussion here will be brief and confined

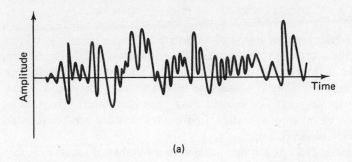

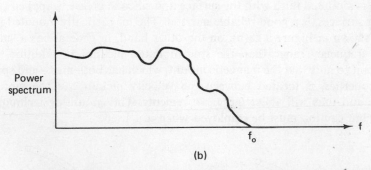

FIGURE 9.26   (*a*) *Random vibration signal; (b) its associated power spectrum*

FIGURE 9.27   *Basic vibration measurement system*

primarily to those components of the vibration measurement system which differ from those used in the measurement of noise.

9.7.2.1 Transducers

In general, the transducers employed in vibration analyses convert mechanical energy into electrical energy; that is, they produce an electrical signal which is a function of mechanical vibration. In the following section, both velocity pickups and accelerometers mounted or attached to the vibrating surface will be studied. Noncontact vibration pickups, used primarily in vibration analyses of rotating machinery, will be discussed in Chapter 10.

9.7.2.1.1 VELOCITY PICKUPS

The electrical output signal of a velocity pickup is proportional to the velocity of the vibrating mechanism. Since the velocity of a vibrating mechanism is cyclic in nature, the sensitivity of the pickup is expressed in peak millivolts/cm/s and thus is a measure of the voltage produced at the point of maximum velocity. The devices have very low natural frequencies and are designed to measure vibration frequencies that are greater than the natural frequency of the pickup.

Velocity pickups can be mounted in a number of ways; for example, they can be stud-mounted or held magnetically to the vibrating surface. However, the mounting technique can vastly affect the pickup's performance. For example, the stud-mounting technique shown in Figure 9.28(a), in which the pickup is mounted flush with the surface and silicone grease is applied to the contact surfaces, is a good reliable method. The magnetically mounted pickup, as shown in Figure 9.28(b), on the other hand, in general has a smaller usable frequency range than the stud-mounted pickup. In addition, it is important to note that the magnetic mount, which has both mass- and spring-like properties, is located between the velocity pickup and the vibrating surface and thus will affect the measurements. This mounting technique is viable, but caution must be employed when it is used.

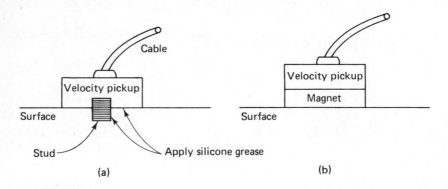

FIGURE 9.28 *Two transducer mounting techniques: (a) stud-mounted pickup; (b) magnetically held velocity pickup*

The velocity pickup is a useful transducer because it is sensitive and yet rugged enough to withstand extreme industrial environments. In addition, velocity is perhaps the most frequently employed measure of vibration severity. However, the device is relatively large and bulky, is adversely affected by magnetic fields generated by large ac machines or ac current-carrying cables, and has somewhat limited amplitude and frequency characteristics.

9.7.2.1.2 ACCELEROMETERS

The accelerometer generates an output signal that is proportional to the acceleration of the vibrating mechanism. This device is, perhaps, preferred over the velocity pickup, for a number of reasons. For example, accelerometers have good sensitivity characteristics and a wide useful frequency range; they are small in size and light in weight and thus are capable of measuring the vibration at a specific point without, in general, loading the vibrating structure. In addition, the devices can be used easily with electronic integrating networks to obtain a voltage proportional to velocity or displacement. However, the accelerometer mounting, the interconnection cable, and the instrumentation connections are critical factors in measurements employing an accelerometer. The general comments made earlier concerning the mounting of a velocity pickup also apply to accelerometers.

Some additional suggestions for eliminating measurement errors when employing accelerometers for vibration measurements are shown in Figure 9.29. Note that the accelerometer mounting employs an isolation stud and

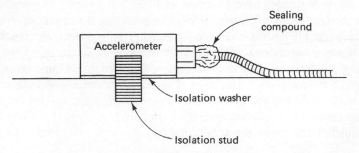

FIGURE 9.29 *Mounting technique for eliminating selected measurement errors*

an isolation washer. This is done so that the measurement system can be grounded at only one point, preferably at the analyzer. An additional ground at the accelerometer will provide a closed (ground) loop which may induce a noise signal that affects the accelerometer output. The sealing compound applied at the cable entry into the accelerometer protects the system from errors caused by moisture.

The cable itself should be glued or strapped to the vibrating mechanism immediately upon leaving the accelerometer, and the other end of the cable, which is connected to the preamplifier, should leave the mechanism under test at a point of minimum vibration. This procedure will eliminate or at least minimize cable noise caused by dynamic bending, compression, or tension in the cable.

Vibration measurements can be affected by background noise. In fact, this can be a significant problem in some industrial plants. A quick check of the relative magnitude of the problem can be obtained by mounting an accelerometer on a nonvibrating surface in the immediate area of the vibrating mechanism and measuring the vibration induced by the background noise. The induced vibrations caused by the background noise should be less than one-third of the actual vibration obtained on the mechanism under test in order to obtain reasonably accurate results. The reader is encouraged to compare this philosophy with that of noise measurements made with background noise present.

The piezoelectric accelerometers are perhaps the most widely used devices. They have a wide operating temperature range, nominally $-100°C$ to $+250°C$, and an output that typically ranges from 1 mV/g to 30 V/g, where g is the acceleration due to gravity. In contrast to the velocity pickup, the devices are designed to function accurately below their natural frequency. The high-frequency limit is typically one-fourth of the accelerometer's resonant frequency, and the low-frequency limit is usually determined by the interconnecting devices. For example, a piezoelectric accelerometer with a 40-kHz resonant frequency will have a usable frequency range that extends from the low-frequency limit to about 10 kHz.

The acceleration is a function of displacement and the square of frequency (i.e., the units are cm/s^2), thus the accelerometer is extremely sensitive to high-frequency vibrations. Therefore, extreme care must be exercised in mounting this transducer. The manufacturer's instructions must be carefully followed to ensure proper operation. The manufacturer's literature indicates that these transducers are available in a wide range of sensitivity, frequency, environmental, and mounting characteristics, and in general the smaller the accelerometer, the lower the sensitivity and the wider the useful frequency range.

There is more than one sensitivity of interest when dealing with accelerometers. The charge sensitivity, measured in picocoulombs/g, and voltage sensitivity, measured in millivolts/g, are important depending upon whether the device is used with charge-measuring or voltage-measuring electronics. In addition, the transverse sensitivity, which is the sensitivity of the device to acceleration in a plane that is perpendicular to the main accelerometer axis, should be less than 3% of the main axis sensitivity at low frequencies.

Although accelerometers are manufactured for useful operation in an extremely wide variety of applications, one must be careful to select a device that not only has the proper electronic characteristics, but the proper environmental characteristics as well.

9.7.2.2 Preamplifiers

The second element in the vibration measurement system is the preamplifier. This device, which may consist of one or more stages, serves two very useful purposes: it amplifies the vibration pickup signal, which is in general very weak, and it acts as an impedance transformer or isolation device between the vibration pickup and the processing and display equipment.

Recall that the manufacturer provides both charge and voltage sensitivities for accelerometers. Likewise, the preamplifier may be designed as a voltage amplifier in which the output voltage is proportional to the input voltage, or a charge amplifier in which the output voltage is proportional to the input charge. The difference between these two types of preamplifiers is important for a number of reasons. For example, changes in cable length (i.e., cable capacitance) between the accelerometer and preamplifier are negligible when a charge amplifier is employed. When a voltage amplifier is used however, the system is very sensitive to changes in cable capacitance. In addition, because the input resistance of a voltage amplifier cannot in general be neglected, the very low frequency response of the system may be affected. Voltage amplifiers, on the other hand, are often less expensive and more reliable because they contain fewer components and thus are easier to construct [8].

Manufacturers supply a variety of preamplifiers suitable for normal vibration applications. These devices possess the necessary operating characteristics, such as sensitivity, frequency response, linearity, and phase distortion, and thus are suitable for interfacing the accelerometer with various types of analyzers employed in processing and displaying the vibration data.

9.7.2.3 Processing and display equipment

The instruments used for the processing and display of vibration data are, with minor modifications, the same as those described earlier for noise analyses. The processing equipment is typically some type of spectrum analyzer. The analyzer may range from a very simple device which yields, for example, the rms value of the vibration displacement, to one that yields an essentially instantaneous analysis of the entire vibration frequency spectrum. As discussed earlier, these analyzers, which are perhaps the most valuable tool in a vibration study, are typically either a constant-bandwidth or constant-percentage-bandwidth type of device. They normally come equipped with some form of graphical display, such as a cathode ray tube, which provides detailed frequency data.

If the signal is random, the frequency analyzer can be used to generate the power spectrum of the vibration signal, which indicates the distribution of vibration energy throughout the frequency spectrum. The reader interested in the many ramifications of this approach is referred to the publications [8,10] and the references contained therein.

The type of analyzer that should be employed in a vibration analysis depends upon two things: first, the form of the vibration signal (i.e., is it periodic, random, or does it contain very sharp resonant peaks that will go undetected without the use of a very narrow bandwidth analyzer?), and second, the purpose of generating the data (i.e., a cursory analysis used to design a test or a detailed analysis used for failure prediction and machine design).

A tape recorder is also an important vibration measurement instrument. Much of its importance stems from the fact that it is often indispensable in industrial environments where, for any number of reasons, other sophisticated equipment cannot be used. The data taken are then used for subsequent laboratory analysis.

Finally, it is informative to examine briefly the overall linearity of the measurement system and the factors that control its response.

9.7.2.4 Overall measurement system performance

Ideally, we would like to have a measurement system that has no phase shift and a relative amplitude response characteristic that is flat with frequency, as shown in Figure 9.30(a). If we could build a system with these characteristics, the system would treat every vibration frequency in exactly the same manner, and thus there would be no amplitude or phase distortion. However, a system with these characteristics cannot be constructed with physical components. The typical characteristics for a physically realizable measurement system are shown in Figure 9.30(b). The distortion at low frequencies is typically caused by the preamplifier and its input cable, and the distortion at high frequencies is normally caused by the accelerometer. To minimize the distortion, the measurement system should be operated in the range shown in Figure 9.30(b) between ω_1 and ω_2. In this range the relative amplitude response is constant; there will be no amplitude distortion. The phase characteristic in this region is linear, and therefore all frequency components of the vibration signal will maintain the same relative position in time, producing no phase distortion. The latter point can be explained by means of an example. Consider a vibration signal consisting of a fundamental $\sin(\omega_0 t)$ and its third harmonic $\sin(3\omega_0 t)$. The signal is of the form

$$x(t) = x_1 \sin(\omega_0 t) + x_3 \sin(3\omega_0 t)$$

A plot of the fundamental and third harmonic signals is shown in Figure 9.31. Comparing Figure 9.31 with Figure 9.32 illustrates that if the phase shift of the system is linear, a 30° phase shift of the fundamental corresponds to a 90° phase shift of the third harmonic.

The phase for the fundamental signal is $\theta_1 = \omega_0 t$ and the phase for the third harmonic is $\theta_3 = 3\omega_0 t$. Since the time delay of a signal of the form

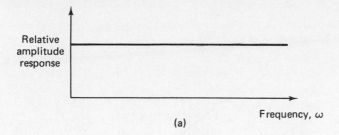

(a)

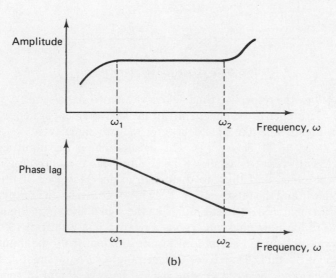

(b)

FIGURE 9.30 *Measurement systems characteristics: (a) ideal amplitude response of a measurement system; (b) physically realizable amplitude and phase response of a measurement system*

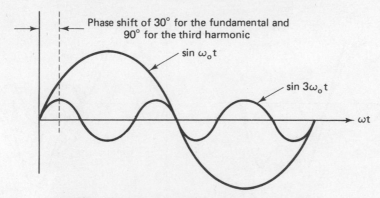

FIGURE 9.31 *Plot of the functions sin $\omega_0 t$ and sin $3\omega_0 t$*

331

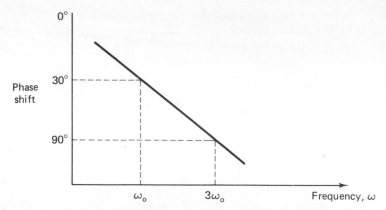

FIGURE 9.32 *Linear-phase characteristic*

$\sin \omega t$ is $t = \theta/\omega$, the respective time delays of the two signals are $t_{d_1} = \theta_1/\omega_0$ and $t_{d_3} = \theta_3/3\omega_0$. Then if the phase shift of the measurement system is linear, as shown in Figure 9.32, the time delays for the two signals will be identical (i.e., $t_{d_1} = 30°/\omega_0 = 90°/3\omega_0 = t_{d_3}$). This means that if the input signal to the measurement system is as shown in Figure 9.31, the output signal will be exactly the same, and thus no distortion exists. However, if the phase shift is not linear and the relative time delay is not constant, the output signal of the measurement system could look like the signal shown in Figure 9.33, where the third harmonic has undergone a relative phase shift of 180°. Note that in this case the output signal looks quite different from the input signal due to phase distortion. Therefore, the measurement equipment should be operated in the range between ω_1 and ω_2 in Figure 9.30 and in general the frequency range of interest should be 10 times the low-frequency limit of the measurement system and one-tenth the high-frequency limit of the system.

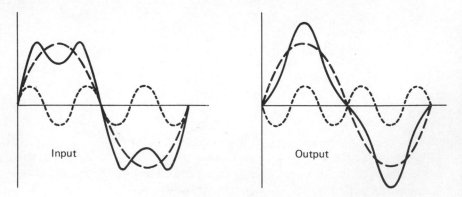

FIGURE 9.33 *Effect of phase distortion*

9.7.2.5 Mechanical impedance and mobility

Perhaps the simplest manner in which to view these two concepts is by means of an analogy with electrical circuits. They are defined as

$$Z = \frac{f}{v} \tag{9.70a}$$

and

$$M = \frac{v}{f} \tag{9.70b}$$

where $f =$ force

       $v =$ velocity

       $Z =$ mechanical impedance

       $M =$ mechanical mobility

The preceding impedance and mobility equations relate the vibrating motion to the force that causes it. In general, the impedance and mobility are functions of both time and space and are actually tensors. However, for simplicity, we shall neglect the space relationships in this discussion.

Although the two analogies are equally valid, the mobility analogy is the most viable. The analogous parameters for the mobility type analog–mechanical and electrical circuits are given in Table 9.2.

TABLE 9.2

Mobility analogs

Mechanical circuit	*Electrical circuit*
Force, f	Current, i
Mass, m	Capacitance, C
Damping coefficient, C_e	Reciprocal of resistance, $\frac{1}{R}$
Spring constant, k	Reciprocal of inductance, $\frac{1}{L}$
Velocity, v	Voltage, e

The two circuits and the equations governing their performance are shown in Figure 9.34. Note the similarities between the two circuits. Since the mass in a mechanical circuit is always referenced to inertial space (i.e., grounded at one end), in effect the two circuits are identical in form. In addition, force is analogous to flow in that whatever force flows into a nodal point, flows

333

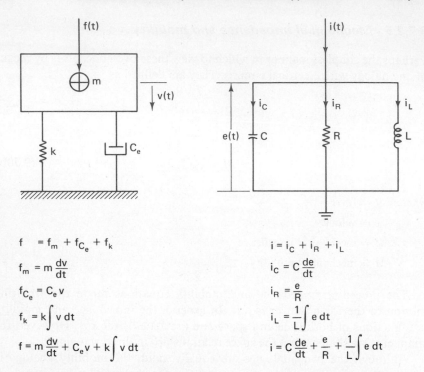

$$f = f_m + f_{C_e} + f_k$$

$$f_m = m\frac{dv}{dt}$$

$$f_{C_e} = C_e v$$

$$f_k = k\int v\,dt$$

$$f = m\frac{dv}{dt} + C_e v + k\int v\,dt$$

$$i = i_C + i_R + i_L$$

$$i_C = C\frac{de}{dt}$$

$$i_R = \frac{e}{R}$$

$$i_L = \frac{1}{L}\int e\,dt$$

$$i = C\frac{de}{dt} + \frac{e}{R} + \frac{1}{L}\int e\,dt$$

FIGURE 9.34 *Mobility-type analogs for mechanical and electrical circuits*

out of it. Similarly, in the electrical circuit, the current that flows into a node is equal to the currents leaving the node. The node points referred to here are simply junctions, such as the machine case at the foundation junction for typical problems in this chapter and a shaft at the bearing-oil film junction for vibration problems of the type considered in Chapter 10. The same type of relationships as those described above exist for the impedance analogy; however, in that case confusion often arises with respect to the system's reference points, and thus that analogy is not considered further here.

The discussion above and the relationships shown in Figure 9.34 are extremely important, not only because the equations used here to describe the vibratory motion are identical to those used throughout this chapter, but in addition it will be shown in Chapter 10 that these equations also describe the vibrations of a rotating shaft.

Let us now examine the importance of mechanical mobility in a vibration analysis. Basically, its importance stems from the fact that it provides a mechanism for estimating (1) the system's overall performance based upon data on the system's components, and (2) the relationship between force and motion within the system. Anyone familiar with electric circuit analysis knows that if the circuit is composed of lumped parameters (i.e., resistors, capacitors, and inductors), it is straightforward to determine the relationships between voltage and current at any point in the circuit. A mechanical circuit is, in general, a distributed parameter system (e.g., a transmission line in electrical systems), and thus the relationship between the system parameters is often quite complex.

The device employed in vibration measurements to determine mechanical point mobilities is called the *impedance head*. A vibration exciter is connected via the impedance head to the mechanical system under test. The impedance head is a dual transducer which measures both the force applied and the resulting acceleration. These two signals can then be used with the proper electronics to determine, for example, the mobility as a function of frequency. For a more complete discussion of some of the ramifications of testing procedures with mechanical impedance methods, the reader is referred to reference [11].

In general, instrumentation is available for essentially any type of vibration analysis. The engineer must be guided in his equipment selection by such things as cost, accuracy and quantity of data, and measurement environment.

9.8 MEASUREMENT OF STRUCTURAL DYNAMICS [12,13,14,15,16]

An important segment of the area of vibration analyses is knowledge of the behavioral mechanisms of structures and their components. Typically, the structure's dynamic characteristics are defined by a transfer function. This cause/effect relationship describes behavior as a function of frequency between two points on a structure. In addition, if we can measure the transfer characteristics of the structure, then the structural dynamics are known and the defining properties of a mode of vibration can be obtained.

Six useful transfer functions are employed in the analysis of structures: compliance (displacement/force), mobility (velocity/force), inertance (acceleration/force), and their reciprocals, all of which are algebraically interrelated. These transfer functions express magnitude and phase as a function of frequency and may be a driving-point or a spatial relationship. The driving-point transfer function expresses the relationship between input and output variables at a single point. The spatial transfer function defines an interrelationship between the variables at two points on the structure (e.g., the displacement at one point caused by the force at another). Both magnitude and phase data are necessary for obtaining modal information; amplitude data alone are not sufficient.

Many practical structures can be characterized by means of linear mathematics. In fact, a multi-degree-of-freedom system can be ideally represented by a series of coupled single-degree-of-freedom systems. The structural dynamics model that will be employed here is the single-degree-of-freedom model used throughout this chapter. The properties of this model were analyzed in detail earlier. If a structure that can be adequately described by this model is excited with a known force and the corresponding motion simultaneously measured, then the resulting transfer function can be used to determine the characteristics of the structure (i.e., inertia, damping, stiffness and resonant frequencies).

Modal analysis determines the dynamic properties of a structure by identifying its modes of vibration. A mode of vibration is a global property of the structure and thus can be excited from any point, with the exception of node points, at which the structure cannot be excited at all. Each mode of vibration that can be found from the transfer functions has a specific natural frequency and damping factor. The resonance associated with each mode is independent of the spatial location and is characterized by a unique distribution of deformation or mode shape across the structure.

If we assume the single-degree-of-freedom model, then the compliance transfer function, for example, is

$$\frac{X(s)}{F(s)} = \frac{1}{ms^2 + C_e s + k} = \frac{a}{s - p_k} + \frac{a^*}{s - p_k^*} \qquad (9.71)$$

where $*$ = complex conjugate

   p_k = pole location in the s plane, defined as $p_k = \sigma_k + j\omega_k$

In general, each complex-conjugate pair of poles corresponds to a mode of vibration. The mode described by equation (9.71) is shown diagramatically in Figure 9.35. Note that there are four important properties of any mode: natural frequency, damping, and the magnitude and phase of the residue. Each mode, which corresponds to a definite deflection configuration such as bending, torsion, etc., has a certain mode shape and is associated with a specific natural frequency. The number of natural frequencies is equal to the number of degrees of freedom of the system.

The residue, which is proportional to the amplitude of the resonance in each transfer function, defines the mode shape. If the phase angle of the residue is zero, or 180°, the mode is said to be a real or normal mode, which is characterized by the fact that all points achieve maximum or minimum deflection at the same instant of time (i.e., all points are either in phase or 180° out of phase). If the mode is complex, then phase angles other than 0° and 180° are possible. This means that node lines are fixed or stationary for normal modes and traveling or nonstationary for complex modes.

The importance of a modal analysis of a mechanical structure stems from the fact that the analysis identifies the dynamic properties of the structure which control its behavior. This knowledge, obtained through the analysis, can be used to predict or control performance.

Although we shall not develop the problem here in this fashion, the reader

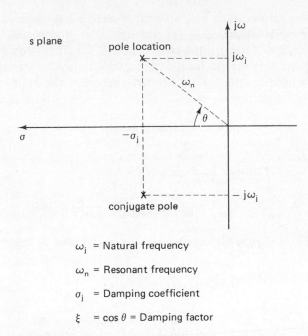

ω_j = Natural frequency

ω_n = Resonant frequency

σ_j = Damping coefficient

ξ = $\cos \theta$ = Damping factor

FIGURE 9.35 *s-Plane representation of a vibration mode*

should realize that a vibration analysis of a structural system is a form of eigenvalue problem of matrix algebra in which the eigenvalues are related to the frequencies of the modal resonances and the eigenvectors to the mode shapes.

The measurement of the transfer function between the point of excitation and various points of interest on the structure may be in magnitude and phase form, in Nyquist plot form, or in the rectangular, (i.e., co-quad) form. The spatial points of interest on the structure should be chosen to ensure that all the modes of vibration are included. In other words, if the points selected are not sufficiently close, some modes may not be detected. The testing methods normally employed for transfer-function measurement are sinusoidal, transient, and random testing. Swept-sine testing was essentially the only technique used for the measurement of transfer functions before the advent of the Fourier analyzer. With this method the energy input at a particular frequency can be high and the excitation force can be accurately controlled. However, this approach has limited low-frequency range, long data acquisition time, difficulty in maintaining accuracy, and problems with nonlinearities and distortion. Transient testing, although easy to implement, typically has limited energy density, frequency resolution, control of spectrum, and it can cause distortion. Random testing, on the other hand, has high-energy density and good control of spectral content, and distortion can be removed through averaging.

The testing methods can be employed to derive a magnitude and phase vs. frequency plot (Bode plot) for the structure under test. From this frequency-response data, the transfer function for the structure (i.e., its structural dynamics) can be identified. In order to illustrate some of the basics of transfer-function identification, consider the following simple example.

EXAMPLE 9.23 Suppose that a simple structure has been tested and that its compliance frequency-response function is shown in Figure 9.36. The form of the magnitude and phase curves indicates that the system is second-order, and therefore the transfer function is of the form

$$\frac{X(j\omega)}{F(j\omega)} = \frac{1}{m(j\omega)^2 + C_e(j\omega) + k}$$

The system parameters for the structure can be identified as follows:

$$\frac{X(j\omega)}{F(j\omega)} = \frac{1}{k} \quad \text{as } \omega \longrightarrow 0$$

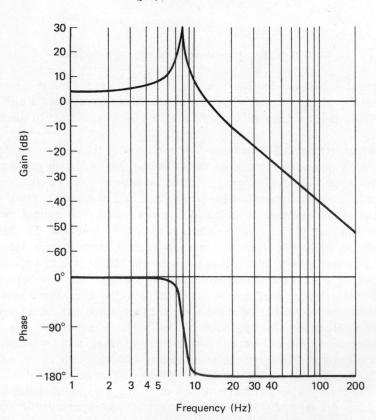

FIGURE 9.36 *Compliance frequency-response function*

Therefore,

$$20 \log_{10} \left(\frac{1}{k} \right) = 4 \text{ dB}$$

$$k = 0.631 \text{ N/m}$$

$$\frac{X(j\omega)}{F(j\omega)} = \frac{-1}{m\omega^2} \quad \text{as } \omega \longrightarrow \infty$$

As an example, let us choose the point at $\omega = 200$ rad/s. Therefore,

$$20 \log_{10} \left[\frac{1}{m(200)^2} \right] = -52 \text{ dB}$$

$$m = 0.0099 \text{ kg}$$

$$\frac{X(j\omega)}{F(j\omega)} = \frac{1}{j\omega C_e} \quad \text{at } \omega = \omega_n = 8 \text{ rad/s}$$

Therefore,

$$20 \log_{10} \left(\frac{1}{8c_e} \right) = 28 \text{ dB}$$

$$C_e = 0.0049 \text{ N-s/m}$$

Note that these parameters completely define the compliance transfer function. In addition, recall from equation (9.35) that

$$Q = \frac{1}{2\xi} = \frac{f_n}{\Delta f}$$

where f_n = natural frequency

Δf = bandwidth (i.e., the distance in frequency between the half-power points)

From Figure 9.36, note that

$$\frac{1}{2\xi} = \frac{f_n}{\Delta f} = \frac{\omega_n}{\Delta \omega} = \frac{8}{8.2 - 7.7} = \frac{8}{0.5}$$

$$\xi = 0.03125$$

Earlier in the chapter we found that MF $= 1/2\xi$ for $X/(F/k)$. Therefore, for X/F, the relationship is

$$\frac{X(j\omega)}{F(j\omega)} = \frac{1/k}{2\xi} = \frac{1/0.63}{2(0.03)} = 26.45$$

And then $20 \log_{10} 26.45 = 28.45$ dB, which is the magnitude of the transfer function at resonance, as shown in Figure 9.36. The actual transfer function used to derive the frequency-response function is

$$\frac{X(j\omega)}{F(j\omega)} = \frac{1}{0.0098(j\omega)^2 + 0.0047j\omega + 0.63}$$

Thus,

$$\frac{X(s)}{F(s)} = \frac{102.04}{s^2 + 0.48s + 64.28}$$

Therefore, the reader can compare the derived parameters with the actual ones for the simple second-order structure with one degree of freedom, which has only one mode of vibration with a natural frequency $\omega_n = 8$ rad/s.

Consider now the two-degree-of-freedom system shown in Figure 9.37 and assume that the dynamic forces $F_1(t)$ and $F_2(t)$ and the damping constants C_{e_1} and C_{e_2} are zero. If the system were excited, the masses would vibrate freely about their equilibrium position and the equations that describe the motion are

$$m_1\ddot{x}_1(t) + k_1x_1(t) - k_2[x_2(t) - x_1(t)] = 0$$
$$m_2\ddot{x}_2(t) + k_2[x_2(t) - x_1(t)] = 0 \tag{9.72}$$

Employing the Laplace transform, the equations become

$$(m_1s^2 + k_1 + k_2)X_1(s) - k_2X_2(s) = 0$$
$$-k_2X_1(s) + (m_2s^2 + k_2)X_2(s) = 0 \tag{9.73}$$

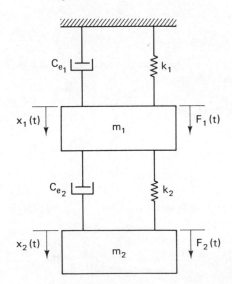

FIGURE 9.37 *Model for a linear two-degree-of-freedom system*

We are interested in the system's modes of vibration, and thus we let $s = j\omega$. Hence,

$$(m_1\omega^2 - k_1 - k_2)X_1(\omega) + k_2X_2(\omega) = 0$$
$$k_2X_1(\omega) + (m_2\omega^2 - k_2)X_2(\omega) = 0$$
(9.74)

The system's natural frequencies are determined from the characteristic equation of the system, which can be written as

$$\begin{vmatrix} m_1\omega^2 - k_1 - k_2 & k_2 \\ k_2 & m_2\omega^2 - k_2 \end{vmatrix} = 0$$
(9.75)

Expanding this determinant yields a quadratic in ω^2 of the form

$$\omega^4 - \left(\frac{k_1 + k_2}{m_1} + \frac{k_2}{m_2}\right)\omega^2 + \frac{k_1k_2}{m_1m_2} = 0$$
(9.76)

and thus the two values for ω^2 are

$$\omega_1^2, \omega_2^2 = \frac{1}{2}\left(\frac{k_1 + k_2}{m_1} + \frac{k_2}{m_2}\right) \pm \frac{1}{2}\left[\left(\frac{k_1 + k_2}{m_1} + \frac{k_2}{m_2}\right)^2 - 4\frac{k_1k_2}{m_1m_2}\right]^{1/2}$$
(9.77)

Only the positive values of ω are of any importance in physically realizable systems, and thus the positive values of ω calculated from the expression above define the two natural frequencies of the system and thus its modes of vibration.

Suppose now that the two natural frequencies determined by the expression above are ω_1 and ω_2. In addition, from the second equation in (9.74), note that

$$\frac{X_2(\omega)}{X_1(\omega)} = \frac{k_2}{k_2 - m_2\omega^2} = B(\omega)$$
(9.78)

Therefore, the relationship that exists between the two amplitudes $X_1(\omega)$ and $X_2(\omega)$ at the two natural frequencies ω_1 and ω_2 is

$$\frac{X_2(\omega_1)}{X_1(\omega_1)} = B(\omega_1)$$
(9.79)

and

$$\frac{X_2(\omega_2)}{X_1(\omega_2)} = B(\omega_2)$$
(9.80)

These two relationships between the amplitudes $X_1(\omega)$ and $X_2(\omega)$ define the mode shapes for the structural system, which correspond to the modes defined by ω_1 and ω_2.

In a manner similar to that shown above, the modes and mode shapes for various structures can be derived from the dynamical equations that describe these systems. For example, if the maximum amplitude is normalized to unity, the first three normal-mode shapes for the flexural vibrations of a uniform simply supported beam of length L are shown in Figure 9.38(a). The normal-mode shapes for the torsional vibrations of a uniform rod of length L free at both ends are shown in Figure 9.38(b).

In our analysis thus far it has been assumed that the system under study is modeled by the linear second-order equation

$$m\ddot{x}(t) + C_e\dot{x}(t) + kx(t) = f(t) \qquad (9.81)$$

In the modal analysis of structures, it is normally assumed that the structure can be adequately described by a set of simultaneous equations of the form

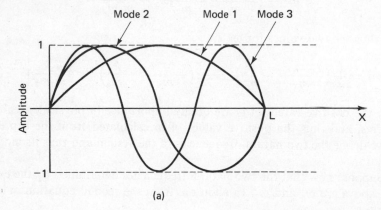

(a)

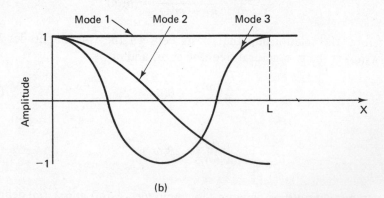

(b)

FIGURE 9.38 *Mode shapes for a beam and rod: (a) flexural vibration normal mode shapes for a uniform, simply supported beam; (b) normal shapes for the torsional vibrations of a uniform rod of length L free at both ends*

shown above. This set of equations can be written as a single vector-matrix equation of the form

$$\mathbf{M}\ddot{\mathbf{x}}(t) + \mathbf{C}'_e\dot{\mathbf{x}}(t) + \mathbf{k}\mathbf{x}(t) = \mathbf{f}(t) \tag{9.82}$$

where, in general, for an n-degree of freedom system $\mathbf{x}(t)$ and $\mathbf{f}(t)$ are n-vectors and $\mathbf{M}$, $\mathbf{C}'_e$, and $\mathbf{K}$ are real symmetric $n \times n$ matrices.

This formulation of the system is commonly referred to as the *state-variable representation* of dynamic systems.

Employing the Laplace transform variable, the equation above can be written as

$$\mathbf{B}(s)\mathbf{X}(s) = \mathbf{F}(s) \tag{9.83}$$

where $\mathbf{B}(s)$ = system matrix

          $\mathbf{X}(s)$ = system state

          $\mathbf{F}(s)$ = forcing function

This equation may be rewritten as

$$\mathbf{X}(s) = \mathbf{H}(s)\mathbf{F}(s) \tag{9.84}$$

where $\mathbf{H}(s)$, the transfer matrix, is given by the expression

$$\mathbf{H}(s) = \mathbf{B}^{-1}(s) = [\mathbf{M}s^2 + \mathbf{C}'_e s + \mathbf{K}]^{-1} \tag{9.85}$$

At this point, recall that the transfer function for the system described by equation (9.81) can be written in general as

$$h(s) = \frac{1}{ms^2 + C_e s + k}$$

$$h(s) = \frac{r}{2j(s - p)} - \frac{r^*}{2j(s - p^*)} \tag{9.86}$$

where   * = complex conjugate

    $r/2j$ = complex residue

    $p = \sigma - j\omega$ and $p^* = \sigma + j\omega$ are the pair of complex-conjugate poles of the transfer function

The inverse Laplace transform of the transfer function $h(s)$ is the impulse response, which is given by the expression

$$x(t) = |r|e^{-\sigma t} \sin(\omega t - \alpha) \tag{9.87}$$

where α = phase angle of the residue

EXAMPLE 9.24 Consider the system in Example 9.23. The transfer function $h(s)$ was

$$h(s) = \frac{102.04}{s^2 + 0.48s + 64.28}$$

$$= \frac{102.04}{(s + 0.24 - j8.01)(s + 0.24 + j8.01)}$$

Therefore,

$$\sigma = 0.24$$
$$\omega = 8.01$$
$$r = 12.74 \underline{|0°}$$

Using Figure 9.39 and employing the graphical residue technique for complex poles which is shown in Appendix B, the impulse response is calculated to be

$$x(t) = (102.04)\left[(2)\frac{1}{(2)(8.01)}e^{-0.24t}\cos(8.01t - 90°)\right]$$

$$= 12.74e^{-0.24t}\sin 8.01t$$

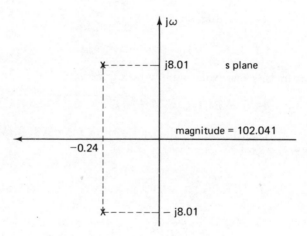

FIGURE 9.39 *s-Plane representation for*

$$h(s) = \frac{102.04}{s^2 + 0.48s + 64.28}$$

For the general n-degree-of-freedom system, the transfer matrix is of the form

$$\mathbf{H}(s) = \begin{bmatrix} h_{11}(s) & \cdots & h_{1n}(s) \\ \cdot & & \cdot \\ \cdot & & \cdot \\ \cdot & & \cdot \\ h_{n1}(s) & \cdots & h_{nn}(s) \end{bmatrix} \tag{9.88}$$

and each element of the transfer matrix is a transfer function of the form shown in equation (9.86). Each pair of complex-conjugate poles corresponds to a mode of vibration for the structure. The poles corresponding to the kth mode are

$$p_k = -\sigma_k + j\omega_k \quad \text{and} \quad p_k^* = -\sigma_k - j\omega_k$$

where σ_k = modal damping coefficient

ω_k = natural frequency

In a manner similar to that shown above for the single-degree-of-freedom system, the impulse response of mode k (i.e., the response obtained if only mode k was excited by a unit impulse) would be

$$x_k(t) = |r_k| e^{-\sigma_k t} \sin(\omega_k t - \alpha_k) \tag{9.89}$$

If α_k is zero, the mode is said to be normal, meaning that all points on the structure reach their maximum or minimum deflection at the same instant.

The transfer matrix completely defines the dynamics of the system, and each element of the transfer matrix is a transfer function.

In general, a complex structure will have many modes of vibration. Our objective in a modal analysis is to determine the modal parameters of the structure, and our method of attack is to obtain this information from the transfer-function data. From the standpoint of modal analyses, transfer functions provide (1) gain and phase data as a function of frequency; (2) frequency, damping, and the amplitude and phase of system resonances (i.e., modes of vibration); and (3) a spatial description of the amplitude and phase at resonances (i.e., mode shapes).

From complex-variable theory, it is well known that the dynamics of the system, defined by its transfer matrix in the s-plane, is completely specified by the frequency-response functions; that is, the values of the system over the entire s plane are completely specified by its values along the $j\omega$ axis, and therefore the frequency-response functions contain all the information necessary to identify modal parameters. In general, the preferred procedure used to obtain the modal parameters is to curve-fit an analytical expression of the form

$$H(\omega_i) = \sum_{k=1}^{n} \left[\frac{r_k}{2j(s - p_k)} - \frac{r_k^*}{2j(s - p_k^*)} \right]_{s=j\omega_i} \tag{9.90}$$

to the measured frequency data so as to minimize the least-squares estimate between the complex data and the complex analytical expression $H(\omega_i)$; that is,

$$E = \sum_{i=1}^{m} [H_i - H(\omega_i)]^2 \qquad (9.91)$$

where $H_i = i$th data point

During the curve-fitting process, the four parameters for each mode (i.e., σ_k, ω_k, and the magnitude and phase of the residue) are extracted, since they are the unknowns in equation (9.90). This method possesses the capacity to accurately identify all modal parameters, even where there is heavy modal overlap.

Modern measurement instrumentation, such as digital Fourier analyzers employing digital parameter identification techniques, has been developed which provides the analyst with a myriad of options for the acquisition and analysis of modal data. The following examples illustrate some of the capabilities of this modern equipment.

EXAMPLE 9.25 The measurement system for determining the modal data for a rectangular plate is shown in Figure 9.40. A load cell measures the input force and the output response is measured with an accelerometer.

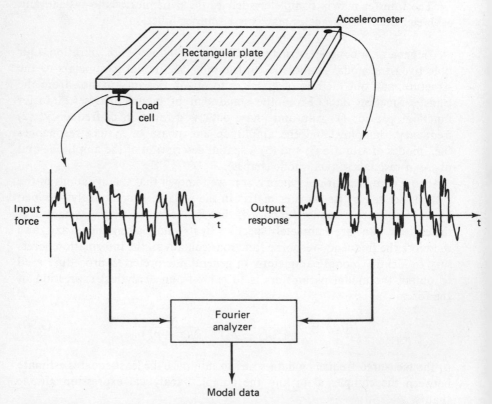

FIGURE 9.40 *Measurement system for determining the modal data on a rectangular plate*

The Fourier analyzer operates upon this information to determine a variety of modal data. For example, a typical measured frequency-response function which illustrates the vibration modes of the plate is shown in Figure 9.41(a). Fifty-five such functions were measured around the plate and stored, and from these data the Fourier analyzer calculated and displayed the animated isometrics shown in Figure 9.41(b)–(f), which illustrate the mode shapes for selected modes.

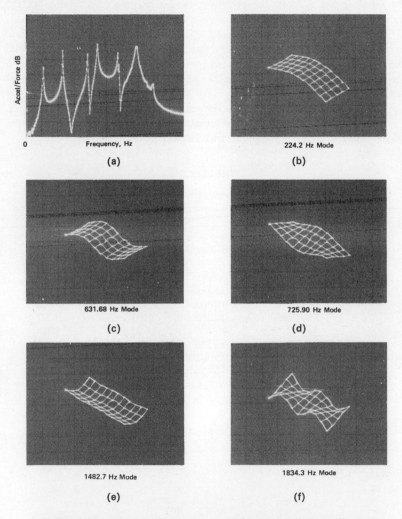

0   Frequency, Hz

(a)

224.2 Hz Mode

(b)

631.68 Hz Mode

(c)

725.90 Hz Mode

(d)

1482.7 Hz Mode

(e)

1834.3 Hz Mode

(f)

FIGURE 9.41 *A typical frequency response function and animated isometric displays of certain vibration modes for a rectangular plate: (a) frequency response function; (b) first longitudinal bending mode; (c) second longitudinal bending mode; (d) second torsion mode; (e) first transverse bending mode; (f) third torsion mode (courtesy of, and reprinted with the permission of, Kenneth A. Ramsey, Hewlett-Packard Corporation, Santa Clara, California and* Sound and Vibration, *Bay Village, Ohio)*

The following example illustrates some of the salient features of a modal analysis.

EXAMPLE 9.26 A set of transfer-function measurements were made on a disk-brake rotor from a foreign car. After the measurements were stored, by the Fourier analyzer, the analytical expression in equation (9.90) was curve-fitted to the measured data as illustrated in equation (9.91). During the curve-fitting process, the complete modal parameters of all vibration modes in the frequency band of interest were extracted. The results of this process are shown in Figure 9.42 and Table 9.3. Note that the analytical expression provides an excellent fit to the measured data, and that the modal parameters—frequency, damping, and mode shape, which are expressed as a magnitude and phase—are completely identified by the Fourier analyzer system.

It is informative to analyze more carefully some of the modal data. Consider mode 6. From Figure 9.42 it can be seen that the amplitude of the mode is 24 dB, which is measured with respect to 1 V. Therefore,

TABLE 9.3*

Modal parameters corresponding to Figure 9.42

Mode	Frequency (Hz)	Damping ratio (%)	RESIDUE	
			Amplitude	*Phase*
1	1099.6194	0.2891	145.5571·	344.7
2	1147.8315	0.7285	212.2571	354.7
3	1355.9260	0.7090	419.2204	352.5
4	1505.1660	0.7013	1031.2451	6.2
5	1583.7939	0.2138	22.7832	340.1
6	2105.0620	0.1726	1138.0957	360.0
7	2927.1890	0.3062	1555.4602	1.5

*Courtesy of Kenneth A. Ramsey, Hewlett-Packard Corporation, Santa Clara, Calif.

$$20 \log \left(\frac{x}{1 \, \text{V}} \right) = 24 \, \text{dB}$$

$$x = 15.85$$

Now, from Figure 9.35,

$$\xi_j = \frac{\sigma_j}{\sqrt{\sigma_j^2 + \omega_j^2}}$$

Since the damping ratio is given as a percentage in Table 9.3,

$$\xi_6 = \frac{\sigma_6}{\sqrt{\sigma_6^2 + \omega_6^2}} \times 100\%$$

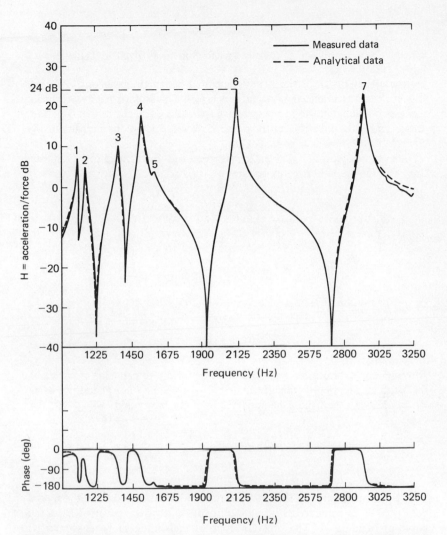

FIGURE 9.42 *Frequency-response function (magnitude and phase) and the corresponding analytical approximation (courtesy of Kenneth A. Ramsey, Hewlett-Packard Corporation, Santa Clara, Calif.)*

Solving the expression above for σ_6, using the data in Table 9.3, (i.e., $f = 2105.0620$ Hz and $\xi = 0.1726$) yields

$$\sigma_6 = 22.82 \text{ rad/s}$$

We can now calculate the magnitude of the transfer function $H(\omega_6)$ using the expression

$$H(\omega_6) = \frac{r_6}{2\sigma_6}$$

where r_6 = amplitude of the residue at mode 6 as given in Table 9.3

Note carefully that r_k is the quantity printed out in the table and not $r_k/2j$ (i.e., the residue is defined as r_k and not $r_k/2j$). This is done for two reasons: it causes our equations to fit the standard form of a co-quad transfer function, and H makes the equation for the impulse response come out mathematically correct.

The expression $H(\omega_k) = r_k/2\sigma_k$ is obtained from equation (9.86) by dropping the second term, to obtain

Thus,

$$H(s) = \frac{r_k}{2j(s - p_k)} = \frac{r_k}{2j(s + \sigma_k - j\omega_k)}$$

$$H(s = j\omega k) = \frac{r_k}{2j\sigma_k} = \frac{r_k}{2j(s + \sigma_k - j\omega_k)}$$

or the magnitude of $H(\omega_k)$ is

$$H(\omega_k) = \frac{r_k}{2\sigma_k}$$

Although this expression is an approximation for the actual value of the transfer function, experience indicates that it is an excellent one. Therefore, using the residue data and the calculated value of σ_6, we obtain

$$H(\omega_6) = \frac{1138.0957}{(2)(22.82)}$$

$$= 24.93$$

That is, the magnitude of the transfer function as calculated from the modal data is 24.93 as compared with a value of 15.85 (24 dB) measured from the curve in Figure 9.42. The obvious question remaining to be answered is: Why is there a discrepancy between these two values? The answer is simple: The discrepancy arises as a result of using digital techniques with finite resolution. The effective resolution for the data shown in Figure 9.42 was 9.76 Hz between data points. Figure 9.43 shows an expansion of the data with no increase in resolution about mode 6. Note from Figure 9.43 that the reading of 15.85 was, indeed, a good guess. However, since the resolution was only 9.76 Hz, the true data simply "fell through the cracks." Table 9.3 indicates that the true natural frequency, as computed by the curve-fitting algorithm, was 2105.0620 Hz. Therefore, if adequate resolution had been employed, the true value of the peak (i.e., 24.93 at a frequency of 2105.0620 Hz) would have been automatically obtained.

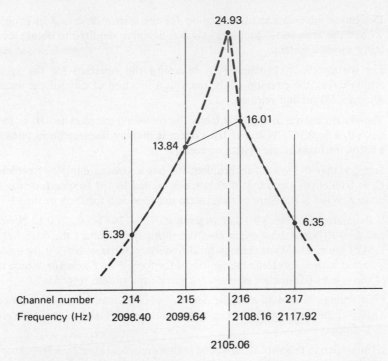

FIGURE 9.43 *Expansion of the data points about mode 6 shown in Figure 9.42 (courtesy of Kenneth A. Ramsey, Hewlett-Packard Corporation, Santa Clara, Calif.)*

The equipment is capable of providing arbitrarily fine frequency resolution; however, it was not used here, in order to illustrate these salient points. Nevertheless, in the absence of very fine resolution, the curve-fitting algorithm was able to extract from the available data the correct residue and natural frequency.

The previous examples illustrate some of the powerful features of modern measurement instrumentation for the extraction of modal data even in the presence of heavy modal overlap, as in the case of mode 5.

PROBLEMS

9.1 The system shown in Figure 9.1 has the following parameters: $W = 50$ N, $k = 0.32$ N/cm, and $\xi = 0.4$. Determine the natural frequency and viscous damping of the system.

9.2 Determine the characteristic equation for the system described in Problem 9.1 and the amount of additional viscous damping required to obtain a critically damped system.

9.3 For the system in Problem 9.1, determine the equation for the system response (i.e., the position of the mass as a function of time) if the mass is displaced 20 cm and released.

9.4 The system shown in Figure 9.1 has the following parameters: $W = 36$ N, $C_e = 0$, $k = 0.3671$ N/cm. If the system is initially displaced and released, what is the form of the system response?

9.5 Suppose that the system in Problem 9.4 has a viscous damping coefficient $C_e = 0.09$ N/cm/s and that the system is excited by the forcing function 4.0 $\sin \omega_n t$. What is the form of the system response as a function of time?

9.6 If the system in Figure 9.1 has parameters $W = 240$ N, $C_e = 0.12$ N/cm/s, and $k = 30$ N/cm, how much does the spring deflect under the dead weight load of the mass? What is the natural frequency of the system? If the system is forced with a harmonic oscillation of the form $F = F_0 \sin 10t$, which one of the system parameters essentially controls the system response?

9.7 Four springs, each with a spring constant of 2.4 N/cm, are used to support a mass. If the springs deflect 1 cm, what is the approximate weight of the mass they support?

9.8 If the system in Problem 9.7 is forced with a sinusoidal signal the frequency of which is twice that of the natural frequency of the system, what is the approximate degree of vibration isolation that is achieved?

9.9 A mechanical system that is to be mounted in a plant is modeled as shown in Figure 9.1 and has the following parameters: $W = 1200$ N and $k = 3.40$ N/cm. If the forcing frequency is $\omega = 10$ rad/s, how much damping will the system tolerate if the percent isolation must be greater than 92%?

9.10 A system containing rotating parts is to be mounted on shock absorbers and springs. The weight of the system per mounting point is 150 N. The forcing function due to an unbalance in the rotating system has an $\omega = 60$ rad/s. The shock absorbers provide a $C_e = 0.7$ N/cm/s. What type of spring (find k and δ_{st}) must be employed to achieve at least 90% vibration isolation?

9.11 A rotating system, together with its inertia base, is mounted on springs. The forcing frequency of the system is 24 rad/s. The system constants are $W_{Mp} = 1000$ N and $k = 40$ N/cm. Shock absorbers are to be added to limit the response to a transmissibility of 3 as the system passes through resonance upon startup. The percent isolation at the operating speed must be at least 80%. Can a shock absorber with a particular C_e be found that will satisfy these constraints?

9.12 Some delicate electronic equipment is to be isolated from a panel that vibrates at 50 rad/s. It is estimated that approximately 90% vibration isolation must be achieved through the use of springs to prevent damage to the equipment. What static deflection is required? (Assume that $\xi = 0$.)

9.13 A measurement system weighing 12 N is mounted on a complex mechanical system. A spectrum analyzer is used to measure the vibration. The spectrum analyzer shows significant vibration peaks at 40 Hz, 80 Hz, and 120 Hz. If the weight of the measuring equipment is equally distributed on four springs that are used to isolate the system, what type of springs are needed to achieve 80% isolation? (Assume that $\xi = 0$.)

9.14 A small engine weighing 60 N operates at 3000 rpm and is mounted on four mounting supports close to some fragile equipment. It is desired to isolate the engine so that it will not induce vibrations in the nearby equipment. Select one of the isolators in Figure 9.11 so that at least 80% vibration isolation will be achieved. (Assume that $\xi = 0$.)

9.15 A machine and its inertia base must be isolated from the floor. The system will load an isolator at 3.5 N/cm². If the forcing frequency of the vibration is 80 Hz and the degree of isolation desired is 95%, which isolator in Figure 9.13 should be chosen? (Assume that $\xi = 0.05$.)

9.16 A vibration-generating mechanical system is placed on a large concrete base. A spectrum analysis of the vibration signal indicates that its lowest-frequency component is 90 Hz. If the isolator will be loaded at 9 N/cm², which isolator in Figure 9.14 should be chosen to obtain 92% isolation? (Assume that $\xi = 0.05$.)

REFERENCES

[1] JONES, D. I. G. "Damping Treatments for Noise and Vibration Control," *Sound and Vibration*, July 1972, pp. 25–31.

[2] SALERNO, C. M. and R. M. HOCHHEISER "Cost Effective Noise Damping for Machinery Quieting," in *Proceedings of Noise-Con '73* Oct. 15–17, Washington D.C., 1973.

[3] RUZICKA, J. E. "Vibration Control," *Electro-Technology*, Vol. 72, No. 2 (Aug. 1963), pp. 63–82.

[4] RUZICKA, J. E. "Damping Structural Resonances Using Viscoelastic Shear Damping Mechanisms," *ASME J. Eng. Ind.*, Nov. 1961, pp. 403–413.

[5] NASHIF, A. D. "Materials for Vibration Control in Engineering," *Shock and Vibration Bull. 43*, June 1973.

[6] LAZAN, B. J. *Damping of Materials and Members in Structural Mechanics*, Pergamon Press, London, 1968.

[7] PERRY, R. E. Sales Manager, Barron/A.S.E., Inc., Leeds, Ala, private communication.

[8] BROCH, J. T. *Mechanical Vibration and Shock Measurements*, Brüel and Kjaer Instruments Co., June 1973.

[9] CHENG, D. K. *Analysis of Linear Systems*, Addison-Wesley Publishing Company, Inc., Reading, Mass., 1959.

[10] RUZICKA, J. R. "Characteristics of Mechanical Vibration and Shock, "*Sound and Vibration*, Apr. 1967, pp. 14–31.

[11] BALLARD, W. C., S. L. CASEY, and J. D. CLAWEN *Sound and Vibration*, Vol. 3, No. 1 (Jan. 1969), pp. 10–21.

[12] RAMSEY, K. A. Hewlett-Packard Company, Santa Clara, Calif., private communication.

[13] RAMSEY, K. A. "Effective Measurements for Structural Dynamics Testing Part-1," *Sound and Vibration*, Nov. 1975, pp. 24–35.

[14] RAMSEY, K. A. "Effective Measurements for Structural Dynamics Testing Part-2," *Sound and Vibration*, Apr. 1976, pp. 18–31.

[15] RICHARDSON, M. and R. POTTER "Identification of the Modal Properties of an Elastic Structure from Measured Transfer Function Data," 20th ISA, Albuquerque, N.M., May 1974, pp. 239–246.

[16] RAMSEY, K. A. and M. RICHARDSON "Making Effective Transfer Function Measurements for Modal Analysis," 22nd Annual Meeting of The Institute of Environmental Sciences, Apr. 26–28, 1976, Philadelphia.

BIBLIOGRAPHY

BERANEK, L. L. ed. *Noise and Vibration Control*, McGraw-Hill Book Company, New York, 1971.

BLAKE, M. P. and W. S. MITCHELL *Vibration and Acoustic Measurement Handbook*, Spartan Books, New York, 1972.

CROCKER, M. J. and A. J. PRICE *Noise and Noise Control*, Vol. 1, CRC Press, Inc., Cleveland, Ohio, 1975.

DIEHL, G. M. *Machinery Acoustics*, John Wiley & Sons, Inc., New York, 1973.

FOWLER, D. F. "Instrumentation for Noise and Vibration Measurement," in *Reduction of Machinery Noise*, edited by M. J. Crocker, Purdue University, 1974, pp. 124–139.

HARRIS, C. M. ed. *Handbook of Noise Control*, McGraw-Hill Book Company, New York, 1957.

HARRIS, C. M. and C. E. CREDE *Shock and Vibration Handbook*, Vol. 2, McGraw-Hill Book Company, New York, 1961.

JACKSON, C.   "A Practical Vibration Primer Part 1, "*Hydrocarbon Processing*, Apr. 1975, pp. 161–163.

JACKSON, C.   "A Practical Vibration Primer-Part 2," *Hydrocarbon Processing*, June 1975, pp. 109–111.

JACKSON, C.   "A Practical Vibration Primer-Part 3," *Hydrocarbon Processing*, Aug. 1975, pp. 109–111.

MAGRAB, E. B.   *Environmental Noise Control*, John Wiley & Sons, Inc., New York, 1975.

"Vibration Measurement," IRD Mechanalysis, Inc., Columbus, Ohio, 1972.

10

MACHINE PROTECTION
AND MALFUNCTION DIAGNOSIS

10.1 INTRODUCTION

In Chapter 9 we were concerned with the isolation of vibration so as to minimize the resulting noise. In this chapter the vibration analysis is aimed at diagnosing malfunctions within rotating machines so that these machines can ultimately be protected from failure.

Mechanical defects, even though very slight, cause rotating machinery to vibrate. These vibrations range from those that are very small and essentially insignificant to those that are severe enough to tear the machine apart. Thus, vibration is a good measure of the mechanical integrity of the machine.

A vibration analysis of the type alluded to above is extremely useful. It can be used to detect trouble long before it becomes a serious problem, and thus should be an integral part of a preventative maintenance program. It is important in acceptance testing, on the one hand, and can also be useful in sales and product promotions, on the other. Quality control is another area in which vibration analyses may play a fundamental, yet paramount role.

Before we attack the techniques for analyzing vibrations in rotating machinery, let us explore some of the fundamental mechanisms that cause vibrations.

10.2 CAUSES OF VIBRATION

In general, vibration occurs because a force, either internal or external, is present to excite it (i.e., motion is always equal to force divided by restraint). Vibration is also a function of the system parameters (e.g., damping) and the relationship between the machine's running speed and its natural frequencies. Although there are numerous causes of vibration, those which occur most frequently are due to such things as unbalance, excitation of resonances due to a lack of system damping, misalignment, looseness, and the like. We shall discuss a number of the important vibration generators in order to obtain an understanding of how they are produced. Knowing how they are produced, we can devise methods for detecting them. This information then yields the data required for the reduction or elimination of these vibration phenomena.

10.2.1 Unbalance

History seems to indicate that perhaps the single most dominant cause of vibration is unbalance. This phenomenon is characterized by the existence of an unequal weight distribution in a part about its centerline of rotation. Although there are many reasons for the existence of unbalance in rotating machinery, a common one is the buildup of improper tolerances when many machine parts are assembled. This case, as well as other cases of unbalance, can be represented or modeled as shown in Figure 10.1(a). Note that in this figure all contributions to an unequal weight distribution are represented by one point, called the *heavy spot*. The actual amount of unbalance, measured in units such as gram-centimeters or ounce-inches, is equal to the product of the weight of the unbalance in grams or ounces and the distance of the unbalanced weight (heavy spot) in centimeters or inches from the rotating centerline, as shown in Figure 10.1. Note that the unbalance is also equal to the product of the weight of the disk in grams and the distance from the center of rotation to the center of mass of the disk in centimeters. It is the centrifugal force of the unbalanced weight that causes vibration, and this force is proportional to the square of the rotating speed.

It is important to note that when the heavy spot depicted in Figure 10.1 exists only in a single plane (i.e., a thin rotor disk), the unbalance is termed *static*. This term is used to describe an unbalance condition which can be simply detected by placing the rotor shaft on knife edges. Then, if a heavy spot does exist, the rotor will rotate so that the heavy spot becomes the nadir of the rotor. *Dynamic* unbalance, on the other hand, exists when there is an unbalance condition in more than one plane. For example, suppose that a shaft contains two thin disks which are located at opposite ends of the shaft. In addition, suppose that the two thin disks have the same amount of unbalance but are located 180° apart. In this case, placing the shaft on knife edges will give no indication of an unbalance condition. However, if the shaft is rotated, the centrifugal forces created by the unbalanced disks will cause the shaft to wobble and rock. Since this rocking condition can only be detected when the shaft is spinning, the unbalance is called dynamic. Thus, two thin disks separated on a shaft may be used to represent unbalanced conditions in two machines which are coupled to form a single shaft.

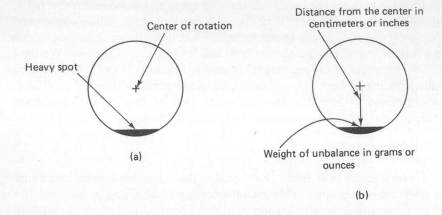

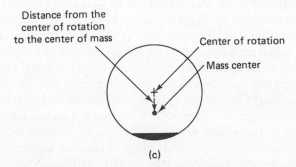

FIGURE 10.1 (*a*) *Representation of unequal weight distribution by heavy spot;* (*b*) *and* (*c*) *measures of unbalance*

10.2.2 Rubs in Rotating Machinery

A *rub* in a rotating machine is in general any type of load on the shaft, such as a steady-state preload, that will prevent the shaft from moving in some direction. For example, an internal rub may be caused by differential expansion in which one end of the machine heats up faster than another. Rubs are more predominant than one might initially expect, and they occur in both the radial and axial directions. Gas turbines and other nonrigid case machines are particularly susceptible to rubs. A vast variety of symptoms occur from rubs, because of the wide variety of reactions that occur in different rub situations.

Rubs are generally classified in two types, *partial* and *full* [1]. Partial rubs occur first and may quickly lead to a full rub. One type of partial rub is like a hit-and-bounce mechanism. If this hit-and-bounce action increases due to an increase in certain system parameters, such as unbalance or bearing clearance, then the rotating machine may attain the violent full-rub condition. This condition occurs very suddenly and catastrophically, leading in most cases to an instantaneous destruction of the machine. Herein lies the importance of the rub phenomena.

10.2.3 Misalignment

Another important source of vibration in rotating machinery, and one that is in some industries the most frequent offender, is that of misalignment. Single-shaft misalignment can exist if the bearings through which the shaft passes are not properly aligned with the shaft. Two shafts may be misaligned in either of the two ways illustrated in Figure 10.2 or a combination of them (i.e., offset at an angle). Misalignment always acts as a unidirectional preload on the rotor. Because of this, its tendency is to create double-frequency vibrations. However, if this preload negates the effect of another deliberate preload, the symptoms and actions of unbalance or oil whip may occur.

In rotating machinery trains, both cold and hot alignment are used

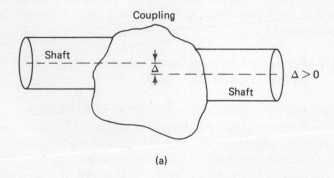

(a)

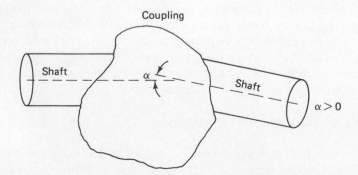

FIGURE 10.2   (a) Offset misalignment; (b) angular misalignment

[2–10]. Cold alignment means that the shafts of a train are positioned at rest so that they will be aligned during actual operation. Hot alignment checks the shaft movement when the system is running at operating speed and temperature.

There are many items that affect alignment: piping that is connected to the trains, as, for example, in turbomachinery, machine supports and foundations, machine casings, thermal relationships between shafts to be aligned, and the like. Alignment of rotating machinery has become very important in recent years because of increased use of high-speed machines. Aligned machines last longer, reduce downtime, and thus minimize operating costs. In addition, it has often been shown in practice that severe misalignment will ultimately cause machine failure.

10.2.4 Looseness

Looseness may lead to severe vibration problems in rotating machinery. It is caused by conditions such as loosely tied-down foundations, loose bearing restraints, excessive bearing clearances, and the like. This looseness problem simply fuels any fire caused by misalignment and unbalance. For example, any small amount of unbalance can cause large vibrations in a loose environment.

10.2.5 Excitation of Resonances due to Insufficient Damping

A good rule of thumb with rotating machinery is that "any poorly damped resonance will sooner or later get excited." Oil whip is one example where a resonance and a forcing function cooperate to damage machines. The most common resonance excited in machinery is the translational balance resonance (first critical). Lack of damping of this resonance can cause twofold problems. First, damage may occur due to excessive amplitude when transversing this resonance. Second, any number of maximum-amplitude-seeking forcing functions can occur to excite this resonance (e.g., oil whirl). For this reason it is important to check the damping of resonances not only while transversing them but under actual operating speeds and conditions as well. This will ensure safe, reliable operation of the machine under normal and abnormal conditions.

10.2.6 Oil Whirl and Oil Whip

Oil whirl [1,11] is a hydrodynamic instability that occurs in bearings at some "onset" speed and is characterized by the generation of some tangential force proportional to the radial deflection of the rotating shaft. This whirling

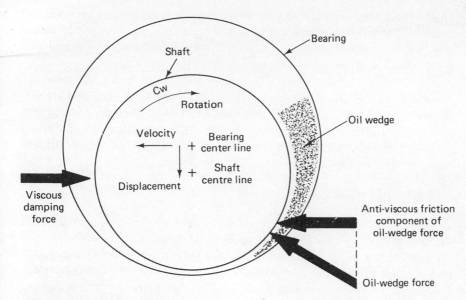

FIGURE 10.3 *Exhibition of oil whirl (courtesy of Donald E. Bently, Bently Nevada Corporation, Minden, Nev.)*

motion of ever-increasing amplitude is limited only by the nonlinearities of the system, which ultimately limit deflections. The position of the shaft inside the bearing is as shown in Fig. 10.3, and the vibrations produced are caused by forces within the oil film. Essentially, energy is wasted in dragging the oil around inside the bearing. This phenomenon is normally produced by light loading and excess bearing clearance. The vibration frequency is approximately 43% of the operating speed; that is, the velocity of the oil film is approximately 43% of the shaft surface speed, and thus the shaft rides on the oil film in the same manner that a surfboard rides a wave.

Oil whirl can also be intermittent. For example, Bently [1] uses as an example a bearing that heats up due to an oil whirl, then changes the bearing clearance and oil viscosity; this, in turn, stops the whirling. Once the bearing cools, the conditions for onset of the whirl are again established and the cycle repeats.

If an oil whirl excites the natural translational balance resonance of the rotor, *oil whip* is said to exist. This phenomenon usually exists within a few percent of the oil-whirl speed and is much more severe. Oil whip can be quite violent and is usually the onset of machine destruction, because the non-synchronous whirling induces alternating stresses, which then lead to fatigue failures and also rubs.

Although the causes of vibration discussed here are by no means complete, they do represent some of the most important and frequently encountered phenomena and thus serve to introduce and provide the basis for the material that follows.

10.3 BASIC ROTOR DYNAMICS

In general, rotors are classified as *rigid* or *flexible*. If the motions are completely described by what occurs at the center of mass, then the rotor is said to be rigid. In the analysis that follows, however, we shall assume a flexible rotor system and will discuss the basic concepts of its critical speeds, system response, and stability [12].

10.3.1 Modeling [12]

A number of modeling techniques have been applied to the analysis of the flexible rotor system. However, we shall be concerned here with only the most basic phenomena and hence will model the flexible rotor-bearing system in a manner that is simple enough to be easily understood, but rigorous enough to illustrate some important basic concepts. In general, a flexible rotor system consists of a rotor, disks, support bearings, foundations, and housings. We shall model this system as a linear system with lumped parameters and will not include such things as different types of bearings, foundations, torsional gyroscopics, axial effects, and the like. Because of its importance, we assume that the forcing phenomenon is a steady-state excitation caused by mass unbalance. The system's restraining forces are the elastic force, the external and internal damping forces, and the force due to inertia.

From Figure 10.4 we note that the position vector from the origin to the center of mass of the disk is given by the expression

$$\mathbf{r} = (y + a \cos \omega t)\mathbf{e}_y + (z + a \sin \omega t)\mathbf{e}_z \tag{10.1}$$

Note that a is the distance from the center of the shaft to the disk center of mass. The acceleration of the mass center is then

$$\ddot{\mathbf{r}} = (\ddot{y} - a\omega^2 \cos \omega t)\mathbf{e}_y + (\ddot{z} - a\omega^2 \sin \omega t)\mathbf{e}_z \tag{10.2}$$

The equation of motion for the system is then

$$m\ddot{\mathbf{r}} = \sum_{j=1}^{n} \mathbf{f}_j \tag{10.3}$$

where the right-hand side of the equation represents the sum of forces acting on the system. The three forces (i.e., $n = 3$) acting on the system are the elastic force $\mathbf{f}_1 = \mathbf{f}_k$, the external friction force $\mathbf{f}_2 = \mathbf{f}_e$, and the internal friction force $\mathbf{f}_3 = \mathbf{f}_i$. The elastic restoring force $\mathbf{f}_k$ is given by the expression

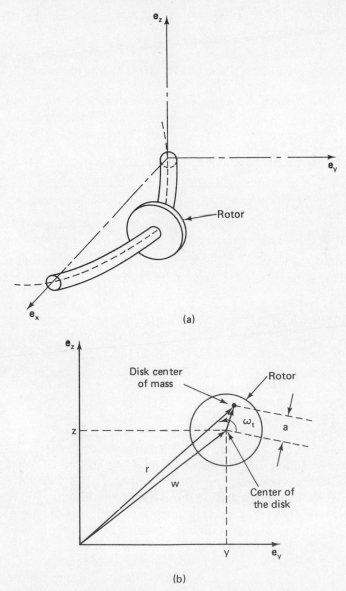

FIGURE 10.4 *Simple model for the flexible rotor system: (a) flexible rotor geometry; (b) projection of the rotor in the yz plane*

$$\mathbf{f}_k = -k(y\mathbf{e}_y + z\mathbf{e}_z) \tag{10.4}$$

The external friction force $\mathbf{f}_e$ is given by

$$\mathbf{f}_e = -C_e(\dot{y}\mathbf{e}_y + \dot{z}\mathbf{e}_z) \tag{10.5}$$

and the internal friction force $\mathbf{f}_i$ is

$$\mathbf{f}_i = -C_i[(\dot{y} + \omega z)\mathbf{e}_y + (\dot{z} - \omega y)\mathbf{e}_z] \tag{10.6}$$

Substituting equations (10.2), (10.4), (10.5), and (10.6) into (10.3) yields

$$\begin{aligned} m[(\ddot{y} - a\omega^2 \cos \omega t)\mathbf{e}_y &+ (\ddot{z} - a\omega^2 \sin \omega t)\mathbf{e}_z] \\ &= -k(y\mathbf{e}_y + z\mathbf{e}_z) - C_e(\dot{y}\mathbf{e}_y + \dot{z}\mathbf{e}_z) \\ &\quad - C_i[(\dot{y} + \omega z)\mathbf{e}_y + (\dot{z} - \omega y)\mathbf{e}_z] \end{aligned} \tag{10.7}$$

Equating coefficients of $\mathbf{e}_y$ and $\mathbf{e}_z$, we obtain the coupled equations

$$\begin{aligned} m\ddot{y} + (C_e + C_i)\dot{y} + ky + C_i\omega z &= m\omega^2 a \cos \omega t \\ m\ddot{z} + (C_e + C_i)\dot{z} + kz - C_i\omega y &= m\omega^2 a \sin \omega t \end{aligned} \tag{10.8}$$

If we now define $w \equiv y + jz$, where $j = \sqrt{-1}$, the imaginary or 90° operator, then multiplying the second equation in (10.8) by j and adding it to the first equation yields

$$m\ddot{w} + (C_e + C_i)\dot{w} + kw - j\omega C_i w = m\omega^2 a e^{j\omega t} \tag{10.9}$$

The equations in (10.8) or (10.9) form our simple model for the system dynamics of the flexible rotor system.

10.3.2 Critical Speed

Critical speeds correspond to resonant frequencies of the system and are identified by natural frequencies and forcing phenomena. The system natural frequencies are, in general, a function of mass, elasticity, external force, and torque as well as a function of rotor speed due to gyroscopic effects. If a periodic component of the forcing function approximates a natural frequency of the system, a resonant condition will exist. If this condition exists at some specific speed, that speed is called a critical speed.

Critical speeds correspond to the eigenvalues of the equations that define the system dynamics. The undamped natural frequency of the system is obtained by solving the homogeneous equation for the system dynamics where $C_e = C_i = 0$:

$$m\ddot{w} + kw = 0$$

Using the Laplace transform, we obtain

$$(ms^2 + k)W(s) = 0$$

The characteristic equation is then

$$ms^2 + k = 0$$

and its solution is

$$s = \pm j\sqrt{\frac{k}{m}}$$

and therefore the undamped natural frequency is $\omega = \sqrt{k/m}$. If the internal friction is zero (i.e., $C_i = 0$) but external friction is present, the characteristic equation becomes

$$ms^2 + C_e s + k = 0$$

The form of the solution of this equation is well known from Chapter 9. It should also be remembered that this damping is controlling only at "resonance" (i.e., where $\omega = \sqrt{k/m}$).

A rapid transition of the running speed through a critical speed tends to restrict the whirl amplitudes, but a slow transition of the critical speed will allow large amplitudes to develop. If the running speed is equal to this critical speed, the rotor will exhibit large deflections, thus damaging internal seals, and the forces transmitted to the bearing may cause bearing failure, all of which may ultimately lead to rotor failure.

The critical speeds of a simple lumped-parameter system such as that being analyzed here can be obtained quite easily. However, in large practical systems, straightforward analytical techniques are simply not mathematically tractable, because the equations to be solved are too complex. Under these conditions one must resort to a numerical technique such as the Rayleigh method, Holzer method, Prohl–Mykelstad method, and others [13].

EXAMPLE 10.1 Assume that a rotor consists of a 1200-lb disk suspended on a 2-in.-diameter steel shaft and mounted between bearings 48 in. apart. Neglecting the distributed mass of the shaft, the static deflection and thus the critical speed can be determined.

SOLUTION It is well known from mechanics that the static deflection is given by the expression

$$\delta_{st} = \frac{WL^3}{48EI}$$

where W = weight of the disk = 1200 lb or 5337.6 N, since the weight of the shaft is neglected

L = span between bearings = 48 in. = 121.92 cm

E = modulus of elasticity = 30×10^6 psi or 20.69×10^6 N/cm² for steel

$I = $ moment of inertia of the shaft $= \pi d^4/64 = 0.785$ in.$^4 =$ 32.67 cm^4

The static deflection is then

$$\delta_{st} = \frac{(1200)(48)^3}{(48)(30 \times 10^6)(0.785)} = \frac{(5337.6)(121.92)^3}{(48)(20.69 \times 10^6)(32.67)}$$

$$= 0.1174 \text{ in.} = 0.298 \text{ cm}$$

Then as shown in Chapter 9 [e.g., equation (9.44)], the critical speed is

$$\omega = \sqrt{\frac{g}{\delta_{st}}}$$

$$= \sqrt{\frac{386}{0.1174}} = \sqrt{\frac{980.7}{0.298}}$$

$$= 57.34 \text{ rad/s}$$

$$= 547.56 \text{ rpm}$$

10.3.3 System Response

In general, one is interested in the rotor response to both internal and external disturbances, and a number of numerical techniques exist for calculating the response [13]. However, we shall be concerned here only with the response of the system shown in Figure 10.4 to harmonic excitation at the rotor speed frequency (i.e., mass unbalance). We shall calculate the unbalanced response (i.e., the rotor dynamic whirl amplitudes resulting from the mass unbalance) by examining the solution of the two equations in (10.8). Assuming that the shaft does not undergo any flexure and hence $C_i = 0$, the equations in (10.8) become

$$m\ddot{y} + C_e\dot{y} + ky = m\omega^2 a \cos \omega t$$
$$m\ddot{z} + C_e\dot{z} + kz = m\omega^2 a \sin \omega t \qquad (10.10)$$

The exact solutions of these equations can be derived using the techniques outlined in Appendix B. However, the general form of these solutions will be

$$\boxed{\begin{aligned} y(t) &= A_1 e^{-\alpha t} \cos (\beta t + \phi_1) + Bm\omega^2 a \cos (\omega t + \phi_2) \\ z(t) &= A_2 e^{-\alpha t} \cos (\beta t + \phi_3) + Bm\omega^2 a \cos (\omega t + \phi_4) \end{aligned}} \qquad (10.11)$$

Note that the first term in the two expressions above contains a decaying exponential, and hence these terms can be neglected after a finite amount of time. The second term in each expression represents a steady-state circular whirl. Since the forcing functions in equations (10.10) are sinusoidal and the equations are linear, we know that the steady-state response will be sinusoidal at the same frequency, and therefore all that need be determined is the amplitude and phase. Hence, using the method of undetermined coefficients and assuming a sinusoidal solution, we can compute the amplitude factor B for the circular whirl to be

$$B = \frac{1}{\sqrt{(k - m\omega^2) + \omega^2 C_e^2}} \qquad (10.12)$$

It is straightforward but tedious to show that the whirl amplitude becomes a maximum when the rotational speed is

$$\omega = \omega_n \frac{1}{\sqrt{1 - \frac{1}{2}(C_e/\omega_n)^2}} \quad \text{rad/s} \qquad (10.13)$$

where $\quad \omega_n = \sqrt{k/m}$

Only when the damping term is actually zero does the critical speed correspond exactly to the natural translational balance resonance frequency ω_n. Note also that equation (10.13) indicates that, in general, the presence of damping increases the system resonant frequency.

It is informative to examine the system response to a mass imbalance from another, and yet parallel, point of view. The equation of motion for the system is

$$m_u r_u \omega^2 \sin \omega t = m \frac{dx^2}{dt^2} + C_e \frac{dx}{dt} + kx \qquad (10.14)$$

where $\quad k$ = spring coefficient of the system

$\quad C_e$ = viscous damping coefficient

$\quad m$ = mass of the rotor

$\quad m_u$ = unbalanced mass

$\quad r_u$ = distance to the unbalanced mass from the center of the rotor

Note carefully the contrast between this model and that used in the discussion above, where, for example, a represents the distance from the center of the rotor to the rotor center of mass. It was shown in Section 9.2.3 that the steady-state solution of equation (10.14) is

$$x(t) = |x| \underline{/\phi}$$

$$= \frac{m_u r_u \omega^2}{\sqrt{(k - m\omega^2)^2 + (C_e \omega)^2}} \left| -\tan^{-1}\left(\frac{C_e \omega}{k - m\omega^2}\right) \right. \qquad (10.15)$$

Thus, the magnitude of the position vector (i.e., amplitude of the motion) is $m_u r_u \omega^2 / \sqrt{(k - m\omega^2)^2 + (C_e \omega)^2}$, and the phase angle is $-\tan^{-1} C_e \omega / (k - m\omega^2)$. A plot of this function is shown in Figure 10.5.

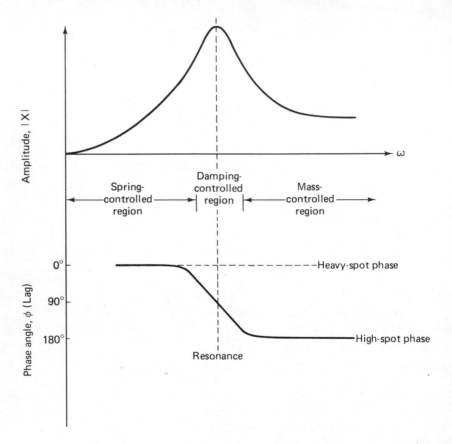

FIGURE 10.5 *System response to an unbalanced mass*

An examination of Figure 10.5 will indicate what physically happens at the first critical, which we refer to as the *translational balance resonance*. Note that the forcing function is proportional to the square of the rotational speed ω^2. Hence, we would expect that the vibration amplitude would simply increase as the speed increases. However, it is well known that the actual response looks as shown in Figure 10.5. This actual response prompts the question: What happens at the translational balance resonance to limit the system response? To answer this very fundamental question, we need to examine the equations for the steady-state response. Note that this response is determined by the amplitude factor given by equation (10.15). At low speed the shaft reacts directly to the unbalanced force and the response is determined by the spring constant k, since the other terms in the amplitude factor are small. Note that the phase angle for small ω is 0°. As the speed is increased, the response peaks because a resonance condition is established where $k - m\omega^2 = 0$, and thus the response is limited in this critical speed range only by the damping term. The phase lag is now 90°. It is in the speed range above the translational balance resonance that the rotor exhibits a fascinating phenomenon. At this higher speed the response is determined primarily by the mass-acceleration term, which reacts 180° out of phase with the unbalanced force. This means that the shaft reacts to an unbalanced force above resonance by moving in a direction opposite to that force, and this, then, is the mechanism by which the system balances itself and thus limits the response.

For example, suppose that the mass unbalance, called the heavy spot, is located at three o'clock on the rotor as measured from the driving end facing the driven load and that the system is rotating counterclockwise. At low speed the heavy spot and the high spot, the point of maximum vibration amplitude, are the same. As the speed increases toward the translational balance resonance, the phase angle of the high spot begins to lag that of the heavy spot until at resonance the high-spot phase lags the heavy-spot phase by 90°. The high spot is then at six o'clock. As the speed is increased further, the high-spot phase will soon begin to lag the heavy-spot phase by 180°, and then the high spot is located at nine o'clock. Thus, the high spot has moved diametrically opposite the heavy spot, thus moving the rotor over to compensate for the unbalanced mass. This action is the *self-balance mechanism*. This concept will be demonstrated in the examples that appear later in the chapter.

D10.1   A rotor that consists of a massive disk suspended on a steel shaft is mounted on bearings. The static deflection of the rotor is measured to be 0.45 cm. Determine the critical speed of the rotor.

Ans.   445.79 rpm.

EXAMPLE 10.2   Suppose that we are given a 1000-lb or 453.6-kg rotor with an unbalanced weight of 0.1 lb or 45.36 g located at a distance of 10 in. or 25.4 cm from the center of the rotor. From equation (10.15) it can be seen that for high values of ω above the translational balance resonance, the phase angle is 180° and the amplitude of the response is

$$x(t) = \frac{m_u r_u}{m}$$

Using the values above,

$$x(t) = \frac{(0.1 \text{ lb})(10 \text{ in.})}{1000 \text{ lb}} = \frac{(0.04536 \text{ kg})(25.4 \text{ cm})}{453.6 \text{ kg}}$$

$$= 1 \text{ mil} = 0.00254 \text{ cm}$$

Therefore, to correct for the unbalanced weight, the center of gravity of the rotor had to move over 1 mil, or 0.00254 cm, in a direction 180° from the location of the unbalanced weight. Note that this value corresponds to the magnitude of the tail of the amplitude response curve shown in Figure 10.5. Note also that the parameter a employed in the previous analysis corresponds to $m_u r_u / m$.

The general form of the amplitude response of a rotor system is shown in Figure 10.6. The two balance resonances correspond to the shaft motions illustrated in Figure 10.7. However it is important to note that this amplitude response is critically dependent upon the rotor-bearing configuration. For

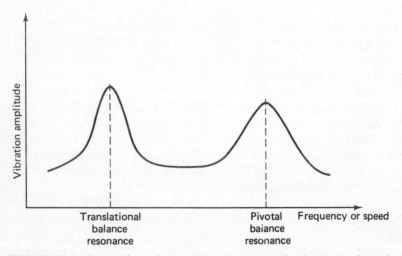

FIGURE 10.6 *General form of the unbalanced response for the system shown in Figure 10.4*

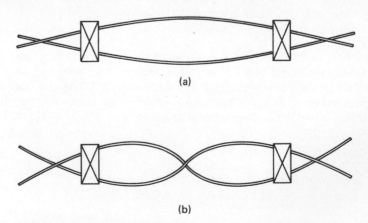

FIGURE 10.7 *(a) Translational and (b) pivotal shaft motion*

example, some systems may have only a translational or pivotal balance resonance, but not both; and in some systems the pivotal balance resonance occurs before the translational balance resonance. This point will be re-examined later in the chapter.

It is important for the reader to note that the previous analysis, although simple, presents the important characteristics of a flexible rotor system through the translational balance resonance.

D10.2　A large rotor is observed to be causing an unbalance in a machinery system. The rotor is measured and found to have an unbalanced mass of 75 g. This mass is located 35 cm from the center of the rotor. The mass of the rotor is 750 kg. What should be done to correct the unbalance? Assume ω to be very large for purposes of simplification.

Ans.　Apply sufficient weight at 180° from the unbalanced mass to move the center of gravity 0.0035 cm.

10.3.4　Stability

Instability in a flexible rotor-bearing system can be caused by such things as the excitation of any undamped resonance, internal friction, asymmetry in the rotating system, the action of an oil-lubricated journal in its bearing (e.g., oil whip) and nonlinearities within the rotational system. Kimball and Hull [14] and Newkirk [15] noted that internal friction could cause shaft instability at operating speeds above the first critical. In addition, we will show later that the transition from stability to instability is dependent upon the ratio C_e/C_i.

If the system dynamics are described by an nth-order linear differential equation with constant coefficients, then the characteristic equation for the system is of the form

$$s^n + a_n s^{n-1} + \ldots + a_1 s + a_0 = 0 \tag{10.16}$$

The roots of this equation define eigenvalues of the form $s = j\lambda$, where λ may be real or complex and thus yields a response of the form $Ae^{j\lambda t}$.

In order to determine the system's stability, we can employ what is known as the Routh–Hurwitz stability criterion. This technique, although easy to use, is very powerful and has been used for many years in the stability analysis of linear control systems. Since λ may in general be complex, the terms that define the system response will be of the form

$$\boxed{Ae^{j\lambda t} = Ae^{j(\lambda_R + j\lambda_I)t} = Ae^{(-\lambda_I + j\lambda_R)t}} \tag{10.17}$$

where　λ_R and λ_I = real and imaginary parts of the eigenvalue respectively

Equation (10.17) may be rewritten as

$$Ae^{j\lambda t} = A(e^{-\lambda_I t})(e^{j\lambda_R t}) \qquad (10.18)$$

In this form we note that if the imaginary part of one of the eigenvalues is negative, this component will cause the response to grow exponentially with time and thus indicates a condition of instability.

One form of the Routh–Hurwitz stability criterion states that if we write the characteristic equation (10.16) in the form

$$(a_0 + jb_0)\lambda^n + (a_1 + jb_1)\lambda^{n-1} + \ldots + (a_n + jb_n) = 0 \qquad (10.19)$$

then a necessary and sufficient condition that the imaginary part of all eigenvalues is positive, is that the set of determinants generated below are positive.

$$-\begin{vmatrix} a_0 & a_1 \\ b_0 & b_1 \end{vmatrix} > 0$$

$$\begin{vmatrix} a_0 & a_1 & a_2 & 0 \\ b_0 & b_1 & b_2 & 0 \\ 0 & a_0 & a_1 & a_2 \\ 0 & b_0 & b_1 & b_2 \end{vmatrix} > 0$$

$$(-1)^n \begin{vmatrix} a_0 & a_1 \ldots a_n & 0 \ldots 0 \\ b_0 & b_1 \ldots b_n & 0 \ldots 0 \\ 0 & a_0 \ldots a_n & 0 \ldots 0 \\ 0 & b_0 \ldots b_n & 0 \ldots 0 \\ 0 & 0 & a_0 \ldots a_n & 0 \ldots 0 \\ 0 & 0 & b_0 \ldots b_n & 0 \ldots 0 \\ & & \cdot \\ & & \cdot \\ & & \cdot \\ 0 \ldots 0 & & a_0 \ldots a_n \\ 0 \ldots 0 & & b_0 \ldots b_n \end{vmatrix} > 0 \qquad (10.20)$$

To demonstrate the use of this stability criterion, let us consider equation (10.9), which describes our simple model. The characteristic equation is

$$s^2 + \frac{C_e + C_i}{m}s + \frac{k - j\omega C_i}{m} = 0 \qquad (10.21)$$

Substituting $s = j\lambda$ in equation (10.21) and comparing the result with equation (10.19) we find that

$$a_0 + jb_0 = -1$$

$$a_1 + jb_1 = j\frac{C_e + C_i}{m}$$

$$a_2 + jb_2 = \frac{k}{m} - j\frac{\omega C_i}{m}$$

and therefore

$$a_0 = -1, \quad b_0 = 0$$

$$a_1 = 0, \quad b_1 = \frac{C_e + C_i}{m}$$

$$a_2 = \frac{k}{m}, \quad b_2 = -\frac{\omega C_i}{m}$$

Substituting these parameters into the determinants of equation (10.20) yields

$$-\begin{vmatrix} a_0 & a_1 \\ b_0 & b_1 \end{vmatrix} = -\begin{vmatrix} -1 & 0 \\ 0 & \dfrac{C_e + C_i}{m} \end{vmatrix} = \frac{C_e + C_i}{m} > 0$$

and

$$\begin{vmatrix} a_0 & a_1 & a_2 & 0 \\ b_0 & b_1 & b_2 & 0 \\ 0 & a_0 & a_1 & a_2 \\ 0 & b_0 & b_1 & b_2 \end{vmatrix} = \begin{vmatrix} -1 & 0 & \dfrac{k}{m} & 0 \\ 0 & \dfrac{C_e + C_i}{m} & \dfrac{-\omega C_i}{m} & 0 \\ 0 & -1 & 0 & \dfrac{k}{m} \\ 0 & 0 & \dfrac{C_e + C_i}{m} & \dfrac{-\omega C_i}{m} \end{vmatrix}$$

Expanding the determinant in minors about the first column yields

$$= -1\left(-\frac{C_e + C_i}{m}\frac{C_e + C_i}{m}\frac{k}{m} + \frac{\omega^2 C_i^2}{m^2} \right) > 0$$

$$= -\left(\frac{C_e + C_i}{m} \right)^2 \frac{k}{m} + \frac{\omega^2 C_i^2}{m^2} > 0$$

Therefore,

$$\left(1 + \frac{C_e}{C_i} \right)^2 \frac{k}{m} - \omega^2 > 0$$

or

$$\left(1 + \frac{C_e}{C_i}\right)\sqrt{\frac{k}{m}} - \omega > 0$$

Once again this equation indicates that internal friction can cause instability at rotating speeds above the first critical.

It is important to note that although the Routh–Hurwitz criterion is a very useful and straightforward technique for determining system stability, it provides no other information such as the particular eigenvalues that cause instability or their values. This information, in general, must be determined by numerical integration and approximation techniques.

10.4 INSTRUMENTATION

In order to obtain the data for the protection and malfunction prediction of rotating machinery systems, a number of measurements must be made. A typical basic measurement configuration is shown in Figure 10.8. The various

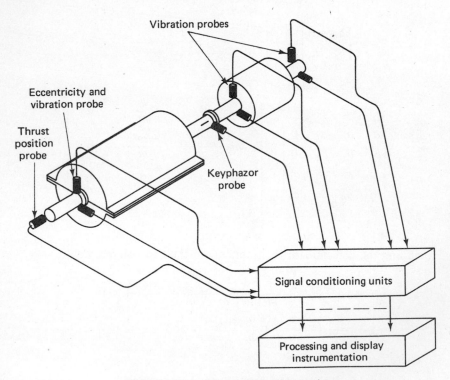

FIGURE 10.8 *Machine protection system*

probes, together with the signal conditioning equipment and the processing and display instrumentation, represent a typical machine protection system. Let us now examine some of the components of this configuration.

10.4.1 Transducer

The transducer employed in the machine protection system shown in Figure 10.8 is the noncontacting eddy-current probe. The probe contains a small coil of wire in its tip, which is excited with a radio frequency signal to produce a magnetic field. The probe employs the magnetic field to operate as a gap-to-voltage transducer with an output that is inherently nonlinear. Thus, it is used with a signal conditioning unit which linearizes the transducer output so that the total transfer function for the probe and signal conditioning unit is of the form shown in Figure 10.9. Note that in the range from about 30 to 80 mils or 0.6 to 2.1 mm, the system is linear with a slope of approximately 200 mV/mil or 8 V/mm. This output signal is then used to drive various types of processing and display instrumentation.

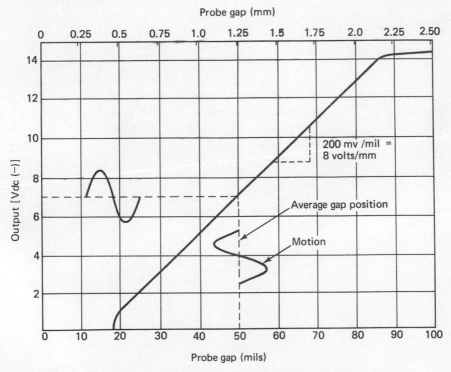

FIGURE 10.9 *Transfer function for probe and signal conditioning unit*

10.4.2 Processing and Display Equipment

10.4.2.1 Vibration monitor

This unit is used to indicate the radial motion of a rotating shaft. The input signals to the vibration monitor are obtained from two probes that are mounted at the same lateral position on the shaft, typically horizontally and vertically, or at 45° in both directions from the vertical. These units normally contain a meter which indicates the highest peak-to-peak vibration displacement of the two probes; alert and danger set points which control lights and relays when the vibration reaches preset-unacceptable levels, and connections for tape or chart recorders and oscilloscopes.

10.4.2.2 Eccentricity monitor

This position monitor is used at operating speed, to obtain an accurate indication of the average shaft position within a bearing. It also is very useful during startup. Thermal bows or bent shafts can be detected by meter fluctuations when the shaft is rotated by a turning gear (1 cpm). Otherwise, the device is very similar to the position monitor discussed below.

10.4.2.3 Axial-position monitor

This unit is used to monitor the average axial (thrust) position of the shaft. The input to this device is derived from a probe that is mounted axially within 1 ft of the thrust bearing to observe the shaft position as shown in Figure 10.10 and provides reliable warning to prevent an axial rub caused by

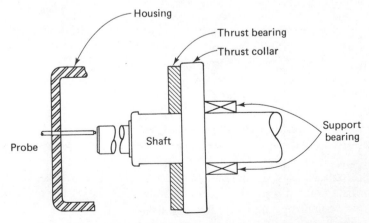

FIGURE 10.10 *Use of a thrust position probe to measure the average gap between probe and shaft*

thrust bearing failure. This unit contains the same type of meter, set points, and connections as those previously described.

In general, the monitors described above can be obtained in units suitable for simply mounting in a control panel or in weatherproof or explosion-proof housings.

10.4.2.4 Tachometers

An accurate indication of shaft speed is very important in rotating machinery protection systems. The input to the tachometer may be derived from a wide range of transducers, including eddy-current probes and magnetic and optical pickups. The eddy-current probe is a good basic speed transducer because this gap-to-voltage converter provides an excellent pulse signal from the notch on the shaft that is essentially independent of the shaft speed.

Tachometers are typically digital because the raw speed information consists of pulses obtained from a machine-mounted pickup and thus is basically digital in nature. In addition, digital tachs can provide greater accuracy and readability under normal field conditions.

10.4.2.5 Oscilloscope

It is important to note that data such as eccentricity, thrust position, and speed can be adequately represented by one or two numbers. However, there is much information needed for diagnostic work, such as the radial motion of a shaft, which cannot be described by a simple set of numbers. Much of this information, as well as many other forms of pertinent data, can be easily displayed for analysis on the oscilloscope. For example, if the vertical and horizontal probe data are fed into the vertical and horizontal inputs, respectively, on the scope, the scope will display the manner in which the shaft is moving within its bearing constraints. A typical display of this type is shown in Figure 10.11. The figure displays the whirling motion of the shaft and is referred to as the shaft *orbit*. Because the orbit provides a graphical picture of the manner in which the shaft is actually moving, it is extremely useful in rotating machinery protection systems.

Note that the orbit shown in Figure 10.11 contains a blanking mark. This spatial reference point is obtained from the keyphazor probe shown in Figure 10.8. This once-per-turn marker can be obtained by using a probe to observe a notch or projection on the shaft or by using an optical pickup to observe a strip of reflective tape. This angular position measurement of the radial motion of the shaft is very important in both malfunction diagnosis and rotor balancing.

The ability of the oscilloscope to simultaneously correlate two machine parameters is also a useful feature. For example, if the information from two

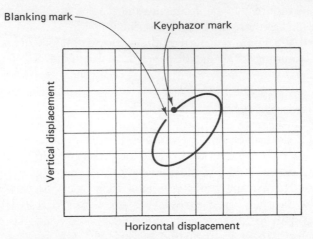

FIGURE 10.11   *Orbit displaying the whirling motion of the shaft*

probes, one mounted on the inboard bearing and the other mounted on the outboard bearing of a shaft, is used to drive the traces of a dual-trace scope, then a translational or pivotal mode of the shaft motion can be easily detected. Finally, a camera, which mounts directly on the oscilloscope, can be used for recording and documenting the data displayed by the scope.

10.4.2.6 Tape recorders

A tape recorder serves a very important function in the diagnostic analysis of rotating machinery, that of storage. Perhaps the most useful tape recorder is the FM (frequency-modulation) type. FM recorders are available with multiple tracks and have the capability of recording data that range from slow-roll information to high-speed radial vibration. It is also useful for obtaining data on intermittent fast transients and startup or rundown. If an on-site analysis is not necessary, then the records taken in the plant can be replayed at a subsequent time in the laboratory, where other data-processing and diagnostic equipment is readily available.

10.4.2.7 Filters and phase meters

Filters are electronic circuits that possess the capability of passing certain frequencies or bands of frequencies while rejecting others. Filters are used in the separation of data in order to provide for a more lucid presentation of the data and to enhance the data analysis. It is important to remember two facts when employing filters; first, filters always throw away some data to enhance other data; and second, filters may distort information, and this distortion may be to either amplitude or phase.

Several types of filters are used for machinery analysis: band pass, which passes one frequency band; band rejection, which rejects one frequency band; low pass, which passes all frequencies below the cutoff frequency; and high pass, which passes all frequencies above the cutoff frequency. These filters are all available as fixed filters, which are set to operate on one frequency or band of frequencies; tunable filters, which are variable frequency devices; or automatically tunable filters, which are capable of automatically following a reference frequency (e.g., the keyphazor).

When a phase measurement is combined with an automatically tuned band-pass filter, a vector-filter phase meter is obtained. Such an instrument can accurately measure the amplitude and phase of a motion with respect to a reference signal. This type of instrument can be used for a wide variety of diagnostic measurements and is therefore very useful in rotor balancing, resonance and mechanical impedance studies.

10.4.2.8 Real-time analyzers

These instruments are normally available in one of three forms: fast Fourier transform (FFT) processors, discrete Fourier transform (DFT) processors, or time-compression analyzers, which are essentially hybrid analog/digital processors. These devices provide a spectrum presentation (i.e., vibration amplitude versus frequency plot) for a particular system under specified running conditions. This presentation displays vital diagnostic data such as the maximum vibration amplitude, and resonance phenomena, such as critical speeds and structural resonances. The manner in which a machine's vibration signature changes with operating speed is also easily obtainable. Like the oscilloscope, these analyzers can be used to correlate the signals from two probes at each end of a rotating machine to detect such things as translational or pivotal motion. When used with a tachometer to provide an angular reference, the analyzers can also be used for balancing. In general, real-time analyzers are extremely useful for quick cause-and-effect analyses of rapidly changing phenomena.

10.4.2.9 Computer systems

The use of the computer to control or augment the diagnostic instrumentation adds tremendous power and flexibility to the equipment. The computer in conjunction with peripheral devices can simultaneously monitor many system functions. Some parameters, such as radial motion and thrust position, must be sampled very fast so that any transient motion within the machine that is potentially damaging will not go undetected. On the other hand, the sampling rate for parameters such as temperature and speed can be relatively slow, since these parameters do not change rapidly. The data collected can be stored and/or processed using various software packages to provide such features as startup analysis and control, emergency shutdown procedures, the machine's operation history, various diagnostic routines, and data for predictive and preventative maintenance.

10.5 DIAGNOSTIC ANALYSES

It has been found in practice that the use of proper monitoring and diagnostic instrumentation coupled with a basic knowledge of machinery behavior will prevent, or at least limit, most accidents involving rotating machinery. When the instrumentation is properly installed and carefully calibrated, it can be used to predict malfunctions caused by a gradual degradation of machine performance, such as a gradual increase in mass unbalance or bearing clearance, and to diagnose the cause of transient forces which act on the machinery and trip the shutdown mechanisms. The use of a tape recorder in these instances provides the instrumentation with the capability to correlate the process operating parameters with those of other similar machines to determine the source of the transient force, and hopefully to provide the data necessary for design changes that will prevent these malfunctions.

The utilization of instrumentation systems for machine protection and malfunction diagnosis will now be presented. This topic is extremely broad and hence we will limit our discussion here to some of the more important and frequently encountered problems. Specifically, we shall treat the subject of rotor balancing in some detail because of the relatively high frequency of occurrence of this phenomenon in rotating machinery. In addition, we shall briefly examine the detection of rotating-machinery rubs, misalignment, looseness, and oil whirl.

10.5.1 Rotor Balancing

Rotor balancing can be done in a number of ways, which range from the crude pencil method of balancing to the use of sophisticated, specially designed balancing machines. The former method is illustrated in Figure 10.12. If the machine is running at slow speed and has large deflections, the pencil will mark the heavy spot on the rotor. Weight should then be added to the shaft at the point 180° opposite the pencil mark. This approach, although not very satisfactory, does illustrate the basic idea of balancing.

10.5.1.1 Instrumentation

The instrumentation system used for balancing as shown in Figure 10.13 is essentially a subset of the machine vibration protection monitoring system

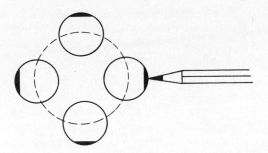

FIGURE 10.12 *Orbital path of the shaft center line*

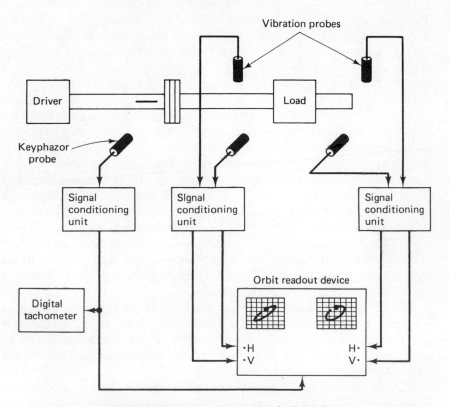

FIGURE 10.13 *Instrument system for balancing*

shown in Figure 10.8. This system is very simple and yet accurate. Although the system can be augmented or partially replaced by other sophisticated devices, it represents a practical and viable scheme for balancing a system of the type shown in Figure 10.13.

The two pairs of vertical and horizontal probes are normally located toward each end of the shaft, usually just inboard and outboard of the bearings. These displacement sensors are positioned 90° apart, hopefully in the true horizontal and vertical positions, at the same lateral location, and provide the orbit of the shaft at these particular positions. The displacement sensors and signal conditioner are typically calibrated as shown in Figure 10.9. Thus, we normally use the 200-mV peak-to-peak volts per mil of peak-to-peak displacement for calibration. Then, with the oscilloscope sensitivity calibrated at 200 mV/cm, the actual mils of displacement can easily be read directly on the scope graticule as 1 mil/cm, or equivalently 25.4 μm/cm.

The keyphazor probe, which observes a single notch or projection on the shaft, provides both phase reference and speed information required for balancing. This keyphazor signal is used not only to synchronize the scope but also to control the scope intensity (i.e., z-axis input).

The dual beam scope (or its equivalent) is used as the readout device and provides a graphical picture of the balancing procedure.

10.5.1.2 Orbits

If the signal from the vertical probe is fed into the vertical input on the scope and the signal from the horizontal probe is fed into the horizontal input on the scope, the resultant image created on the cathode-ray tube (CRT) represents the orbit (shaft center-line motion) at that particular lateral location on the shaft. Typically, the probes are mounted on the bearing housing and thus the motion indicated by the probes is the relative motion of the shaft with respect to this housing. A typical orbit, which is generated by two vibration signals obtained from the vertical and horizontal probes, is shown in Figure 10.14. This orbit represents the actual motion of the shaft within the bearing. The orbit at any lateral location along the shaft is the path of travel of the center line of the shaft; as such, the orbit is the motion of the high spot at that location.

It is always important to establish at the outset the direction of the orbit with respect to the position from which the machine is viewed. If we always view the rotating machinery from outward the driving end facing the driven load, then a *forward* orbit is one that rotates in the same direction as the shaft.

As indicated earlier, the signal from the keyphazor controls the intensity of the scope's CRT. Thus, when the notch on the shaft enters the keyphazor probe face area, the gap between the probe and shaft is suddenly increased, producing an increase in negative voltage which in turn causes a bright spot on the CRT. When the trailing wall of the notch passes the probe face, the gap is suddenly decreased, which produces a positive burst of voltage, which, in turn, causes a blank spot on the CRT. A typical orbit containing the keyphazor mark is shown in Figure 10.15. Recall that if the probes have a transfer function of 200 mV/mil and the scope has a sensitivity of 200

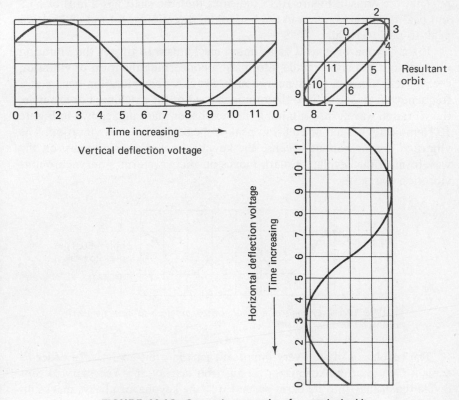

FIGURE 10.14 *Generation procedure for a typical orbit*

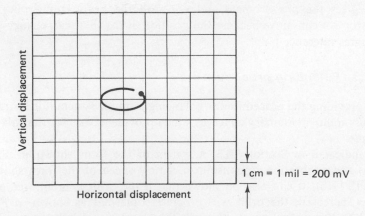

FIGURE 10.15 *Typical orbit pattern containing the keyphazor mark*

383

mV/cm, the orbit in Figure 10.15 indicates that the shaft has 2 mils or 50.8 μm peak-to-peak vibration in the horizontal direction and 1 mil or 25.4 μm peak-to-peak vertically.

If the keyphazor is used as a timing mechanism to trigger the vibration waveform, it can also provide information concerning the form of vibration (i.e., synchronous or nonsynchronous). Synchronous vibration occurs at a frequency that is locked into the running speed of the machine. For example, the vibration waveform containing the keyphazor mark and shown in Figure 10.16 indicates a 2-mil or 50.8-μm peak-to-peak synchronous vibration. The vibration is synchronous because the keyphazor mark is stationary on the waveform. If the keyphazor mark moves on the waveform, nonsynchronous vibration exists.

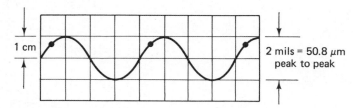

FIGURE 10.16   *Example of a 2-mil or 50.8-μm peak-to-peak synchronous vibration*

The keyphazor plays a very important part in orbit analysis. In order to explain the use of the keyphazor on an orbit display, it is necessary to subdivide the display into its basic parts. First, the keyphazor always marks an exact *angle of rotation*. For convenience this angle of rotation is assigned to be zero degrees. Second, a pair of *X*-*Y* probes totally define shaft center-line motion (orbit) at that particular station. Third, the orbit with keyphazor tells us the exact position of the shaft at the instant the shaft was rotating past zero degrees (keyphazor mark indexed with the keyphazor probe). The keyphazor dot can move because the position of the shaft can change at the zero-degree reference point.

10.5.1.3   Balancing procedure

Before providing the procedure for balancing a simple system it is instructive to discuss some preliminary concepts which provide a basis for the balancing technique.

As indicated in Section 10.3, a system of the form shown in Figure 10.17(a) has a vibration response at a critical speed of the form shown in Figure 10.17(b). It can also be shown that if the orbit is monitored as the speed is increased, the orbit will progress somewhat as shown in Figure 10.17(c) as the system passes through the translational balance resonance

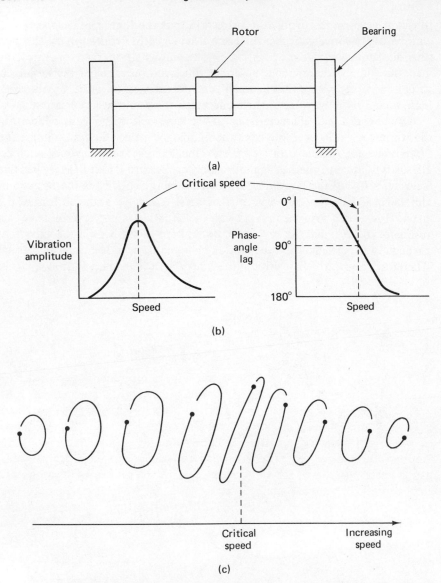

FIGURE 10.17 *Vibration characteristics of a simple rotor system*

(i.e., the first critical). Note the manner in which the machine attempts to balance itself. Note also that the phase shift of the keyphazor is ideally 180°, as shown in Figure 10.17(b) and (c). Thus, the orbits look different depending upon whether the machine is operating above or below the first critical.

Recall that the shaft is displaced by the mass unbalance and since the keyphazor mark shows the exact location of the shaft at the instant when the notch passes the probe, the orbit, which is the high spot on the scope, can be used to locate the heavy spot on the shaft by working back through the mechanical impedance response. This information then provides the necessary data to determine the point at which to apply the corrective weight. It is important to note that in some cases filters may be necessary to eliminate noise on the waveform which can mask the essential information.

The operational characteristics above and below the first critical are shown in Figure 10.18. Note from the figure that below the first critical, the keyphazor dot on the orbit represents the location of the heavy spot on the shaft relative to the bearing where the probes are located (i.e., when the keyway is facing the keyphazor probe, the heavy spot is at the bottom of the shaft). Concurrently, the keyphazor mark is located at the bottom of the orbit. However, above the first critical, the situation is just reversed. We saw in Figure 10.17(c) that the keyphazor dot shifted 180° when the machine went through the first critical and the motion changed from translational to pivotal. Therefore, above the first critical the heavy spot is located 180° away from

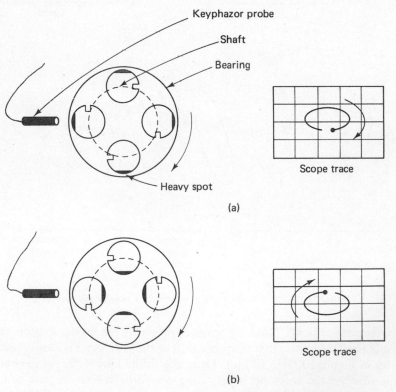

FIGURE 10.18 *Operational characteristics of a simple rotor system (a) below and (b) above the first critical*

the keyphazor mark, as shown in Figure 10.18(b) (i.e., when the keyway is facing the keyphazor probe, the heavy spot is at the bottom of the shaft but the keyphazor mark is at the top of the orbit). Thus, if we stopped the machine and lined up the notch on the shaft with the keyphazor probe, the orbit information would dictate the location of the corrective weight.

If we simply examine the vibration signal from a single probe, we can locate the heavy spot by observing the phase angle, which is defined to be the number of degrees from the timer mark to the next positive peak on the vibration waveform. This is illustrated in Figure 10.19, where 45° after the notch passes the keyphazor probe, the heavy spot passes the vibration probe.

The following examples illustrate many of the concepts discussed earlier. The laboratory model employed in these examples is a Bently Nevada Corporation rotor kit approximately 2 1/2 ft long and 6 in. wide. A small, fractional-horsepower motor drives the system, and rotation is in the counterclockwise direction. An oscilloscope equipped with a camera is used for graphical display, and a vector filter unit designed to measure speed and vibration amplitude and phase is also employed. In the following examples, measurements were made of the vertical vibration amplitude and phase. The

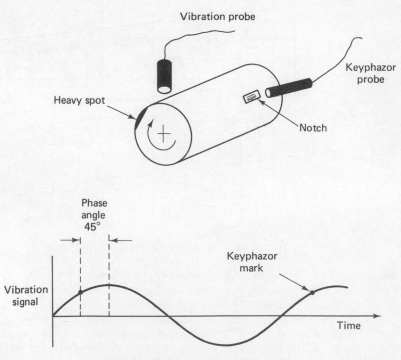

FIGURE 10.19 *Vibration phase angle*

signal conditioning units adjust the signal to match the input to the oscilloscope, as described earlier. A schematic of the laboratory model is shown
in Figure 10.20.

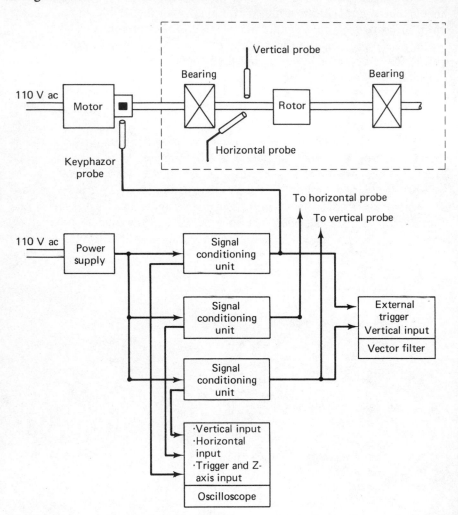

FIGURE 10.20 *Laboratory rotor system model*

EXAMPLE 10.3 Using the rotor-bearing configuration shown within the
dashed lines in Figure 10.20, the laboratory model was used to generate the
curves shown in Figure 10.21 from the vector filter unit. The speed of
the system was varied from zero to 7000 rpm. The curves demonstrate the
performance of the system as it passes through the translational balance
resonance (first critical). In addition, photographs of the orbit were taken
below resonance at 2000 rpm, at the translational balance resonance speed,

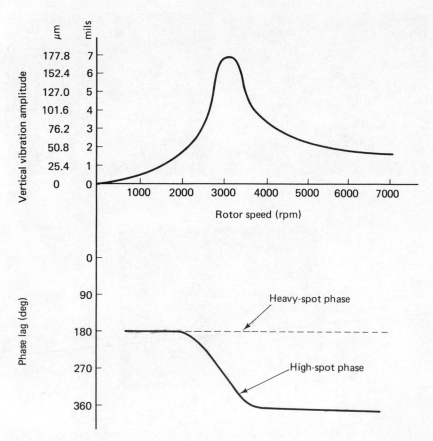

FIGURE 10.21 *Vertical vibration amplitude and phase for the system shown in Figure 10.20*

and above resonance at 5000 rpm. These orbits are shown in Figure 10.22. Each major division of the scope graticule corresponds to 1 mil or 25.4 μm. Note the close agreement that exists between the data obtained from the vector filter unit as shown in Figure 10.21 and the orbits obtained from a camera mounted on the oscilloscope and displayed in Figure 10.22. At 2000 rpm the vertical vibration amplitude, as shown on the orbit, is approximately 2 mils or 51 μm, and the phase angle marked by the keyphazor is approximately 200°. At the translational balance resonance speed, which is approximately 3100 rpm, the vibration amplitude has increased to approximately 7 mils or 178 μm, and the phase angle has shifted approximately 90° from the value obtained at 2000 rpm. The irregularities in the orbit are due to normal machine nonlinearities. Above the translational balance resonance at 5000 rpm, the vertical vibration amplitude is about 2 mils or 51 μm, and the phase angle has now shifted essentially 180° from its position below

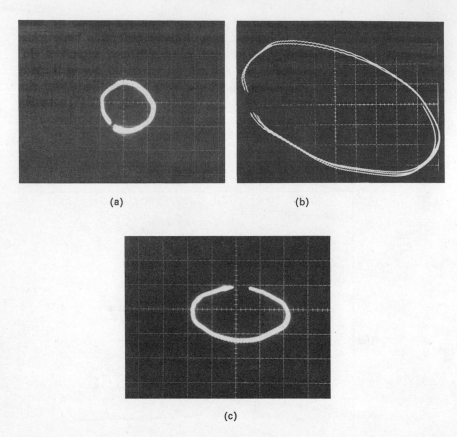

(a)   (b)

(c)

FIGURE 10.22 *Orbits which display the transition through a translational balance resonance: (a) orbit below resonance at 2000 rpm; (b) translational balance resonance orbit at 3100 rpm; (c) orbit above resonance at 5000 rpm (courtesy of Mr. Donald E. Bently, Bently Nevada Corporation)*

resonance at 2000 rpm. Thus, the system behaves as the theory discussed earlier suggests.

The rotor-bearing configuration analyzed in this example is typical of large steam and gas turbines, electric motors and generators, and axial and radial compressors.

EXAMPLE 10.4 The rotor shaft used in Example 10.3 is essentially a beam, and thus its translational balance resonance can be determined by applying an impulse to the shaft and measuring the response. The shaft was hit with a pencil between the rotor and the outer bearing and the response was measured, by means of the vertical probe, with an oscilloscope. This response is shown in Figure 10.23. The time base on the oscilloscope was set for 5 ms/division, and hence it can be seen from the figure that the time between the peaks was 19 ms (i.e., 19 ms/cycle). Since there are 60,000 ms/min,

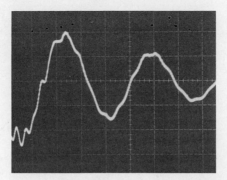

FIGURE 10.23 *Impulse response of the rotor-shaft shown in Figure 10.20 (courtesy of Mr. Donald E. Bently, Bently Nevada Corporation)*

the resonance in cycles/minute can be obtained by dividing 60,000 ms/min by 19 ms/cycle. The result is 3157 cycles/min (i.e., 3157 rpm). This value compares favorably with that obtained in the previous example.

EXAMPLE 10.5   The rotor-bearing configuration in Figure 10.20 is replaced in this example with that shown in Figure 10.24. This configuration is typical of small steam turbines, blowers, fans, and hot and cold gas expanders. This particular rotor-bearing arrangement displays a pivotal balance resonance. The vector filter unit was used to obtain the data shown in Figure 10.25, which illustrates the transition through the pivotal balance resonance. For purposes of comparison, orbits of the motion were obtained below resonance, at resonance, and above resonance. Photographs of the orbits taken at 4000, 9100, and 15,000 rpm are shown in Figure 10.26. The calibration of the oscilloscope was set such that each major division on the graticule corresponds to $2\frac{1}{2}$ mils, or 63.5 μm, of vibration. Once again note the comparison in vibration amplitude and phase between the data shown in Figure 10.25 and that displayed by the orbits shown in Figure 10.26.

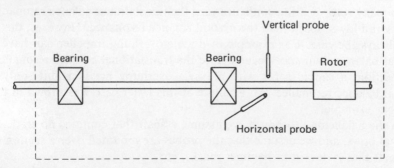

FIGURE 10.24 *Overhung rotor configuration for examining a pivotal balance resonance condition*

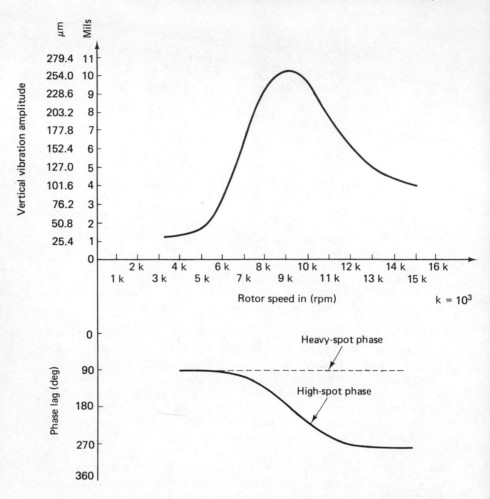

FIGURE 10.25 *Vertical vibration amplitude and phase for the system shown in Figure 10.24*

In many rotor-bearing configurations the translational balance resonance occurs at a lower speed than the pivotal balance resonance. However, this is not always the case. For example, a double-overhung impeller can have a pivotal balance resonance occur before the translational balance resonance. In addition, a simple rotor with a single overhung mass exhibits only a pivotal balance resonance. An example of this latter case was shown in the previous example.

In the balancing procedure, we assume a shaft that contains no static or thermal bows, and we assume that the probes are mounted over a section of the shaft that is devoid of hammer marks, etc.

With the machine instrumented for balancing as described earlier, the balancing technique is as follows. At the selected balancing speed, observe

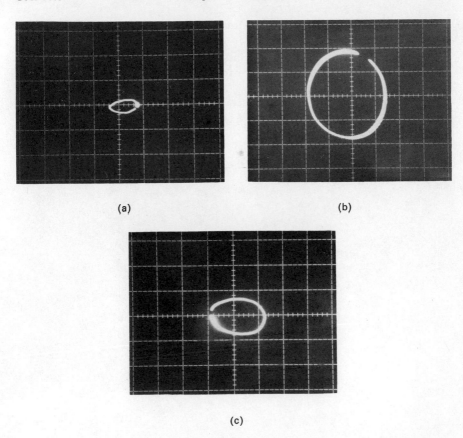

(a)   (b)

(c)

FIGURE 10.26 *Orbits which display the transition through a pivotal balance resonance: (a) orbit below resonance at 4000 rpm; (b) pivotal balance resonance at 9100 rpm; (c) orbit above resonance at 15000 rpm (courtesy of Mr. Donald E. Bently, Bently Nevada Corporation)*

the orbit and the phase angle. If the machine is running below the first critical and the keyphazor dot is at 60°, the machine should be stopped, rolled to line up the notch with the keyphazor probe, and a corrective weight added at 240°. If the machine is running above the first critical, the corrective weight should be added at 60°. An oscilloscope equipped with a camera or a storage scope will greatly facilitate this balancing procedure.

The actual speed at which the balancing is performed is important. Slow-roll balancing is not very useful, but it is better than no balance at all. One should not necessarily balance at the operating speed either, and yet the machine should certainly have good balance at this number of rpms. In fact, the balancing speed will depend upon whether the balancing is being performed to help the machine balance itself at the first or second critical, or both. However, the actual speed chosen should not be near one of these criticals so that the phase shifting caused by these modes does not confuse the motion due to imbalance.

In practice, we find that most orbits are not circular but rather elliptical, and thus we must have a procedure for deriving the true keyphazor angle when orbits are noncircular.

Consider the forward orbit shown in Figure 10.27(a). The step-by step procedure for deriving the true keyphazor angle is as follows:

1. Draw a major axis through the orbit as shown in Figure 10.27(b).
2. Circumscribe a circle about the ellipse using the major axis as diameter as shown in Figure 10.27(c).

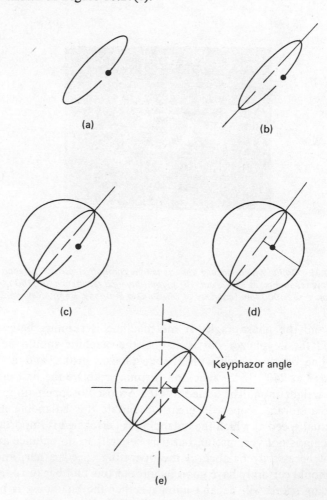

FIGURE 10.27 *Procedure for locating the true keyphazor angle for noncircular orbits*

3. Through the keyphazor dot, draw a line perpendicular to the major axis and extend it radially until it intersects the circle, as shown in Figure 10.27(d).

4. The angle between the vertical and the intercept of the perpendicular line with the circle is then the true keyphazor angle, as shown in Figure 10.27(e).

EXAMPLE 10.6 The system shown in Figure 10.24 will be used to illustrate the balancing procedure. In general, balancing should be performed at a speed that is either approximately $\frac{2}{3}$ or $1\frac{1}{2}$ times the resonance speed. In this example the balancing will be performed at high speed. The initial orbit is that shown in Figure 10.26(c). Since the balancing is performed above resonance, additional weight must be applied at the keyphazor position on the orbit. The first application of a weight at this position results in the orbit shown in the left side of Figure 10.28. It is obvious that this application of additional weight has reduced the vibration somewhat, but it is not sufficient. When additional weight was added in the same position, the new orbit obtained is shown on the right side of Figure 10.28. Note that this last orbit indicates that the system is perfectly balanced.

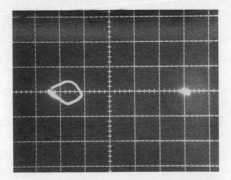

FIGURE 10.28 *Orbits obtained in balancing the system shown in Figure 10.24 (courtesy of Mr. Donald E. Bently, Bently Nevada Corporation)*

EXAMPLE 10.7 The rotor-bearing configuration shown in Figure 10.20 was employed in this balancing example. A larger rotor than that used in the other examples was employed and the translational balance resonance of the system was found to be at 8000 rpm. This system was balanced below resonance at 4200 rpm. The oscilloscope was calibrated for $2\frac{1}{2}$ mils, or 63.5 μm/division, and the unbalanced orbit is shown on the left side of Figure 10.29. Since balancing was performed below resonance, the additional weight to be added to balance the system must be added to the position directly opposite the keyphazor mark (i.e., at the twelve-o'clock position).

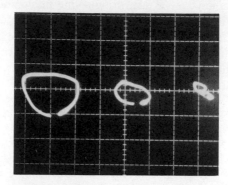

FIGURE 10.29 *Orbits obtained in balancing a system as shown in Figure 10.20 (courtesy of Mr. Donald E. Bently, Bently Nevada Corporation)*

The application of additional weight produced the orbit shown in the center of the figure. This additional weight did reduce the amount of vibration, but it was not sufficient. To correct the unbalanced condition, twice as much weight as was added the first time was applied at the twelve-o'clock position, and a small weight was added at the ten-o'clock position. The resulting orbit is shown on the right side of Figure 10.29. Note that the system is now essentially well balanced. There remains, however, some double-frequency component in the orbit caused by normal machine nonlinearities.

A word of caution is in order when balancing a system. If motions such as $2 \times$ rpm, $\frac{1}{2}$rpm, or 43%rpm are present in a system, the cause of the vibrations should be eliminated before attempting to balance the system. For example, if 5 to 7 mils (127 to 178 μm) of vibration are measured at a bearing that has 6 mils (152 μm) of bearing clearance, any attempt to balance the system without correcting the bearing clearance will not solve the problem but may lead to an oil-whirl or oil-whip condition.

In this discussion of rotor balancing we have treated only the simplest case. No attempt has been made to examine more advanced topics, such as, for example, the use of modal balancing and the influence coefficient method for multiplane balancing of flexible rotors. However, the reader interested in pursuing these topics will find a good discussion and host of references in [16].

10.5.2 Rotating-Machinery Rubs

We have already seen that a rotating system inherently possesses the ability to correct for certain forces, such as mass unbalance. However, it has no mechanism to compensate for rubs. These rubs in rotating machinery can be detected using the instruments employed in balancing. Although there are perhaps an infinite variety of partial rub orbits, in general the orbits typically display such phenomena as loops, as shown in Figure 10.30(a), (b), and (c), and keyphazor jitter, as shown in Figure 10.30(d).

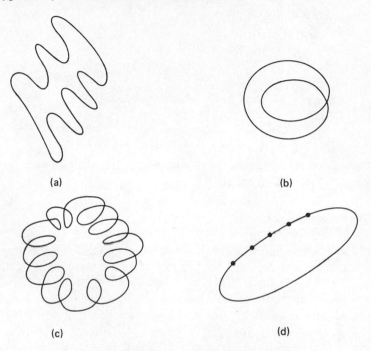

(a)

(b)

(c)

(d)

FIGURE 10.30 *Typical partial-rub orbits*

In sharp contrast to the extremely large number of partial-rub orbits, the full-rub condition always displays the same format, assuming, of course, that the machine actually survives the initiation of this condition. The full-rub orbit results, in general, from a progressive action toward a more unstable condition which begins with a partial rub such as that displayed in Figure 10.30 and progresses through the heavy rub shown in Figure 10.31(a) to the destructive full-rub condition shown in Figure 10.31(b).

Another important facet of the rub condition is the resultant shift in critical frequencies. For example, if the system shown in Figure 10.17(a) develops a rub at the center of the rotor, the point of rub begins to act like another bearing, thus increasing the total spring coefficient and reducing the effective mass. Since the natural resonance of the machines is related to $\sqrt{k/m}$ as shown in Section 10.5.1, this change in system parameters acts to

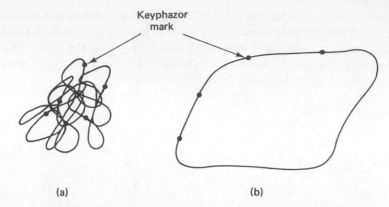

FIGURE 10.31   (a) *Heavy-rub and* (b) *full-rub orbits*

increase the natural resonance speeds of the machine and thus shift the curve in Figure 10.7 to higher rpm's.

10.5.3 Misalignment

Cold and hot alignment techniques were briefly mentioned earlier and a list of references was provided for the interested reader. However, the cold alignment and subsequent "hot check" are useful only as an initial step, since the operating parameters of the system will change with time and load. Thus, a continuous check by means of the protection monitoring system is needed to ensure that any type of misalignment that already exists or develops is detected. Once again, the orbits provide the data for misalignment detection. Orbits of the form shown in Figure 10.32 are typical of misalignment caused by a downward vertical load. In general, machine misalignment is usually indicated by the presence of a 2 × rpm vibration (i.e., vibration that occurs at twice the running speed). This is caused by the bearing curvature working against the preload. A good rule of thumb for determining the frequency of the vibration with respect to the running speed is that this vibration-speed ratio is equal to the number of outside loops −1 or the number of inside loops +1 on the orbit pattern. For example, both orbits in Figure 10.33 illustrate a 2 × rpm vibration.

Bently [1] indicates that a static gap shaft position check coupled with the orbits at running speed provide the pertinent data for troubleshooting misalignment. For example, suppose that a check of the machine shown in Figure 10.13 at standstill is performed with vertical probes mounted at the outlet of the driver and inlet of the load and that these probes indicate that the machine is properly aligned. However, when the machine is operating at

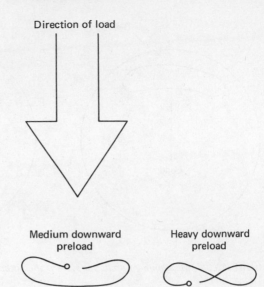

FIGURE 10.32 *Typical misalignment orbits*

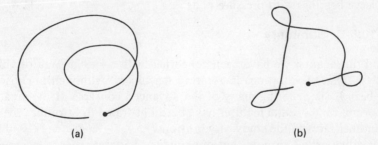

FIGURE 10.33 *Typical orbits for a 2 × rpm vibration: (a) inside orbits; (b) outside orbits*

running speed and full load, the orbits obtained from the probes are as shown in Figure 10.34. Hence, we would conclude that the machine is misaligned.

10.5.4 Looseness and Oil Whirl

Two other causes of vibration that we have discussed are looseness and oil whirl. Typically, a loose housing or loose radial bearing constraint results in radial dynamic motion that is $\frac{1}{2}$, $\frac{1}{3}$, $\frac{1}{4}$, etc., the forcing frequency rate. The ramifications of this condition are treated in some detail by Bently [17,18].

Oil whirl is another forced subrotative speed mechanism. The oil whirl

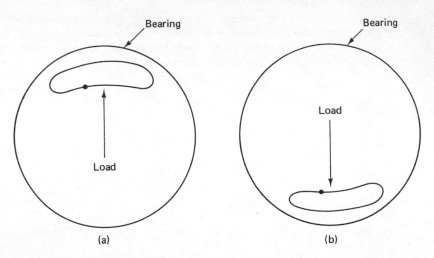

FIGURE 10.34 *Misalignment orbits for system shown in Figure 10.13:* (a) *driver outlet;* (b) *load inlet*

can often be corrected through speed reduction, application of a steady pre-load, or by changing to a different type of bearing, such as tilted pad bearings or sleeve bearings with pressure pads.

10.5.5 Summary

In this chapter we have been concerned with a special area of vibration analysis, that of vibrations in rotating machinery. Although our discussion has been basic, ignoring many of the advanced concepts of the area, many important fundamental features have been illustrated to introduce the reader to another facet of the study of vibrations.

It is important to note that our analysis here has been concerned with very small vibrations with essentially no regard for the resultant noise level. The reason for this is that in rotating machinery, by the time that noise is produced due to vibrations, the machine may be on the verge of self-destruction.

PROBLEMS

10.1 A rotor weighing 3500 N is suspended on a 3-cm-diameter steel shaft which is mounted on bearings positioned 85 cm apart. Neglecting the mass of the shaft, determine the critical speed. Assume that the modulus of elasticity for steel is 20.7×10^6 N cm^2.

10.2 A particular rotor which consists of a massive disk is suspended on a 4-cm-diameter shaft mounted between two bearings. The distance between the bearings is 100 cm. The rotor is observed to have a critical speed of 500 rpm. Assume the modulus of elasticity of the shaft to be 20.7×10^6 N/cm². From this information, determine the weight of the rotor arrangement.

REFERENCES

[1] BENTLY, D. E. "Orbits," Bently-Nevada Corporation, Minden, Nev., 1969.

[2] JACKSON, C. "Cold and Hot Alignment Techniques of Turbomachinery," *Proceedings of the Second Annual Turbomachinery Symposium*, Texas A & M University, College Station, Tex., Oct. 1973, pp. 1–7.

[3] JACKSON, C. "Successful Shaft Hot-Alignment," *Hydrocarbon Processing*, Vol. 48, No. 1 (Jan. 1969), pp. 100–104.

[4] JACKSON, C. "Shaft Alignment Using Proximity Transducers," *ASME Paper 68-PET-25*, Sept. 1968.

[5] JACKSON, C. "How to Align Barrel-Type Centrifugal Compressors," *Hydrocarbon Processing*, Vol. 50, No. 9 (Sept. 1971), pp. 189–194.

[6] JACKSON, C. "Alignment of Rotating Equipment," presented at the NPRA Refinery and Petrochemical Plant Maintenance Conference, Jan. 30–Feb. 1, Houston, Tex.

[7] BARNES, E. F. "Optical Alignment Case Histories,".*Hydrocarbon Processing*, Vol. 50, No. 1 (Jan. 1971), p. 80.

[8] DODD, V. R. "Shaft Alignment Monitoring Cuts Costs," *Oil Gas J.*, Sept. 1972, pp. 91–96.

[9] ESSINGER, J. N. "A Closer Look at Turbomachinery Alignment," *Hydrocarbon Processing*, Sept. 1973, pp. 185–188.

[10] CAMPBELL, A. J. "Optical Alignment of Turbomachinery," *Proceedings of the Second Annual Turbomachinery Symposium*, Texas A & M University, College Station, Tex., Oct. 1973, pp. 8–12.

[11] GUNTER, E. J., JR. "Dynamic Stability of Rotor-Bearing Systems," *NASA SP 113*, 1966.

[12] LOEWY, R. G. and V. J. PIARULLI *Dynamics of Rotating Shafts*, Shock and Vibration Monograph SVM-4, The Shock and Vibration Information Center, Naval Research Laboratory, Washington, D.C., 1969.

[13] ESHLEMAN, R. L. "Flexible Rotor-Bearing System Dynamics I. Critical Speeds and Response of Flexible Rotor Systems," ASME, 1972.

[14] KIMBALL, A. L. and E. H. HULL "Vibration Phenomena of a Loaded Unbalanced Shaft While Passing through Critical Speed," *Trans. ASME*, Vol. 47 (1926), p. 673.

[15] NEWKIRK, B. L. "Shaft Whipping," *General Electric Rev.*, Vol. 27 (1924), pp. 169–178.

[16] RIEGER, N. F. "Flexible Rotor-Bearing System Dynamics III. Unbalance Response and Balancing of Flexible Rotors in Bearings," ASME, 1973.

[17] BENTLY, D. E. "Forced Subrotative Speed Dynamic Action of Rotating Machinery," *ASME Publication No. 74-PET-16*, 1974.

[18] BENTLY, D. E. "Forward Subrotative Speed Resonance Action of Rotating Machinery," presented at the Fourth Turbomachinery Symposium, Texas A & M University, College Station, Tex., Oct. 1975.

BIBLIOGRAPHY

BENTLY, D. E. "Continuous Turbine Supervision and Simplified In-Place Balancing in One Package," *Diesel and Gas Turbine Progr.*, Apr. 1973.

BENTLY, D. E. "Monitor Machinery Condition for Safe Operation," *Hydrocarbon Processing*, Nov. 1974, pp. 205–208.

BLAKE, M. P. "Use Phase Measuring to Balance Rotors in Place," *Hydrocarbon Processing*, Aug., 1967, pp. 127–132.

IRD Mechanalysis, Inc., "Mechanalysis Instruction Manual—Vibration Analysis and Dynamic Balancing," 1961.

JACKSON, C. "New Look at Vibration Measurement," *Hydrocarbon Processing*, Jan. 1969, pp. 89–99.

JACKSON, C. "Balance Rotors by Orbit Analysis," *Hydrocarbon Processing*, Jan. 1971.

KELLER, A. C. "Real-Time Spectrum Analysis of Machinery Dynamics," *Sound and Vibration*, Apr. 1975, pp. 40–48.

SHAPIRO, W. and J. H. RUMBARGER "Flexible Rotor-Bearing System Dynamics II. Bearing Influence and Representation in Rotor Dynamics Analysis," ASME, 1972.

APPENDIX A

CONVERSION FACTORS

To convert	Into	Multiply by	Conversely, multiply by
atm (atmosphere)	mm Hg at 0°C	760	1.316×10^{-3}
	lb/in.^{2}	14.70	6.805×10^{-2}
	N/m^2 (Pa)	1.0132×10^5	9.872×10^{-6}
	kg/m^2	1.033×10^4	9.681×10^{-5}
°C (Celsius)	°F (fahrenheit)	$[(°C \times 9)/5] + 32$	$(°F\text{-}32) \times 5/9$
cm (centimeter)	in. (inch)	0.3937	2.540
	ft (foot)	3.281×10^{-2}	30.48
	m (meter)	10^{-2}	10^2
cm^2 (square centimeter)	in.^{2}	0.1550	6.452
	ft^2	1.0764×10^{-3}	929
	m^2	10^{-4}	10^4
cm^3 (cubic centimeter)	in.^{3}	0.06102	16.387
	ft^3	3.531×10^{-5}	2.832×10^4
	m^3	10^{-6}	10^6
dyne	lb (force)	2.248×10^{-6}	4.448×10^5
	N (newton)	10^{-5}	10^5
dynes/cm^2	lb/ft^2 (force)	2.090×10^{-3}	478.5
	N/m^2 (Pa)	10^{-1}	10
ft (foot)	in. (inch)	12	0.08333
	cm (centimeter)	30.48	3.281×10^{-2}
	m (meter)	0.3048	3.281
ft^2 (square foot)	in.^{2}	144	6.945×10^{-3}
	cm^2	9.290×10^2	0.010764
	m^2	9.290×10^{-2}	10.764

To convert	Into	Multiply by	Conversely, multiply by
ft^3 (cubic foot)	in.^{2}	1728	5.787×10^{-4}
	cm^3	2.832×10^4	3.531×10^{-5}
	m^3	2.832×10^{-2}	35.31
hp (horsepower)	W (watt)	745.7	1.341×10^{-3}
in. (inch)	ft (foot)	0.0833	12
	cm (centimeter)	2.540	0.3937
	m (meter)	0.0254	39.37
in.^{2} (square inch)	ft^2	6.945×10^{-3}	144
	cm^2	6.452	0.1550
	m^2	6.452×10^{-4}	1550
in.^{3} (cubic inch)	ft^3	5.787×10^{-4}	1.728×10^3
	cm^3	16.387	6.102×10^{-2}
	m^3	1.639×10^{-5}	6.102×10^4
kg (kilogram)	lb (weight)	2.2046	0.4536
	slug	0.06852	14.594
	g (gram)	10^3	10^{-3}
kg/m^2	lb/in.^{2} (weight)	1.422×10^{-3}	703.0
	lb/ft^2 (weight)	0.2048	4.882
	g/cm^2	10^{-1}	10
m (meter)	in. (inch)	39.371	2.540×10^{-2}
	ft (foot)	3.2808	0.30481
	cm (centimeter)	10^2	10^{-2}
m^2 (square meter)	in.^{2}	1550	6.452×10^{-4}
	ft^2	10.764	9.290×10^{-2}
	cm^2	10^4	10^{-4}
m^3 (cubic meter)	in.^{3}	6.102×10^4	1.639×10^{-5}
	ft^3	35.31	2.832×10^{-2}
	cm^3	10^6	10^{-6}
microbar (dynes/cm^2)	lb/in.^{2}	1.4513×10^{-5}	6.890×10^4
	lb/ft^2	2.090×10^{-3}	478.5
	N/m^2 (Pa)	10^{-1}	10
Np (neper)	dB (decibel)	8.686	0.1151
N (newton)	lb (force)	0.2248	4.448
	dynes	10^5	10^{-5}
N/m^2 (pascal, Pa)	lb/in.^{2} (force)	1.4513×10^{-2}	6.890×10^3
	lb/ft^2 (force)	2.090×10^{-2}	47.85
	dynes/cm^2	10	10^{-1}
lb (force) (pound)	N (newton)	4.448	0.2248
lb (weight) (pound)	slug	0.03108	32.17
	kg (kilogram)	0.4536	2.2046
lb/in.^{2} (weight)	lb/ft^2 (weight)	144	6.945×10^{-3}
	kg/m^2	703	1.422×10^{-3}
lb/in.^{2} (force)	lb/ft^2 (force)	144	6.945×10^{-3}
	N/m^2 (Pa)	6894	1.4506×10^{-4}
lb/ft^2 (weight)	lb/in.^{2} (weight)	6.945×10^{-3}	144
	g/cm^2	0.4882	2.0482
	kg/m^2	4.882	0.2048

To convert	Into	Multiply by	Conversely, multiply by
lb/ft^2 (force)	lb/in.^{2} (force)	6.945×10^{-3}	144
	N/m^2 (Pa)	47.85	2.090×10^{-2}
slugs	lb (weight)	32.17	3.108×10^{-2}
	kg (kilogram)	14.594	6.852×10^{-2}
W (watt)	hp (horsepower)	1.341×10^{-3}	745.7

APPENDIX B

SOLUTION OF ORDINARY DIFFERENTIAL EQUATIONS
BY USE OF LAPLACE TRANSFORMS

INTRODUCTION

In this appendix we shall discuss elements of the Laplace transform that will aid the reader in understanding the material in the preceding chapters. Only a very few of the Laplace transform characteristics will be examined. However, the techniques presented are simple and extremely powerful procedures which can be effectively employed to solve linear differential equations.

The reader should note as we proceed that in essence we trade a differential equation for an algebraic one when we attack a differential equation via the Laplace transform.

Definition:

If $f(t)$ is a function of t for $t > 0$, then its Laplace transform $F(s)$ is defined by the expression

$$F(s) = \int_0^\infty f(t)e^{-st}\, dt = \mathcal{L}[f(t)] \qquad \text{(B.1)}$$

where s is a complex variable $s = \sigma + j\omega$

The function $f(t)$ is Laplace-transformable for $\sigma > 0$ if

$$\lim_{T \to \infty} \int_0^T |f(t)|\, e^{-\sigma t}\, dt < \infty \qquad \text{(B.2)}$$

406

EXAMPLE B.1 If $f(t) = A$ for $t > 0$, then the Laplace transform of the constant A is

$$\mathcal{L}[A] = \int_0^\infty Ae^{-st}\, dt = -\frac{1}{s}e^{-st}\Big|_0^\infty = \frac{A}{s} = F(s)$$

Hence, one of the Laplace transform pairs is

$$f(t) = A \Longleftrightarrow F(s) = \frac{A}{s}$$

EXAMPLE B.2 If $f(t) = e^{-at}$ for $t > 0$, then the Laplace transform of the exponential function is

$$\mathcal{L}e^{-at} = \int_0^\infty e^{-at}e^{-st}\, dt = \int_0^\infty e^{-(s+a)t}\, dt = F(s)$$

$$F(s) = \frac{-1}{s+a}e^{-(s+a)t}\Big|_0^\infty$$

$$= \frac{1}{s+a}$$

Therefore,

$$f(t) = e^{-at} \Longleftrightarrow F(s) = \frac{1}{s+a}$$

LAPLACE TRANSFORMS OF DERIVATIVES

Given that a function $f(t)$ and its derivative $df(t)/dt$ are both Laplace-transformable, and in addition if the function $f(t)$ has the Laplace transform $F(s)$, then

$$\mathcal{L}\frac{df(t)}{dt} = sF(s) - f(0+) \qquad (B.3)$$

Equation (B.3) can be easily proved by using integration by parts,

$$F(s) = \int_0^\infty f(t)e^{-st}\, dt$$

Letting

$$u = f(t), \qquad du = \frac{df(t)}{dt}\, dt$$

$$dv = e^{-st}\, dt, \qquad v = -\frac{1}{s}e^{-st}$$

Then

$$F(s) = uv\Big|_0^\infty - \int_0^\infty v\, du$$

$$= -\frac{1}{s}f(t)e^{-st}\Big|_0^\infty + \frac{1}{s}\int_0^\infty \frac{df(t)}{dt}e^{-st}\, dt$$

$$= \frac{f(0+)}{s} + \frac{1}{s}\int_0^\infty \frac{df(t)}{dt}e^{-st}\, dt$$

or, in other words,

$$\int_0^\infty \frac{df(t)}{dt} e^{-st}\, dt = sF(s) - f(0+)$$

The term $f(0+)$ is the value of $f(t)$ as $t \to 0$ from the positive side.
In a similar manner, we can show that

$$\mathcal{L}\frac{d^2f(t)}{dt^2} = s^2F(s) - sf(0+) - \dot{f}(0+) \tag{B.4}$$

SOLUTION OF DIFFERENTIAL EQUATIONS

A differential equation that is of primary importance to us is one of the form

$$m\ddot{x}(t) + C\dot{x}(t) + kx(t) = f(t) \tag{B.5}$$

or, equivalently,

$$\ddot{x}(t) + 2\zeta\omega_n\dot{x}(t) + \omega_n^2 x(t) = f_1(t) \tag{B.6}$$

Applying equations (B.3) and (B.4) to equation (B.6) yields

$$s^2X(s) - sx(0+) - \dot{x}(0+) + 2\zeta\omega_n[sX(s) - x(0+)] + \omega_n^2 X(s) = F_1(s)$$

Solving for $X(s)$,

$$X(s) = \frac{F_1(s) + sx(0+) + 2\zeta\omega_n x(0+) + \dot{x}(0+)}{s^2 + 2\zeta\omega_n s + \omega_n^2}$$

or

$$\boxed{X(s) = \frac{F_1(s) + x(0+)(s + 2\zeta\omega_n) + \dot{x}(0+)}{s^2 + 2\zeta\omega_n s + \omega_n^2}} \tag{B.7}$$

Note that in general we can write equation (B.7) in the form

$$X(s) = \frac{A(s)}{B(s)} \tag{B.8}$$

Although we have derived equation (B.7) and thus the form of equation (B.8) using a second-order equation, the techniques that follow are valid for any order function.

Equations with Simple Roots

For equations with simple roots, equation (B.8) is of the form

$$X(s) = \frac{A(s)}{(s + a_1)(s + a_2)\dots(s + a_n)} = \frac{A(s)}{B(s)} \tag{B.9}$$

We assume throughout the discussion that $B(s)$ is of higher order than $A(s)$. Equation (B.9) can be expanded using partial fraction expansions to obtain

$$X(s) = \frac{C_1}{s + a_1} + \frac{C_2}{s + a_2} + \dots + \frac{C_n}{s + a_n} \qquad (B.10)$$

where

$$C_k = \lim_{s \to -a_k} (s + a_k)\frac{A(s)}{B(s)} \qquad (B.11)$$

EXAMPLE B.3 We want to find the inverse Laplace transform of the function

$$X(s) = \frac{10(s + 3)}{s(s + 1)(s + 2)}$$

The roots of the numerator (e.g., $s = -3$) are called *zeros* of the function and the roots of the denominator (e.g., $s = 0, -1, -2$) are called *poles*. A partial fraction expansion of the function yields

$$\frac{10(s + 3)}{s(s + 1)(s + 2)} = 10\left(\frac{C_1}{s} + \frac{C_2}{s + 1} + \frac{C_3}{s + 2}\right)$$

The residues C_1, C_2, and C_3 are obtained as follows. Using equation (B.11),

$$C_1 = \lim_{s \to 0} \frac{s + 3}{(s + 1)(s + 2)}$$

$$= \frac{3}{(1)(2)} = \frac{3}{2}$$

$$C_2 = \lim_{s \to -1} \frac{s + 3}{s(s + 2)}$$

$$= \frac{(2)}{(-1)(1)} = -2$$

$$C_3 = \lim_{s \to -2} \frac{s + 3}{s(s + 1)}$$

$$= \frac{1}{(-2)(-1)} = \frac{1}{2}$$

Therefore,

$$X(s) = 10\left(\frac{\frac{3}{2}}{s} + \frac{-2}{s + 1} + \frac{\frac{1}{2}}{s + 2}\right)$$

Using the transform pairs derived earlier,

$$x(t) = 10(\tfrac{3}{2} - 2e^{-t} + \tfrac{1}{2}e^{-2t})$$

Knowing the form of the inverse transform above we can actually derive the residues graphically by inspection. The procedure for doing this is simply to divide the product of all vectors from the zeros to the pole whose residue is being determined by the product of the vectors from all the other poles to the pole in question. This technique is demonstrated in Figure B.1. From Figure B.1(a) we derive C_1 as

$$C_1 = \frac{3\underline{/0°}}{(2\underline{/0°})(1\underline{/0°})}$$

$$= \frac{3}{2}$$

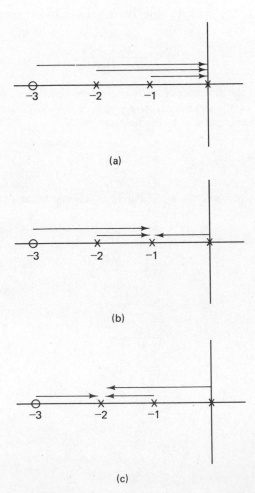

(a)

(b)

(c)

FIGURE B.1 *Example of the graphical residue technique*

From Figure B.1(b) we obtain

$$C_2 = \frac{2/0°}{(1/0°)(1/180°)}$$

$$= \frac{2}{(1)(-1)}$$

$$= -2$$

and from Figure B.1(c),

$$C_3 = \frac{1/0°}{(1/180°)(2/180°)}$$

$$= \frac{1}{(-1)(-2)}$$

$$= \frac{1}{2}$$

Note carefully how the numbers calculated by use of the graphical residue technique compare with those derived above using equation (B.11).

The sign of the residue can be easily determined by simply counting the number of poles and zeros to the right in the s plane of the pole whose residue is being calculated. If the number is even, the sign of the residue is positive, and if the number is odd, the sign is negative.

EXAMPLE B.4 Given the function

$$X(s) = \frac{5(s+2)(s+4)}{s(s+1)(s+3)}$$

The pole–zero pattern is shown in Figure B.2. Using the graphical residue technique outlined above, we can write the inverse transform down by inspection as

$$x(t) = 5\left[\frac{(2)(4)}{(1)(3)} - \frac{(1)(3)}{(1)(2)}e^{-t} - \frac{(1)(1)}{(2)(3)}e^{-3t}\right]$$

$$= 5[\tfrac{8}{3} - \tfrac{3}{2}e^{-t} - \tfrac{1}{6}e^{-3t}]$$

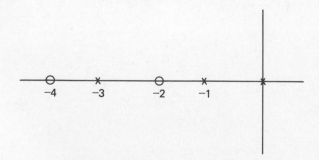

FIGURE B.2 *Pole–zero pattern for Example B.4*

Equation with Multiple-Order Roots

If the function $X(s)$ has multiple-order roots of the form

$$X(s) = \frac{A(s)}{B(s)} = \frac{A(s)}{(s + a_1)^m (s + a_2) \dots (s + a_n)}$$

then the partial fraction expansion of the function must be of the form

$$\frac{A(s)}{B(s)} = \frac{C_{11}}{(s + a_1)^m} + \frac{C_{12}}{(s + a_1)^{m-1}} + \dots + \frac{C_{1m}}{s + a_1} + \frac{C_2}{s + a_2} + \dots + \frac{C_n}{s + a_n}$$

The residues $C_2, C_3, \dots, C_n$ are evaluated as shown in the preceding section. However, the coefficients $C_{11}, C_{12}, \dots, C_{1m}$ require special consideration. They are evaluated by means of the expression

$$C_{1j} = \frac{1}{(j - 1)!} \frac{d^{j-1}}{ds^{j-1}} (s + a_1)^m \frac{A(s)}{B(s)} \qquad (B.12)$$

which is evaluated at $s = -a_1$. Then the inverse Laplace transform is obtained from the pair.

$$\mathcal{L}^{-1}\left[\frac{1}{(s + a)^j}\right] = \frac{t^{j-1}}{(j - 1)!} e^{-at} \qquad (B.13)$$

The following simple example will serve to illustrate the procedure.

EXAMPLE B.5   Given

$$X(s) = \frac{s + 1}{(s + 2)^2 (s + 3)} = \frac{C_{11}}{(s + 2)^2} + \frac{C_{12}}{s + 2} + \frac{C_2}{s + 3}$$

Using equation (B.11), we derive

$$C_{11} = \frac{s + 1}{s + 3}\bigg|_{s=-2}$$

$$= -1$$

$$C_2 = \frac{s + 1}{(s + 2)^2}\bigg|_{s=-3}$$

$$= -2$$

The coefficient C_{12} is obtained from equation (B.12):

$$C_{12} = \frac{1}{(2 - 1)!} \frac{d^{2-1}}{ds^{2-1}} \frac{s + 1}{s + 3}$$

$$= \frac{d}{ds} \frac{s+1}{s+3} \Big|_{s=-2}$$

$$= 2$$

Therefore,

$$X(s) = \frac{-1}{(s+2)^2} + \frac{2}{s+2} - \frac{2}{s+3}$$

Using the Laplace transform pairs discussed earlier, we obtain

$$x(t) = -te^{-2t} + 2e^{-2t} - 2e^{-3t}$$

Equations with Complex Roots

Consider the problem of finding the inverse Laplace transform of a function of the form

$$X(s) = \frac{(s+b_1)(s+b_2)}{[(s+\alpha)^2 + \beta^2](s+a_2)(s+a_3)(s+a_4)}$$

The inverse transform can be evaluated via a partial fraction expansion as demonstrated earlier. However, if complex roots are present, this procedure can be quite involved. To attack this problem, we will first show that

$$\frac{as+b}{(s+\alpha)^2 + \beta^2} \Longleftrightarrow 2Me^{-\alpha t} \cos(\beta t + \phi)$$

is a transform pair with unknowns M and ϕ which are dependent upon a and b. We will then show that M and ϕ can be evaluated by inspection using the graphical residue approach.

Let us suppose that the function above has the pole–zero pattern shown in Figure B.3(a). We will determine only the residues of the complex poles at $-\alpha \pm j\beta$. Using the graphical approach outlined above for simple roots, we obtain, using Figure B.3(b),

$$X(s) = \frac{\dfrac{(z_1 e^{j\theta_2})(z_2 e^{j\theta_6})}{(P_1 e^{j\theta_1})(P_2 e^{j\theta_3})(P_3 e^{j\theta_4})(P_4 e^{j\theta_5})}}{s + \alpha - j\beta}$$

$$+ \frac{\dfrac{(z_1 e^{-j\theta_2})(z_2 e^{-j\theta_6})}{(P_1 e^{-j\theta_1})(P_2 e^{-j\theta_3})(P_3 e^{-j\theta_4})(P_4 e^{-j\theta_5})}}{s + \alpha + j\beta} + \text{all other terms}$$

Note that the coefficient of the second term, which is the residue at $s = -\alpha - j\beta$, is the complex conjugate of the coefficient of the first term, the residue at $s = -\alpha + j\beta$. The equation above can be written in the form

$$X(s) = \frac{Me^{j\phi}}{s + \alpha - j\beta} + \frac{Me^{-j\phi}}{s + \alpha + j\beta} + \text{all other terms}$$

where $\quad M = \dfrac{z_1 z_2}{P_1 P_2 P_3 P_4}, \quad \phi = \theta_2 + \theta_6 - (\theta_1 + \theta_3 + \theta_4 + \theta_5)$

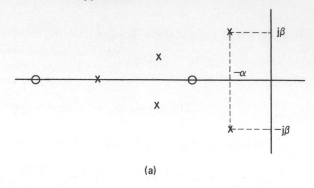

(a)

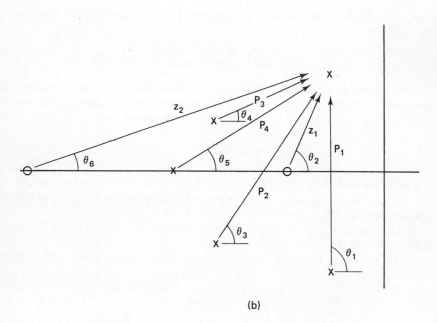

(b)

FIGURE B.3　*Graphical residue approach for complex roots*

Then

$$x(t) = M(e^{-\alpha t + j\beta t + j\phi} + e^{-\alpha t - j\beta t - j\phi}) + \text{all other terms}$$

$$= Me^{-\alpha t}(e^{j(\beta t + \phi)} + e^{-j(\beta t + \phi)}) + \text{all other terms}$$

$$= 2Me^{-\alpha t}\frac{e^{j(\beta t + \phi)} + e^{-j(\beta t + \phi)}}{2} + \text{all other terms}$$

We recognize the term in the brackets as the cosine function and hence

$$x(t) = 2Me^{-\alpha t}\cos(\beta t + \phi) + \text{all other terms}$$

where M and ϕ can be easily evaluated graphically using Figure B.3.

It is important for the reader to note that the application of the graphical residue technique to equations with complex roots is extremely powerful in that the inverse transform can be essentially written down by inspection. The following examples will illustrate the ease with which the technique can be applied.

EXAMPLE B.6 Given

$$X(R) = \frac{s + 2}{s(s^2 + 4s + 5)}$$

The function is shown in Figure B.4(a). The residue for the pole at the origin is calculated using Figure B.4(b), and the residues for the complex poles are

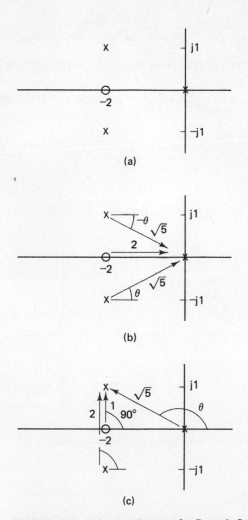

(a)

(b)

(c)

FIGURE B.4 *Pole–zero diagrams for Example B.6*

calculated using Figure B.4(c). The resulting function is

$$x(t) = \frac{2}{\sqrt{5}\sqrt{5}} + 2\frac{1}{2\sqrt{5}}e^{-2t}\cos(t + \phi)$$

$$\phi = (90°) - (90° + \theta)$$

$$\theta = (\pi - \tan^{-1}\tfrac{1}{2})$$

EXAMPLE B.7   If $X(s)$ is of the form

$$X(s) = \frac{x_0(s + 2\xi\omega_n)}{s^2 + 2\xi\omega_n + \omega_n^2)}$$

$$= \frac{x_0(s + 2\xi\omega_n)}{(s + \xi\omega_n + j\omega_n\sqrt{1 - \xi^2})(s + \xi\omega_n - j\omega_n\sqrt{1 - \xi^2})}$$

Then the inverse transform can be evaluated by inspection from Figure B.5 as

$$x(t) = x_0\left[\frac{\omega_n}{\omega_n\sqrt{1 - \xi^2}}e^{-\xi\omega_n t}\cos(\omega_n\sqrt{1 - \xi^2}\,t + \phi)\right]$$

where $\phi = \theta - \pi/2$

$\theta = \cos^{-1}\xi$

so that

$$x(t) = \frac{x_0}{\sqrt{1 - \xi^2}}e^{-\xi\omega_n t}\cos(\omega_n\sqrt{1 - \xi^2}\,t - \pi/2 + \cos^{-1}\xi)$$

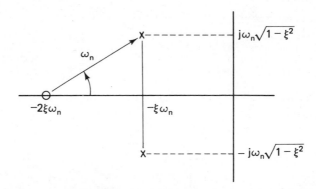

FIGURE B.5   *Pole–zero diagram for Example B.7*

APPENDIX C*

DETAILED SUMMARY OF NOISE CONTROL ACT OF 1972†

The Noise Control Act of 1972 charges the Environmental Protection Agency with the primary responsibilities for controlling environmental noise. The major provisions of the Act passed by the 92nd Congress on October 27, 1972, are summarized section by section below.

SECTION 1. SHORT TITLE

The Act may be cited as the "Noise Control Act of 1972."

SECTION 2. FINDINGS AND POLICY

Findings

That inadequately controlled noise presents a growing danger to the health and welfare of the nation's population and that the control of major sources of noise requires national uniformity of treatment.

Policy

It is the policy of the United States to promote an environment for all Americans free from noise that jeopardizes their health or welfare by (1) establishing a means for effective coordination of federal research in noise control, (2) authorizing the establishment of federal noise emission standards for products distributed in com-

*Prepared by Charles E. Hickman, Southern Company Services, Inc., Birmingham, Alabama.
†Public Law 92–574, 92nd Congress, H.R. 11021, October 27, 1972.

merce, and (3) providing information to the public respecting the noise emission and noise reduction characteristics of such products.

SECTION 3. DEFINITIONS

The Act defines "administrator" as the administrator of the Environmental Protection Agency. Other definitions listed are "person," "product," "ultimate purchaser," "new product," "manufacturer," "commerce," "distribute in commerce," "state," "federal agency," and "environmental noise."

SECTION 4. FEDERAL PROGRAMS

(A) Federal agencies shall

1. administer the noise control programs within their control,
2. comply with federal, state, interstate, and local noise regulations unless exempted by the president, and
3. consult with EPA in prescribing standards or regulations respecting noise.

(B) EPA shall

1. coordinate the programs of all federal agencies relating to noise research and noise control, and
2. publish periodic reports on the status and progress of federal activities relating to noise research and noise control.

SECTION 5. IDENTIFICATION OF MAJOR NOISE SOURCES; NOISE CRITERIA AND CONTROL TECHNOLOGY

EPA shall

1. develop and publish criteria by July 27, 1973, with respect to the kind and extent of all identifiable effects on the public health or welfare which may be expected from different quantities and qualities of noise,
2. publish information by October 27, 1973, on the levels of environmental noise required in defined areas and under various conditions to protect the public health and welfare with an adequate margin of safety,
3. compile and publish reports by April 27, 1974
   (a) identifying products which are major sources of noise, and
   (b) providing information on techniques for control of noise from such products, and
4. review and revise from time to time any criteria or reports published under this section.

SECTION 6. NOISE EMISSION STANDARDS FOR PRODUCTS DISTRIBUTED IN COMMERCE

(A) EPA shall publish by April 27, 1974, proposed noise emission standards for products identified as major sources of noise in the following categories:

1. Construction equipment.
2. Transportation equipment (including recreational vehicles and related equipment).
3. Any motor or engine (including any equipment of which an engine or motor is an integral part).
4. Electrical or electronic equipment.

(B) Manufacturers of each new regulated product shall

1. warrant that such product is designed and built to meet the regulation at the time of sale, and
2. shall *not* pass on to the dealer the cost of such a warranty.

(C) No state or political subdivision thereof may set noise emission limits different from those promulgated by EPA, but the states retain the right to regulate the use, operation, or movement of products.

SECTION 7. AIRCRAFT NOISE STANDARDS

(A) EPA shall conduct a study of:

1. the adequacy of Federal Aviation Administration flight and operational noise controls,
2. the adequacy of noise emission standards on new and existing aircraft (including recommendations on the retrofitting and phaseout of existing aircraft),
3. the implications of identifying and achieving levels of cumulative noise exposure around airports, and
4. the additional measures available to airport operators and local governments to control aircraft noise.

(B) The FAA retains the right under the provisions of Section 611 of the Federal Aviation Act of 1958 to prescribe and amend standards for the measurement of aircraft noise and sonic boom. However, EPA can question FAA standards which, in the opinion of EPA, do not protect the public health and welfare.

SECTION 8. LABELING

EPA shall by regulation designate any product which

1. emits noise capable of adversely affecting the public health or welfare,
2. is sold wholly or in part on the basis of its effectiveness in reducing noise

and shall require the manufacturer of such product to provide the consumer with the noise level the product emits or its effectiveness in reducing noise, as the case may be. The methods and units of measurement shall be specified.

SECTION 9. IMPORTS

The Secretary of the Treasury shall, in consultation with EPA, issue regulations to carry out the provisions of this act with respect to new products imported or offered for importation.

SECTION 10. PROHIBITED ACTS

After the effective date of the noise regulations, the following acts are prohibited.

(A) Manufacturer:

1. to distribute any new product which does not conform with the regulations
2. to distribute any new product without proper labeling

(B) All persons:

1. to remove or render inoperative any noise abatement device
2. to use a product after noise abatement device has been removed or rendered inoperative
3. to remove any noise information label
4. to import any new product in violation of regulations.

Under certain specified conditions, EPA may exempt certain products for a specified period of time.

SECTION 11. ENFORCEMENT

Any person who violates the prohibited acts of Section 10 shall be subject to a fine of up to $25,000 per day of violation, or by imprisonment for up to one year, or by both. The penalty may be doubled for subsequent violations. The district courts of the United States have jurisdiction to restrain violations of this act.

SECTION 12. CITIZEN SUITS

Any person may commence a civil action on his own behalf against any person (including the United States) who is alleged to be in violation of any noise control requirement or against the administrator of EPA or FAA for alleged failure to perform any nondiscretionary duty under this act. Legal details are included in the act.

SECTION 13. RECORDS, REPORTS, AND INFORMATION

Each manufacturer of a product to which noise emission standards or labeling requirements apply shall

1. maintain records of compliance,
2. permit EPA to have access to records, and
3. make available to EPA assembly line products.

Although trade secrets shall be considered confidential, such information may be disclosed to federal officers or committees of the Congress. Violation of this section is punishable by a fine of up to $10,000, or by imprisonment of up to six months, or by both.

SECTION 14. RESEARCH, TECHNICAL ASSISTANCE, AND PUBLIC INFORMATION

To further knowledge concerning noise and its effects, EPA is authorized to

1. conduct research and finance research by contract,
2. provide technical assistance to state and local governments, and
3. distribute the information to the public.

SECTION 15. DEVELOPMENT OF LOW-NOISE-EMISSION PRODUCTS

Under the provisions of this section, EPA shall

1. certify "low-noise-emission products" (defined as: products which emit noise in amounts significantly below the levels specified under Section 6 regulations),
2. establish a Low-Noise-Emission Product Advisory Committee, and
3. periodically test the certified products.

Conditions for the procurement of certified products by the federal government are presented.

SECTION 16. JUDICIAL REVIEW; WITNESSES

A petition for review of action of the Administrator of EPA or FAA may be filed only in the United States Court of Appeals for the District of Columbia Circuit. EPA may issue subpenas for the attendance and testimony of witnesses and the production of relevant papers for the purpose of obtaining information to carry out the act.

SECTION 17. RAILROAD NOISE EMISSION STANDARDS

EPA shall publish by July 27, 1973, proposed noise emission regulations for surface carriers engaged in interstate commerce by railroads. Any standard or regulation shall be promulgated only after consultation with the Secretary of Transportation. State and local governments are prohibited from adopting different standards except when necessitated by special local conditions and then only with the approval of EPA and the Secretary of Transportation.

SECTION 18. MOTOR CARRIER NOISE EMISSION STANDARDS

Provisions of this section are nearly identical to those of Section 17, except the standards are for motor carriers engaged in interstate commerce.

SECTION 19. AUTHORIZATION OF APPROPRIATIONS

Appropriations authorized to carry out this act are as follows:

Fiscal year ending	Amount
June 30, 1973	$ 3,000,000
June 30, 1974	6,000,000
June 30, 1975	12,000,000

ANSWERS
TO ODD-NUMBERED PROBLEMS

ANSWERS

TO ODD-NUMBERED PROBLEMS

1.1 (a) $f = 1720$ Hz; (b) $p = 0.45$ Pa.

1.3 $p = 2.83$ Pa.

1.5 (a) $Q = 1.86$; (b) $Q = 2.82$; (c) $Q = 5.75$.

1.7 $W_t = 0.01$ W.

1.9 (Omitted).

2.1 61, 75, 82, 83, 80, 91, 84 phons.

2.3 PN = 156.65 noys, 113.09 PNdB.

2.5 PN = 93.6 noys, PNL = 105.64 PNdB.

2.7 (a) $L_{pt} = 99.8$ dB; (b) $L_t = 96.4$ dBA; (c) $L_t = 97.1$ dBB; (d) $L_t = 99.2$ dBC.

2.9 (a) shouting (+); (b) normal voice; (c) raised voice.

3.1 $D = 2.17$.

3.3 $L_{10} = 97$ dBA, $L_{50} = 90$ dBA, $L_{90} = 84$ dBA.

3.5 $L_{NP} = 103$ dB_{NP} [eq. (3.5)], $L_{NP} = 102$ dB_{NP} [eq. (3.6)].

3.7 $L_{dn} = 75.28$ dBA.

4.1 $L_{p_{machine}} = 87.7 \approx 88$ dB.

4.3 7.1 ft $\times$ 10.25 ft or 2.18 m $\times$ 3.13 m.

4.5 $L_s = -81.4$ dBV/μbar.

4.7 (a) $f_1 = 22.3$ Hz, $f_2 = 44.5$ Hz, bw = 22.2 Hz; (b) $f_1 = 177$ Hz, $f_2 = 354$ Hz, bw = 177 Hz.

5.1 $L_{w_{low}} = 95$ dB, $L_{w_{mid}} = 111$ dB, $L_{w_{high}} = 116$ dB.

5.3 $B_f = 900$ Hz, $L_w = 84$ dB.

5.5 $L_w = 112$ dB [eq. (5.5)], $L_w = 108$ dB (Table 5.2).

5.7 $L_w = 105$ dB [eq. (5.8)], $L_w = 105$ dB (Table 5.2).

6.1 $\delta' = 1.08 \times 10^{-4}$ J/m³.

6.3 $\delta' = 1.056 \times 10^{-5}$ J/ft³ or 3.73×10^{-4} J/m³.

6.5 $L_p = 40$ dB.

6.7 $L_p = 105$ dB.

6.9 $L_{p_{10ft}} = 98$ dB.

6.11 $L_{p_{TOTAL}} = 105$ dB.

6.13 $\bar{L}_p = 90$ dB, $w = 0.525$ W.

6.15 $L_p = 73$ dB.

6.17 $L_{p_{40ft}} = 105$ dB.

7.1 $L_{p_3} = 80$ dB.

7.3 $L_{p_B} = 109$ dB.

7.5 $L_{p_{LARGE}}$ (near the wall in the large room) $= 69$ dB.

7.7 $L_{p_2} = 65$ dB.

7.9 $L_{p_2} = 75$ dB.

7.11 $L_{p_2} = 71$ dB.

7.13 $\bar{\alpha} = 0.796$.

7.15 TL $= 33$ dB.

7.17 L_{p_A}(without barrier) $= 105$ dB, L_{p_A}(with barrier) $= 97$ dB, IL $= 8$ dB.

8.1   (a) NRC $= 0.293$;   (b) NRC $= 0.763$;   (c) NRC $= 0.628$.

8.3 $\bar{\alpha}_2 = 0.317$.

8.5 $f_{BP} = 180$ Hz.

8.7   (a) Att. $= 14$ dB;   (b) att. $= 14$ dB;   (c) att. $= 14$ dB;
         (d) att. $= 14$ dB.

8.9 $m = 35.6$, $L = 2.256$ ft, $48 \leq f \leq 191$ Hz.

8.11 $A = 0.3376$ m².

9.1 $\omega_n = 2.5$ rad/s, $C_e = 10.22$ N/m/s.

9.3 $x(t) = 0.218e^{-t} \sin(2.29t + 66.42°)$ m.

9.5 $x(t) = 0.144 \sin(3.16t - \pi/2)$ m.

9.7 $W = 9.6$ N.

9.9 $C_e = 81.62$ N/m/s.

9.11 Yes, $C_e = 2.56$ N/cm/s.

9.13 $k = 322.26 \, \text{N/cm}.$

9.15 Isolator B.

10.1 $\omega = 42.4 \, \text{rad/s} = 405.5 \, \text{rpm}.$

INDEX

INDEX

INDEX

431

M

Machinery noise, 129 (*see* Noise sources)
Malfunction diagnosis, 356
  looseness, 360
  misalignment, 359
  oil whirl and whip, 360
  resonances, 360
  rubs, 358
  unbalanced, 357 (*see also* Diagnosis
    instrumentation)
Masking, 53-54
Mass law, 187
Microphones, 77
  calibration, 84
  ceramic, 78
  condensor, 77, 78
  crystal, 78
  dynamic, 78
  electret condensor, 78
  frequency range, 78
  sensitivity level, 79
  stability, temperature and humidity, 81-83
  windscreen, 81
Mufflers, 249
  expansion, 255
  Helmholtz resonator, 258
  prefabricated, 262
  reactive, 255

N

National Environmental Policy Act of
    1969, 58
Natural frequencies, 273
NC curves, 45-47 (*see* Noise criteria curves)
Noise Control Act of 1972, 61, 417
Noise control criteria, 58
Noise control regulations, 58, 68
Noise criteria curves. 45-47
  NC curves, 45
  noise criteria range, 47
  PNC curves, 46
Noise exposure:
  acceptable, 60
  octave bands, 93, 96
  Walsh-Healey Act, 59
Noise Pollution and Abatement Act of
    1970, 58

Noise reduction (NR), 191
  hood or enclosure, 205, 208
  wall, 190
    certain common, 197
    inside, 196
    outside, 198
Noise reduction coefficient, 239
Noise sources, 66, 67, 86, 114
  appliances, home, 132
  compressors, air, 126
    centrifugal, 127
    reciprocating, 127
  fans, 116
    axial, 117, 118, 119
    blade passage frequency, 117
    centrifugal, 117, 119-121
    induced draft, 122
    propeller, 117, 121, 122
  gears, 116
  machinery:
    building equipment, 129
    construction equipment, 129
    loudspeakers, 116
    motors:
      diesel, 116
      electric, 116
    pumps, 116, 124
    turbines, 116
Noise survey, 85
Noys, 42
NR, 191
NRC, 239

O

Occupational Safety and Health Act of 1970,
    59-61
Octave bands, 91, 93, 96
  conversion to A-weighted level, 167 (*see
    also* Filters)
OSHA, 59-61
Oval window, 33

P

Pascals (Pa), 8
Perceived noise level, 41-45

α	DATE DUE		
JY 19 83			
MAY 1 2 87			
JUN 2, 87			
JUN 9 '87			
JAN -5 '87			
JAN 2 6 '88			
FEB 16 '88			
MAR 8 '88			
APR 1 7 '90			
MAR 2 0 1995			
JUN 1 2 1995			
GAYLORD			PRINTED IN U S.A

The Gardens of Provence and the French Riviera

Library of Congress Cataloging-in-Publication Data

Racine, Michel, 1942-
    The gardens of Provence and the French Riviera.

    Translation of: Jardins de Provence, Jardins de la Côte d'Azur.
    Bibliography: p.
    Includes index.
    1. Gardens—France—Provence-Alpes-Côte d'Azur.
I. Binet, Françoise. II. Boursier-Mougenot, Ernest.
III. Title.
SB466.F82P76713   1987   712'.0944'9   87-16928
ISBN 0-262-18128-2

First MIT Press edition, 1987
© 1987, Edisud, 13090 Aix-en-Provence
English translation © 1987, Edisud

Published in France under the title
Jardins de Provence, and *Jardins de la Côte d'Azur*,
© 1987 Edisud, Aix-en-Provence.

The Gardens
of Provence
and the French Riviera

Michel Racine, Ernest J.-P. Boursier-Mougenot,
and Françoise Binet

Translated from French by Alice Parte, and Helen Agarathé

The MIT Press
Cambridge, Massachusetts
London, England

The Gardens of Provence and the French Riviera
by Michel Racine, Ernest J.-P. Boursier-Mougenot, and Françoise Binet

With contributions by:

Gérard Cadel: The Alpine garden of the Lautaret.
Roderick Cameron: La Fiorentina and Le Clos.
François Cayol: Illustrations.
Noël Coulet: The medieval gardens.
Sylvain Gagnière: The papal gardens in Avignon.
Hélène Homps: Fruit parks and landscaping of the "Mexicans" in the Alps.
Marcel Kroenlein: The Jardin Exotique in Monaco.
Camille Milliet-Mondon: The winter gardens of Cannes.
Tobie Lou de Viane: Val Joanis.
Martine Mattio et Jean-Pierre Olive: Illustrations and plans.

Anne-Marie Lapillonne was the coordinating editor for this book.

This authored work is an extension of a survey of gardens noteworthy for their scenery, historical and/or botanical interest commissionned by the Ministère de l'Equipement, du Logement, de l'Aménagement du territoire et des Transports, for which Anne-Marie Cousin served as project supervisor. The survey was carried out by a team headed by Michel Racine from the Délégation régionale à l'Architecture et à l'Environnement Provence-Alpes-Côte d'Azur, under the direction of Michèle Prats and Yves Lassaigne.
The garden map, and most of the garden plans were executed for the survey, between 1981 and 1983.
An extension of this survey was commissionned in 1985 by the Ministère de la Culture, direction du Patrimoine, for the gardens of the French Riviera.
The text and photographs represent the authors' own views. The Association pour l'Art des Paysages et des Jardins (ARPEJ) coordinated this work from 1983 to 1987.

Acknowledgements:
Marquis Olivier d'Albertas; Mr. Gabriel Alziar; Marquis Henri de Puget de Barbentane; Count Simon de Bendern; Mr. Henri Besson; Mrs. Jean-Claude Bottin; Mr. Jean-Claude Bouilly; Mrs. Madeleine Cambuzat; Mr. Roderick Cameron; Mr. Jacques Casalini; Mr. and Mrs. Pierre Champin; Mrs. Fleur Champin; Mrs. Cécile Chancel; Mr. and Mrs. Pierre Chiesa Gautier-Vignal; Count and Countess Jean de Clarens; Mrs. Nerte Dautier; Mr. and Mrs. Régis David; Mrs. Joëlle Déjardin; Baron Double; Mrs. Marius Durbec; Mr. Bruno de Fabri; Mr. Charles-Alexandre Fighiera; Mr. Henri Fisch; Mr. Claude Fournet; Mr. Paul Franck; Mr. François Fray; Mr. and Mrs. McGarvie Munn; Mr. Marcel Gaucher; Mrs. Béatrice de Germay; Mr. R. Girard; Mr. Marcellin Godard; Mrs. Anne Grüner-Schlumberger; Mrs. Guévrékian; Mrs. Francine Guibert; Mrs. E. Hatemian; Mr. Bertrand de la Haye Jousselin; Mr. Charles de la Haye Jousselin; Mrs. Anne Hugues; Mrs. Noëlle Icardo; Mrs. Ladan-Bockairy; Mrs. Christine Laurant; Mrs. Jacqueline Martial-Salm; Mrs. Émilie Michaux; Mr. Graeme Moore; Mrs. Monique Mosser; Mrs. Jacqueline Neve; Mrs. Claudette Nicolaï; Mr. Antony Norman; Mr. Ocelli; Mr. Georges Pascal; Mr. Peraud; Prince Louis de Polignac; Mr. André Pons; Mr. Laurent Puech; Mrs. Agnès Rastit; Mr. Remusat; Mr. André Rousseau; Mr. Jean-Pierre Rozan; The Duchess of Sabran Pontevès; Mrs. Michèle Salmona; Countess Sanjust di Teulada; Mr. Louis de Saporta; Mr. and Mrs. A. de Saporta; Mrs. Pierre Schlumberger; Mr. Jean Sechiari; Mr. Settimione; Mr. R. Touzet; Mrs. Anne Troisier de Diaz; Mrs. Danièle Veran; Baron and Baronness Guillaume de Vitrolles; Baronness Louise de Waldner; Mr. William Waterfield; Count Jacques de Wurstenberger.
The curators of the following libraries and museums: the Municipal Library, Fréjus; the Société d'Études Scientifiques et Archéologiques of Draguignan; the Municipal Library, Nice; the Inguebertine Library, Carpentras; the Arbaud Museum, Aix-en-Provence; the Chamber of Commerce and Industry Museum, Marseilles; the Château-Gombert Museum, Marseilles; the Bouches-du-Rhône Departmental Archives; the Var Departmental Archives; the Marseilles Municipal Archives; the Museum of Fine Arts, Toulon; the Municipal Library, Toulon; the Draguignan Museum.

Contents

Advertising

....

A garden designed and executed by a gifted gardener has aesthetic delights for all five senses. It is a space charged with meaning, a space that embodies an ideal or fuses ideas. It gives a faithful portrait of the gardener, revealing his intimacy with the site and surroundings and his ideas on the natural world. A garden is a system to be read, a living architectural composition representing the tastes and practices of a given society, full of allusions to the other arts: literature, painting, sculpture, and music.

In preparing this book, we visited countless gardens in Provence and on the French Riviera. We hope indeed other garden-lovers will share our sense of excitement and discovery as they read this book. We selected over one hundred outstanding examples from different periods to demonstrate the full scope of landscape creation in this remarkable region. Through our descriptions of the extant gardens, we wish to evoke the relationship between gardener and site, that special bond with the earth, rocks, sun, and water inherent in the physical or imaginative aspect of a design.

As we set out to review the most important periods in garden history, we saw it necessary to undertake two detours to afford the reader insight into the nature of the garden in the south of France. Thus this book first deals with the importance of the ordinary family vegetable plot, typical of this region, which has been perpetuated from one generation to the next for centuries without major alterations. The second detour is a historical account of gardening in this region, from the Middle Ages to the present day.

Many different people have inhabited and traversed this region in the course of history, and local gardens naturally reflect this diversity. Yet what twists of fate brought the trends from Italy, the Loire Valley, the Ile-de-France, England, Spain, and other distant lands to fuse or coexist in Provence and on the French Riviera? How was typical coastal land transformed into the Côte d'Azur, and how did it develop a garden type that became characteristic of this geographical region? The answers are in this book.

We hope that through the photographs the reader will be able to delight in the sensuous beauty of the gardens we describe. Many of these gardens have been well-kept secrets. Others are very recent, and still others may vanish unless rapid measures are taken to preserve them. The reader can pick and choose his own itinerary through these gardens.

Through the photographs and illustrations we also hope to remember these gardens as beautiful landmarks expressive of their time and of the intelligence and sensitivity of their creators. In cases where plans did not exist, we engaged artists to make measured drawings to establish the specific garden designs. In addition we interviewed several landscape designers and gardeners who allowed us insights into their intentions and techniques.

We dedicate this book to all garden-lovers. We urge them to read it on site, to experience the vitality of these creations firsthand.

....

La Bargemone, gateway.

Garden map

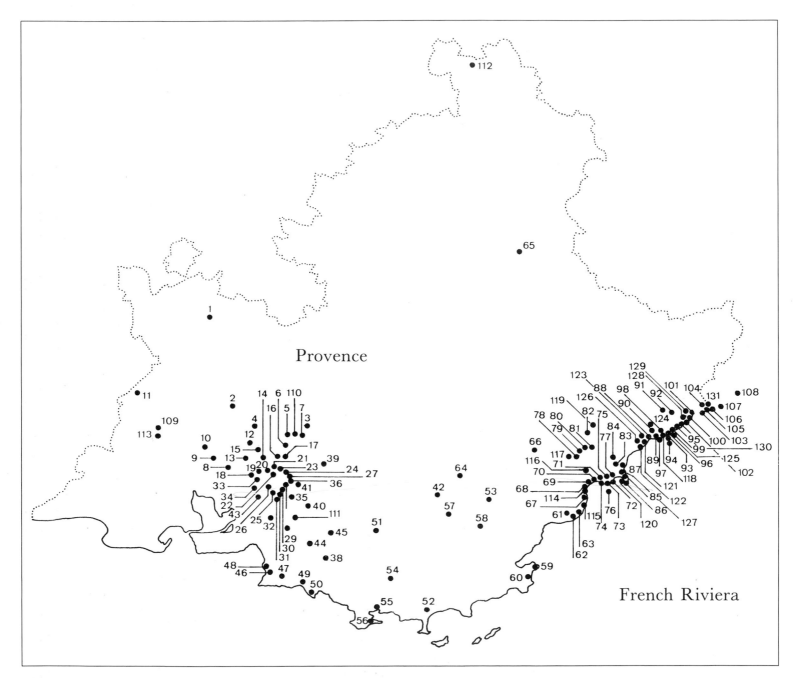

Provence

French Riviera

Garden map: one hundred thirty-one from over three hundred selected in our survey.

#	Name	Existant	Vestiges	Vanished	17th century	18th century	1820-1914	1914-1950	1950-1986	Of botanical interest
1	Jas Créma	●						●	●	
2	Les Quatre Sources	●							●	●
3	Pradines	●			●	●				
4	Lauris		●		●	●	●			
5	Ansouis	●			●	●	●			
6	Val Joanis	●			●	●	●			●
7	La Tour d'Aigues		●	●	●	●	●			
8	La Barben	●			●	●	●			
9	Saint-Pierre-des-Canons	●			●	●	●			
10	Barbentane	●			●	●	●			
11	Lamanon		●		●	●	●			
12	La Roque-d'Anthéron		●		●	●	●			
13	Aiguebelle	●			●	●	●			
14	Beaulieu	●			●	●	●			
15	Le Seuil	●			●	●	●			
16	Arnajon	●			●	●	●			
17	Fonscolombe	●			●	●	●			
18	La Bargemone	●			●	●		●		
19	Beaupré	●			●	●	●			
20	Bourgogne	●			●	●	●			
21	Saint-Simon	●			●	●	●			
22	La Brillanne		●			●				
23	La Gantèse		●			●				
24	Saint-Hippolyte		●			●				
25	Hôtel d'Espagnet	●			●					
26	Pavillon Vendôme	●			●			●		
27	La Violaine	●			●					
28	La Gaude	●			●	●				
29	La Mignarde	●			●	●				
30	Lenfant	●			●	●				
31	Trimont	●			●	●				
32	Albertas	●			●	●				
33	Le Verger	●	●		●					
34	La Bougerelle	●			●					
35	Pavillon de la Torse	●			●					
36	Bouteille	●			●					
37	Château de l'Arc	●			●			●		
38	Gémenos et Saint-Pons	●	●		●					
39	Jouques	●	●		●					
40	Châteauneuf-le-Rouge	●	●		●					
41	Saint-Marc-de-Jaumegarde	●			●					
42	Entrecasteaux	●			●					
43	Galice	●				●				
44	Roquevaire	●				●				
45	Le Moulin Blanc	●				●				●
46	Parc Borély	●					●		●	●
47	La Bastide Blanche	●					●			
48	Le Jardin de Gardini	●						●		
49	Le Mas Calendal	●	●					●	●	●
50	Le Mugel	●					●		●	
51	Tourves		●				●			
52	Parc St-Bernard - Villa Noailles (Hyères) Ste-Claire-le-Château	●					●	●		
53	Trans			●		●				
54	Belgentier		●	●						
55	Jardin botanique de Toulon		●		●	●		●		
56	Tamaris		●			●				
57	Le Thoronet	●								
58	Astros	●				●				
59	La Moutte	●				●			●	
60	Château Camarat	●				●	●			
61	Villa Magali	●					●			
62	Villa Claudine		●							
63	Leïs Messugues			●						
64	Une fondation en Haute-Provence	●						●		
65	Barcelonnette	●				●	●			
66	Château de Beauregard	●	●		●					
67	Château de la Napoule	●	●					●		
68	La Croix-des-Gardes	●						●		
69	Villa St-Georges - La Rochefoucauld	●					●			
70	Villa Eléonore-Louise		●				●			
71	Ch. des Tours - Vallombrosa	●					●			●
	Villa Rothschild	●					●			●
72	Villa Camille-Amélie	●					●			●
73	Champfleuri	●					●		●	●
74	Villa les Violettes	●					●			
	Villa Winslow	●					●			
75	Villa Saint-Priest	●					●			
	Villa La Cava	●					●			
76	Le Grand Jardin	●				●				
77	Villa Domergue	●						●		
	Isola Bella		●					●		
78	Villa Victoria	●						●		
79	Villa Croisset		●					●		
80	Villa Nora	●						●		
81	Château de Malbosc	●						●		
82	Château de Gourdon	●				●			●	●
83	La Bastide du Roy	●					●			●
84	La Chèvre d'Or	●						●		●
85	Jardin Thuret	●					●			●
86	Eilen-Roc	●					●			●
87	L'Altana	●	●					●		
88	Cimiez (16th)	●						●		
89	Parc Vigier	●					●			●
90	Valrose	●					●			
91	Le Vignal	●						●	●	
92	La Palaréa	●			●	●				
93	Villa Ile-de-France	●					●	●		
	Villa Sylvia		●				●			
94	Maryland	●					●			
	Rosemary		●				●			
95	Jardin exotique de Monaco	●						●	●	●
96	La Fiorentina et le Clos	●					●	●	●	
97	Villa Espalmador	●					●	●		
98	La Léopolda	●						●	●	
99	Villa Sallés		●				●			
100	Torre Clémentina	●					●			
	Villa Cypris	●					●			●
101	Val Rahmeh	●					●	●	●	●
102	Jardin du casino de Monte-Carlo	●					●			
103	Fontana Rosa		●					●		●
104	Les Colombières	●						●		
105	Le Clos du Peyronnet	●					●			●
106	Jardin Grimaldi (Bennet)	●				●				●
107	Jardin Hanbury	●					●			●
108	Villa Charles-Garnier (Bordighera)	●					●			
109	Saint-Paul-de-Mausole	●								
110	Sannes	●						●		
111	Château de l'Arc	●			●	●				
112	Jardin du Lautaret	●					●	●	●	●
113	Pavillon de la Reine-Jeanne	●					●	●		
114	Rêve d'Or	●					●	●	●	
115	Château de Théoule	●						●		
116	Yakimour	●							●	●
117	Villa Noailles (Grasse)	●							●	●
118	Les Terrasses de la darse	●							●	●
119	Grand Jardin de Vence (16th)			●						
120	Les Cocotiers	●					●			●
121	Villa Furtado-Heine		●				●			
122	Le Palais de marbre	●					●	●		●
123	Villa Bermond	●					●			●
	Villa Peillon	●							●	●
124	Parc Chambrun	●					●			●
125	La Berlugane	●					●			
126	Les Tropiques	●					●			
127	Château de la Garoupe	●					●			●
128	La Paloma	●					●	●		
129	La Serre de la Madone	●						●		●
130	Eden Hôtel	●	●						●	
131	Bella Vista	●					●			

The G. Grammont garden, Saint-Tropez,
by Charles Camoin.

About Paradise

The art of gardening
in Provence

To citizens selecting a place for a country house, the Roman architect Vitruvius advised: "Pay attention to the quality of the air, the climate, and convenience of the location, and choose a healthy place: it should not be a low, swampy area where the infected breath of poisonous animals corrupts the air and breeds many humors and diseases; neither should it be too high up, for fog and severe winds prevail at high altitudes.[2]

Today irrigation and drainage systems have transformed the uncultivated hills and low-lying marshes of Provence into hospitable land. Yet anyone who undertakes to build or design a garden should keep in mind that Provence has only recently become in a sense "a paradise." Although Provence is blessed with a mild climate, the land is unyielding and unforgiving toward all who, unschooled in its ways, attempt to make it obey them. Provence is really a region of great contrasts in its topography, climate, and plant life; even its inhabitants are strong-minded!

. . . .

The Vaucluse Fountain, by Jean-Antoine Constantin.
Petrarch was perfectly aware that he had enhanced the reputation of the "very illustrious spring of La Sorgue, famous on its own merits for many years and now even more so after my long stay there and through my lyric verses."

One thousand and one microclimates

Several regions of the world are characterized by a Mediterranean climate—which is to say, by hot, dry summers with at least one month of protracted drought, cool, moist winters, and two springlike intermediary seasons. True, the climate is mild in the immediate vicinity of the coast, thanks to the regulating effect of the sea (it retains heat in winter and exerts a cooling effect in summer), but the mildness is relative; the temperature can change from $+15°C$ to $-5°C$ in a few hours. On the same day plants in one spot of a garden can freeze while those in another are spared, depending on the lie of the land. There are many microclimates existing side by side in Provence. This is due to a combination of factors: a multifaceted topography that produces various types of exposure to the sun, and conditions inland that are accentuated as one travels away from the sea. Some generalizations can be attempted, but they hold true to varying degrees depending on circumstances. Southern slopes ("adrets"), sheltered from the violent mistral wind, are warmed by the sun in winter. In the summer they may get breezes if located high enough, but they tend to get too hot. Cooler air can be found on plateaus or northern slopes ("ubacs"). Still lower down the slopes and in the enclosed valleys, a temperature inversion occurs: the cool air remains at ground level, trapped there by a layer of warm air.

Park, Le Muguel, La Ciotat. A large garden protected by rock, warmed by the sea in winter.

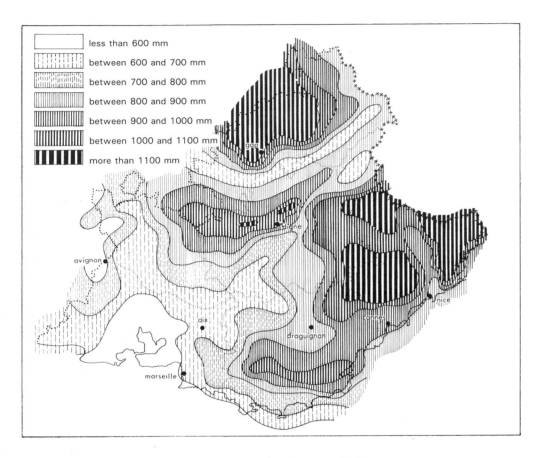

less than 600 mm
between 600 and 700 mm
between 700 and 800 mm
between 800 and 900 mm
between 900 and 1000 mm
between 1000 and 1100 mm
more than 1100 mm

gap

sisteron

avignon

aix

draguignan

cannes

nice

marseille

Urbanization has also produced microclimates. It is warmer and rains more in cities than in the country, and the urban air is more heavily loaded with dust and gases.

There are in fact infinite ways in which favorable microclimates are created in Provence, even on a very small scale (around a low wall, a bush, or the stump of a tree). One example is the way vegetation is used for its heat-retaining properties, as a plant cover to prevent the ground from freezing. All these microclimates give rise to a complex network of microclimates, each influencing the others around it. The art of gardening in Provence is based first and foremost upon a judicious use of its microclimates.

. . . .

Annual precipitations in the Provence-Alpes-Côte d'Azur region.

One of the last violet fields in the Alpes-Maritimes, Tourette-sur-Loup.

Where there is mistral, there is sun

"Oh sun! Without whom nothing would be the same!" Edmond Rostand

The principal climatic asset of the region is the high number of hours of sun, an average of 2,750 hours a year in Provence; 3,000 hours between La Ciotat and Saint-Tropez, the most southerly part; and 2,710 on the Côte d'Azur, as compared with 1,650 in the Parisian region.

This asset, however, together with an incomparably luminous sky swept by the wind, cannot be exploited unless a gardener knows how to domesticate the effects of this same cold wind which blows one hundred days a year. Certainly the mistral carries away the clouds, but it also carries away trees and chills and dries the leaves. For gardeners and fishermen alike, the days when the mistral blows are "bad weather" days—hence the importance of shelter. To discover where the good spots are to be found in Provence, follow the curve traced by this prevailing wind, and from the Rhône valley to the Italian border, note what stops it on its course: a mountain, the façade of a well-oriented house, a gar-

den wall, a fence, a row of cypresses, or a thick hedge, each creating a microclimate in which a garden can thrive.

Situated at the end of the mistral's journey, sheltered by the mountain mass of the Maures and the Esterel, and with an average of only five days of mistral a year, the Côte d'Azur is an enormous garden whose most privileged part is to be found between Villefranche and Italy. But the person who does not know how to find these sheltered corners can cross the whole of Provence without finding the least trace of one, getting more and more aggravated with each gust of wind. "Just when I thought I was going to be able to explore the charms of Provence; up till now, I've only seen the mistral," complains Stendhal during one of his visits to the Midi.[3]

Yet going into a well-sheltered garden on a day when the mistral is blowing is like entering an oasis and leaving the desert behind—a sensuous experience enjoyed

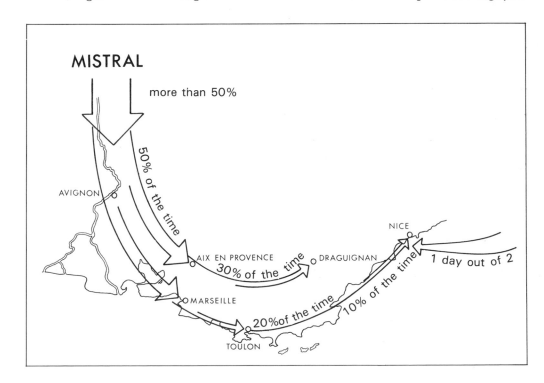

Mistral *and east wind.*

by Gustave Flaubert in 1840 when he discovered the famous botanical garden of Toulon, which unfortunately no longer exists: "The *mistrao* was blowing in Toulon; we were blinded by dust. But, once inside the garden, perhaps because of the shelter of its walls, the air became calm . . . Under tall, shady trees, near a grassy bank, two or three convicts were at work in the garden; one could hear their chains dragging on the sand, but they had no guard, no sergeant, no warders. While their fellow convicts were in the prison lifting up beams, nailing planks on the ribs of boats, working with iron and wood, they could hear the sound of the wind in the palm trees and in the aloes; in this garden there are strangely shaped reeds from India, banana trees, agaves, myrtle, cacti, all the lovely plants from unknown countries where tigers leap and serpents coil, and where brightly colored birds perch and sing. I felt that it must warm their hearts to be continually surrounded with these plants, the silence, the shade, all the leaves both large and small, the murmur of the little ornamental ponds, and the spray of the fountains. It's cool under the trees and warm in the sun; the wind stirs the trellised vines, the air is full of the scent of jasmine, of honeysuckle, of flowers whose names I ignore, but whose perfume is such that one's heart feels vulnerable and full of love. Water lilies spread themselves over the springs; reeds leaned over from all sides. The wind had flattened the shrubs and was rustling the tops of the palm trees, two palm trees of the kind called Royal; they are at the end of the garden, and so beautiful that I understood why Xerxes had fallen in love and had placed rings and necklaces round the neck of one of them as though it were his mistress. The uppermost fronds fell in gently curving sheaves, and the *mistrao* blew them together, making a sound unlike that made by any tree of this country, while the trunk stayed calm and motionless, like a woman whose hair alone moves in the wind."[4]

Greenhouses, Vallauris.

Saint-Rémy, protecting orchards against the mistral.

Sheltered corners in winter

Certain places protected from the wind have the reputation of being warm and sheltered—good corners in winter because of a natural or man-made shelter where all the warmth of the sun is concentrated. They are easily spotted in towns or villages; they are the habitual meeting places for the local people who know how to combine conversation with comfort. There in the hills, one frequently finds almond trees in flower at Christmas, or even in January, and the early crops of vegetables and fruit that are greatly prized by connoisseurs. In most cases making good use of these warm sheltered spots has entailed very hard work. The terraces were built by individual farmers, whereas the irrigation canals and their smaller channels were the result of communal effort. The traditional vegetable garden in Provence has always modeled itself on the combination of protected terraces and irrigation canals.

The practice of building terraces, know as *faïsses, restanques*, or *bancaous* in Provence, is said to have been introduced to the region during the Crusades. From the twelfth to the late nineteenth and early twentieth centuries, when the plains were rendered productive by draining or irrigation, the terraces occupied a privileged position in the countryside of Provence, ascending even the steepest slopes of the hills as the local population increased. Laid over a substructure of stone in such a way that heat stored during the day is released at night, they create microclimates that enable vines and olive trees to be productive far beyond their natural habits, and they have led to the development of very profitable enterprises, such as the cultivation of caper plants on the hills near Gémenos, Cuges, Auriol, and Roquevaire.

The Grilles district, Lauris.

Terraces near Grasse, the Saint-Jean district.

A fragile environment: a challenge for the imagination. The skillful planting of fruit trees, grapevines, olive trees, grains, vegetables and flowers in juxtaposition. This was what the typical Mediterranean agricultural landscape looked like before drainage and irrigation methods conquered insalubrious plains.

The luxury of shade

In Villars, a small village in the Vaucluse, the public bench is movable. The villagers move it back and forth depending on whether they want to be in the sun or in the shade. With the recent mass of vacationers seeking the summer sunshine, one has forgotten the value attached to shade by those who, in Provence, put their comfort first. Flaubert was among them: in Marseilles the sun for him was "such a beautiful but terrible thing" that he took "to the sordid little streets" in order to avoid it. In Toulon he appreciated lunching in a country house "in a large garden full of shade, with tall Provençal canes, and cool drives."[5]

In Marseilles at about the same time, Stendhal also says, "I spend my leisure hours at Belle-ombre, on the banks of the river Huveaune. My well-being is surely due to the fact that in this part of the country shade is a necessity." Shade was apparently so sought after that it had market value: "Lots of country houses boast of having shade, and For Sale notices often state 'Country house with shade,'" notes Stendhal.[6]

Shade is not complete darkness, but modulated light, an essential component of Mediterranean gardens and also agricultural fields. Just as in Sumeria the gardeners used the shade of certain trees to shelter their crops, so today agriculturists or gardeners of the Comtat[7] make use of various degrees of insolation to retard or advance their crops.

Greenhouses, Antibes.
Projections of white or pink lime and use of tinted glass prevent plants from being scorched by intense sunlight. Plastic is used to obtain double thermal insulation in winter; previously, cloth or old newspapers served the same purpose.

A bastide in the Pinchinats valley, Aix-en-Provence. Drawing by François Cayol.

Each tree has its own specific shade characteristics, though very little is known about the effects of the play of light and shade on vegetation. Yet from Pliny the Elder stating that "the shade of walnut trees, pine trees, spruces, and fir trees is undoubtedly a poison for all that it touches"[8] to Stendhal who also says about walnut trees "corn cannot grow in their shade," the notion of the inherent harmfulness of shade has long been accepted. We have always known, as a local market-gardener so candidly put it, "Without sun everything's a flop." We also know that the growth of certain plants can be slowed down at certain stages by too much light: "When you thin out the the leaves of tomato plants or when you tie up the sweet peppers, do this on a sunless day."[9]

A better understanding of the importance of the energy of light for agriculture or forestry had to wait for the nineteenth century. All the varieties were then classified as light-loving or shade-loving species, depending on whether or not they needed a lot of light in the early stages of their growth.

According to recent research[10] the growth of plants is conditioned by the amount of solar radiation they receive and by their internal biological rhythms; these vary according to species, age, and season.

Nothing much is really known about the effects of shade on the human organism. If one knows by experience that the shade of the Aleppo pine is not very cool and that of the walnut tree too cold, it may be worth-while to do research into old wives' tales and try to assess them scientifically.

Shade has always been as much sought after as the sun, not only for the comfort it brings to plants and bodies but also for the pleasure it gives to the eye. Shade combined with the transpiration of plants creates cool conditions. Studying gardens in the Midi should teach us how their inventors made the most out of the sun's rays, created cool places, increased or diminished the importance of a design by taking advantage of the unique natural lighting. The gardener in Provence is as skilled in his use of the sun as in his use of shade. Rows of trees, groves, isolated trees, garden architecture, trelliswork—all of these form relatively compact bodies that enable a gardener to play with light rays, the effects of light filtering through shade, twilight, and semidarkness.

In Provence and on the Mediterranean coast, gardeners have delighted in combining water, shade, and fragrance. At dusk belladonna petals open, and masses of flowers send out their perfume, but when night falls on a Provençal garden, perhaps Henri Bosco expresses it best: "These mysterious places seen in the light of day lose their magic power, and their rich emotional impact is lost before its time . . . The shadows spread naturally over the park and the bittersweet scent of the pines drifted down from the hill Nothing stirred, and yet everything was animated."[11]

. . . .

Shade under the plane trees, garden path,
La Torse pavilion, Aix-en-Provence.

Water here and there

In Provence, more than anywhere else in France, water works miracles. Pierre de Gasparin, the nineteenth-century agronomist and politician, originally from Orange, once characterized it thus: "If everywhere in France one measure of earth and one of water produce twice the yield, in Provence one of earth and one of water produce four times as much, because the sun also plays its part." There is some truth to this formulation, provided that one can control the water when it takes the form of violent rainstorms and sudden floods. One too often forgets that it rains just as much, if not more, in Marseilles as in Paris; the only difference is that the same quantity of water falls here in a much shorter time. As soon as the rain touches the ground, most of it disappears into the subsoil or evaporates with the first rays of the sun or the first gusts of the mistral.

Keeping the soil wet is not the easiest task for a Provençal gardener. Water where it is not rare is unequally distributed in Provence, and so is it valued that it is referred to in at least one-third of the local place names. Whole regions have as their water supply only the cistern for the house and the reservoir for the garden. The reservoir, which is an important element in the layout of Mediterranean gardens, has catchment areas for capting rainwater and trenches or rills that lead to the household cistern, thus ensuring a supply for the dry summer days. In Provence, before the irrigation systems of the twentieth century, the well was the most frequent source of water. Even today the person who knows where to dig to find the trickle of water hidden somewhere underground possesses a power even the most learned geologist cannot claim. In Provence the word for water-diviner is almost the same as that for wizard: *sourcier/sorcier*. The man who knows how to interpret the vibrations of his divining wand or the oscillations of his pendulum holds a certain fascination. He is a mastermind of profound understanding and appreciation of the land he stands on.

The underworld and countryside meet at the well. Indeed, the prosperity of the market-gardeners of the Comtat, Ollioules, or Hyères is founded on a very complex underground architecture. The wells in these areas are connected to a veritable labyrinth of galleries built by generations of country people, up to 80 meters from where the water is found.[12] Deep wells were dug during years of drought to get as much groundwater as possible. These wells were dug by the whole family.

Basin, La Roque-d'Anthéron. Drawing by François Cayol.

Prior to the installation of pumping systems, water was brought slowly to the surface by means of a large cupped wheel or *noria (pouso rago, puits à roue)* driven by a mule; the mule was sometimes blindfolded to keep him from getting giddy. From there the water went to the garden reservoir. The water that poured into and ran out of the reservoir represented life for the garden, the animals, and the grass.[13] In Provence today most of the watering is ensured by water from the mains, although the traditional method in Mediterranean gardens of irrigation by gravity-feed is still in practice. The water, capted from a spring or a canal, is run off slowly so as to irrigate as large a surface as possible, but however slowly it descends, descend it does and is lost. Despite this drawback this system of irrigation still has its adherents. For instance, in Marseilles the water from the canal runs over 40 kilometers of hills around the town before terminating its journey at the sea. All along its route the main canal feeds the trenches or rills that supply the various market gardens and the old country houses of the Marseilles burgeoisie of the last century.

Each market-gardener takes his turn for the water

An irrigation drain, coolness under the pines, Ansouis.

Engraving from l'Histoire critique des pratiques superstitieuses, *by Pierre Lebrun, 1750.*

According to the Abbot Gondret in *Moeurs et coutumes des habitants du Queyras,* Jacques Aimard, from Saint-Véran, Hautes-Alpes, invented the diviner's rod.

and has certain hours for watering regulated by a very strict schedule: once a week on a cetain day, whatever the season, and at a certain time not necessarily corresponding to the biological rhythm of the plants. Such practice is unheard of in the Comtat. As one market-gardener said, "Proper watering is one of the skills of gardening; it depends on how much sun the land gets, the kind of land, whether there's the mistral; what land drinks more than others; and one has to take into account the weather, the plants ..." In the Comtat, legend has it that it takes three generations to learn how to control the use of water. The gardeners know that plants themselves help to maintain the humidity in the soil and that in sufficient quantity they can exert a very powerful pull on the groundwater.

The art of vegetable gardening in the Midi came into being because of the difficulties in resolving the problems of incomptatability between the fragility of the plants and the brutality of the climate. There is a perpetual contest in which a lively imagination is as important as actual gardening skills, and an open mind an essential, for there is no place for preconceived ideas.

When Jules Michelet in his monumental history of France was comparing this region with the rest of the country, he stressed Provence's unique social organization which encourages individual merits: "Even on the farming estates serfdom never became as established as elsewhere in France. The peasants were their own liberators and conquerors of the Moors; they alone were capable of cultivating the steep hills and narrowing the beds of torrents. Only free, intelligent men could master such a difficult country." [14]

Mechanical device controlling a spillway at the Béal de la Paroisse, *Auriol. Drawing by Claude Cordoléani, architect.*

Watering the garden, at the foot of the Alpilles.

The evergreen country

The coast of Provence, the summer paradise of today, the winter paradise of yesterday, has been thought of as a paradise of a third kind: the evergreen country. Most of the vegetation here is nondeciduous, which gives rise to this notion. The legend of an evergreen paradise, enjoying a permanent spring, the symbol of eternal life, very often occurs in Mediterranean civilizations. The fabulous garden of Alcinoüs invented by Homer is a good example of the Mediterranean type of evergreen country.

If men from the north were slightly hasty in attributing the permanent green of the Provençal countryside to the existence of an eternal springtime, it is nevertheless true that Mediterranean vegetation, especially in France, has a remarkable biological rhythm. Unlike the twofold biological rhythm that is normal for most of the vegetable kingdom in the world, the vegetation here is subjected to a fourfold rhythm. Plant growth slows down in winter during the cold dormant period and in summer during the dry dormant period. The vegetation has two periods of growth, one in spring and the other in autumn. The permanent green *sempervirens* are rather a dark green that sets off the tender spring shoots. The texture of the evergreens, the most widespread plants, is thick and tough; the spines, or needles, enable these plants to withstand the dryness of summer.

Borély Park, Marseilles.
Jean de La Fontaine alluded to the triumph of the green trees of Provence over winter and the elements:
"Are we, said he, in Provence,
What expense of ever-green trees
Triumphs here over the inclemency
Of the North wind and of Winter?"

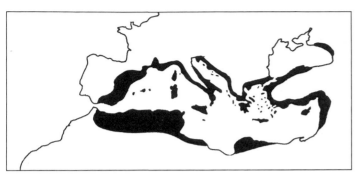

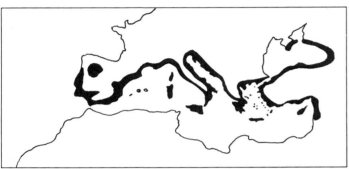

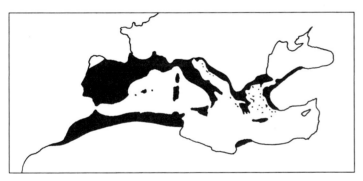

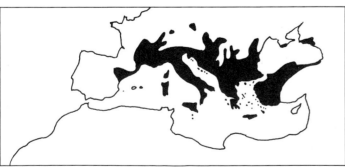

Indigenous or otherwise, vegetation has often been used to define the various parts of the French Mediterranean region. The Midi was long described as a region of olive trees. In the nineteenth century the eastern coastline was referred to as the region of orange trees: ''The region of orange trees is a succession of localized climates whose mildness is due to the foothills which from Toulon to Nice form, like protective giants, a barrier to the wind ... the sea also lends its warming influence to these enchanted shores.''

La Flore pittoresque[15], published in the midnineteenth century, represented the present administrative region of Provence-Alpes-Côte d'Azur as one native entity: ''The region known as 'the olive tree region' is divided into six zones. The *orange tree zone*: here one finds the orange tree, the mandarine, the Cassia or mimosa farnesus, the palm, the long-leaved acacia, the Spanish jasmine, the eucalyptus, the American or false pepper tree. The *Aleppo pine zone*: this zone is characterized by the Aleppo pine, the olive tree, the medlar, the mastic tree, the strawberry tree, the cork oak, the holly oak, the myrtle, and the Phoenician juniper; it rarely exceeds an altitude of 500 meters. The *mulberry tree zone*: situated between 500 and 1,000 meters in altitude; here one meets with the mulberry tree, vines, the Austrian or Russian oak, the sweet chestnut tree, the Norwegian pine, the beech tree, the fir tree, the box tree, broom, sessile-leaved cystus, the fustet, and the coronillas. The *subalpine zone*: both pastureland and forest, with a maximum altitude of 1,800 meters; here one finds beech, fir, spruce, mountain pine, yellow gentian, mountain tobacco, martagon lily, and the aconite. The *alpine zone*: up to 2,500 meters; here there is an abundant supply of mountain pine, spruce, larch, Siberian pine, green alder, and also pastureland in in which very few annual flowers grow; finally, from 2,500 meters and higher, there is the *zone of the eternal snows* ... ''[16]

. . . .

Plantations of Pinus balepensis, Pinus pinea, Quercus ilex, Quercus pubescens.

Garden, La Chèvre d'Or, Biot.

"The Fish Pond," fresco by Matteo Giovanetti, 1343. Palais des Papes, Avignon.

History of the gardens of Provence

Medieval gardens

Treatise on surveying by Bertrand Boysset, 1350-1415.

Gardens open to the sky

Little is known about the gardens of the Middle Ages in France, including Provence. Courtyards and walled gardens of monasteries and castles remain an undiscovered domaine. Enclosed behind a maze of ramparts, dark lanes, and walls, they were the only clear spaces, the only openings where the spirit could feel free to fly away. Enclosed gardens open but to the sky were holes where one could come to drink in light, observe the growth of a few plants, the path of the sun, and the progression of the planets. They were the wells joining the town to the cosmos. The only contact with nature for the monk was the cloistered garden, walled off on all four sides, where medicinal plants were grown. These medicinal plants represented the essence of plant lore of that day—a purified image of the outside world.

In the castle gardens a wider variety of plants was cultivated. Apart from the usual medicinal herbs one could find there aromatic herbs and fruit trees, as in the garden of the Château de Laupie in the Drôme,[17] designed by its noble owner on his return from the Crusades. He brought back numerous plants from his travels and, inspired by all that he had seen, created a garden sheltered from the wind and onlookers by high walls—a garden capable of sustaining his dreams of the Orient.

Cloister, Abbey of Le Thoronet.

Cloister, Saint-Paul de Mausole, near Saint-Rémy-de-Provence where Vincent van Gogh lived during the year preceding his death. Drawing by Albert Laprade.

Cloister, Saint-Trophime, Arles.

Orts, gardens of necessity

Are there any ruins or old documents that give us an idea of what the ordinary medieval garden in the south of France was really like? Contemporary administrative records and cartographers of plant life give very little information on the contents of present-day gardens, so it is hardly surprising that neither drawings nor records of legal or financial transactions pertaining to medieval gardens have survived. As with present-day gardens the produce of the medieval garden was not part of an organized commercial effort, nor did it have an official place in the economy of the country. Nevertheless, historians have tried to sift through the numerous surviving manuscripts to obtain at least a vague idea of what a medieval garden consisted. Noël Coulet managed to reconstruct the garden plans of Aix-en-Provence between 1350 and 1450. It seems clear from her work that as with many medieval towns, a ring of gardens surrounded Aix-en-Provence during the fourteenth century, and their produce helped to feed the citizens. These gardens, known in old Provençal as *orts* ("irrigated gardens," from the Latin *hortus*), were grouped around wells, fountains, and *beals*, small canals that supplied the mills (or *paradous*). The well-known term *jardin de paradou* has nothing to do with a potential garden of paradise. It gets its origin from an association formed between gardeners (*ortolans*) and millers that was based on the sharing of available water, and this often caused problems if not actual conflict. The *ortolans* were neither peasants nor craftsmen; they had a social status equal to that of millers and tailors. They used the domestic refuse of the town as fertilizer and sold the vegetables they grew directly in the streets of the town. Inside the ramparts the town could be said to be afloat within a coronet of walled gardens.

At the end of the fourteenth century in the *ville champêtre*, which was actually a transitional area between the town and the country, many gardens (*casaux*) were planted on the sites of houses destroyed during the peasant rebellions. Here notables of the town and craftsmen often had their gardens side by side. The garden was an integral part of both the countryside and the town, where it was surrounded by walls. The garden of the Middle Ages in Aix produced a rather limited variety of vegetables. From May to November one could find spinach, onions, pumpkin, melons, leeks, cabbages, and turnips. In addition there were aromatic herbs and fruit such as plums, apples, peaches, figs, pomegranates, cherries, grapes, pears, walnuts, and hazelnuts. In those days the butcher's shop opened only once a week,[18] and this gives a fairly good picture of the diet of the period.

Treatise on surveying, by Bertrand Boysset, 1350-1415.

The delight of
the kitchen garden

. . . .

It would be wrong to think that the more pleasurable side of gardening did not exist in medieval gardens. For the *ortolans*, the professional gardeners, hard work surely did not mean that they could not from time to time take a moment to enjoy the fresh air. The notables, whose houses were separate from the irrigated gardens, or *orts*, were able to stay in touch with the earth through these gardens and the vegetables growing there. They came also for the pleasure of touching, smelling, seeing, and gathering flowers or even daydreaming or meditating, something impossible to do in the small dark lanes of the town. We know from surviving documents that in 1427 the merchant Guillaume Aymeric kept for him-

self not only "the crops of his fruit trees and the grapes of his vines, but also the *roses from the rosebush in the middle of the garden* and those that grow at the foot of the vines." A somewhat earlier document—a contract drawn up in 1410 between the brothers Arnaud, lords of Châteauneuf-le-Rouge, and their farmer—forbade fishing "*in the fish pond which is in the middle of the orchard* in the shade of the fig tree and the apple tree." It also stipulated that the farmer "allow at all times free access to their brother-in-law, Jourdain Brice, a local magistrate, whenever it may please him to enter the garden for a stroll [*pro se spaciando*]."[19]

. . . .

Emilie in the Garden. Boccacio, "The Story of Thesus." Attributed to the School of Avignon.

The papal gardens in Avignon

"The Bird Charmer," fresco by Matteo Giovanetti, 1343. Palais des Papes, Avignon.

The papacy installed itself in Avignon in 1309 to escape the struggles prevalent at that time in Italy. Seven popes reigned there until Gregory XI's move to Rome in 1377, but it was only in 1334 that work was started on the actual papal palace. Thanks to the accounts kept by two of the gardeners, Michel Brun and Jean Pierre, and also to the findings of several archaeological excavations, it has been possible to obtain a fairly precise idea of the layout of the gardens of the Palais des Papes. Inside the ramparts and just below the pope's bedroom and private apartments, the garden of Benedict XII was embellished with a large, richly sculpted fountain, or *griffon*. The water was drawn up from the garden well by an elevating mechanism to a cistern, and the garden was watered from this cistern. The lawn around the fountain, the *pratellum*, was cut regularly. This ornamental garden stretched north as far as the kitchen garden, situated by the kitchens. Sheltered from the formidable mistral by the fortress and its ramparts, the garden received only the morning sun. It was a cool perfumed garden. Roses and beds of violets grew there. A large amount of the available space was taken up by vines trained over trellises or other supports. All the sweet-smelling and medicinal plants were cultivated there: marjoram, sage, fennel, and borage. In the large kitchen garden various vegetables were grown, including white and green cabbages, spinach, chard, leeks, and parsley. There were also trees in this garden. The account books mention the pruning and the removal of branches and also tell anecdotes about the garden, such as how the fountain had to be repaired owing to damage by the pope's young nephews, or how the well had to be completely drained in 1371 because a cat drowned in it. At the foot of the ramparts a second garden was planted and walled in during the reign of Clement VI. From 1372 the mechanism for raising the water was a wheel with willow baskets, and this enabled the orchard to be watered as well. Helped by both male and female day laborers, the gardeners carried out the manuring, the sowing, and the weeding. They also had to look after the large collection of animals, mainly deer, though at times there were a lioness, a wild cat, several bears, a camel, a wild boar, a parrot, cranes, and ostriches.[20]

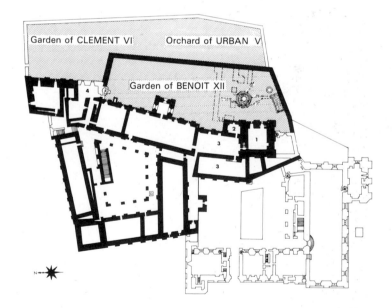

Plan, gardens, Palais des Papes, after S. Gagnière.

1. The Pope's bedchamber - 2. Studium - 3. Private apartments - 4. Kitchens.

Searching for new ways—
from Petrarch to King René

The enclosed medieval garden was an urban garden. The gardens outside the city walls always belonged to people who were freer in spirit than those of their time and who had the means for living outside the town. The poet Petrarch in many ways was a forerunner of this breed. To be able to work in peace and solitude, he retreated often to the Vaucluse where he kept what would nowadays be known as a second residence. As Petrarch noted in his letters, no place but the Vaucluse offered him such an escape from worldly distractions and such pleasure.[21]

There, besides writing poetry, he took up gardening. He consulted the writings of Virgil for guidance in the planning and maintenance of his garden. In a letter written in 1336 at Fontaine de Vaucluse, he talks not only about his love of gardens but also about his ideas on gardens which owed none of their inspiration to the gardens of his era: "I have created two marvelous gardens. One of them is for contemplation. It is full of shade and dedicated to Apollo. It overhangs the riverhead and finally loses itself in rocky outcrop where only birds can go. The other garden is close to the house and is less austere; this I have dedicated to Bacchus. Its situation is quite out of the ordinary, it's right in the midst of a rushing river. One gets to it by crossing a rock bridge. Under the rock arch there is a natural grotto were the light of day can never penetrate."[22]

In this garden Petrarch planted many laurels, symbol of glory of fire. This plant, dedicated to the sun since antiquity, also reminded him of the name of Laura, with whom he was deeply in love. Petrarch's interest in nature was not confined to his garden. He is equally

Petrarch.

Vaucluse fountain.

famous for a rather unusual exploit for his era—climbing Mt. Ventoux. His description of this adventure reminds one of the Grand Tour invented by the English four centuries later, in particular of travel accounts of picturesque journeys to the Alpes, Provence, and later the Côte d'Azur. But without looking so far forward, Petrarch had a considerable influence on his contemporaries and succeeding generations. By awakening interest in the buried world of antiquity, and by advocating, like his great friend Boccacio, a free love of nature, he prepared the way for the blossoming of the gardens of the Renaissance.

King René, who installed himself in the Midi from 1442 to 1480, can be considered as the first great Provençal gardener to follow Petrarch. Although he has often been wrongy credited with the invention of gardens that extend residences into the landscape by a sequence of terraces, known as the Italian-style garden, careful research shows him to have been an enlightened man, a great lover of gardens, but not one to anticipate the innovations of the Italian Renaissance. An evocation of his life-style in Provence makes clear his ever-growing desire for a wider contact with nature, which was somewhat restrained by the need for security in those days.[23] Faced with the impossibility of extending the grounds of his castle, René increased the number of his residences, particularly in the countryside, and traveled constantly from one to another. He had at least ten in Provence, enclosed by walls.

He concerned himself with developing a synthesis between house and garden that showed itself particularly in the increased size of windows and in the construction of pavilions. His garden in Aix-en-Provence had no less than four pavilions that were made habitable, enabling their occupants to spend more and more time in the garden. His preference for country houses, called *bastides*, was probably due to his stay in Italy, where he also picked up the idea of having a menagerie in his garden. The garden in Aix had at one time an elephant and a dromedary. Architectural elements, plants, and water were all used to effect a transition between the dwelling place and nature, giving rise to multiple trellises, arbors, hedges, and palisades, as well as fountains which became very popular. The demand for these new scenic effects soon created new professions, but at that time the necessary techniques were not always fully mastered.

We are still a long way from the accomplishments of the artists and craftsmen of the Renaissance garden. Some of the problems encountered by René in his garden in Angers bear witness to this. After several years of failure during which his fountain engineers never managed to get the water to flow at the right level, he was obliged to abandon his projects.

King René.

The garden of the ''bastide''

. . . .

For almost three centuries the garden in Provence remained a *bastide*, or country-house, garden. The *bastide* (the principal house) was situated at the center of a farm among farm buildings, and it imposed a very specific architecture on the countryside. Like the Italian villa, the *bastide* betokened the ever-growing encroachment of the town into the countryside and the desire to combine the advantages of a farm and its farmlands with those of a country estate and its gardens.

. . . .

Carpet of narcissi and formal grove, Beaupré, Saint-Cannat.

The origins of the bastide: the conquest of new territories

The appearance of the first bastides coincided with the increase in population of the thirteenth century. They were fortified edifices of noblemen built outside of a town or a village at strategic posts—near busy roads, at tollgates, or at high vantage points. Around the fifteenth century certain *bastides*, such as the Tour de Sabran, began to engage in farming, and so fulfilled the double function of farm and residence. In the mid-sixteenth century, when the restructuring of agricultural lands in Provence began, the *bastide* made another transition from fortified residence to open county residence, and thus it remained unchanged until the mid-nineteenth century.

The agricultural development of the region of La Crau became possible in 1541 with the construction of the Craponne canal, and in 1611 the Pierrelatte canal rendered the same service to the region of the Comtat. Whether as a simple real estate investment, a source of revenue from the land, or as a place for experimenting with new gardening techniques, the *bastide* ensured for its owner a pleasant haven during summer heat waves, popular uprisings, or epidemics. The *bastide* was the most prominent building on the farm, and it constituted the nerve center for the management of the farm. Its stately architecture, however, and its gardens set it apart from the actual farm, and it brought a different culture to the countryside.

"Queen Jeanne's" summerhouse,
Les Baux.
Built around 1575 in the royal garden for
Jeanne de Quinqueran, the governor's wife.

Olivier de Serres: project for a landscape and a global exploitation of the land

In his famous work, *Le Théâtre de l'agriculture et Mesnage des Champs*, first published in 1600 and reprinted twenty times during the seventeenth century, Olivier de Serres expounds his liberal ideas of how the countryside could be intelligently administered as one entity. His work presented a comprehensive program for the economic and aesthetic development of the whole Midi region.

A native of southern France, Olivier de Serres (1539-1619) visited the botanical gardens of Pisa, Padua, Florence, Genoa, and Montpellier, which then represented the new centers of research and information on the vegetable kingdom. This was at about the time when new specimens from Italy were being introduced into the kitchen gardens of the Midi, notably pumpkins, aubergines, artichokes, and lettuces known as "romaines." De Serres, profiting from what he had learned on his excursions to Italy, experimented with new methods of cultivation and with the growing of new plants. He propagated the white mulberry in France and also new agronomical methods for improving the crop yield.

The garden surrounding his house at the center of his farm in Pradel in the Ardèche occupied a privileged position. The farm itself was quite standard, but the garden was reserved for experimenting with and perfecting new techniques, including ploughing, hoeing, raking, harrowing, and rolling, and for improving the soil. The garden was a combination of kitchen garden, medicinal garden, orchard, and pleasure garden, where strawberries grew alongside flowers.

Essentially rural in character, the garden was laid out along a sanded alley which provided both easy access and a beautiful place for a stroll. The alleys, bordered by small irrigation trenches and stone pools, passed through the various garden plots. The pleasurable aspect of the *jardin de plaisir* was increased by the agreeable sights of a cut-flower garden, known as a *jardin bouquetier*. In the medicinal garden, de Serres had built a miniature mountain like the one he had seen in the botanical gardens of Montpellier. This was actually a very much reduced model of a mountain with a great variety of soils and exposures to the sun.

One would like to know exactly how this *jardin bouquetier* was laid out, for Olivier de Serres was a friend of Claude Mollet, gardener to Henri IV and one of the first to introduce the embroidered parterre into France. Did de Serres embellish his farm in Pradel with flower beds and the other more precious elements he recommends in his famous work—statues, columns, pyramids, obelisks and other sculptures in porphyry? One can imagine this as being so. Yet, for de Serres, the main purpose of his garden was to grow food. Pleasure was of limited importance; as the plan of his garden suggests, only one of the four plots in the kitchen garden was reserved for flowers.

LIEV SIXIESME,
DV
THEATRE D'AGRICVLTVRE·
ET
MESNAGE DES CHAMPS.

DES JARDINAGES, POVR AVOIR
des Herbes & Fruits Potagers : des Herbes & Fleurs
odorantes: des Herbes medecinales, des Fruits
des Arbres : du Saffran, du Lin, du Chan-
ure, du Cuesde, de la Garence, des Char-
dons à draps, des Rozeaux en suite,
la maniere de faire les cloisons
pour la conservation des
Fruits en general.

Gg iij

Olivier de Serres' Théâtre d'agriculture et mesnage des Champs.

A villa in Provence: the Peiresc's house in Belgentier, general view, mideighteenth century. Engraving by Israël Sylvestre.
The square symbolizes land and control of space. It is used to master the landscape. In the Belgentier garden, like many Tuscan gardens—la Petraia, l'Ambrogiana, Castello—each component is separate from

Belgentier: a seventeenth-century Italian villa in Provence

In the garden of the first large *bastide* in Provence, the residence of Claude-Nicolas Fabri de Peiresc in Belgentier, the relation between the kitchen and pleasure parts of the garden was exactly opposite to that of Olivier de Serres's garden. Of the four plots in the garden, only one was reserved for vegetables; the other three comprised a pleasure garden. They were planted with orange and lemon trees about banks of greenery in the middle screened by myrtle trees. After studying for three years in Padua, Rome, Naples, and Venice, and discovering the architecture and gardens of the Italian Renaissance, Peiresc returned to France full of ideas for the renovation of his family home in Belgentier.

Planned as a villa, this *bastide* is surmounted by a belvedere, and its main façade opens onto the principal axis of the garden from a horseshoe-shaped staircase leading down to an ornamental pool and a fountain. Embroidered parterres enhance the grounds of the main house. Down the sides two large arbors provide shady walks. Two canals from the Gapeau River, which flows freely through the surrounding countryside, pass through the grounds perpendicularly to the main axis.

Peiresc, who was one of the greatest humanists of his time, completed this masterpiece toward 1623. It antedated the *bastides* in Aix-en-Provence by several decades and was visited by many important people. Its conception marked a turning point in the evolution of the *bastide* and its garden. Although Peiresc collected his plans of Italian gardens directly while traveling in Italy, he used French artisans who had their own interpretations of ideas and innovations coming from Italy. He corresponded widely with other scholars, and among his correspondents was Jacques Boyceau, a well-known theoretician of gardening, whom he contacted not only to get advice for his own garden, which had to be "laid out with embroidered parterres and edgings," but also

for the archbishop of Aix-en-Provence's garden which needed "a summary design for the parterre."

Why were the gardens of the *bastides* so influenced by Italy? Provence's fascination with Italy was as much due to its geographical proximity, its Roman past, and the Italian training of the chief gardeners and sculptors as to the desire to emulate the gardens of Italy.

the others. A grid divides up the space and creates a pattern; the feeling of order is reinforced by the cubic volume of the bastide.

Claude-Nicolas Fabri de Peiresc, 1580-1637.

From Aix-en-Provence to Marseilles: two different "bastide" landscapes

In the eighteenth century for a parliamentarian of Aix-en-Provence or a merchant from Marseilles wanting to improve his social standing, the possession of a country estate was essential. Anyone of consequence had to have a countryseat. In short, owning land was a social requisite. Families already enjoying a certain social position often took the names of their new estates. Thus the family Boyer d'Eguilles decided to change their name to Boyer de Fonscolombe, the name of the castle they had just purchased. Among the rising social classes, those who were known by the epithet of "newly rich" had to attribute their name to one or several places and had to own several dwellings, all bearing the same name: the private town house, the *bastide* in the country, and even a lodge between town and country. Such was the pressure on the middle class in the vicinity of Aix-en-Provence, Marseilles, and the small towns of the region that often travelers and writers mentioned this feature.

Each of these smaller regions, however, has its own character. The *bastides* in the Aix-en-Provence countryside provide a focal point otherwise lacking in this open country where the only interruption between fields and wooded hills are the hedges. "The country estates, the vines, the olive trees, and numerous wealthy homes make this district one of the most delightful around," notes Michel Darluc at Puyricard.[24] In Provence garden walls are still rare. They are sometimes used as a frame for an entrance gate, or to close off a more private corner of a garden, but they are never used to surround a property. In contrast, in Marseilles a great number of *bastides* are completely walled in, creating a semiurban landscape. The same is true of Genoa and other Medi-

Bastides *near Aix-en-Provence: Sannes and Bourgogne in their natural surroundings.*
Many of the *bastides* near Aix have survived thanks to continued farming activity and their proprietors' love of the land.

terranean harbortowns, where labyrinths of lanes and blind alleys bordered by high walls ensure the safety and privacy of their citizens. To protect their *bastides* from nosy people and prowlers, the owners sometimes surrounded their property with walls that rose as high as two meters. Consequently along the lanes of the outlying areas of Marseilles the view is blocked on either side by a continuous wall bordering the fields and the country villas.

. . . .

A landscape of walls, Marseilles'
bastides.

Maze of country roads and dead-ends not far from Marseilles. Sainte-Marthe, Saint-Just in 1864.

View of the Sainte-Victoire peak, from the Les Pinchinats road.

The stage setting for aristocratic power: the "bastide" and pavilion gardens

In the second half of the seventeenth century urbanism and architecture opening into nature were all the rage in Aix-en-Provence. Town houses were built whose gardens were as large in surface area as the area of the house. A public promenade, the *cours*, was created, known today as the Cours Mirabeau, situated in a luxurious residential area; this attests to the city's high degree of accomplishment in the development of real estate[25] and urban comfort. The *bastide* garden became more of a pleasure garden and a stage where aristocrats could show off their life-style.

Rank, arrogance, and ostentation, these were the rules of the game for improving social standing under Louis XIV, and they found a particularly favorable setting in Aix-en-Provence. It was because of the social jousting among the nobility of the sword, the cloth, and the upper classes, and the fascination exerted by the Court on the whole of the cultivated classes, whose constant rivalries stimulated artistic creativity, that Aix had the reputation of being "the best example of the French baroque after Versailles." Early in the century the middle classes had taken advantage of the new technical, scientific, and artistic discoveries; they had invested in the countryside and spread their influence from *bastide* to *bastide* over whole areas surrounding the towns. Around midcentury, coinciding with the reign of Louis XIV, the pleasure gardens really began to multiply. In the eighteenth century the increase in ground rent led to an enormous increase in the construction of *bastides* all over the Aix countryside. Nobles, members of the legal profession, high dignitaries of both the government and the church, and members of the upper class gradually formed a class of landowners, all with their own *bastides*. They were neighbors, they socialized together, and they tried to outdo each other in the creation of their architecture and garden schemes.

Grille, Albertas garden.

Hôtel d'Espagnet, Aix-en-Provence.

Basically, there evolved a new emphasis on the garden, now called a *jardin de propreté*. This garden, which was sepatate from the working farm, had the distinctive purpose of showing off the main house. It was the most sophisticated place on the grounds; it was there that the high culture of the town expressed its conception of the countryside and shaped its ideas on nature. It was there that the countryside was inflicted with the upper classes' conception of order.

The residential architecture, the gardens, the landscaping, all played a very important role in reinforcing the social position of the owner. They established him in society. Likewise, the large open space in front of the house, the terracing and leveling of the ground, the straight lines of planted trees, a gateway, a sculpture, distinguished the landscape with the imprint of the new owner's refined taste. In this period the *bastide* garden came under the twofold influence of Italy and Versailles.

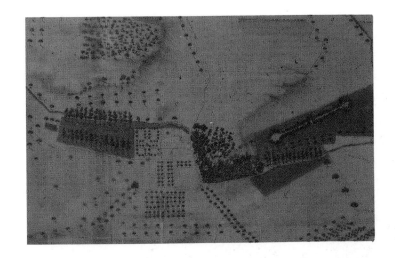

Plan, Beaulieu garden. Late eighteenth century.

Saint-Pierre des Canons, staircase descending to terraced garden.

Geometrically designed flower beds and a very Italianate use of artificial slopes were popular. Terraces and monumental staircases were used to raise the viewpoint and provide a downward perspective onto some focal point, often a circular ornamental pool but sometimes a visual contrast to the ordered design of the foreground and the vast panoramic view in the background. Such was the case with the gardens at Montjoli (ca. 1675), Saint-Pierre des Canons, and at Beauregard at the end of the century. This approach to garden design continued into the early eighteenth century at the Château du Seuil, Entrecasteaux, and at the Trimont Pavilion, but there the vistas were less grandiose.

To reproduce what they had seen in Versailles, and in the Parisian area, certain owners must have employed professional landscape designers. Very few names have come down to us, however. Many may have tried to lay out their own gardens, following guidebooks such as those published in 1739, J.-F. Blondel's *Maison de plaisance* or Louis Liger's *Nouvelle maison rustique*. For landowners looking for models and practical advice, the treatise of Dézallier d'Argenville, which was reprinted several times during the eighteenth century, was a godsend. This work was published for the first time in 1709, and many French gardens owe everything to it. The book examined in detail beautiful gardens and provided measurements for all sorts of motifs as well as a section on hydraulics. Dézallier regretted that so few French authors dealt with the art of gardening. In his book he makes clear from the start that he does not share the opinions of La Quintinie, de Tournefort, or Olivier de Serres. For Dézallier, the pleasure garden, or *jardin de propreté* was "a garden . . . in which one seeks principally a rational disposition and all that pleasing to the eye, such as flower beds, groves, and grassy lawns adorned with porticos, trelliswork shelters, statues, staircases, fountains, and waterfalls." Excessive formalism in the treatment of nature and trees trimmed into unnatural shapes had no place in his theory of gardening.

Château de l'Arc.

A later edition of his book introduced a gardening style that became more pronounced during the Regency and the reign of Louis XV, the opening out of the garden into the surrounding landscape.

The "ha-ha," a wide and deep ditch shored up on one side to retain the earth and giving a view over the countryside, was already part of Dézallier's schemes.

Beaupré, aerial view.

La Torse pavilion, garden.

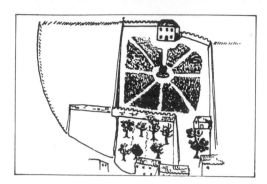

As Dézallier explains, "These hillsides are called *sauts de loup* or ha-has because the view they reveal surprises one . . ." The ha-ha, however, was not really justified in *bastide* gardens; many new owners preferred the traditional motifs for viewing the Provençal scenery. An embankment or a terrace was sufficient to discourage unwanted visitors while permitting a harmonious transition between the garden and the surrounding countryside.

The pavilion garden was very different from the garden of the *bastide* in both its use and design. Surrounded by walls, for it is just outside the town limits, the pavilion garden provided a setting for formal social gatherings; it was an extension of the drawing room organized round an architectural focal point. The gardens of the pavilions of Vendôme, Lenfant, Trimont, and La Torse are good examples of this type of garden. But the pavilion garden was unlike the garden of a *bastide* where one could become more intimate with the dreams and aspirations of the master of the house, as evidenced in his garden design, and where the entertainment was less formal.

A judicious simplicity marked the classical façade of the *bastide*. The design of the exterior contrasted with the baroque extravagance and refinement of the interior decorations and with the gardens where the imagination could run riot. The parterres were laid out along the axis, often an alley in front of the house extending to the terrace on which the house was sited, and they transformed the central part of the estate into a veritable showplace. Many an owner took fancy in the use of geometric motifs, fountains, ornamental pools, and illusions that made the rural countryside beyond seem like part of the *bastide*'s grounds. Shady lateral walks, groves, *tèses*, nymphaea, grottos, and mazes could be gradually discovered during a leisurely stroll through the garden. There were comfortable spots for a quiet rest and meditation and theatrics that encouraged a laugh. As the eighteenth century advanced, the more the garden of the *bastide* became a place where one could enjoy in comfort a refined style of country living.

Before visiting Provence, one must remember that all that is left of the splendid gardens now is the architecture. The spirit of the baroque garden—which was an endless play with optical illusions, water effects, gilded decorations, theatrical masques, ephemeral constructions, and costuming and makeup—can now be seen only by the imagination.

. . . .

Pavillon Vendôme garden, Aix-en-Provence. Drawing by Cundier.

Terrace, Montjoli, Aix-en-Provence, circa 1680.

Summerhouse, garden. Open-air salon at the city's edge.
Pavillon Lenfant, Aix-en-Provence.

The alley and the avenue

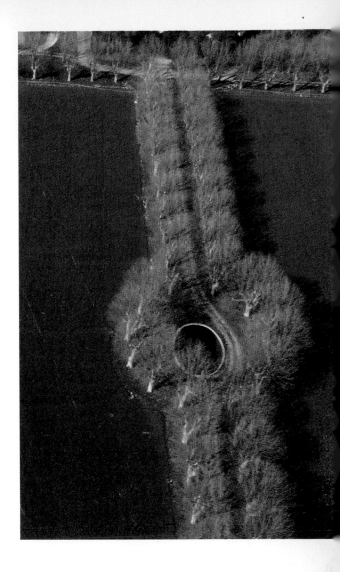

Visitors do not come upon the gardens of a *bastide* by accident. A long, broad stretch of road warns them well in advance of the approach, even in these days of the motorcar. Varying in length from several hundred meters to several kilometers, these avenues enabled *bastide* owners to leave their mark on the Aix countryside and enhanced the importance of their estates. Taken together, these avenues are an extension of the order found in the idealized geometric designs of the *bastide* gardens. They are also the rural equivalent of the wide boulevards constructed in cities during the reigns of Louis XIV and Louis XV, of which the Cours Mirabeau in Aix-en-Provence is an example.

The avenue takes the visitor in hand as soon as he nears the property, and its shade procures for him a summertime pleasure that one has to have enjoyed at least once to appreciate it. To savor this pleasure, one should walk slowly along the fifteen hundred meters of road that leads to the *bastide* of Saint-Hippolyte near Venelles. It is on a really hot sunny day, when the whole countryside seems flattened by the light, that one can best appreciate the delight of entering an avenue shaded by plane trees, holly oaks, parasol pines, horse chestnuts, and elms. (The avenues bordered by plane trees leading to Saint-Simon, the Gantese, the Violaine, and Bourgogne provide other fine examples.)

The alley, however, is part of the garden, and it leads from the gate to the *bastide* itself; it can be edged with box or laurels. It takes the visitor past flower beds, past a succession of views, down a fine gravel path, which once would have been carefully maintained by daily raking.

La Gantèse, garden path.

Garden path leading to the bastide, *drawing by François Cayol.*

Albertas, chestnut double avenue.

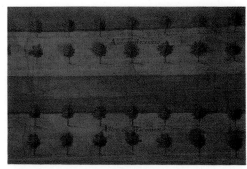

The gateway

Rarely of modest proportions, the gateway marked the end of the alley, the entrance to the estate. A low wall or a bank could flank it, but enclosing walls were rare. Thus, standing in the countryside practically unexpectedly, the gateway had a purely symbolic role to play. The decorative ironwork enabled the owner to display his initials or coat of arms against the sky (as at Châteauneuf-le-Rouge, Albertas, and Sannes). It also celebrated the notion of passage, so prevalent in the art of the Baroque period, which found in gateways an ideal place for expressing extravagant motifs. Sometimes the *jardin de propreté* was framed by a wall or a low wall, in which case the *bastide'* garden was conceived as a veritable stage (Bourgogne, la Bougerelle, or even Arnajon illustrate this feature).

Ansouis.

Jas Créma.

Bourgogne.

La Bourgerelle.

Le Verger.

Albertas.

Stairways and terraces

Today it seems as though the terraces have always been part of the countryside, so one never stops to think of the enormous work necessary to create a terraced garden, given the technology of the seventeenth century. Yet, despite practical limitations and the exorbitant cost of terracing, the owners of *bastides* were determined to clear spaces, to re-level their land, and to construct retaining walls and stairways. There are very few *bastides* without a terrace. As a status symbol it could hardly be bettered. The purpose of the terrace (and sometimes there was only one terrace) was to set apart the main house from the other buildings. Its surface was covered with gravel, and every day the cleaning of the garden began with the raking of this fine gravel around the foundations of the *bastide*. Different raking patterns were made daily. In this way the presence of the visitors and the movements of the day were recorded. No one could approach the *bastide* without being heralded by the crisp crunching of gravel underfoot.

From its position on the terrace the *bastide* surveyed the whole countryside, and in turn could be seen by it. A terrace was therefore never positioned immediately in front of the main house; something, if only a parterre, always separated them. The sides of the house, however, were often planted with large trees that provided shade and framed the composition.

The edge of the first terrace would invite visitors to go a bit farther, to the succeeding terraces. Through the balustrade they could glimpse the landscape below, which they could discover upon descending the stairway.

La Bargemone, secret garden above the lawn.

Staircase in the archbishop's gardens, Jouques.

Arnajon, stairway, terrace, and pool.

La Gaude.

La Mignarde, edge of the terrace.

The vitality of water in Provence

The idea that water can be used visually, as a design element, takes on a very special significance in Provence. In Provence, more than elsewhere, water refreshes the spirit. To understand what a fountain or an ordinary reservoir in a Provençal garden really represents, one has to have experienced the relative scarcity of water in Provence during the dry season. One has to appreciate the enormous effort involved in transporting this vital fluid from where it is to where it is not. All the canals, aquaducts, pumps, siphons, rams, channels, and irrigation trenches were constructed for the purpose of stocking water in reservoirs, cisterns, or catch basins. Provence does not have romantic lakes or meandering river streams. Lakeshores, wide estuaries, or marshs are just not part of the Provençal scenery; they are replaced by a complex network of small and large works all destined to ensure the supply of water. This network passes through rocky hills to link all the small refuges of the springs and valleys. It is rare in Provence for a farm or *bastide* to monopolize a water source the way the Roman farms did. Most of the time—before the conduits bring water to the plants and coolness to people, before they stimulate the imagination with sophisticated tricks, jetting out of the shadows of an alley or spraying unsuspecting onlookers via small pipes hidden in the dark sculptured walls of a grotto— the water is aerated in the sun. The *bastide* garden often has two reservoirs, one at an upper level for watering the garden and another on lower ground for agricultural purposes. Between the two reservoirs lies a succession of cascades, pools, waterspouts, and fountains supplied by buried canalizations.

Bourgogne, reflecting pool.

La Torse pavilion.

Saint-Pierre des Canons.

Saint-Hippolyte, fountain.

La Gantèse, pool in garden alley.

Château Lenfant.

The parterre

In Provence, according to surviving documents, embroidered parterres were laid out around the *bastides*. The plan of the Jardin d'Albertas, drawn up in 1751, is revealing for the principles it obeys from the works of Dézallier and Blondel. On the second terrace, along the sides of three fountains grouped in a triangle, along a stretch of lawn marking the central axis, two parterres are laid out surrounded by flower beds and yews trimmed into conical shapes. Each parterre is made up of embroideries that are symmetrical in their general outlines but differ in the detail of motifs used, including foliated designs, volutes, trefoils, and fleurons of grass, flowers, and colored sand.

Today few owners have the means or the desire to restore the embroidered designs in the manner of the seventeenth and eighteenth centuries. There is both the cost of upkeep and of transporting plants from garden nurseries which has been diminishing in number. The parterres that have best survived the changes in ownership and varying fashions are the boxwood parterres. Their upkeep is limited to two trimmings a year. Boxwood can border a garden section surfaced in gravel (as at Ansouis), a lawn, or a central pool of water (as at Bourgogne, the Lenfant Pavilion, or La Barben). However, it is rare to find box as carefully looked after as that of La Gaude or La Violaine. The garden parts are no longer laid out in sand as in the olden days, and the box has grown to obliterate the fine design; this is a problem Dézallier d'Argenville remarked on in his treatise. Yet, although box tends to overgrow, and sometimes its contours even to merge, the effect is a aesthetically pleasing to the contemporary eye. Who today would want to uproot the box of Beauregard or Ansouis so as to replant the parterres?

Parterre, Albertas park, Bouc.

Detail, lateral parterre planned by Embry
in 1770 for the château Borély.

Parterre, summer palace of Monseigneur du
Bellay, Draguignan.

The "tèse"

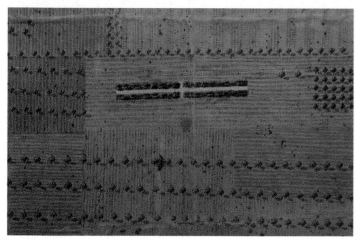

Hunting is one of the great pleasures of living in the country, and it went hand in glove with owning a *bastide*. Catching small birds in nets in certain dense woods suitable for this pastime was a custom practiced since the Middle Ages. In Italy, the creators of villa gardens planted carefully selected trees and bushes in such a way that they provided a place for a pleasant promenade as well as a place where this form of society game hunt could be pursued. John Evelyn described such a plantation laid out with a narrow alley especially designed for this kind of hunting in his diary when he stayed at the Villa Aldobrandini near Rome in 1644. "Here in the alleys," he observed, "they have hung large nets to catch woodcocks."

The *tèse* (in Provençal, *tèso*), derived from the Latin *tendere* ("to stretch"), is the Provençal equivalent of these alleys. A net was stretched across an alley planted with thick, berrying bushes and trees that met overhead to form a kind of tunnel. Down the middle ran an artificial stream. The birds, which were attracted by the berries and the water, were flushed out of the trees by the hunters, and as they flew down the alley, they precipitated themselves into the net. The hunt of course was not very exhausting; it was rather like hunting in an outdoors drawing room. The *tèse* thus had the combined purpose of accommodating a refined form of hunting and a pleasant stroll; it was certainly one of the most extraordinary landscape creations in Provence, and its presence in the *bastide* garden provides a most revealing insight into the real life in the Provençal countryside up to the midnineteenth century. It was part of the program of any self-respecting *bastide*, as is shown by the wonderful description given by the Comte de Villeneuve in the inventory of the Bouches-du-Rhône in 1829: "The *tèses* are plantations of trees and shrubbery that bear berries attractive to birds. They are planted in two rows 150 to 200 meters long and 10 meters thick. In the center is a small trench containing a trickle of water; sometimes there is a walkway in the center and

Tèse, *Beaulieu, late seventeenth century.* Each end terminates in an outdoor room where a stone basin collects rainwater for birds to drink.

Tèse, *La Tour d'Aigues, late eighteenth century.*

then the trenches are disposed on either side. Half-way down there is a space in the alley of about 2 to 3 meters, and it is here that the net is hung in a vertical position. This net has a very fine mesh with pockets, and the birds get caught in there and cannot escape. The actual hunt consists of walking slowly down the alley from where it begins to where the net is stretched, and then throwing handfuls of gravel or earth into the trees so that the birds are frightened and fly out, precipitating themselves into the net. One catches mainly passerines with this kind of hunting. The trees selected for these alleys are hazel, nutmeg, arbutus, dogberry, mastic, and spindle. Close to the net two screens of the common osier *(Salix viminalis)* are grown and their fine branches and thick leaves help to conceal the net from the birds.''

At some *bastides*, such as at Beaulieu or La Tour d'Aigues, the *tèse* was planted in the field; at others it was an integral part of the gardens, as in the gardens of Albertas at Bouc and at La Gaude. In either location the *tèse* retained its narrow, tunnel-like aspect which was essential for catching the birds.

The original meaning of the *tèse* has sometimes been confused to the point where it has been taken for an avenue or alley, especially the entrance alley, which is often planted with a mixture of trees—elms, plane trees, or horse-chestnut trees for the tall-growing varieties and laurels for the low-growing variety. Although there is some resemblance, the alley plantation is not as wide, and its foliage is less dense.

Inside the tèse *at La Gaude.*

Tèse*, Albertas.*

Woodlands, groves, the orchard, and the kitchen garden

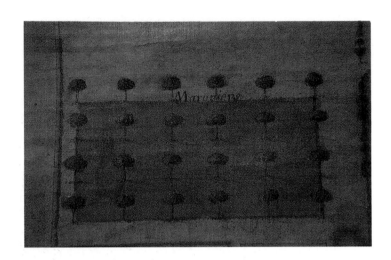

The woodland area around the *bastide* is often not very large. It is rather the meadows that take up most of the landscape. The thick greenery of the groves serves to screen the sun and to frame the view of the garden. And their height sets off the parterres.

Woodlands are an essential element of a *bastide*; they provide coolness in the summer. Often situated to the northwest, they protect the garden from the mistral and also provide firewood and a stock of wood for the carpenter. A woodland or a grove planted just for shade is often found in the quincunx at one side of the house. The species vary, but the principal varieties are pine, cypress, holly oak, and the common oak. From the point of view of color, the selection seems always to obey the principles laid down by Dézallier d'Argenville: "To be effective, all trees and shrubs must be of of a dark green variety, verging on black." At the end of the eighteenth century the woodlands increased in importance. They became a place to explore, a place for outings. More care was given to their layout. Alleys were laid in star formations, crossing diagonally to multiply the vistas (as at La Tour d'Aigues, Trans, and La Violaine). In the romantic period that followed, the straight alleys were replaced by winding paths (as at Fonscolombe, Barbentane, and Moulin Blanc).

The orchard and the kitchen garden are often particularly well tended and walled in to protect the crops from the wind and pilferers.

Albertas, chestnut groves.
The scenic garden groves frame the view on either side of the second terrace.

Beaupré, regular stand of plane trees.

Decoration and sculptures

A few names of sculptors and their workshops in Aix-en-Provence are known to us from contemporary documents, such as Thomas Veyrier, Jean-Claude Rambot, and Jean Turreau of the seventeenth century, and Jean Pancrace Chastel of the eighteenth century. Many noteworthy creations remain anonymous, although, overall, the abundant sculpture of the eighteenth-century *bastide* gardens was very unequal in quality.

The sculpture and the decorative stonework were unique for each garden. They were adapted to whatever theme was to be represented. Antiquity, vegetation deities, the four seasons, all served as pretexts for placing in the landscape sculptures of men at work or at war, women unveiling themselves, or children discovering the beauties of nature.

Bourgogne.

La Mignarde, head of a dog.

La Mignarde.

La Mignarde, oak leaves on base of the Nemean lion.

A rococo garden and
an experimental garden:
La Tour d'Aigues

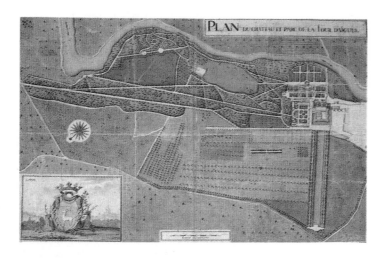

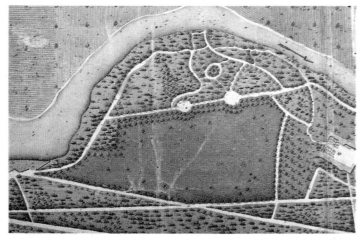

The château of La Tour d'Aigues is well known as a masterpiece of French Renaissance architecture. Although the garden of the château was completely destroyed in 1792, its plan, which has been preserved in the Bibliothèque Méjanes, reveals a very typical garden of the Regency period, a garden type that persisted until the end of the reign of Louis xv. The first garden was laid out to the northeast of the château by the dukes of Lesdiguières in the midseventeenth century. Over the years they bought up the adjacent land, diverted the road between Aix-en-Provence and Forcalquier, and created an enormous park enclosed by walls four meters high, completely breaking with traditional rural practice.[26] They terraced the ground around the château. The long terrace dominated the ornamental French-style parterre, with its rolling grounds and its big half-moon shaped ornamental basin, and overhung the Eze valley countryside.

When in 1719, Baron Jean-Baptiste Jérôme Bruny, a newly enobled merchant from Marseilles, bought the château, he found a garden that had been abandoned for the past thirty years. He restored the garden and added vast sections of land to it. He began by replanting the parterre and creating green bowers at either end of the terrace. In the park he worked out the alignments characteristic of Regency gardens. The design of the seventeenth century was replaced by a system of diagonal and radial alleys that crossed the park and divided it geometrically, though not in symmetric units. The design multiplied the vistas and controlled the view of the landscape from any fixed point.

Bruny drained the pond to the northwest of the château and excavated on its site an immense canal, 32 meters wide and running 350 meters long, framed by two alleys planted with young elms and a grove. In the area between the garden and the humid countryside, he laid a labyrinth of winding paths that evoked an English-style garden. In a hidden part of this elaborate scheme, entered by an ever-narrowing path, was a surprise: monkeys, chameleons, several species of African

Plan for the park, La Tour d'Aigues, 1783.

Park, La Tour d'Aigues, detail.

pigeons, turtle doves, pheasants, deer, gazelles, wild Corsican mountain sheep, and a menagerie of other exotic animals.[27] At the foot of the château, to the South east where there were once rented-out market gardens, he sited an ornamental, experimental kitchen garden. The kitchen garden was not the least surprising part of Bruny's undertaking. It was designed by him, and it occupied the best spot on the grounds in terms of exposure to the sun. But the experimental aspect of this garden was due to a renewal of interest on the part of the ruling classes in the improvement of garden produce.

A convinced Physiocrat, an esteemed correspondent of the Académie des Sciences in Paris, winner of the gold medal of the Société Royale d'Agriculture in 1785 for his introduction to Provence of a breed of sheep with extra fine wool, angora goats, and various trees and useful plants, the last owner of the château planned a garden whose produce the eighteenth-century English agriculturist Arthur Young enthusiastically described as the "most exquisite and advanced for the seasons." In the greenhouse could be found jasmine from the Cape, India, and Azores, a variety of African geranium, numerous euphoriba, Mexican cacti, mesembryanthemums, yuccas from Canada, atoemuria, several varieties of solanum, and aloes. In the garden, among the local species, were various exotic trees such as the papiriferous Chinese mulberry, mimosas, mountain ash, maples, alpine cytisus or false ebony tree, and the robinia or false acacia. In the park were planted cedars, thuyas, privets, and numerous other trees.[28] In the herb section of the garden were "the most beautiful plants of Provence, of the Levantine, the Pyrenees, and Cayenne," and the selection excited the admiration of the botanist Darluc, who decreed "any interested person or any conoisseur who comes to Provence must visit La Tour d'Aigues."[29] Bruny managed to interest the garden expert Brother Gabriel in his enterprise and engaged him as steward of the gardens.

A forerunner of the experimental gardens of the nineteenth century, the orchard and the kitchen garden of La Tour d'Aigues, which were once muddy meadowland, constituted the most cared-for section of his land. The kitchen garden was bordered by fruit trees, and its rectangular beds were positioned around pools of water that also followed an orthogonal plan. But this "laboratory" where experiments with new techniques of plant cultivation took place was not sacrosanct: it did not escape the revolutionary violence of 1792. Provoked by the owner's arrogance, the villagers made no distinction between the gardens and the château, and they ransacked them both.

. . . .

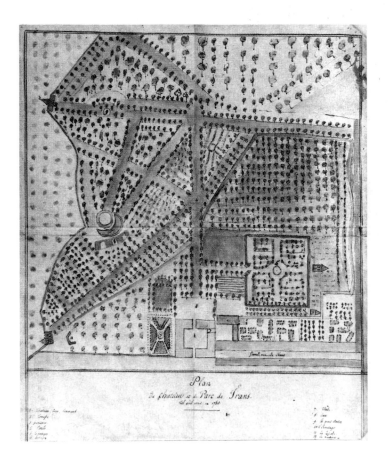

Park, Trans, 1780.
Like at La Tour d'Aigues, that at Trans (Var) included a walled ornamental orchard, a parterre, a *tèse*, and alleys crossing diagonally to form a star that leads to ornamental constructions.

The landscape garden in Provence

Although landscape parks inspired by English and Chinese examples became widespread in France toward the end of the eighteenth century, there were only a few parks created in Provence. Of these very little remain today. Among Provence's few enlightened residents, only the Comte de Valbelle at Tourves, the Marquis d'Arbaud at Jouques, the Baron de Castille at Castille, and to some extent the Marquis d'Albertas at Saint-Pons and the Baron Bruny at La Tour d'Aigues managed to accommodate park gardens. The topography of Provence made it difficult to import models inspired by the English or Chinese countryside without a few modifications.

While in England the landscape park evolved from a new approach to the science of growing plants, in France it went hand in hand with a need of escape from social and economic turmoil to an idyllic setting or rural utopia.[30] This gave rise, on the one hand, to the ornamental, nonworking farm, the *ferme ornée*, and, on the other, to the model commune projects of the Physiocrats, who drew their inspiration from a very idealized vision of Chinese agriculture. Thus, whereas exceptional agricultural research and experiments took place at La Tour d'Aigues, the dairy farm at Tourves produced nothing but a very fine example of a humorous and imaginative use of architecture. The dairy, designed simply to evoke a rural setting, was a kind of *ferme ornée*. Cl. Watelet commented on it in his *Essai sur les jardins*; he found at Tourves a welcome setting, ideal for a country stroll, where visitors could enjoy "a country-style meal consisting mainly of milk and fruit."[31] The main objective at Tourves was to transform the owner into an actor in a pastoral scene.[32] The physical decor had the definite function of giving substance to the illusion; the farm had no economic function. In order to really enjoy this rural Eden, and not to feel foolish, the pleasure-seekers had to have some semblance of authenticity in their surroundings. The countryside became in a sense a theater with actors ready to go on stage, having surrounded themselves with every possible

Park, Collongues.

stage prop to give credibility to the character they had chosen to play.[33]

In this sybaritic program new refinements were introduced for enjoying the simple pleasures of the countryside, and the whole of the landscape was used by the owner for his pleasure. The whole of the park in Tourves was organized into a series of environments with an emotive use of natural elements and many small structures that were completely foreign to the local surroundings. Over the years this type of construction could easily have been forgotten or destroyed, for there is almost nothing left of this park in Tourves, nor the picturesque park of Jouques except a pinewood which ressembles any other woodland.

Only the park of Saint-Pons in Gémenos has survived, thanks to the vitality of nature, stimulated by an abundant water supply, and to a protective site. Jean-Antoine Constantin, a Provençal landscape painter, born in 1757, enjoyed working there. Saint-Pons was one of the first parks in France to seek inspiration directly from nature, and not the design studio. Constantin's work at Saint-Pons offers us the rare opportunity of comparing the eighteenth-century vision with our own.

Park, Lamanon.

View of Saint-Pons. Painting by J.-A. Constantin.

From picturesque to gardenesque: the search for regular irregularities

At the beginning of the nineteenth century, the ruling classes in Provence were caught up in the large-scale European movement to assimilate the picturesque model in their gardens. The incentive came from the north. Innumerable catalogs appeared containing plans of existing gardens with directions for carrying these out. Professional landscape architects, influenced by the Beaux-Arts school and the Académie, made their reputation by a return to plans and formal designs. "The art of constructing gardens today has reached such a pitch of perfection that it should be placed on the same footing as the liberal arts," a prominent Marseilles citizen, M. Croze-Magnan, declared in 1813. Landscape architects and authors of prescriptive works abounded. J.-Ch. Krafft's *Plans of the Most Beautiful Gardens in France, England and Germany*, published in 1809 in three languages and dedicated to the architects and connoisseurs of gardens, proposed pell-mell, "edifices, monuments, and constructions beautified by the use of all styles of architecture such as Chinese, Egyptian, Arabic, Moorish . . ."[34]

The popularity of the picturesque garden gave rise to numerous publications that enjoyed a tremendous success. J.-G. Grohmann's *Recueil d'idées nouvelles*, published in both German and French, was reprinted three times between 1799 and 1810, and it sold equally as well in Florence as in The Hague, Paris, Budapest, and St. Petersburg. Lalos's work, *De la Composition des parcs et des jardins pittoresques*, was reprinted five times between 1817 and 1832, and Gabriel Thouin's work, *Plans raisonnés de toute espèce de jardins* (1819), was also very much sought after. The appeal of these publications was to the urban middle classes of Europe, for they pro-

posed models of picturesque gardens adapted to much smaller plots of land than those hitherto utilized.

With the picturesque garden, the landscape park, which had provided large landowners with a means of relating to nature, became reduced to a symbolic and standardized form. The past was evoked by an accumulation of references to bygone days, mainly through the use of the Gothic style, and faraway places were evoked through exotic details.

In 1832 the landscape architect and journalist J.-C. Loudon qualified as "gardenesque" the very style he had himself popularized, a style whose purpose was to incorporate the contemporary developments in botany and horticulture into the art of landscaping.

Fonscolombe, park.

Collongues, park.

The gardens of the Marseilles "bastides" in the nineteenth century

The northern gardenesque style was not readily accepted by the urban middle class in Provence; it caused much reflection and discussion. The president of the Académie in Marseilles, though open-minded about these innovations, also wanted to safeguard the local heritage. In 1813 he outlined his thoughts on the subject in his *Essai sur les jardins pittoresques, convenables au territoire de Marseille*. He noted that for some time exotic trees had been successfully planted and that only northern trees such as larch, birch, and certain fir trees had not taken kindly to their new habitat. He then reviewed exotic and local plants in Provence, analyzing their structure, shape, texture, intensity, and hue. He drew up a landscape plan for the Marseilles area. Deploring the stony aspect created by the endless walls built to protect the small local properties, he went as far as to suggest they be demolished and relaced with hedges composed of Mediterranean species such as hawthorn, burning bush, or sea purslane, which would both ensure the owner's privacy and provide the countryside with attractive greenery.

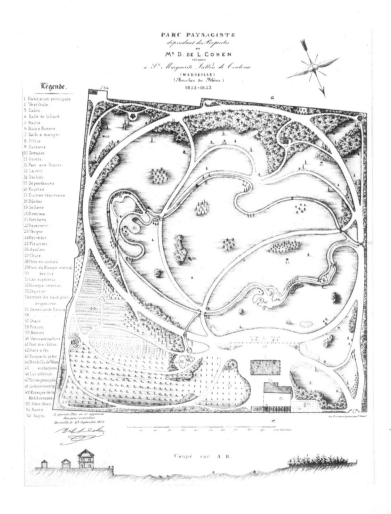

Member of the Marseilles bourgeoisie at home in his bastide.
Château d'Astros, bastide *in the Var belonging to a member of the Marseilles bourgeoisie.*
Landscape park, Marseilles.
One of the landscaped parks designed by F. Duvilliers (architect-engineer-landscape gardener) for his Marseilles clients. Today the park is called *La Ferme Blanche* and belongs to the city of Marseilles.

Parisian landscape architects and Marseilles' rock builders

The wealthy middle class, wanting to be in the forefront of fashion, summoned professional landscape architects from the capital. Others made do with the local experts. The gardens of the Second Empire, designed to provide the family with healthful walks, seem very uninspiring today. It is true that this aesthetic has reemerged in many recently created public gardens.

The banality of private gardens, however, was lessened by artful rock creations, thanks to a new material, cement. Thus masons who were also gardeners and sculptors developed between 1845 and 1914 a new speciality, a new profession dating back to the Renaissance grottos. As long as one had a few artistic pretensions, one could construct mountains instead of moving them and could use the plastic qualities of cement to imitate nature or any conventional building material. These *rocaille* ("rockwork") specialists transformed stony hills into phantasmagorical rocks, pavilions into fabulous castles, and expressed in concrete the impatient dreams

Imitation window in a summerhouse, Marseilles.

Rusticated door.

of the middle class. Cement applied in several layers and then colored enabled all kinds of visual deceptions: fashionable rustic constructions, where imitation wood, imitation fissures, or imitation tree trunks were placed side by side with the real thing, thus enhancing the illusion. In the gardens of country houses the *rocaille* specialists enjoyed a certain freedom of expression. In this privileged place the dreams of two cultures could meet; on the one side, the craftsman with his folk culture and, on the other, the new middle class, filled with classical ideas and ready for industrial and colonial conquests. The numerous signed pieces from this era show the desire of a group of craftsmen to assert their individuality as a challenge to industrialization. Their pretended oversights—a hat left on a nail, a coat hanging from a branch—are like a reassuring wink at us: everything is fine with them, they have just gone off for a drink and will come back when the garden regains its sense of humor.[39]

Fanciful structure on grotto in Marseilles, Montolivet.

"Forgotten" gardener's hat on tree stump, Marseilles.

Imitation roots on a terrace, La Ciotat.

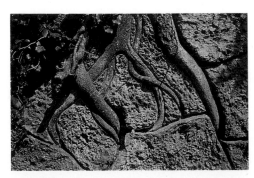

Fruit parks and landscaping of the "Mexicans" in the Alps

Between 1830 and 1930 many people emigrated to Mexico from the Barcelonnette valley in the Alps of Haute Provence.[40] Those who had made their fortune and returned to this region had not forgotten their rural roots. In the parks around their "dream Mexican villas" these rich adventurers began by planting the best fruit trees available as well as enormous kitchen gardens with vegetables and aromatic herbs.

Today we see how cleverly orchestrated these private gardens of Barcelonnette were in the juxtaposition of the agricultural garden, which was in fact the most important element of the estates, and the showy parks. The park stands out by its remarkable variety of fruit trees. The kitchen garden is often half the size of the park (a surface of 5,000 square meters in a park of 10,000). The landscaping scheme is centered around the task of protecting the vegetation; such is its importance. The villa's itself, with its high, enclosing walls, offers shelter for apricots, raspberries, or even a few rare pear or plum trees planted at the foot of the wall.

The importance of the rustic house or farmhouse in the scheme of the "Mexican villa" of the Barcelonnettes is easily understood. Besides providing a barn for the hay and stables for the animals, it also served as a lodging for the gardener, who was indispensable for the maintenance of such a sizable park and garden. The access to this rustic house was carefully specified, as this estimate by a horticultural specialist brought specially from Grenoble shows: "For the border of the alley leading to the rustic house, we would advise the following plants for the shady side: ivy and periwinkle. For the sunny side, two-colored couch-grass, canary grass, fescue, plumbago . . . Rosebushes also make strong and attractive borders." The park, for its part, was planted with the most subtly perfumed, and the most beautiful and extremely rich vegetation made its appearance in the valley: Judas tree, sophora, giant sequoia, Nordmann's fir tree, catalpa, Japanese copper beech, Austrian pine, acuba, magnolia, yucca, and jasmine.

The most ambitious of the Barcelonnettes employed the Italian landscape architects, the Mangiarotti brothers, who created the public parks of Genoa. Certain owners, anxious to modernize and beautify their park, employed a horticulturist-landscape gardener from Cannes to restore their park, fifteen years after its creation.

The design remains faithful to conventional, middle-class standards and provides a circuit both landscaped and picturesque, dotted with all sorts of constructions: water towers, grottos, small *rocaille* shelters, artificial lakes, and especially summer houses. The wide majestic alley leads right to luxuriant terraces and pillared porches. Colorful verandas form transitional gardens between the house and the park, an intermediary space between the lost paradise of Mexico and the paradise regained of the Alps. On the stained glass windows flourishing orange trees and pomegranates of cut glass can flower year-round while outdoors snow may fall.

. . . .

Stained glass, Barcelonnette.

The botanical passion

In the eighteenth century botanical explorers returning from the East found in the region between Marseilles and Toulon an ideal place for acclimatizing new species. The climate enabled the plants to recover their strength after the privations of the long sea voyage. In 1762, Jean-Baptiste Fusée-Aublet, appointed botanist to the king, created a garden in Salon where he cultivated the "most rare and useful plants brought back from the two Indies [India and the West Indies], thus giving them the time to grow stronger and be able to sustain the journey to Paris, where it was then easier to acclimatize them."[41] A few of the plants acclimatized in Salon spread from there throughout Provence and then into the surrounding areas—for example, Cape jasmine, Azores jasmine, Indian heliotrope, the Virginian tulip tree, Virginian catalpa, Cape heather, and other now-common plants.[42] The botanical gardens created in 1775 in Aix-en-Provence, where Lieutaud and then Darluc worked, had a rather brief existence; before they disappeared during the French Revolution, they contributed many new varieties of plants to local *bastide* gardens.

The naval botanical garden in Toulon, intended to supply the doctors with medicinal plants, was created in 1786. The naval officers regularly brought back seeds and plants from warmer climes, and thanks to this supply, the gardens rapidly became well known and supplied the whole region with exotic plants, some of which became naturalized. This is the case of the Japanese medlar brought back to France from Canton in 1787, which has propagated widely. We know, for example, that in 1820 in the botanical garden of M. Filhe, known as the Jardin de Flore, plants such as *Phoenix dactylifera, Plumbago capensis, Acacia melanoxylon*, and *Cassia tomentosa* were grown beside plants from the botanical garden of Toulon. This botanical garden was also a very pleasant place to stroll in. Flaubert wrote in 1845: "If I lived in Toulon, I would go every day to the botanical garden." Despite this admirable sentiment, the garden was transferred five years later far from the public eye to Saint-Mandrier.

The botanical garden of Marseilles played an important role in the distribution of new plants, although Flaubert found it ugly compared with the one in Toulon. In 1827 the director offered seeds, graftings, and cuttings to all interested persons. At this time there were over a hundred *bastides* with heated conservatories, whose owners had been caught up by the craze of the century—the passion for botany. As the Comte de Villeneuve observed, "From this praiseworthy and useful cooperation not one year passes without the acclimatization of several exotic plants, and one can say... all the plants are of foreign origin, starting with the plants brought from Greece and Italy by the Phocaeans and the Romans, and acclimatized by them, and continuing with the plants from the West Indies and India brought

Greenhouse, Marseilles.

over by cargo loads, and which every year are taken from the botanical garden and planted along the walks of the town.''[43]

In cultivating new species, the amateur botanists also expanded their knowledge of the local flora. Countless amateurs scoured the countryside in search of plants to study and collect. A few became famous, such as Jean-Louis Castagne, the mayor of Miramas, A. Lions, a solicitor in Marseilles, Félix Salze, Ludovic Lègre, a Marseilles lawyer, Antoine Honoré Roux, street-porter in Marseilles, and Reinaud de Fonvert, former counselor to the Imperial Court. In the Second Empire, when flowers returned in force to Provençal gardens, amateurs and professionals formed associations and published new reviews. The Horticultural Society of the Bouches-du-Rhône was created in 1846, and the *Horticultural Review* of the Bouches-du-Rhône in 1854. Horticulture, like botany, became a social phenomenon. More eclectic than ever, the garden of the Second Empire gave priority to plant collection over composition. Tropical or northern, exoticism was the word of the day. The botanist in his paradise saw the plants one by one. To the joy of discovering a succession of microcosms in an Eden of labels was added the pleasure of naming each plant and of reciting the list.

. . . .

Villa Bermond, Nice.

Villa Winslow, Cannes.

Château des Tours, Cannes.

"Like a perfumed woman:" The point of view of the literati

A change in philosophy and life-style

In the second half of the eighteenth century, the French Mediterranean coast, not yet christened the "Azure Coast," began to attract visitors, and its reputation began to spread. Before then, people traveling between France and the Italian states or Rome, such as Mon-

taigne in 1580, generally crossed the Alps at the Mont Cenis pass. The sea route was not an attractive alternative, for pirates were legion, and the coastal road linking Nice and Genoa was not only so primitive that carriages could not use it, but was often cut off when the Var River overran its banks. With the advent of the ideology of nature and development of natural sciences, it became fashionable to compare customs, institutions and religions. The picturesque ideal, rooted in the work of Nicolas Poussin, Claude Lorrain, and their followers, found a perfect vehicle in landscape gardening, in England and later in France. A new code of aesthetics was emerging, accompanied by a surge of curiosity about other lands and cultures. The tenets of classicism fell out of favor as, in literature and the arts, representations of rural and wild landscapes became the fashion. Eighteenth-century intelligentsia cultivated an inner sensitivity and proclaimed an intense urge to experience new ways of living which could only be assuaged by travel. They wanted to follow, to the letter, Jean-Jacques Rousseau's maxim that "the observation of all species is not based on books, but on personal experience."[1] Intellectuals wanted to see the world, motivated either by scientific curiousity or pleasure. Earlier travelers had condemned the "abominable isolation" of the Alps, looking upon them merely as an obstacle between themselves and Turin, Vienna, or Venice. Now, however, the new breed of traveler was in raptures at the exalted sensations experienced in the Alps. They went to Switzerland in search of climatic novelty, venturing farther abroad than ever before in hopes of discovering a panacea for all their ills in change of scenery and air. In this general context of changing ideas and practices, some travelers stopping off at Nice and Hyères found them similar in landscape and climate to faraway places already discovered by other, more daring travelers, but much more conveniently located.

The Antibes harbor. After Ozanne, 1776.

View of Antibes, where palms first appeared. Early nineteenth century.

Naturally beautiful sites: a premonition of change

What could be more scenic than Nice, set at the foot of a rock, with a curve of evergreen hills behind it and, stretching out before it, a bay reflecting the blue of the sky? It seemed as if nature had fashioned it to please landscape connoisseurs. Moreover, the town was surrounded by "landscape pictures" bathed in white light: slopes, hills, capes, beaches, and the Esterel in the background. Soon, it was noticed that the Alps and Alpine foothills warded off cold winds, making a "greenhouse" of this south-facing coast;[2] in short, this was a perfect place to winter. In about 1750 Nice began to attract the English. Just prior to the French Revolution, at a time when Nice's population was 15,000 inhabitants, an official survey showed that 100 foreign families were in residence from October to May, either in the new residential areas or *bastides* near the town.

Precedents could be found for this vogue. In 1560 Marguerite of France, sister to Henri II, wedded Emmanuel-Philibert, Duke of Savoy, and was received with great pomp in Nice, located just at the border of his estates. In his account of these mid-January nuptials, Marguerite's escort, Chancelor Michel de l'Hospital,[3] sang the praises of this town, "caressed by the west wind," mentioning "the flowers in the gardens." Before regaining the transalpine part of Savoy, the duke and duchess spent the winter months, "less harsh there than anywhere else," in the château on the rock cliff overlooking the town. Four years later, Charles IX toured Provence with his court, and was received at Hyères.

The queen mother, Catherine de Médicis, dazzled by orange and palm groves, asked the king to build her a "royal house surrounded by gardens," but political and religious strife led to the cancellation of this project.[4] A century later, when the aristocracy began to make heavy use of perfumes to render social relations all the more pleasant, their practiced sense of smell quickly detected the aromatic richness of Mediterranean plants. Antoine Godeau,[5] a man of letters who later became Bishop of Grasse and Vence, compared Provence to a *gueuse parfumée*, or a woman wearing scent, because its charm helped compensate for having to leave his Parisian coteries. The beauty of the landscape moved him to cry out, in poetic enthusiasm:

O fields, O fields of Grasse, O fertile hills,
O cultivated rock, O silvery springs,
O myrtle, O jasmine, O orange groves ...

Saint-Hospice bay.

A. Beaumont, *Voyage historique et pittoresque du Comté de Nice*, 1787.

In a more affected tone, the Marquise de Sévigné, in a letter to her cousin Coulanges, described the siege of Nice by the troops of Louis XIV in 1691. She expressed concern that the besiegers, including her grandson Grignan, would have their heads turned by the scent of the *fascines* (pieces of wood used to reinforce military earthworks), which were "all in orange, laurel-rose, and pomegranate wood! All they fear is being excessively perfumed!" She added, referring to Nice: "Never has there been such a beautiful landscape, nor one so delicious."[6]

The Nice coast.
A. Beaumont, *Voyage historique et pittoresque du Comté de Nice*, 1787.

Fruits in winter

Early in the eighteenth century, Vauban arrived in the Midi region on a tour of inspection of fortifications and strongholds. In his report on the ramparts of Saint-Paul, he included a description of the countryside, which gives historians an idea of what kind of local farming methods prevailed at the time: "The land is covered with grapevines, olive trees, and fig trees; one often notes these three types of plant in the same field, set in three separate rows with wheat planted between them. Thus the same piece of property yields wheat, olives, and figs." The general commissioner of fortifications made no effort to conceal his admiration for this "land, where one finds beautiful oranges of all species in summer or winter, standing right out in the wind, a sight to be seen nowhere else except in Hyères..."[7] At this time, exotic plants were virtually unknown, and the orange tree was perceived as a little miracle: always green, prodigal with flowers, fruit, fragrance, and savor. It embodied refinement, and its mere presence confirmed the presence of a garden. The introduction of the orange tree coincided with the creation of the oldest gardens known today. Let us recall that the citron, a citrus fruit indigenous to northern India, was already being planted in Greece: in the third century B.C., the natural scientist Theophrastus alluded to it in his *Plant Studies* as the "Median or Persian fruit." The orange, which originated on the shores of the Bay of Bengal, was acclimated in Africa and Sicily by the Moors around the year 1000. The Crusaders saw orange trees in Palestine. The local introduction of the orange tree seems to date from the fifteenth century, when Genoese emigrants settled on depopulated Provençal estates in Biot,[8] although it may have been introduced from other countries at the same time, since, in Hyères, it was referred to as the "Portuguese orange tree."

Maltese orange.

A. Risso and A. Pointeau, *Histoire naturelle des orangers*, 1818.

Bergamotte mellarose *with double flower.*

A. Risso and A. Pointeau, *Histoire naturelle des orangers*, 1818.

Garden composition in its initial stages

Erste Aufsicht von dem Seehaven zu Antibes Premiere Vue de le Port d'Antibes en Provence

The oldest garden on the Riviera is located in Nice, on the hilltop of Cimiez, where the Gallo-Roman town of Cemenelum used to stand. The garden plan was drawn up by monks, who began to build their abbey in 1546; it is claimed that this plan has not been modified since that time. Large alleys set at right angles define the plots that used to be cultivated by the "brothers." In Vence, the *Grand Jardin* esplanade lies on the former site of a high-walled garden, built in about 1580 by the Barons of Villeneuve, whose fame has been transmitted by the *félibres*, writers devoted to preserving the Provençal language. After the end of the sixteenth century, the aristocracy of Nice had large *bastides* built on their estates in the neighboring hills. At the time these were referred to as *maisons civiles* to distinguish them from military or religious establishments, which, up until then, had been the only type of edifice whose fortifications were independent of urban defenses. In Grasse the house later known as Villa Fragonard was one of the first seventeenth-century mansions built outside the surrounding walls of the town. These country houses were sometimes decorated with great care inside; outside, they had a walled garden or terrace with a well, fountain, or basin and regular orange parterres, even on estates engaged in cultivating oranges. Alleys shaded by vine-covered pergolas and flower beds were also typical of this type of formal garden, which generally covered only a small proportion of these vast properties. One is tempted to see in this a persisting tendency of feudal architecture, which made use of very limited open-air spaces in fortified edifices. Naturally, garden size depended to a great extent on topography. In the early seventeenth century the châteaux in Cagnes and Gourdon underwent considerable architectural transformation. At Cagnes, superposed galleries were added, and the triangular patio was caught between walls projecting into the town. At Gourdon, located outside the village, the terraces were extended by means of enormous barrel vaults built up on a rock base to create a tiered garden planted with cordons of clipped box. The

Orange and fig picking near Antibes in the eighteenth century.

Grasse.
Parterre planned for the Hôtel Théas de Thorenc, built in 1690, which later became the Hôtel Court de Fontmichel in 1774.

orange tree "adorned with a pleasant green which evokes an image of perpetual springtime,"[9] was to be found in the gardens of seventeenth- and eighteenth-century town houses built by prominent families. In Grasse seven of these little gardens remain, almost all of which have retained the orthagonal layouts of their alleys: three are within the confines of the old ramparts, and the others are close by. These aristocratic gardens were hidden from view and failed to come to the notice of writers traveling in the vicinity. Abbot Jean-Pierre Papon, traveling in the winter of 1778, stopped to rest in Grasse. In his guidebook,[10] he noted that gardens were every where: "South of the town are meadows and gardens scattered with all sorts of flowers fed by the life-giving waters of mountain sources. When in bloom, the orange, lemon, and citron trees give off a lovely fragrance which mingles with that of Spanish jasmine." Several years earlier, another wandering man of letters, the Scot Tobias George Smollett, had expressed his astonishment upon discovering the countryside around Nice near the stream running along the old town. Smollett claimed that, from the ramparts, he could almost believe himself under a spell. He thought this stretch of country much like a garden, full of green

trees laden with oranges, lemons, citrons, and bergamots, a scene that he found delicious to behold. He went on to add that all kinds of green peas, lettuces, and excellent vegetables were ready for the picking. He observed that the shrub roses, massed carnations, buttercups, anemones, and asphodels showed a vigor and scent not to be found in English flowers.[11] These little gardens, a combination of orchard, vegetable plot, and flower garden, lay on the outskirts of towns and were a normal part of everyday life, however extraordinary they might seem to newcomers. A Swiss physician, Johann Georg Sulzer, in a description dating from about 1775,[12] contended that gardens in the middle ground of a panoramic view could induce the observer to believe that gardens and natural environment were as one. In 1782, almost a century after A. Godeau, the writer Antoine Léonard Thomas, in descriptions of his rambles in the countryside near Nice, spoke of a certain duality in the landscape's power of attraction, seeing "everywhere the blend of untouched and cultivated Nature, mountains which are gardens and others which bristle with rocks, pine and cypress, and, in the far distance, the snow-covered peaks of the Alps . . ."[13]

The route de France *in Nice. J. Defer,*
1865.

History of the gardens of the French Riviera

Picturesque excursions into the land of Acclimatization

When Stephen Liégeard found a name for the coast stretching from Marseilles to Genoa,[14] winter residents were flocking to the region, but its legend had already been some time in the making. Years before, its mild climate had been recognized as a sort of Garden of Eden for plant-lovers. The Hyères basin, which receives almost twice as many hours of sun annually as the Paris basin, provided ideal conditions in which plants shipped from overseas to large local harbors could adapt to the Mediterranean climate. In the early nineteenth century different species native to intertropical regions from the botanical garden in Toulon were planted in private gardens, such as the Filhe garden and that of the Count of Beauregard. But the first attempts at naturalization were made much earlier. François-Emmanuel Fodéré mentioned that, in 1640, a Hyères man by the name of Arène introduced "twenty species of orange tree, thirty-one species of lemon tree, palm trees, sugarcane and many exotic plants which had not previously been known."[15] We have already seen that when Charles IX visited the town in 1564, orange and palm trees were in cultivation. A sugar-maker named Gabriel had presented the previous monarch, François II, Charles' brother, with a scheme to raise sugarcane in Hyères.[16]

In Nice, Jean-François Bermond, upon his return from the Antilles, planted a garden with species native to the New World, such as the cotton plant, indigo plant, colocase, banana tree, guava tree... On December 11, 1796, the head engineer in the Alpes-Maritimes *département* certified that he had seen, growing out in the open, "sugarcane plants at all stages of development, next to plants from all corners of the earth, grown for their usefulness or for pleasure," concluding that the soil was being cultivated "in a manner both interesting to the naturalist and useful to the Republic."[17] Officialdom and private citizens joined forces to introduce new species. The Empress Josephine, nostalgic for her native Martinique, played a part in recreating the landscape of the Côte d'Azur. In 1804 she had several different plants indigenous to New Holland sent from her greenhouses at Malmaison to the administrator of the Nice gardens,[18] asking that he assists "in her plan to naturalize a multitude of exotic plants in French soil." At that time Antoine Risso was preparing his *Histoire naturelle des orangers* and planting a collection of citrus trees on his estate in the Saint-Roch district. His *New Travelers' Guide to Nice* came out at that time, evidence of the influx of foreigners who came to spend what, in 1787, illustrator Albanis Beaumont christened the "season."[19] The English preferred the new residential area off the route de France, and those who had villas built on lots between this road and the sea were able to display their gardening prowess.[20] In 1822 work began in Nice on the transformation of the hill where the château stood into a public park; the landscape model was selected, showing the impact of English ideas. Other English families were attracted by the tranquility of Hyères, where they found homes that had been available for rent for decades. Yet another British colony sprang up around Cannes, "discovered" in 1834 by Lord Brougham. Châteaux, villas, and hotels

Stephen Liégeard.

were built, along with their gardens. Cannes' first real estate developer was Sir Thomas Woodfield who, in 1837, began to build large mansions surrounded by parks, which he intended to sell to fellow Englishmen. One example was Château Sainte-Ursule, subsequently called Château de Tours and Vallombrosa. Villa Rothschild, facing the Vallombrosa park, was built at a later date, and provides another example of the typical nineteenth-century English garden in this area. In Nice, about 1840, the Count of Pierlas planted a landscape botanical garden on his Saint-Maurice estate. The Chambrun and Liserb parks today are all that remain of this park. Cedars, araucarias, and palms standing all throughout this vicinity give some indication as to the park's original size. Other Nice families relandscaped *bastide* gardens in keeping with the taste of the day; the Count of Cessole and banker Honoré Gastaud had their villa gardens redone. The eclecticism of the age was revealed in these gardens' irregular bodies of water, waterfalls, pavilions, and "ruins"; they also occasionally featured greenhouses, a novelty introduc-

ed in the nineteenth century. The Villas Peillon and Bermond, well-known for their adjoining picturesque parks, were rented by the Imperial family of Russia, and Tsar Alexander II received Napoléon III at Villa Peillon in 1864. In the same year, the Count of Vigier succeeded in naturalizing the first palm, *Phoenix canariensis*, on his estate near the harbor; this species would later spread from Hyères to Italy and become a hallmark of the Riviera. This was the name used in Italy to designate the coast stretching from Bordighera to La Spezia, where mountains come into abrupt contact with the sea; it was the term used by Smollett in 1765.[21] In his account of the voyage from Nice to Rome, he related that, on the other side of the Cape,[22] lay a superb land cultivated like a garden, whose plantations reached up the slopes to the hilltops. He reported that this area was scattered with villages, châteaux, churches, and villas, and that all along the Riviera the same type of landscape predominated, except in areas bare of construction and agricultural activity. A distinction was initially made between the Riviera di Ponente and

Plant study, watercolor.

Plant study, watercolor.

On the route de la Grande Corniche *above Monaco.*
F. Benoist, *Nice et Savoie*, 1864.

the Riviera di Levante, respectively west and east of Genoa. In the nineteenth century the term ''Riviera'' came to include the French coast between Menton and Toulon.

During the nineteenth century plant-lovers were not alone in noticing this stretch of coast between Marseilles and Genoa. Gradually, it became the mecca of writers, musicians, journalists, painters, architects, and landscape architects in search of inspiration and of the idle rich in search of perpetual springtime. It also attracted real estate developers sniffing out new ways to make money, engineers working hand in glove with local entrepreneurs, and doctors, who were just discovering the beneficial effects of the Mediterranean, sunshine in winter and a wealthy clientele. A cult was growing up around the ''Azure Coast.'' Yet traveling was still an uncomfortable and occasionally hazardous occupation. The new aesthetes seeking picturesque landscapes had to put up with a great deal along the way. In 1837 Stendhal declared in Toulon: ''In hopes of encountering beauty, I put up with the fatigue of the diligence and poor lodgings.'' Gardens were vital to this quest for beauty. Flaubert traveled to Italy in 1840, and visited one garden after another: he went through the botanical gardens in Marseilles, Toulon, and Saint-Mandrier-sur-Mer and orange groves in Hyères. He mentioned the first cacti growing out in the open in the Esterel, the poor taste of gardens in Antibes and Nice, a charming hotel garden, and a large garden of superposed terraces in Nice.

In 1861 the coast had not yet been ''invaded'' by visitors. The *Revue de Nice* could still applaud tolerant ''proprietors and farmers who allow people whom they do not know onto their property,'' or recommend outings to ''the Gairaut Valley, carpeted with violets along its entire length, the Count of Pierlas's villa, so richly adorned with rare trees and shrubs, the beautiful home of the Count of Cessole, where vast reflecting pools mirror green pines or the residence of Baron Van Zuylen, who possesses one of the loveliest date trees in the area and a Judas tree much admired by visitors in April.'' It also mentioned the Villa Girard, farther away toward the Var River, ''its olive trees underplanted with flowers, where wheat, green peas and beans are cultivated with grapes, and also the Saint-Aubin estate, whose artichokes, asparagus and strawberries are familiar to anyone who has wintered in Nice.'' Visitors noticed the similarity between these properties and country estates. But the hordes were on their way, and the Riviera prepared for the onslaught. Students in architecture at the Ecole des Beaux-Arts in Paris were asked to treat a question on the ''country house in the south of France'' as early as 1857. This may have been an invitation to provide a French solution to a problem raised by Prosper Mérimée. In 1856, in Cannes, Mérimée, inspector of historic monuments, bitterly protested against the way the coast was being colonized by the English, ''settled here as if in conquered territory.'' Moreover he abhorred the architecture they favored: ''They have built fifty villas or châteaux, each more extraordinary than the last, and deserve to be impaled upon the architecture which they have brought to this area.''[23] Mérimée had an inkling that the trickle would soon swell to a flood and feared what each fresh trainload of passengers would bring.

Cap Ferrat, watercolor.

The coming of the railroad: exotic resort gardens

"Here instead of châteaux, there are elegant country houses called villas. Instead of parks, gardens—but oh, what gardens!" Arthur Mangin, 1867

In 1861 the railroad reached Cannes. In 1869 Nice was linked to the rail network, and Vintimille in 1872. Travelers no longer needed to be hardy pioneers. Guides were soon published recommending scenic rail excursions, and new tourist resorts began to spring up. A string of railway stations provided ready access to this area, which had not yet named the "Azure Coast."[24]

Visitors could descend at any number of railroad stations, each in a marvelous corner of this dreamland, far from city smoke and troubles, midway between colonial empires and the fogs of northern Europe. This was the heyday of the seaside resort. Hyères and Nice had been drawing winter residents since the mideighteenth century, and Cannes had already been put on the map. Saint-Raphaël was next to become the rage; this was engineered by its mayor-architect, Félix Martin, and by Alphonse Karr, who was also instrumental in developing the cut-flower industry. In about 1865, Karr's idea of sending packages of cut flowers to Paris by rail was put into practice.[25] Before long, whole wagonloads were being sent to meet Parisian demand. Later Emile Ollivier succeeded in attracting crowds to Saint-Tropez, and Michel Pacha did the same for Tamaris. The idle rich of Europe customarily migrated from place to place with a weather eye on the barometer, in search of eternal springtime. For them, the so-called "land of orange groves" between Hyères and Menton turned out to be a winter paradise. Members of high society professed to idealize travel and in particular travel to Oriental countries. Here, they could look out over the Mediterranean and allow their imaginations to take flight. This explains the decidedly exotic tone of some houses and gardens to be found on the "Azure Coast." An exotic landscape grew up, characterized by an eclectic assortment of architectural styles. A Moorish villa might very well be built next to a home in the *troubadour* (a nineteenth-century neomedieval) style. Lavish residences complete with parks and gardens were built all along the coast. In Cannes alone, 250 were built between 1840 and 1870. In 1887 Guy

Lord Brougham's Villa Eléonore-Louise, Cannes.

Stained glass window of the winter garden, at the Riviera Palace, Beausoleil.

Antibes and the railway line.

de Maupassant traveled by boat between Antibes and Saint-Tropez and saw "countless villas amid the shrubbery all along this endless shore, looking like white eggs laid during the night in the sand, rocks, and pine forests by monstrous birds from those snow-covered peaks visible beyond the coast."[26] He went by Cap d'Antibes, noticing a "prodigious garden between two bodies of water where the most beautiful flowers in Europe grow . . . more villas and, at the tip of the point, Eilen-Roc, the lovely, whimsical house which visitors come from Nice and Cannes to see." Today's visitors to Eilen-Roc might find it difficult to believe that a century ago this vast wooded park was nothing but a pile of rocks by the sea. To create flat ground around the villa designed by Charles Garnier, thousands of tons of earth were brought in. This flat surface was not typical of the region; in fact, only a gigantic effort could produce it. Most gardens were designed with the natural terrain in mind, as Edouard André, the great nineteenth-century French landscape architect, confirmed: "Gardens in this region are very rich in plant life and are totally different in design from those laid out on a flat surface or regular slope. There are sharp variations in level obliging one to structure the garden space by means of terracing . . . Sometimes terraces are large enough to be treated as parterres; otherwise one must resort to using winding paths up the slope, punctuated with terraced landings."[27] Up until 1900 landscape gardens predominated on the Côte d'Azur, with a special emphasis on irregularities, rare plants, and flowers. The palm was a choice collector's item because of its size and its power to suggest exotic lands. Gustave Thuret introduced varieties of palm at the botanical gar-

Michel Pacha's summerhouse, Tamaris.

Michel Pacha.

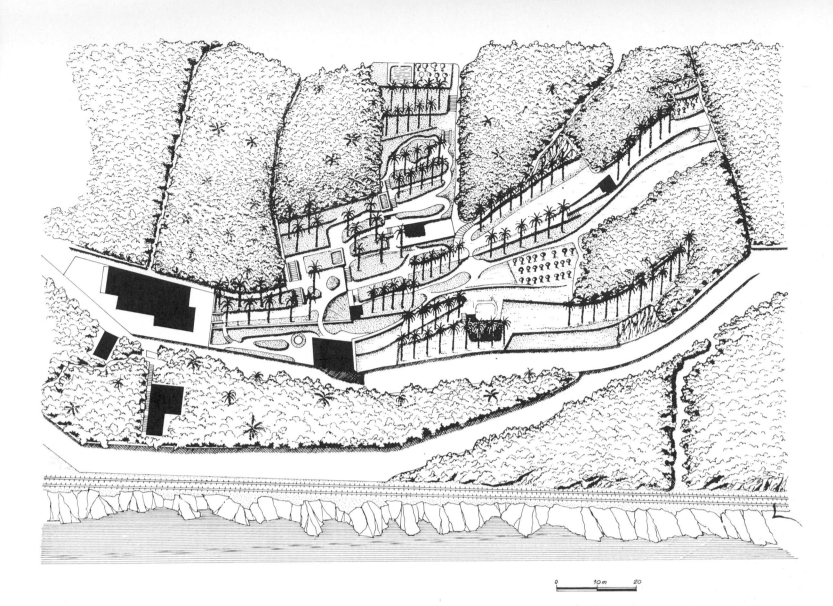

Plan, the Charles Garnier Villa.

The Charles Garnier Villa, Bordighera,
Italy.
Edouard André, *Parcs et jardins*, 1879.

den which he established in 1856 in Cap d'Antibes. At Villa les Cocotiers in Golfe-Juan, the Count of Eprémesnil owned more than 5,000 plants including 62 species of palm. In his tropical garden in the Grottes de Sainte-Hélène district in Nice, Dr. Robertson-Prochowsky started a collection in 1892 that eventually featured 125 varieties. At the end of the nineteenth century there was a trend in favor of succulents. Connoisseurs living between Nice and Menton found themselves in the place in France benefiting from the greatest amount of natural protection from the cold. In about 1895 botanist Robert Gosselin cultivated cacti among subtropical rarities in the Colline de la Paix park in Villefranche-sur-Mer; simultaneously, the head-gardener in Monaco, Augustin Gastaud, was putting together extensive collections of succulents that were later conveyed to the rocky site of the celebrated Jardin Exotique. Another specialty was the cultivation of exotic food crops, in which Arpad Plesch took a deep interest. Plesch worked in Beaulieu-sur-Mer, at La Léonina, located in a sheltered site called the Petite Afrique. Between 1937 and 1974 he was able to naturalize wild or

horticultural varieties of plants bearing comestible fruits, seeds, pods, stems, or roots, in addition to aromatic and medicinal plants and those used in making condiments. In 1950 Miss Campbell began planting a botanical garden on her Menton property, Val Rahmeh, acquired in 1966 by the Museum National d'Histoire Naturelle.

The mild climate helped spread "botanical fever" among European high society wintering on the coast. Hothouses were not absolutely necessary here, for many exotic species could be grown in the garden for contemplation on one's daily walk. Deciduous species fell out of favor because many gardeners were winter residents; they preferred subtropical plants, which were green all year round. "These gardens often have an Oriental air to them," noted E. André in 1879,[28] describing villas such as Bordighera which belonged to Charles Garnier, architect of the Paris Opera House. The villa overlooked the water from a height about 50 feet above the railroad tracks. Wide ramps wound up a steep slope, through a small terraced palm grove. The palm trees were presented like pieces of sculpture; one's gaze was meant to travel from bottom to top. The same is true of a column among the palms which had been recovered from the wreckage of the Tuileries. Prickly pear and agave added an Oriental touch, and vine-covered pergolas provided shelter from the summer heat. Botanical explorers used exotics to remind them of their far-flung travels, juxtaposing plants from the distant erstwhile territorial possessions of old Europe.[29] But "exoticism" was not synonymous with "jungle." In turn-of-the-century gardens on the Côte d'Azur, the object was to create domesticated tropical luxuriance, appropriate for viewing on one's daily promenade. Landscape architects belonged to the "public garden school" and tended to see gardens as a means of dressing up lots adjacent to roadways. Members of this school, when commissioned to do private landscaping, often produced gardens very like a public square. The garden alley was a key element in garden planning due to the central role played by the daily promenade and to proprietors' desire

The Monte-Carlo gardens designed by Edouard André.
The mixed-garden style combined the French love of geometry with an emphasis on the curved line.

to be able to show visitors around in a carriage. Visiting gardens was a social pastime on the Côte d'Azur. Each town organized competitions between property owners recognized for their role in beautifying the township. In 1887 Stephen Liégeard noted that specific gardens were always open to the public; the Villa Bermond in Nice, for instance, had orange and lemon groves that could be "freely visited at any hour, without the slightest supervision." In Menton, there was the Bennet garden, where the "doors opened each morning and remained open until one o'clock. One did not have to ask for permission or put up with the inconvenience of a gateman or gardener on watch ... A plaque in Carrara marble bid one a cosmopolitan welcome: *Salvete, amici!*" Naturally, free access was far from being the general rule. In La Mortola, near Vintimille, Sir Thomas Hanbury only delivered an admission pass when one applied in writing. And Liégeard went on to say of the plant-lover owning the garden at Villa Camille-Amélie

in Cannes, that "once the lock was undone and the door open" his hospitality was "prodigal," his nomenclature "relentless."

At the turn of the century some gardens were planned with an eye for architecture. Some gardeners collected stone elements much as others collected plants. Caroline Miolan-Carvalho, the favorite singer of Charles Gounod, planted a garden in Saint-Raphaël which contained a total of 43 fragments of the Tuileries, whose fragments were sold and scattered in 1882. Guidebooks of the time highly recommended a tour of her garden. Amid botanical rarities stood statuary, pediments, columns, capitals, and other fragments of what had been one of the finest examples of the French Renaissance style. She gave some fragments to friends, to place in their gardens, and others to municipal authorities, to place about public buildings where they may still be seen today.

Gradually, there evolved a Côte d'Azur style for winter gardens. This unique style was characteristic of both time and place, of the landscape of the entire coast as it looked in the late nineteenth century. This typical landscape already contained a scattering of white houses against the dark greens of pine, yew, and privet, grey-greens of agave, eucalyptus, and tamarisk, and bright hues of the bougainvillea, mimosa, and wisteria.

. . . .

Colonnade from the Tuileries, Villa Magali, Saint-Raphaël.

The winter gardens of Cannes

The English were Cannes' first winter residents, and the city owes its nineteenth-century gardens to their efforts. Many Europeans patterned their gardens after the English landscape model; the natural style was not only adapted to various types of site and climate but also to the dictates of local fashion. In Cannes, gardens tended to be planned with great emphasis on naturalness of form and used as sites for experimentation with plants.

Gardens were deliberately asymmetrical in layout and relied on the curved line to define planted areas. Lawns, parterres, massed shrubbery, and stands of trees crowned hilltops or descended into valleys depending on the contours of the land; there were usually many banks and slopes to work with. A complex system of garden paths and avenues provided a variety of itineraries, enabling one to admire winding streams, waterfalls, harmonious lakes, pieces of sculpture, colonnades, ruins, grottos, rockwork, and so forth. The local variant of the naturally landscaped garden veered toward the exotic, as increasing numbers of local gardeners sought to introduce rare plants. In Cannes and elsewhere on the Riviera the hallmark of nineteenth-century gardening was the integration of previously unknown plants into garden design. Landscape gardeners, acquiring a taste for the exotic, developed a collector's interest in seeking out the most attractive and curious plants available. Gradually, a large number of new species selected for beauty of form, singularity of foliage, color of blossom, and subtlety of scent was acclimated. The winter season was lengthy, extending from the mild autumn typical of the Mediterranean climate through spring. Winter gardens thus tended to concentrate on plant species with indeciduous foliage and those flowering between November and March. A land survey done between 1883 and 1886 clearly shows the great diversity of existing garden plans. Several were in the English style, including the gardens at the Hôtel des Pins, Villa Alexandra, and Villa Madrid, but the majority combined the classical revival and natural styles. Generally

Rockwork, Villa Rothschild, Cannes.

Stream, Villa Rothschild, Cannes.

speaking, the residence was placed on the property's highest ground to afford a view over the garden and take advantage of whatever other views could be captured. In fact, a terrace was often put in to raise the ground floor to a higher level. Moreover the empty space of the terrace was used to highlight architectural features, while effectively separating house and garden. To provide an appropriate foundation, the site was banked or the terrace supported by arcades. The latter solution was advantageous in that the lower space could be converted into a grotto or open-air room, as at Villa Saint-Priest. Pleasure gardens with a terrace, or several consecutive terraces, could be divided into two different areas: a formal garden of symmetrical composition, with geometric flowering parterres laid out on the terrace or terraces, and a "walk" throughout the rest of the property, where natural garden scenes could be carefully cultivated. The structured space and free space would be connected by a series of staircases, straight or semicircular, set along the garden axis or laterally. In the case of combination gardens, the formal area was generally much smaller than the natural part. A few gardens were entirely geometrical; landscape gardeners faced with a narrow lot found this solution convenient. The pleasure garden was not the only cultivated area on the Cannes estate. Often kitchen gardens and orchards were planted off to one side, near the water tanks and outbuildings; generally, they were hidden behind a row of trees or set in a walled enclosure.

In the case of landscape and combination gardens, the architectural design of the house had no bearing on the garden plan. One might find a Gothic revival mansion amid Mediterranean terracing and an Italianate, neoclassical, or Moorish villa in a landscape setting. Garden furnishings and monuments—Chinese and Moorish pavilions, ruins recalling antiquity or the Middle Ages—evoked faraway, exotic lands. In this type of decor, designed to stimulate the imagination as well as the physical senses, the dream element played a significant role.

Park, Villa Rothschild, Cannes.

The most prestigious Cannes gardens were open to the public, and tourist guidebooks of the time did not fail to devote a chapter to their luxuriant vegetation. In one guidebook Dr. Valcourt[30] mentioned the garden at Villa Larochefoucault, the former Château Saint-Georges, as being particularly fine. This garden, created in 1852 and relandscaped in 1860, included several varieties of tall tree native to Australia: araucarias, filaos with dark green spindly branches, grevilleas remarkable for the beauty of their long bipennate leaves, tree ferns, Egyptian papyrus, Chinese camphor trees, and coconut trees. In 1858 the Duke of Vallombrosa purchased the Château des Tours and had the Villa Vallombrosa garden landscaped at that time. It featured a collection of bamboo and palm, in addition to a variety of mimosa (*Acacia baileyana*) of spectacular dimensions. There were also aralias from Central America, with glossy, colored leaves, araucarias, and ficus. At Villa Alexandra the garden's reputation was largely due to its flowering shrubs, and to its azaleas and camellias in particular. Among the cork oaks were planted

taxodium, or bald cypress, whose leaves turn red in the fall. At Villa Camille-Amélie, created in about 1883, many rare species previously unable to grow outside of glasshouses flourished in the open: varieties of flowering-top cactus, Chinese palm, Central American philodendron, Abyssinian banana tree, and South African aquatic plants. The Garden of the Hesperides was also much admired by winter residents; it dated from 1850, and was located on a ten-acre tract out on the point of La Croisette. Here one found many varieties of citrus trees, whose fruits were preserved in brandy or crystallized, in addition to arbutus and pomegranate trees. The main difficulty faced by gardeners was the shortage of water for sprinkling. The canal

de la Siagne finally reached Cannes in 1868, but prior to that time, water had to be transported and stored with care to avoid evaporation. Each garden had its own water storage and irrigation systems vital to the upkeep of the plantings. In the Cannes area the groundwater supply kept the moisture level constant in many places; for many species this encouraged growth. Moreover the soil structure was characterized by formations of gneiss, Jurassic limestone, and weathered rock; the surface layer was extremely fertile, despite its arid appearance.

Gardeners in Cannes gradually succeeded in getting delicate cold-hating exotics to grow out in the open. Initially, efforts with plants from Australia, New Zealand, and the Cape of Good Hope were the most successful, for the climate and soil type more or less corresponded to their native environment. Eventually, species from Africa, the Americas, and Asia were introduced.

Among the most celebrated gardeners of the age was Gilbert Nabonnand, horticulturist and rose-lover, who made his permanent home in Golfe-Juan in 1860. Before then he had worked in Avignon as head of a well-reputed floral establishment. In 1865, at the request of Lord Brougham, he planned the park and rose-garden at Villa Eléonore-Louise. His success was such that several proprietors commissioned him to do their parks. Nabonnand set about developing the cultivation of imported species, while continuing his experiments with roses. At about this time, the first specimens of eucalyptus, mimosa, and palm were brought to Cannes.

In the 1860s he redesigned the gardens at the Villas Alexandra and Larochefoucault, and created those at Villas Rebecca, Hollandia, Passiflora, Florida, les Violettes, as well as those at the Château de la Bocca and the Hôtel Pavillon. A school devoted to the horiculture

Summerhouse, Villa La Cava, Cannes.

of exotics sprang up around him, composed of many of Cannes' horticulturists and several English gardeners. Stabile, Maccary, Gourdelon, Wortham, Roubion, Précelle, Bidon, Crotta, and Desgeorges strove to outdo one another. Denery received an award for his show of sugarcane varieties from La Croisette. Opoix, too, won a prize for pineapples cultivated at the Duke of Vallombrosa's château. Vergalet won acclaim for cotton plants grown at the Garden of the Hesperides, Gerlowsky for his clusters of mature dates, Debionne for his banana trees bearing bunches, and Perrin, head of Lord Brougham's gardening staff, for his discovery of the forcing principle applicable to the mimosa.

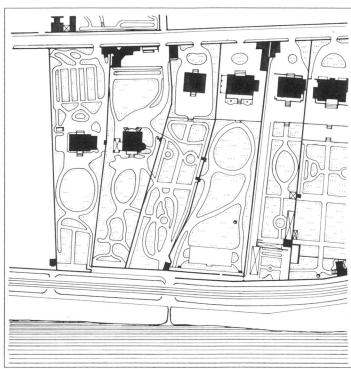

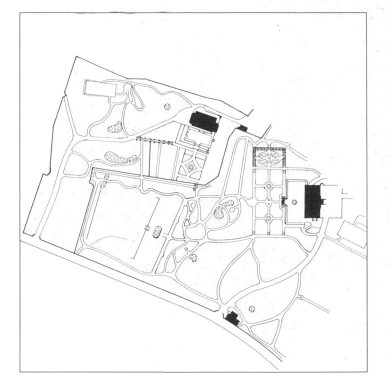

Araucarias and palms, Villa Rothschild, Cannes.

Plan, La Croisette, Cannes, showing a number of irregular gardens.

Plan, Villa Clémentina and a neighboring villa, Cannes.
Two examples of the mixed garden.

English gardens and gardeners: Italian and Andalusian dreams garden of Harold Peto

At the end of the nineteenth century the tide finally turned in favor of a more intuitive, yet ordered eclecticism, and once again the impulse for change came from the English. Before becoming interested in the Riviera and seeking out its rustic style, English and American residents thought of it as the perfect place to invent a compromise between two tendencies: the latest English school, the so-called "wild garden" school,[31] which advocated the association of different perennials in mixed borders for a pictorial effect as well as gentle shapes and vistas of color, and the school in favor of architectural forms borrowed from Roman, Italian, and Andalusian gardens. Thus the opposition between formalism and the natural style was reconciled in a combination of the trim Mediterranean and sophisticated English natural styles. The work of garden-architect Harold Peto (1854-1933) exemplifies this new movement. Up until 1892 he worked for Ernest George at various English palaces and country houses. Peto left his partner, who exacted from him a promise that he would not practice in Great Britain before 1907, and Peto turned toward the Riviera. In ten years' time he had created or contributed to least eight major gardens on the Côte d'Azur, including three on Cap Ferrat: Villa Sylvia, strongly influenced by the Villa Médicis in Fiesole, for the American, Ralph Curtis, in 1902, and Villas Maryland and Rosemary. Peto's reputation on the coast spread rapidly. He was called upon to extend the garden of the king of Belgium in Saint-Jean-Cap-Ferrat. In 1910, in Cannes, he composed the Isola Bella garden, with its colonnaded oval pool, that at the Villa Sallés-Eiffel, enhanced by statuary and a great loggia giving onto the sea, and the Bella Vista cloister and *jardin d'eau*.[32] Open pavilions, arcades, and hemicycles were often taken as a point of departure or arrival of vistas created in these gardens. At the Villa Maryland, Peto sought to strike a harmonious balance between the garden and Italian Renaissance house. He designed a series of gardens, effecting subtle transitions between the Roman garden, Italian flower garden, and the

Villa Maryland, Cap Ferrat.

Harold Peto in his garden, Itford Manor.

Rosemary.

garden of wild flowers along a green vista. Peto went counter to French tradition, which dictated that the most ornate parts be located close to the house and that the design should grow simpler as one drew away from the house.[33] Peto went back to the layout used at the Villa Rosemary, in which orange, olive, and cherry orchards near the house opened up onto parterres adorned with flowers that in turn gave onto very sophisticated natural gardens. The Roman garden is not intended to strike archeological chords but is rather playful, seeming to treat antique fragments from the perspective of an English watercolorist. Witness the airy grace of ivy clinging to the pergola's Doric columns, a garden path bordered with softly shaded rosemary, and truncated columns placed at intervals. The patio was Italianate in feeling but featured the colors of Spanish ceramics. A winding spring walk in green turf was bordered with trees underplanted with carpets of spring flowers: tulips, hyacinths, narcissus, and snowball bushes. The natural landscaping intentionally highlighted the budding green of spring foliage in the background.

Previously, many Riviera garden planners had looked to the Italian model for inspiration. Among these Italianate compositions were the Villa Arson garden in Nice, done in 1810 in the eighteenth-century baroque style, with sculpture, porticos and grottos; La Fiorentina in Cannes, an 1880 garden by Sir Goldsmid, where a parterre set in marble dominates a series of terraces traversed by a canal; and in Beaulieu, the La Berlugane garden belonging to the Count Gautier-Vignal, dating from about 1900, whose compartmentalized design featured outdoor ''rooms,'' pools, and fountains.

. . . .

Bella Vista.

Isola Bella in 1984.

Isola Bella in 1920.

From vestiges to theme gardens: gardens from around the world

In 1912 Danaé Vagliano moved into Champfleuri, her house in Cannes; she had all the palm trees in the park taken out to make way for what would gradually become seven gardens within a garden, each in a different style, set at the base of a wide sweep of lawn. This was typical of a transition period between 1900 and 1914, which marked the end of nineteenth-century exoticism and exaggerated curves and ushered in the geometric garden based on local plants and materials. New sites were found, and sports became popular. A fresh wave of well-to-do newcomers to the coast turned away from the old Riviera trend. Gardeners no longer took pride in mixing themes but relied on theme juxtapositions. This new brand of ordered eclecticism, as opposed to the former mosaic-type eclecticism, lent itself to the theme of garden history. One no longer collected plants, but styles, and one's garden was no longer a plant catalog but a garden catalog or, in some cases, an encyclopedia. Specialized literature of the time reveals that landscape gardeners wished to conduct a historical review, so to speak, while many owners desired nothing more than a garden for show which would display their worldliness and knowledge of garden history. Among reference books published then were the Benoit and Bouchot history of French landscaping (1908), a recapitulation of four centuries of traditional French gardens by Maumène and Duchêne (1910), a work on two hundred years of English gardening by Maumène and André (1911), and a study of various gardening styles, covering European and Oriental styles, by Fouquier and Duchêne.

The Moorish garden, Villa Cypris.
R. Maïnella, garden architect.

During this period, Raffaële Mainella, landscape painter turned landscape gardener, was hard at work at Villa Cypris and Villa Torre Clementina in Cap Martin. He combined long, winding alleys and archaeological vestiges typical of the Riviera style with irregular free spaces characteristic of gardens done between the World Wars. The axes were defined by immense flights of steps sweeping downward, accentuating the descending movement of the ground and enhancing the view of the sea, or else by small enclosed geometric gardens along winding paths. Mainella established an interplay between these two systems and drew attention to contact points with visual markers such as the large mosaic and the bridge at Villa Cypris. History and globetrotting were telescoped into a very small space and time frame: in the course of a walk, one came across a small Greek theater, a coral-strewn Japanese garden, a Venetian garden monument, and a Moorish garden. At the turn of the century the Riviera could be mapped out in terms of its "colonies": the English flocked to Cannes, Nice, and Menton, the Russians to Nice and Villefranche, while tropical-plant connoisseurs settled in localities on either side of the Italian border. Several gardens celebrated the high points in world garden history. At Champfleuri in Cannes, and at Villa Ile de France in Saint-Jean-Cap-Ferrat, Japanese, Florentine, Spanish, English, Moorish, and French gardens were landscaped. Another approach was the theme garden: water gardens, rock gardens, single-color gardens, rose gardens, or regional gardens such as Champfleuri's Provençal garden.

. . . .

Lou Sueil, Parterre de broderie *and loggia.*
A. Duchêne, garden architect.

English country revival and neo-Mediterranean gardens

By the end of the nineteenth century the principle of the landscape garden had become incongruous in England. The countryside had changed in character: properties were smaller enclosed spaces owned by a new bourgeoisie formed during the industrial revolution. Between 1880 and 1914 ideas adapted to this new landscape had been put forward by talented gardeners like Gertrude Jekyll, landscape painter, photographer, and journalist, and architects including Sir Edwin Luytens and Inigo Triggs. The new garden was patterned after the English cottage garden, while featuring a highly structured and impressionistic approach. A bond between garden and site would be created by means of working a central idea into a geometric design with simple lines. Indigenous plants and materials would be used to establish the layout and the color scheme. The objective was to find an intelligent blend of formalism and natural surroundings at a time when urbanization was devouring the remaining countryside; this school of thought sought the "natural" element in traditional plant life, materials, and forms. At the same time light Italian or Oriental touches would add zest to the overall effect. This new approach was popularized in many prewar books and magazines. It surfaced on the Riviera either because this literature was sent directly to English and American families living on the coast or because many French landscape architects had come to assimilate the new English concepts. After the First World War there was a general trend in favor of adapting the ideas of the English school to the Mediterranean environment. Costly prewar techniques of pricking out and repotting shrubs were abandoned, and there developed a new respect which amounted to veneration in the case of certain local plants. The cypress became the object of a cult, not only because it was a typical feature of the local landscape and a reference to the area's Latin heritage but also because it was, like long-lived Mediterranean trees such as the olive and carob trees, looked upon as a symbol of traditional values and stability by a bourgeoisie shaken by the

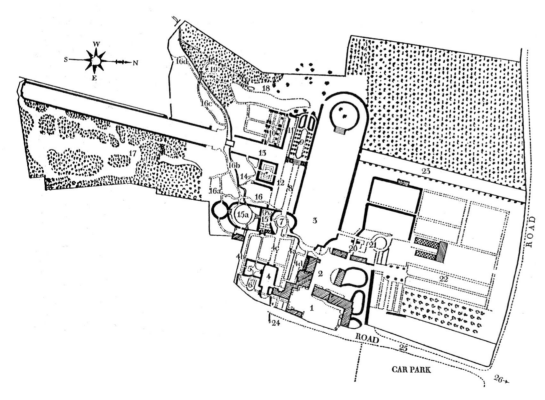

Plan, Hidcote Manor.

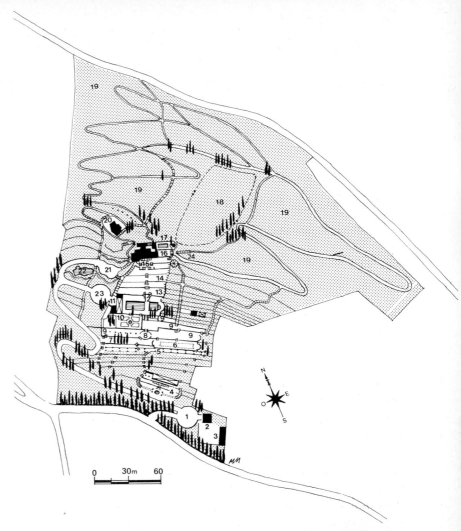

winds of change. The cypress supplanted the palm as
the most-preferred tree; it was planted alone, in hedges
to define spaces or the layout, and used to reinforce
garden axes. In 1901 the cypress was the principal component in the main vista at La Léopolda in Villefranche,
the garden designed for the king of Belgium by Ogden
Codman. Deciduous trees were once again planted near
the house. Plants indigenous to the Mediterranean,
maquis and *garrigue*, were no longer removed from a plot
to make way for other plants. In fact, if native plants
were lacking, gardeners went out of their way to introduce them. These included heather, cistus, myrtle, trailing arbutus, mastic trees, alaternus, broom, juniper,
rosemary, lavender, and asphodels, to which the gardener then added bulbs and perennials.[34] This repertory afforded gardners a full palette of shades with which
they could create an impressionistic effect, while high-
lighting geometric patterns through the use of contrasting colors. Garden ambiance was also changed by the
use of climbing plants; they provided a "binding ele-

Plan, La Serre de la Madone.

The white garden, Hidcote Manor,
designed by L. Johnston.

ment," supple lines and color on pergola columns, wall façades, and terrace balustrades, or even on a tall tree to be "sacrificed."

Two Americans were prominent among local garden-lovers. Little remains of Sainte-Claire-le-Château, Edith Wharton's natural and flower gardens in Hyères, described in *Gardening in Sunny Lands*,[35] but Lawrence Johnston's Serre de la Madone garden in Menton still exists. Like Edith Wharton, he belonged to a segment of American high society utterly fascinated by the Europe portrayed by novelist Henry James. During his childhood in Paris he became familiar with many museums and châteaux in the Ile-de-France area. After university studies in England and extensive travels, he started, in 1907, to plan his first garden at Hidcote Manor, in Gloucestershire. Combining his reminiscences of structured Italian, French, and Elizabethan gardens with the natural tendency, he originated "an absolutely new style"[36] that influenced generations of landscape gardeners. His last garden, La Serre de la Madone, is a masterly exercise in composition, although it is little known on the Côte d'Azur. Instead of focusing on the view of the water, Johnston laid out a garden on a coastal valley slope with a stylized central terraced part and outlying areas that gradually blended into the natural site. Although many botanical rarities have been removed, varieties of large plants remain which Johnston himself brought back from botanizing trips to South Africa and Asia.

. . . .

Watercourse, Buscot, designed by H. Peto.

The coming of the automobile: gardens and local color

In the 1920s any garden-lover with an automobile could set off on expeditions to find new garden sights. Hilltop locations with views of the sea thus became more accessible. Even if one did not have direct access to the water, one could create a garden focusing on the role of sunlight. In the past people had sought out shady areas to escape the burning rays of the sun, but now the only thing that mattered was acquiring a golden tan. The "sun garden" was included in the repertory of André Vera.[37] The golden tan became a status symbol, in the way that top hats and parasols had previously been emblems of class. Local color became fashionable. Around 1928 Juan-les-Pins launched the new era of summer vacationing and the sun craze with the proclamation of an "official" seaside bathing season. The casinos in Juan-les-Pins and Cannes were packed with yachting aficionados and social butterflies. To an increasing extent, gardens had to accommodate to a new type of life-style in which sports occupied an important place. Sun terraces were readily integrated into garden design, and swimming pools could be disguised as a reflecting pool or basin, but it was not an easy proposition to lay out a garden with a tennis court. Some landscape gardeners, such as J.-C.-N. Forestier and Achille Duchêne, decided to place the court at the heart of the garden.

Generally speaking, the size of gardens shrank to more modest proportions in the postwar period. It became more and more difficult to find good gardeners. The idea of the maintenance-free garden had its advocates. Yet, on the Riviera, there seemed to be more large gardens than ever. There were quite a number of gardens exceeding twenty-five acres, although for sheer size one could not find anything to rival with the garden of Alice de Rothschild in Grasse, which occupied about 335 acres. New estates often swallowed up several small farms. Ironically, vanishing farmland was often replaced by what one might call "rural revival" gardens. Ironically enough, this regional revival style commemorated the very life-style that had vanished owing to its own advance.

Terra-cotta jar.
R. Charmaison, *Les jardins précieux*, 1919.

Trompe-l'œil garden, Les Colombières, by Ferdinand Bac:
"I invite you to take a journey for which you will need neither wheels, train compartments nor automobile engines..." Henri de Régnier, preface, *Les jardins précieux*, R. Charmaison, 1919.

Mediterranean gardens

The Côte d'Azur is the perfect place to study landscaping trends that prevailed between the World Wars. In the wake of the English, Russians, and Germans came the Americans, Greek shipbuilders, Oriental potentates, and the very wealthy French. The Riviera came to resemble nothing so much as one vast garden. Some of these new garden planners were millionaires; others were gifted gardeners. In some happy instances the millionaire was a gifted gardener. Ferdinand Bac once said: "When one does not belong to any profession in particular, nothing prevents one from becoming a landscape gardener." Some of the most remarkable gardens of this period were produced by enlightened amateurs who became great gardeners, such as Raffaële Mainella, Ferdinand Bac, Lawrence Johnston, Edith Wharton, and the Viscount Charles de Noailles. Whether prominent in the arts or in industry, cultivated gardeners such as Vicente Blasco Ibañez, Jean-Gabriel Domergue, Maurice Maeterlinck, and Francis de Croisset used the garden as a medium to give clear expression to their imagination. Many professionals also did outstanding work at this time; among the more prominent were Achille Duchêne, Harold Peto, Jean-Claude-Nicolas Forestier, Gabriel Guévrékian, Octave Godard, and Jacques Greber. Each contributed to the development of the "Mediterranean garden." The Mediterranean was a vast confluence of cultures which allowed gardeners to make use of a wide range of references. One could go back to an austere classicism or ride the wave of contemporary trends. One could weave an Oriental or Arabic theme into the design or follow the regional model of the so-called "Côte d'Azur garden." This model gained international reknown after being featured in the press and at the 1925 International Decorative Arts Exhibition. Fast-moving celebrities attracted to the Riviera made both Mediterranean and Côte d'Azur garden styles a succès fou. The Riviera's climate, which combined the climatic conditions of Italy and Provence, was the ideal place to create gardens blending both influences. The strong contingent of

Challenge: to lay out a garden in a garrigue *environment scorched by fire and blazing sun and exposed to wind. Solution: a skillfull reproduction of the native* maquis.

Recent wild garden.

English gardeners developed what might have seemed unexpected, a feeling for local character. But when it came to creating "wild gardens" with a heavy emphasis on perennials, the English had stood unchallenged since William Robinson's stand in favor of this type of garden in 1871. Another clear influence could be detected in Gertrude Jekyll's impressionist gardens, landscaped in about 1880 to accompany the neo-rural architectural designs of Sir Edwin Lutyens.

The Côte d'Azur was a crossroads, too, for different schools of thought. Varying degrees of symbiosis could often be detected. The intuitive and pragmatic English approach was often juxtaposed with the highly ordered French approach as many gardeners tried to strike a balance. English landscape gardeners sometimes tried a more structured approach, as witness Lawrence Johnston's La Serre de la Madone garden, whereas the French were known upon occasion to employ painterly techniques, among them Jean-Claude-Nicolas Forestier at the Bastide du Roy, where color and formal composition are strongly related.

Fescue lawn simulating a pool of water in a dry garden.

Staircase bordered with summer flowers.

Ferdinand Bac

Ferdinand Bac referred to himself as the instigator of a ''Renaissance of Mediterranean gardens.'' His tendency to make heavy use of mythological references can be taken as an illustration of what Bachelard called the ''Nausicaâ complex,'' in other words, ''nymphs recollected from Latin and Greek classes at school, transported to the countryside.'' Ferdinand Bac, born in 1859, the presumed grandson of Jérôme, king of Westphalia and brother to Napoléon Bonaparte, was indeed a man of the nineteenth century. Yet he made a name for himself as an innovater at the turn of the century. In his search for form, color, and new materials, Bac went back to what he had seen on his travels: ''I had accumulated piles of annotations on the art of building as practiced in Latin countries. Those very buildings which were modest in scale and simple in design remained engraved in my memory as being the most rational of all. From all this memorabilia, which had lain dormant for so long, sprang the renewal of an architectural formula in which one will find everything but the hand of an architect . . . This is the architecture of sentiments . . . an art which builds upon all our nostalgic reminiscences of places where we would have liked to set up our tent and remain, uplifted by Beauty and strengthened by Simplicity . . . There is a special ease to Mediterranean grace—as if it would fit into the palm of one's hand.'' Bac started out in 1912 with the Villa Croisset in Grasse, where gardens of Italian and Spanish inspiration formed, with the cloister, a significant landscaped ensemble. He went on to work in a similar, ''antiexotic'' vein at the Villa Fiorentina, in Cap Ferrat, which belonged to the Countess of Beauchamp. His masterpiece is at Les Colombières in Menton, where he created a garden itinerary. He wanted to ''create a bouquet of travel memories . . . and bring them to life in order to relive the past . . . The idea was not to produce servile copies of Italian villas, but to take inspiration from the Italian experience, Spanish mystery, and from the admirable Orient, the true father of landscape gardening whose forms are infinitely inter-

Bust of Ferdinand Bac, Les Colombières, Menton.

Two sketches for the Villa Croisset, Grasse, by Ferdinand Bac.

Some of these projects were carried out at Les Colombières in Menton. In each garden space, a theme evoked one of F. Bac's fantasies: the Pink Torrent, the Monastery at the Bottom of the Chasm, the Gate of Nostalgia, the Tower of Things Delectable, the Red Garden, the Gate of White Ibis, the Spring with the Black Grille, the Yellow Summerhouse, the Autumn Garden, the Shell of Water, the Cruel Garden, etc.

twined with our classic art and which landed on this coast many years ago." The composition of the Les Colombières house and garden is characterized by what Bac called the "submission to the majestic panorama," in other words, by the way the natural advantages of the site are taken into consideration. He treated nature with reverence instead of looking upon it as a conquest. He wished to "venerate Nature, render hommage to her as miraculous." The landscape and clearly defined garden areas became a veritable pantheon of deities. Bac retained an existing grove of olive trees and divided up the garden into small geometric subgardens by means of clipped cypress hedges. Using the vegetation or architectural elements (columns or arcades of various garden monuments), he created frames to capture garden views. Bac would sometimes present part of the surrounding landscape for viewing, but elsewhere a single element—a tree or a rock—would be the chosen focus. He built a bridge and laid out a path in celebration of

The Roman bench and the gate to the
allée des Jarres, *gouache by F. Bac.*

a thousand-year-old carob tree. Another bridge was built to honor a cypress tree, and a staircase for an olive tree. At Les Colombières the house is treated with equal attention. Ferdinand Bac's pictorial approach was strongly influenced by the 1920s style. All around the patio a frieze of heros and goddesses may be read as a personal mythology. In the other rooms and bedrooms the artist portrayed his own odyssey to the Orient, Attica, Venice, and the Generalife gardens in Grenada. Enclosed spaces were used to focus more intensely on the travel theme. The "odyssey" covered all the walls of the patio. The dining room, treated as an exterior space, was designed as a window garden, able magically to transform the landscape: the steeples in Menton could be viewed as towers, campaniles, or minarets, depending on one's vantage point. The handling of the perched-site location and travel-memory theme was brilliant, the overall effect particularly conducive to a flight of imagination.

The French Mediterranean revival school: an architectural approach

In the early twentieth century the regional revival style appeared in France, perfectly exemplified by Villa Arnaga, owned by Edmond Rostand. Tournaire designed the Basque revival house and classic French garden in 1903, effecting a transition between asymmetrical architecture and symmetrical garden with strategically placed flower-covered pergolas and beds of massed flowers near the house. On the Côte d'Azur Harold Peto, with his Tuscan revival experiments between 1902 and 1910, and Ferdinand Bac were trailblazers, showing the way to a return to local architectural forms, ordinary materials, Mediterranean vegetation, and the "total elimination of ornament."[38] In the wake of the prewar theme gardens, some of which were precursors to the new "decorative arts gardens," and after Ferdinand Bac's initial project at Villa Croisset (1912), several landscape gardeners were instrumental in promoting the concept of the Mediterranean garden. Among them were J.-C.-N. Forestier and André Véra,

Villa Croisset, Grasse, designed by F. Bac.

who was a fervent partisan of the regular garden. Both relied heavily on the universal language of the classical—especially Roman—heritage, and on the use of cultural fragments picked up in new territories just opening up to travelers under the impetus of colonial expansionism. Architects, landscape gardeners, and interior decorators cumulated influences: to that of the Roman country villa, they added references to Spain and countries across the Mediterranean that had come under the French protectorate. In short, a vast new field of creative experimentation was opening up. J.-C.-N. Forestier gradually built up a solid reputation. At the 1925 International Decorative Arts Exhibition in Paris and, again, at the 1929 International Exhibition in Seville, Forestier presided over the Garden Show. Earlier, he had begun to attract attention for his work in Morocco, when, around 1912, he was with Maréchal Lyautey, and then later in Spain. In 1920 a collection of his notebooks was published,[39] including gar-

den plans and sketches, a history of the Mediterranean garden, models for vine-covered arbors and terraces inspired by the Roman country villa, and vast projects intended for the most prominent residents of the favored "climate of the orange tree." Other landscape gardeners, such as Octave Godard, strived to formulate gardening rules specific to the south of France. Godard, a student of Edouard André, was originally from Picardy but had settled in Nice. He wrote *L'Art des jardins dans le Midi*,[40] trying, like his colleagues, to reconcile a return to French classicism with the new emphasis on preserving local and rural characteristics. Terracing, espaliered trees, and massed flowers were organized in geometric patterns that ultimately led to the sea. At the Palais de Marbre in Nice, La Paloma in Monaco, and the Château de la Croix des Gardes in Cannes, Godard rigorously applied his principles as he created neoclassical seaside settings. Between 1918 and 1922 a new garden type had developed in response

The Spanish garden, Villa Croisset.

to new needs generated by advertising and the automobile a new interest in outdoor exercise and a desire to own a portion of that horizon between sky and water. In consequence the Mediterranean garden, and the Côte d'Azur type in particular, complete with pergolas, flights of steps, terraces, *opus incertum* paving, reflecting pools, and faience tiling, was adopted by gardeners all over France.

At the 1925 Exhibition in Paris, a number of pavilions made specific allusions to the Mediterranean. The most popular garden of all, landscaped by Messrs. Delmas and Lavergne could be found in the Maritime Alps Pavilion. Jostling crowds made their way through the Garden of Provence and Tunisian Patio. Ferdinand Bac wrote a commentary in *L'Illustration* in which he declared: "Clearly, inspiration has come from the Midi and from our colonies."

The pergola.
Alpes-Maritimes Pavilion at the International Exhibition of Decorative Arts, 1925.

A formal French garden,
painting by A. Devambrez.

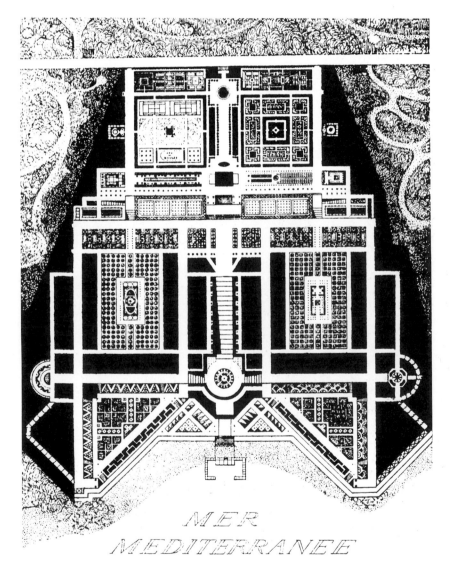

MER
MEDITERRANEE

General view and plan, André Vera's sun garden, 1919.
Tunnel arbor flanked by paved and planted beds. Wide staircase with low, broad treads. Line of cypresses. Mimosa groves enclosed by clipped yew.

The mediterranean garden and its attributes

Terraces and patios

The terrace had been known in earlier centuries as the *espace de propreté*, and functioned as the base for the typical Provençal *bastide*. It now became an extension of the living space, a sort of paved living room lighted by dazzling sun, and was one of the preferred architectural forms of the new outdoor-oriented life-style. The formerly cultivated terraces, or *restanques*, were used by landscape gardeners as a focal point; this type of traditional feature was an integral part of the Mediterranean garden. A landscape gardener who wished to keep up with the latest trends had to find ways to work the sun, Midi, and Mediterranean themes into his landscaping. The sun dial came back into favor. Various cloister ensembles created at the Résidence in Sainte-Maxime by R. Lallemant and the Charles de Noailles villa in Hyères by R. Mallet-Stevens extended the living space out into the sun-drenched outdoors and used a portico to frame a panoramic view. Elsewhere could be found large and theatrical balconies looking out over the water. Yet another solution was the small open-air theater; the Mediterranean sea, after all, provided the best of all possible backdrops.

Raquel Meller's terrace, Villefranche.

Paving, Villa Nora, Grasse.
Léon Lebel, architect.

Terrace, La Croix des Gardes, Cannes.
O. Godard, garden architect.

The pergola

This type of arbor, intended to support climbing plants and flowers, was first used in France in the early nineteenth century. It was brought back to France by painters and architects returning from Italy in the early nineteenth century, and was used later for picturesque country houses, naturally christened "villas." The pergola effected a transition between the terrace and the landscape. Many of the pergola models[42] designed by Jean-Nicolas Durand in 1809 used pergolas for this purpose. At the same time an art collector named Valentin sponsored the reconstruction of the town of Clisson in the middle of Vendée, far north of the Riviera, in an Italian style. He himself proposed building an astonishing villa with a gazebo and surrounded by pergolas. Previously, we mentioned the pergolas at the Charles Garnier villa, Bordighera, but these were different in the sense that they were built during the eclectic period, at a time when local architecture was not fashionable.

In the twentieth century the pergola became all the rage. It could be used simultaneously as a reference to ancient Rome, the Renaissance, and Provence. It provided refreshing shade for those who wanted it, while being picturesque and decorative. Trailing flowers and climbing plants created a dappled sun-and-shade effect when trained across well-spaced columns. Sometimes the pergola was no more than a flower-covered portico designed to frame the view.

Stairway-pergola, Champfleuri, Cannes.

Pergola in the Grand Jardin, *Ile Sainte-Marguerite, Cannes.*

Summerhouse-arbor, seaside garden.

The axes:
alleyways, steps, and canals

The Mediterranean garden took its place within a larger current prevailing during the period between World Wars I and II, based on a renewal of interest in symmetry. Yet landscape gardeners no longer systematically used the house and the line of the steepest slope to determine the axis of the garden. They wanted to set up a series of framed vistas, often lateral. Garden symmetry no longer possessed an inner structure which, as Pascal once said, reflected "the face of man, in a symmetry of width, rather than height or depth."[43] Nor was it still employed as a form of control over the landscape: symmetry had evolved into a means of creating new vistas on a large or small scale. Vistas could open up onto the infinity of sea or sky, and encourage the projecting of one's spirit toward those mirror surfaces. They could also narrow the view to center on garden detail, encouraging private thoughts in the privacy of the garden landscape.

On the coast, landscape gardeners frequently used straight alleys to produce the desired vistas. To handle the problem of steep slopes, they resorted to a system of tiers, as witness Léon Lebel's path at the Villa Nora in Grasse, with its broad steps paved with cobbled mosaics, bordered with orange trees and palms, or the cypress-lined walk at the Villa Espalamador in Villefranche by Jacques Greber. In the Villa Altana park in Antibes and the Château de Malbosc garden in Grasse, Greber used canals, connecting small pools set in *opus incertum* paving, to define the axis of sloping vistas

The Spanish basin, Les Colombières, Menton.

Staircases, Fontana Rosa, Menton.

and transversal terraces. Slopes were often quite steep, and the "one-hundred-steps" theme became so widespread that a Parisian statesman finally quipped that staircase duels were taking place on the Riviera.

Steps sweeping down to the sea had been used in prewar theme gardens, but now they had become veritable monuments, such as the one at the Château de la Garoupe, or the A. Riousse staircase at Château Camarat near Saint-Tropez. At the Château Théoule, one of the most original works of the period, tiny gypsofilia were planted in cracks between the paving stones, playing off soft petal textures and the mineral quality of natural local stone. Subsequently this technique was employed at the Villa Yakimour belonging to the Aga Khan in Le Cannet.

. . . .

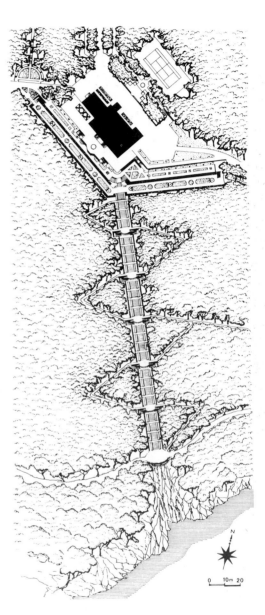

Watercourse.
H. Peto, architect.

Château de Théoule.

Plan, Château de Camarat, Saint-Tropez.
A. Riousse, architect.

The colors of the sun

We have seen that there was a ground swell of interest in favor of Mediterranean references, local materials, stone, ceramics, and colored and white stuccos; inspiration came from both classicism and colonized cultures, as well as from efforts to cut gardening costs.

The 1925 Exhibition was the showplace for the new tendancies. "One could not help succumbing to the charm of materials such as these pretty golden pebbles that go into cement like almonds into a spice cake, or the use of ceramics."[44] Ferdinand Bac, in his commentary on the Exhibition, confessed that he had an aversion to the gravel used in picturesque gardens and the growing influence of the English school on French landscape gardeners. In gardens of the Maritime Alps, "the novelty of using marble scraps for paving has created something of a sensation here. Perhaps it will succeed in driving out those horrible gravel paths which torture one's feet and irritate the ear. The Maritime Alps Pavilion has rendered a great service to French gardens by popularizing this method, widely used in England and Holland, which I have personally been trying to reintroduce into the Mediterranean garden style since 1912." Ceramics were becoming popular, and to meet local demand, ceramics work shops sprang up in Biot, Vallauris, Aubagne, and other localities. The decorative use of terra-cotta pots brimming with flowers, exemplified in the allée des Jarres at Les Colombières, reinforced a Mediterranean blend of Italian, Provençal, and Spanish influences.

Villa Croisset, Grasse, painting by André Boursier-Mougenot, 1935.
"I discovered the shade of red for this surface finish in Carpaccio's paintings," F. Bac.

Les Colombières.

The allée rouge.
R. Charmaison, *Les Jardins précieux,* 1919.

Modernist gardens

In the 1920s modernism swept through postwar Europe and shook up the world of landscape gardening. Preconceived ideas, color nuances, and delicate shade effects under the trees were thrown out and replaced by stark black-and-white contrast and geometrics. Nature was subordinated to graphics; gardens became a few amoeba-shaped curves around buildings planted in the landscape without the slightest transition. Natural surroundings were totally overwhelmed by architecture. Modernists paid little attention to gardens, but landscape gardeners were still busy trying to find solutions to problems caused by high maintenance costs and land prices. Gardens were shrinking in size. Some people talked about a "maintenance-free" garden. The rules of the game had changed: Le Corbusier put terraced gardens up on roofs, whereas J.-C.-N. Forestier and Albert Laprade, reinterpreting newly "discovered" Andalusian gardens, suggested that flowerless gardens be used to highlight architectural design. André Vera's whitewashed surfaces emphasized the brightness of the sun; he declared that reflecting surfaces would create the impression of being "spanking new." Softening touches were provided by climbing plants: roses, honeysuckle, wisteria, and jasmine. The 1925 Exhibition attracted architects, landscape gardeners, and specialists of the plastic arts, who all shared ideas, with the exception of a few renegades, like Gabriel Guévrékian, a member of the avant-garde circle. Guévrékian turned his back on traditional inspiration and the rustic garden. His bold garden design, centered on water and light, was a revolution in terms of architecture and the plastic arts.

Guévrékian brilliantly resolved the spatial difficulties of small and triangular plots, the most challenging shape of all to work with. He refrained from forcing conventional right-angle and circular shapes into this cramped space and found a solution conceived in three-dimensional terms. A myriad of triangular facets set on different planes were arranged throughout the garden, creating very special visual dynamics. This was in effect a cubist garden. Guévrékian managed to open up startling vistas on a very ordinary, small plot. These planes set at oblique angles directed one's gaze up and over the surrounding walls toward treetops, in several different directions. The color orchestration and violent contrasts were reminiscent of Orphism,[45] and of Robert Delaunay's work in particular. The ensemble bore the imprint of Oriental decorative arts and their optical effects. Art critics were scandalized by this departure from normal standards, but Charles de Noailles, a discerning amateur of modern art, was dazzled by this display of artistic virtuosity and commissioned Guévrékian to do a similar garden for him. The plot was once again triangular, and it lay between his villa at Hyères, built by Robert Mallet-Stevens,[46] and the garden of his friend, Edith Wharton. In giving these two artists a free hand, Noailles engineered the creation of a masterpiece of avant-garde landscape gardening. This was probably the only cubist garden in the world, with the exception of the short-lived garden that had served as the model.

Garden of Water and Light.
G. Guévrékian, International Exhibition of Decorative Arts, Paris, 1925.

Garden-lovers amateurs and occasionally professionals keep the torch burning

Work began on Viscount Charles de Noailles' second great garden in 1947, but this time he himself took a large hand in its design. It showed a return to traditional forms and subtle plant associations, as well as sophistication in its evocation of ambience and sense-pleasing effects. This was the age of Italian gardens as seen by the ever-active English school of landscape gardening. The Riviera was no longer exclusively inhabited by winter residents. In fact most residents lived there all year round. For flower-lovers, "gardening became an exercise in prolonging" springtime and searching for plants suited to local sites and microclimates.[47] Charles de Noailles occasionally consulted outside experts and commissioned architects to execute projects, such as the stairway for an outdoor alcove by Emilio Terry, architect and designer of surrealistic works, but he gradually gained confidence in his own talent as a gentleman-gardener. His personal changes in style reflect general postwar trends in landscape gardening. He took part in the creation of several gardens, and in time, his love of plants and gardening experience led him to be one of the foremost garden authorities in Europe. Noailles, patron of the arts and consultant to French National Museums, became a botanist and wrote a book on plants used in Mediterranean gardens with Roy Lancaster, curator at the celebrated Hillier House in Great Britain. Noailles was English on his mother's side and had special relationships with great English and American gardeners and plant-lovers living on the coast, among them Lawrence Johnston, Russell Page, Roderick Cameron, Miss Campbell, and Basil Leng. In his Hyères garden notebook all the technical terms were in English. This is hardly surprising, because after 1945 only the English and isolated groups elsewhere in Europe maintained and progressed in the skills

Color study in guarrigue *garden.*

and lore fundamental to the art of gardening. Members of this "garden society" traveled a good deal and often exchanged plants. When Lawrence Johnston died, his friends inherited rare specimens. They often consulted each other on gardening matters. On one occasion Russell Page suggested that Charles de Noailles try a sequence alternating four red-purple with one white *Cercis siliquastrum* (Judas trees) all along the length of a flowering tunnel.[48] Long discussions were held on all kinds of subjects: what would one have to do if such and such a cypress produced too much shade at maturity, and so forth. And they took their time in making garden-related decisions. Russell Page kept Marcel Boussac at the Eden Hotel in Cap d'Ail waiting for two years before he finally set to landscaping the hotel garden. Page only made up his mind after many visits, when he began to "glimpse the possibilities of the site."

Tracing and replanting a Japanese checkerboard pattern.
Santolina and white sand. La Chèvre d'Or garden.

Parterre of aromatic plants.

Rockwork and Mediterranean plants.

Garden at Cap Martin.

Beauregard, gateway.

Gardens of Provence
and the French Riviera

Jas Créma, birth of an Eden

The garden of Jas Créma is only seven years old. Nothing in this part of the Vaucluse could have led one to imagine an Eden in the midst of vines. It was created from nothing toward the end of 1979 by an English-woman, Mrs. de Waldner, who was forced to leave her garden at Mortefontaine in the Ile-de-France. Like the Vicomte de Noailles, creator of the Parc St.-Bernard in Hyères, and Roderick Cameron, creator of a Jardin Neuf in the Vaucluse, she is part of the pioneering spirit that has come to regenerate the Provençal garden.

"There was nothing here in the beginning." When she recounts the genesis of her garden, one can imagine the lunar landscape of the "beginning," when bull-dozers transported the cubic meters of earth brought by truckloads. Several drillings from seventy to ninety meters in depth were carried out, and several sources of water were located. Then gradually the plan of the garden was made, using a white powder to mark the design.

The terraces were built, and then the first plantations were established, mainly with plants from England but also from South America. Stretching out in the plain and dominated by two vertical landmarks, one far and one near, Mt. Ventoux and the medieval Château du Barrou whose profiles form and admirable background, this garden is perfectly at home in its site.

A simple row of cypresses, including a few young Flor-entine varieties, forms the southernmost boundary, and a plantation of olive trees separates the garden from the

Broderie *border.*

Around the house.

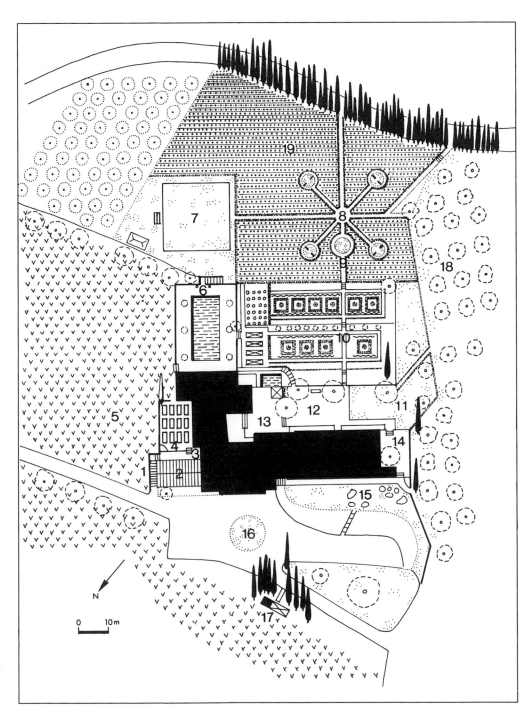

Plan

1. Entrance staircase - 2. Greenhouse -
3. Garden path with iris borders - 4. Herb
garden - 5. Grapevines - 6. Pool - 7. Lawn -
8. Lavender and rambler roses - 9. The
elephant - 10. Formal garden - 11. Lawn -
12. Main terrace with plane trees - 13. Small
terrace with aviary - 14. The west terrace -
15. Mount planted with irises, lawn -
16. Circular drive - 17. Aviary - 18. Olive
trees - 19. Lavender.

Herb garden.

Iris clump before rural landscape.

*Rows of lavender bordered with rosemary,
flowered arches.*

surrounding vines on the western boundary. "Take what the country offers you. Choose in the countryside the most interesting backgrounds and distant vistas, and try to preserve either the constructions or the plantations already in place, all that can enter into the composition of your picture." Such was the advice of the Marquis de Girardin in 1777 in his treatise entitled *The Composition of Landscapes*. For an owner, making the countryside part of the garden, playing with the backgrounds and the distant views, joining them to the garden by means of a succession of different planes, is his way of appropriating what does not really belong to him, of enlarging his field of vision. The upright stances of the cypresses, the profile of the hills, and the château are all in harmony with the horizontal lines of the garden laid out in successive levels. Unlike enclosed gardens, this one offers an opening onto a vast scenic panorama. In addition it establishes a link between it and the rural land that surrounds it by means of intermediary planes planted with the traditional plants of the Mediterranean countryside: olive trees, irises, cypress hedges, and lavander. Thus the useful and the agreeable are joined in this garden which seems to have "no end."

But let the visitor go through the small garden door, the one that invites him to contemplate, like the stone owl above the door, a small "priest's garden" surrounded by walls and set against the side of the house opposite the conservatory (this was created by Alexander Fabre and contains a collection of *Passiflora*). The garden is made up of small regular beds full of aromatic herbs and sweet-smelling plants. A row of earthenware jars containing lemon trees edges one of the house.

As the visitor moves toward the main garden, he discovers the south façade of the house. From the terrace of the *bastide*, the garden descends from one terrace to another. The first and second terraces are planted with beds of *Pittosporum* enlivened by wrought-iron baskets. In a circular space from which alleys lead off in a star

A field of lavender designed to link the parterre and the surrounding landscape.

Broderie *border.*

Detail, parterre.

formation, one comes upon the form of an elephant. Its ironwork carcass was formerly covered by yew and must have been a remarkable piece of topiary art. Unfortunately, the yew died. Mrs. de Waldner replaced it with banksia roses, a beautiful evergreen climber. Elephants, owls, and horse heads accent various points along the axes. Each alley is terminated by a double arch covered with roses. Below the terraces are beds of lavender laid out geometrically. This sea of lavender, its rows undulating in the wind, seen through the curves of the arches, is a restful sight. How wondrous it must be when the lavender is in flower! In June the garden is full of flowers of all descriptions: Chapeau de Napoléon, Apricot Nectar, Blessing, Golden Shower, to mention but a few. This garden is life. Perhaps for its owner who says she wanted to break with the north, it is a new lease on life. The basic shape is prominent and full of promise for the garden in the future.

Unlike many historical gardens, today abandoned, Jas Créma is cared for every day by its creator. It has not suffered the distortions that time and a succession of owners often inflict. The garden lives because its owner is alive. This is what characterizes a new garden. Often owners of new gardens ask to remain anonymous.

· ·

Geometric parterre and lavender parterre.

Quatre Sources, present-day romanticism

I was luckily on my own when I first visited this garden. I discovered the garden of the Quatre Sources shrouded in the morning mist, my steps accompanied by the light-hearted rhythm of the overture of *Don Giovanni*, faultlessly whistled by the music-loving gardener. From the underwood covered with dew to the terraces soaked in the morning sun, along paved paths, and up hidden stairways, I had the feeling I was discovering a new universe, where each planet sent out its own perfume, its own message or myth; wild mint, lavender, rushleaved broom, and dead leaves combined their fragrances. One should know the language of scents to understand this garden.

Roderick Cameron had already proved his talent as a gentleman gardener when he arrived in Ménerbes in 1975 from Cap Ferrat. The Côte d'Azur had become too crowded in summer, so he had sold La Fiorentina and started looking for a calmer site where he could create a new garden with his friend Gilbert Ocelli, also a keen gardener. "We were immediately very taken by the idea of creating a garden in a wood," explained Ocelli, "we started the garden before the house ... we were so eager ... we started by building walls, stairways, cutting down trees, making glades in the middle of the wood, tracing paths. At first we tried rather sophisticated plants, but after three years we learnt only to set plants which could resist both the cold and the lack of water in summer. It's a very harsh climate here for a garden, especially a spring garden, when one goes down into the underwoods and there are thousands of bulbs everywhere, dwarf irises from Turkey, crocuses, also dwarf tulips from Turkey ..."

Around the house there is an interior garden, planted uniquely with herbs from Provence and sage, santolina, and lavender.

"We put most of the gardens near to the house, because it's so marvellous, from spring onwards, to be in the house and almost in the garden at the same time, with all the doors open, and music playing ..." Very gradually the garden reveals itself. First of all, there is

Terrace.

Restanque *and flowers.*

Walking through the woods.

the formal garden against the house, the terrace of box hedges, and the silver garden with santolinas, buddleias, ragworts, and lavender, a whole mixture of silvered plants. Next, down three rounded steps, the visitor arrives at a path, "we had to have it paved because grass would not grow in the shade of the wood."

At the end of the alley, the statue of a pensive young man turns one's steps toward a collection of rockroses, surrounded by myrtles and ceanothus, or toward the Japanese garden, or along a country path where there is a statue of a hunter leaning on his rifle. One can then return via a series of small stairways edged with a collection of boxwood trees with different colored leaves. On one of the intermediary terraces, Roderick Cameron has erected a memorial to *Enid, his beloved mother, Countess of Kenmare, one of the beauties of her time.*

Higher up on the last terrace, an obelisk flanked by two urns, interrupts the perspective. Under the obelisk lie the ashes of the friend who showed this place to Roderick Cameron. Roderick Cameron wanted his garden to be inhabited by all those most dear to him. Unfortunately, this master landscaper has since died. "I wanted to create a romantic garden," he had confided to me.

. . . .

Urn in grey garden.

Walled garden off the kitchen.

Young man daydreaming, lower part of garden.

Hunter at rest between garden and landscape.

Stone vase in woods.

Ansouis, suspended paradise

From a distance the castle situated above the Aigue valley seems very impressive, but nothing in this mass of stone set upon a series of terraces and overhanging the village leads one to suspect the existence of a garden. One has to climb up to the castle to discover its gardens. Enclosed for centuries behind high walls, the castle which has been owned by the Comtes de Sabran since the thirteenth century, started opening out toward the south in the sixteenth century when a first series of narrow terraces was constructed. At the beginning of the seventeenth century, Marc-Antoine d'Escalis began transforming the castle, and his son continued restructuring it and created a beautiful façade that opens onto the top terrace. On the three sunny sides of the promontory, the terraces surrounding the castle were transformed into box tree gardens and balcony gardens.

Garden of Paradise, general view.

Edge of secret garden.

The complicated designs are pleasing to the eye, and one can then look over the balustrade to enjoy the boundless horizon of the immense countryside. On fine days one can see for miles, as far as the Alps covered with snow.

It is not known when the box trees were planted, but they are certainly several centuries old and have undergone a variety of treatments over the years. The different gardens were lovingly restored by the Duchesse de Sabran Pontevès and her husband. Whatever the season, each parterre has the same unchangeable design. To the east is the cool garden known as the Jardin de Peiresc, where the embroideries of yesteryear have become architectural figures covered with vegetation. The box trees have outgrown the geometric plan, and the original scheme has been transformed into a flourishing topiary where the full curves have overpassed the hollows. To the south there are a few box trees shaped like giant chessmen who stand guard over the entrance to the castle, reducing the visitor to the size of a humble pawn. The most agreeable of these box-tree terraces, the best-proportioned one in relation to the high walls surrounding it, is the one known as Paradise; an ideal garden for meditating as the sun sets below the horizon. The vast countryside surrounding these terraced gardens makes them seem smaller than they really are; their location among the ruins of a castle anchored to a rock gives the impression of the land being arid. The lower garden is not completely separate since one has a glimpse of it on arrival. To go down there from the castle is not a major expedition, but it does take some time to reach and enjoy it.

The water gushes forth from a small woodland and winds down among the pine trees to the nymphaeum, a large rectangular pool, and from there to a strange pentagonal ornamental pool surrounded by box trees which, when seen from the height of the castle, looks like a coat of arms thrown down onto the lawn.

Garden of Paradise, detail.

"De Peiresc" garden.

Large basin, pentagonal basin.

Terrace in front of château entrance.

Val Joanis,
a pleasure-kitchen garden

Val Joanis is first and foremost a vineyard, and so parts of it are very clearly separated. The gardens and the buildings sit on a small hill. The main building, which has two wings, was built with walls and parapets that enclose, on the eastern side, a large square courtyard paved with cobblestones (an exercising ground) and decorated with numerous large earthenware jars containing evergreen bushes. The west façade, in contrast, rises above a small valley that is accessed by three terraces, one level with the house and shaded by tall Aleppo pines, an intermediary terrace in the Italian style with four immense urns, a large bed planted with santolinas, old-fashioned roses, and other aromatic plants, and a lower terrace reserved for the swimming pool. On the south side of the hill there is a shady garden containing indigenous or imported shrubs and a few bamboos growing in a wet hollow. The northern side of the hill is very steep, and it has been left in its natural state, embellished with its original flora.

On the eastern side, and overlooking the vineyard buildings, tiers of grass, olive trees, and cypresses give the impression of a miniature Tuscan countryside. The retaining walls are planted top and bottom with all sorts of vegetation, climbers, small bushes mainly grey and blue in color, rambling rosebushes, and lagerstroemias which also line the lower alley.

General view, kitchen garden.

Flowering berceau.

A bit apart from the rest and well exposed toward the south is a very large garden that extends over three levels; part of it is reserved for cut flowers and rosebushes. A very long arbor closes it toward the east, and the lowest level is planted with Greek plane trees. Box trees and yews give it a very traditional look. Everywhere there is exuberant growth despite the strict design.

. . . .

Shade garden next to the house.

La Barben,
between stone and water

Perched on a rock, the château of La Barben emerges from a sea of greenery like a dream castle. There, the imprints of all ages are found: medieval foundations, embossed stonework of the Renaissance, baroque stairways, and crenellations worthy of a nineteenth-century operetta. It overhangs one of those sinuous rivers in the south of France. On the approach to the château, the sonorous, tumultuous outpour of the Touloubre River can be heard.

The garden of La Barben is always cool and fresh; it is sited on a long terrace bordered on one side by the sloping approach to the château, and on the other by the river below. The terrace, bathed in sun and water, and sandwiched between the stony mass of the château and the abundant vegetation of the riverside woods, is a studied reply to the woodland gallery following the path of the water. At one end its turf stretches as far as the nymphaeum situated at the eastern wall. A few

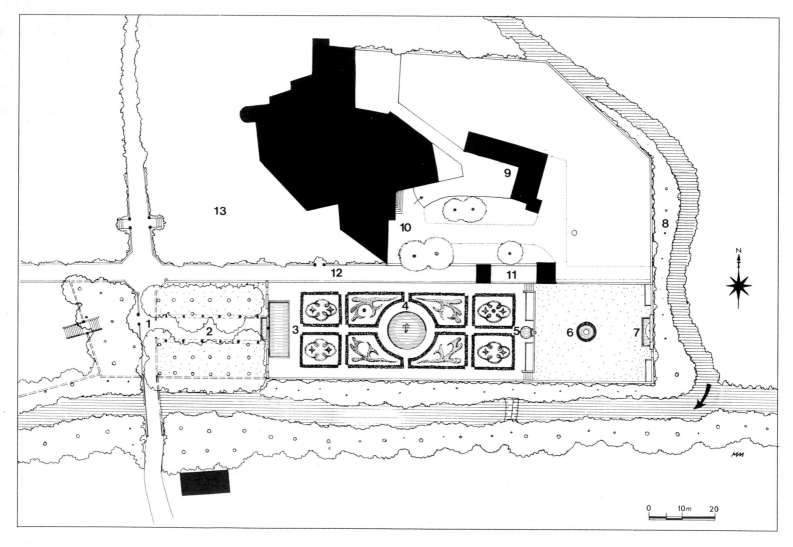

Garden plan.
1. Park entrance - 2. Romantic garden - 3. Large rectangular pool - 4. Large round pool in center of box parterre - 5. Small pool - 6. Piece of sculpture on the lawn - 7. Pool and grotto - 8. Ripisylve - 9. Outlying buildings - 10. Château - 11. Entrance tower - 12. Access ramp over arcades (greenhouses, today) - 13. Woods.

The Comte de Forbin, 1777-1844.

steps down, and on an axis with a reflecting pool, are flower beds edged with boxwood trees. Witnesses to nineteenth-century taste, a few rare trees have been planted amid the beds, notably a sequoia and a Cembro pine, which from the depths of the valley rise up as high as the château's crenellated walls. On the western side two high symmetrical walls, each with a door giving access to the garden, flank a great rectangular pool. From the outside this very original layout permits a visual access to the garden. The eye follows the main axis, is stopped by the pool with its still waters, and then discovers the whole length of the terrace beyond. The effect is reinforced by the vista formed by the alley leading from the road to the garden. The vista is not only framed by the massed vegetation of the two lateral paths leading to the garden doors but also by the subtle artifice of a double row of columns cut off at the level of the ground, suggesting a clearing away of architectural impediments for the exclusive benefit of the landscape.

The château.

General view, parterre.

Visual access to the garden.

The baroque terraces of Barbentane

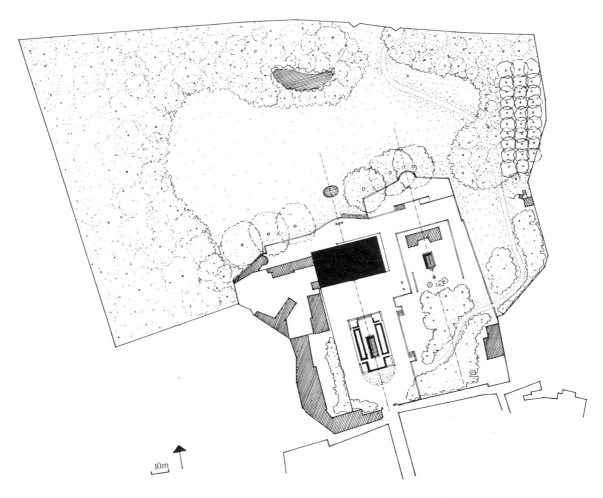

Paul-François de Puget, first consul in Aix-en-Provence in 1693, started building the Château de Barbentane in 1654. It would be interesting to know what the garden of this very Parisian-style château was like, as it was one of the few to be built during the seventeenth century. Unfortunately, the garden which was finished in the eighteenth century was transformed into a picturesque park at the beginning of the nineteenth century, and toward 1880 an orangerie was erected on the site of the large formal parterre which was situated to the east of the castle. An enormous cistern, built under the main entrance alley leading to the castle, bears wit-

Plan, Barbentane garden.

Barbentane terrace, north view.

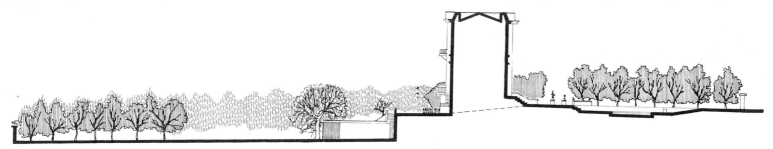

ness to the importance of the waterworks for this former parterre. The sculptures that once decorated it have been repositioned around a large ornamental basin situated in the main quadrangle: one is of Hercules standing over the carcass of the Nemean lion, and is signed by the Parisian sculptor Louis-Claude Vasse and dated 1755; another is a Pomona, sculptured by Bertrand, an artist from Avignon. A shady gallery decorated in the exotic style of the times was installed in the castle's basement on the eastern façade, and facing it at the end of the terrace is a smoker's pavilion. The northern side of the garden with its large terrace overlooking the picturesque park remains today a most spectacular sight. The enormous branches of the plane trees, planted at the foot of the terrace two centuries ago, today twist their serpentine coils around the carved-stone cornucopias and the provocative female-breasted sphinxes that adorn the balustrade. The sphinxes are dated 1776 and signed by Buffardin, an Avignon artist.

. . . .

Sectional view, garden.

Barbentane terrace, east view.

Fruit basket.

Aiguebelle, as its name implies

Beautiful and good is the water of Aiguebelle, transforming this corner of Provence into a territory coveted since the beginning of time. Is this the reason for the enclosing walls? The land around the *bastide* in the countryside of Aix-of-Provence is open to all; Aiguebelle is therefore the exception that proves the rule. As soon as one passes through the gateway, the imposing mass of the *bastide* and its outbuildings dominates the scene from afar, set on the long terrace as though on the deck of a ship anchored in the midst of enormous lawns. The whole is reminiscent of a Venetian countryside transported into Provence. Is it because of the size of the meadows, the flat and wet countryside? Emphasized by a green fringe of water-logged meadow, the terrace looks like a quay emerging from grass. It is a quay. This surprises all who approach the *bastide*, for along the whole length of the terrace, an enormous expanse of water replaces the usual parterre. The deco-

Nymphaeum and surroundings.

Nymphaeum.

rative motifs are replaced here by the shimmering shadows of the trees and the moving reflection of the *bastide*. The water and its changing images serve as a natural mirror for the baroque interior. Set against the terrace, as though it were the wall of a panoramic stage, a giant in white stone seems to tear himself out of his recess, and like the statue of the "Commendatore," comes to meet the foolhardy visitor. It is true that he has no voice, but his presence alone is enough to transform the fields around into a vast opera where the farmworker and the visitor are both expected to know their roles by heart. The master's eye tirelessly surveys the progress of the work in the fields. On approaching the ornamental pond, there is a metamorphosis; the statue of the Commendatore suddenly changes into a statue of Neptune. More friendly than impressive, an athlete seems to pose there, lightly draped, relaxed, with a slightly clouded gaze, letting a dolphin leap away from between his legs. At the edge of the canal, a small boat awaits the traveler; it is surrounded by enough water so as not to appear ridiculous. One can dream, and even row a little to go on dreaming a bit further on. At the level of the terrace a gateway separates the main living quarters from those of the farmer. One has to go around the left side of the *bastide* to get down to the second expanse of water, whose staging is more private than the first. It is a place for a delightful walk. Bordered by an alley planted with plane trees, the great rectangular pool is fed on each side by a lion's head from which water pours into a shallow basin. One of them is dated 1717. The alley leads past a nymphaeum viewed through an arch, a true woody theater, and the nymph there with, of course, a fountain flowing at her feet, represents both an image of the spring and the landscape. In a leisurely manner, and for the greater pleasure of those fanciers of ambiguity, this naiad has for three centuries been covering herself with a veil.

. . . .

Nymph dressing.

Neptune, place guardian.

Arnajon, a baroque stairway over the Durance

At the end of the seventeenth century Valfère consisted of a simple hunting pavilion, situated at the top of a superb terraced garden. Jacques Le Blanc, who was treasurer of France for Provence, bought the property in 1674, and he probably supervised the extensive work carried out in the shaded garden above the Val de Durance. The garden, which faces east, had a simple composition whose essential idea remains to this day.

On his death in 1692, the estate passed to another family but was bought back by his grandson in 1750. The extension of the *bastide* toward the west to obtain a southern exposure for the façade dates from the end of the eighteenth century.

On a sunny, shadowless day in July, the grandson of the present owner welcomed me with a grand gesture. As if to give me the keys to the garden without an excessive explanation, he opened the water valve of the fountain on the first terrace. Immediately the water overflowed the fountain's basin and rapidly flooded half of the terrace before disappearing in the direction of the garden.

Seen from the perron of the horseshoe-shaped stairway are two large rectangular pools which constitute the focal point of the composition. The first crosses one's line of vision like a luminous canal, a springboard projecting one's gaze toward the distant landscape of the Luberon. It is situated on the longest terrace, the third from the top. The second closes off the lower part of the garden, and the level of this reflecting pool invites the gaze to pass beyond the woods in the direction of the horizon, toward the Alpine sky.

An idea of the extent of the initial project can be had by glancing at the fields that border the garden, even if it is impossible to calculate the number of days' work carried out by terrace builders, fountain engineers, and sculptors to transform the hill and its *restanques* into three large terraces, impeccably traced out, leveled, and embellished with vast ornamental pools and multiple fountains. The principal axis, descending from the house and connecting the three terraces, gives rise to

Aerial view, Arnajon.

Gateway.

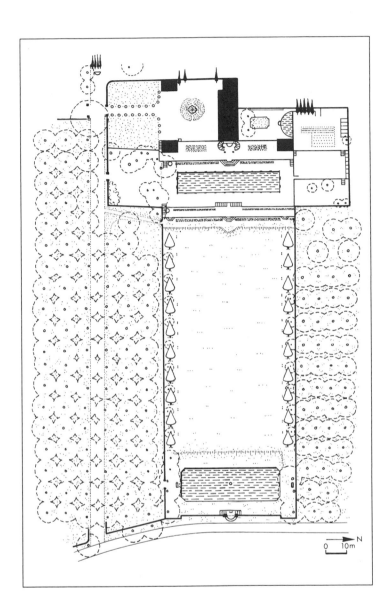

Plan.

Large upper pool and bastide.

Arnajon terraces.

Large pool and view of the Luberon.

a veritable exercise of style in stairway construction. At the end of each flight of steps of the horseshoe-shaped stairway, there is a wall fountain framed by two enormous urns. Devoid of a balustrade so as not to interrupt the view toward the large ornamental pool, the second stairway widens out at the bottom. Two straight flights meeting at a small perron, enable one to descend the steep gradient between the last terrace and what was, without doubt, the orchard. One might imagine that the gateway at the bottom of the garden served as an entrance to the estate, but the steps make the passage of any wheeled vehicle out of the question and do not even seem wide enough for horses. Probably the gateway's only purpose was to prompt the respect of those who passed along the path on the other side.

On going back up toward the *bastide*, the right-hand pigeon house, rare of its kind in Provence, has a surprise in store for the inquisitive visitor. The visitor enters the small building by a low door. At the precise moment that he imagines that he is going to step into a pile of pigeon-droppings, he finds himself to his amazement confronted with the unexpected decor of a *salle de fraîcheur*. As his eyes become accustomed to the gloom, he perceives pleasant-looking caryatids holding baskets of flowers made up of thousands of white shells and colored sand. Hidden among the concretions are small pipes, which in bygone days used to spray the surprised visitor with water. After the heat and dust of the road, or after the fatigue of the hunt, people searching for a cool place would come here to rest and to dream a little in the semiobscurity. In the midst of this octagonal room a cold pool with a bench permits the visitor to cool off from the waist down, while giving free rein to his reveries inspired by traces of strange arabesques on the ceiling.

. . . .

Monumental vase, horseshoe staircase.

Caryatid, salle fraîche.

Fonscolombe,
the landscaping tidal wave

In 1715 Honoré Boyer, counsellor and secretary to the king, acquired the property of Fonscolombe with its small château situated in the midst of a large plain abounding in streams and springs. Around 1720 the château was demolished and replaced by a *bastide*. The project had to deal with the flatness of the land, and had to find some means of making the garden stand out from the surrounding meadowland. In order to separate the new construction from the countryside, a single terrace was built at the foot of the south façade leading to the garden by a double flight of steps. To effect a refined transition between the house and the garden, large urns in the rococo style and two female sphinxes were placed on the balustrade. The eighteenth century parterre was very probably established slightly below the level of the present-day one. But the garden was completely reorganized at the beginning of the nineteenth century and integrated with a landscape park. The *jardin de propreté* disappeared under an immense lawn which came up like a lake to the foot of the terrace. One has the impression that there has been a flood: the only objects emerging from the all-prevailing green are the two large urns placed on pedestals that marked the limits of the *jardin de propreté* and the branched fountain decorated with three dolphins in the center of the parterre. Curiously enough, the basin of the fountain is scarcely higher than the lawn, clearly indicating that the level of the ground has been changed.

New varieties of trees were planted in the park. A bald cypress throws its dark shadow on a swan gliding in the canal. In the middle of this pool, a statue of Hercules seems vainly to be seeking a better site. It is probable that the statues, of which certain have been attributed to Chastel, have been moved several times and that their surroundings have been modified. The memorable fountain of the faun giving water to a young Bacchus stands out today against a hedge of blue cypress,

Aerial view.

a setting of rather questionable effect. The park and its winding walks almost certainly laid out in an existing wood. Along one walk, small edifices in an early nineteenth-century style were erected: a column in Egyptian granite set upon a pedestal and a cenotaph, both of which are surmounted by urns.

Near the house an iron gate with lateral pillars topped with iron balls separates the main courtyard from the road. The gateway affords access to the courtyard, and from there one can go up to the main house, enter the offices and outbuildings, or go down to the rose garden. To get to the rose garden, one has to leave the graveled courtyard, go along a magnificent ornamental pool with a fountain, pass by two pigeon houses with glazed tiled roofs, and finally go through a small wrought-iron gate. Here one finds several statues, survivors of the landscaping upheaval among which are statues representing summer and winter. In the past they must have adorned the South Garden and were surely placed here for safety. If their new role is to watch over the roses, they seem, all things considered, to have adapted themselves remarkably well.

. . . .

Cenotaph in kitchen garden.

Walled garden, north of bastide.

Bust and stone dove.

Pool, north of bastide.

La Violaine,
an invitation to come down
into the garden

Neither the recent buildings that surround Aix-en-Provence, nor the earthquake that in 1909 divested the *bastide* of its ornaments, or the two and a half centuries that separate us from the moment it was created, have really affected the Parc de la Violaine. On arriving at the foot of the house after a drive down the very long alley of plane trees, there seems, at first, nothing to see. There is no terrace or parterre in front of the house. One has to turn away from the building to discover the garden.

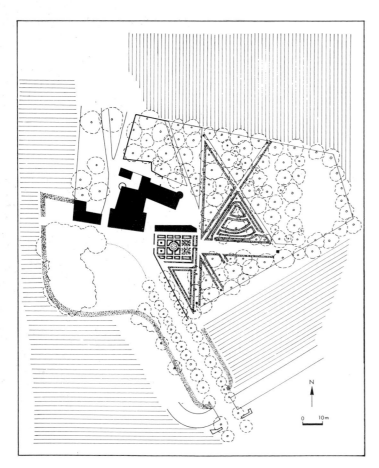

Aerial view.

General plan.

Invitation to walk in the woods.

A clearing beckons you to one side, bordered by an orangery whose whiteness contrasts severely with a marvelous parterre of box tree, dating in all probability from the mideighteenth century. M. de Violaine bought the estate in 1720. The original design has been somewhat enlarged inasmuch as the curves have become wider, though the plant growth has always been kept under control in accordance with a strict plan. Like the designer trying to find the perfect curve, the successive gardeners have respected and mastered the box tree's growth.

Upon looking closer, one sees that the parterre has been slightly inclined, like the Italian stage, so that it can all be seen at the entrance point, since the terrain did not permit the construction of terraces. The parterre proffers an invitation to stroll in the surrounding woods. Three alleys connecting in an equilateral triangle open vistas onto the surrounding countryside. Each alley leads to a statue that provides both a focal point and a terminal point to the park, outside of which is the agrarian landscape. At the heart of a small woodland a box-tree maze gives one the illusion of being lost in a huge forest.

Parterre in a clearing.

Box parterre, detail.

Plan, box parterre.

Box parterre, general view.

La Gaude, the perfect "bastide" garden

Thanks to the loving care of its owner Mme de Vitrolles, La Gaude, a *bastide* garden in the small valley of Les Pinchinats near Aix-en-Provence, was able to rise from its ruins. Very little is known about the history of this estate. Certain names follow one another. Charles Albert Pisani renovated the property in 1750, then the Lubières family took up possession, followed by the Arlatan de Lauris family. In February 1955 the Baron and Baronne de Vitrolles inherited La Gaude which was then in an abandoned state. Mme de Vitrolles, the daughter of the Duc de Castries, is an artist. She spent her early years in the garden at the Château de Castries, near Montpellier, a garden that was designed by Le Nôtre. As with the garden of her childhood, Mme de Vitrolles, found at La Gaude a garden that had to be restored.

Today this garden really deserves a leisurely visit. After entering and passing through the high gateway, one comes upon a long alley of box and spindle trees, in the shade of horse-chestnut trees. At the end of this alley is the private part of the estate marked by a low wrought-iron gate. This is the threshold of the garden. Once it has been crossed, you come to a raised terrace on which the house with its two outlooks is situated. The principal façade is turned toward the distant rolling countryside, where in the horizon rises Mt. Sainte-

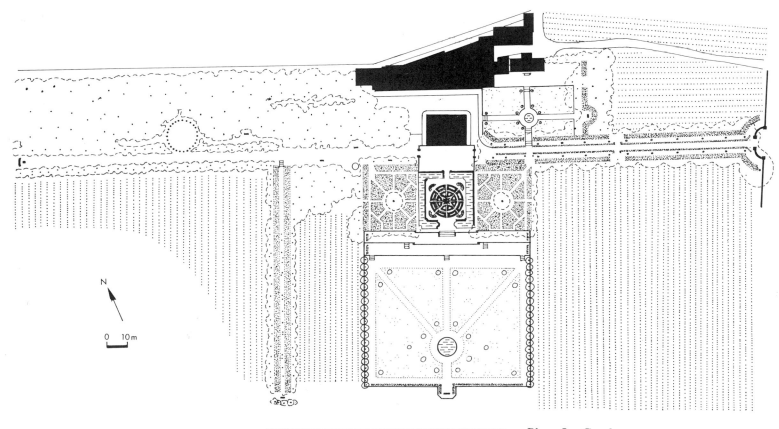

0   10m

Plan, La Gaude.

Bastide, *view from last terrace.*

Victoire. In the foreground are the successive terraces of the garden linked together by small symmetric stairways. Seen from above, the *jardin de propreté*—whose parterre consists of topiary motifs in box, the oldest of which are at least two hundred years old—appears framed by a splendid canal; the box-tree parterre is really an island joined to the terrace by two small bridges, marking the axis of the view. The motifs were traced with a compass as was traditionally done for the labyrinths of the sixteenth and seventeenth centuries.

Although simplified and symmetricalized, the parterre bears a strange ressemblance to the famous Stra labyrinth at the Villa Pisani near Venice, designed in the mideighteenth century. It so happens that the renovator of La Gaude in 1750 was from Pisani; doubtless he tried in this way to leave his mark on the countryside of Provence. Three-quarters of the box-tree parterre were remodeled by the present owners who found

it in a pitiable condition. For eighteen years Mme de Vitrolles trimmed them with ordinary hedge-clippers; then her husband bought her an electric pair which changed her previously artistic gesture into a simpler task.

There are two ways to look at this property. The best viewpoint is of course from the central balcony. Looking down, the succeeding terraces are but tiers that lead down to the country scenery. But looking up at the house and garden from the lowest terrace is a breathtaking sight.

Beyond the box-tree parterre, another terrace was created. Mme de Vitrolles felt that the perspective was too short, and to extend her garden, she added a terrace, which she set off by a grassy lawn and a fountain, with alleys leading off in the shape of a goosefoot. At La Mignarde, a neighboring property, statues fill the garden; decorating the graden at La Gaude are Andalusian

Aerial view.

Parterre, view from bastide.

vases and animal sculptures attributed to Chastel. "To each his own taste," the Baronne de Vitrolles commented.

A promenade round the garden at La Gaude has many other surprises in store. Copses, green bowers, former round threshing floors also used for harvest balls, and shady woodlands also await the visitor.

On either side of the *jardin de propreté* two symmetrical copses screen a small clearing in the shade of a horse-chestnut tree. Far from inquisitive eyes one can rest a while on one of the stone benches in a private, shady, and protected place.

On approaching the pinewood, one discovers what must be the best-kept *tèse* in Provence. One can still hunt there. Then one follows a long alley lined with thickset trees that leads to a circular space known as the *salle verte* ("green room") surrounded with benches. The layout somewhat resembles that proposed by Dézallier d'Argenville in his treatise on gardening of the early eighteenth century. The land has been "shaped," but generally speaking, art gives way to nature. Contrasts are created between the open parterres and the copses that screen the view. Not all is seen at once, and the curiosity of the visitor is aroused as he comes to discover the garden little by little.

The main façade is for show and dreams of greatness. But La Gaude has also another side; a much more modest façade gives onto rustic buildings covered with climbing plants. For although La Gaude is a showplace, it is also a place for work and production. The vineyard, and the well-reputed wine it produces, is the moving force of the estate. Today the owner has to assume a double personality, that of an aristocrat and that of a director of an enterprise. To the east, beyond the agricultural buildings, two greenhouses frame a chapel with pilasters, vermiculated masonry, and a wooden balustrade. A recently made *parterre* with a fountain sets off the greenhouses and the façade of the chapel. La Gaude is an exemplary garden; it is unforgettable.

Parterre, detail.

Sea dog on terrace edge.

The ball of La Mignarde

General view, garden.

No, you are not imagining it. It actually is a meat pie and a piece of puff pastry sculpted over the stone pilasters at the entrance gate that welcomes you to La Mignarde.

Enriched by the art of his pastry-cook father, Sauveur Mignard did not try to deny his origins when in 1768 he took it into his head to transform his country house into a countryseat. The sumptuous interior decoration had its counterpart in the garden outside, although the layout of this disappeared during the modifications carried out at the beginning of the nineteenth century. A space remains. As with many country houses in the region of Aix, the successive modifications and periods of neglect have left intact the space of the garden, whose limits are marked by drives lined with plane trees and a terrace that is divided down the center by a succession of ornamental pools. La Mignarde is certainly blessed with the architecture of its garden, fields of red earth, the pine forests on the hills, Mt. Sainte-Victoire set upon the Cengle plateau, and especially the miraculous existence of two dozen statues that lined the principal axis of the garden.

Except for a slender shepherd carrying a lamb on his shoulders and a powerful Hercules overwhelming the lion Nemeus, the statuary consists of seductively veiled women, some standing and some in enamored poses, but all partaking of the joys of nature. What better scenery could be found for the backdrop of the hectic love affairs of the Princess Pauline? During the summer of 1807 while, around her, plans were being devised to separate her from her Provençal lover, Auguste de Forbin, La Mignarde was the setting for their reunion, thanks to its new owner, J. B. Ray, chief commissary of the emperor's armies, who was unable to refuse anything to the princess and was even forced to extend hospitality to the two lovers at the same time as to her husband, Prince Camille Borghese, who found himself in a very difficult situation. As recalls his companion A. C. Thibaudeau, ''The Prince exclaimed that his wife was

Sauveur Mignard.

Bastide, *general view.*

very lucky to be the Emperor's sister; otherwise he would have given her a beating she would not have forgotten in a hurry. This was at six o'clock in the morning, we were outside enjoying the fresh air just under the windows of the Princess whom he presumed was in arms other than those of Morpheus."[1]

Two centuries after the creation of the garden, La Mignarde's sensuality does not seem at all diminished. If you feel slightly giddy after a visit to La Mignarde, it's absolutely normal—just the aftereffects of the waltz round the garden that the statues have forced you to dance. Each one, set upon its pedestal, invites you to look up at it and to go round it at least once, so that your passage through the park resembles a large curve embellished with a small loop for each statue. The statues have certainly changed places since the first garden, but at the ball of La Mignarde, you're the one dancing. What visual delight! So many women as to dizzy the mind! The eighteenth-century sculptor Chardigny must have sketched all the beautiful young women in the Aix region to give us this comely group.

To try to identify Summer, Flora, or Venus among these statues is an uncertain enterprise. Mythological references are less important than their direct gaze on the natural world. The statues are there wrapped in wet draperies or garlands of flowers. Insects and snails crawl over them. In deep communion with the countryside, they are supremely indifferent to our presence.

. . . .

Lenfant,
the drawing-room garden

Although the original disposition and harmonious proportions of the main façade of the Lenfant Pavilion are still discernible to an expert eye, the building no longer has the characteristics of a pavilion because an additional story was constructed in the nineteenth century. One has to mentally block out this attic story, which rises beyond the surrounding tree tops, to appreciate the relationship that once existed between the small immaculate pavilion and its vast garden. The pavilion was built for Simon Lenfant, war commissary and general treasurer of France, in 1685. It was situated close to the town, and its sumptuous garden served as an outdoor extension of the first-floor drawing room, which had a rich baroque interior.

It would be a mistake to judge the garden merely by its box edgings, ornamental pools, and lines of trees. One has to try to imagine the complex embroidered parterres inside the rigorous geometric outlines of box, which are all that remains today. The intricate motifs of embroidered parterres and the geometric schemes of the small lawns with red-and-yellow sanded paths were reflected in the house in the profusion of plaster curves and the ceilings painted by Van Loo.

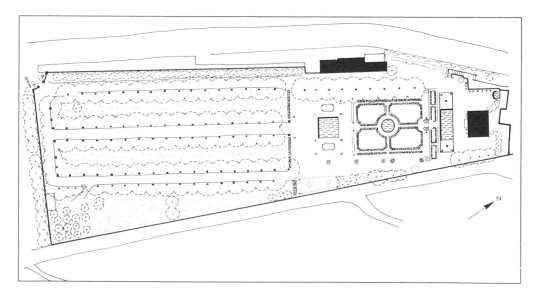

The Lenfant pavilion.

Plan, Lenfant garden.

Simon Lenfant, engraving by Cundier.

The box edgings run parallel to the principal façade and each gives a visual impression of being a green step to a shallow terrace. In this way the garden imparts an illusion of height to the pavilion. A fine horseshoe-shaped stairway frames the fountain against the terrace. The head of a dolphin spouts water into a shell, and from there the water pours out in a sheet, submerging two cupids and a gargoyle. A double row of box fulfills the role of a balustrade at the edge of the terrace, and it prevents the stairway from assuming its real function, that of connecting the parterre with the terrace. The unfortunate visitor who tries to go up the stairway will have to come down again frustrated, having got-

ten as far as the edge of an ornamental pool blocking his path. He will have to console himself by walking round one of the superb *pot-à-feu* whose petrified flames are the most eloquent expression of baroque art and its love of all that is elusive and ephemeral.

The charm of the garden of Lenfant also lies in the balance achieved between the space open to the sky, the terrace and the parterre, and the covered space, the space in the shade, separated only by a box border. An alley of lime trees and two counteralleys of horse-chestnut trees, surrounding the two parts of the open rolling grounds, provide cool refuges from the strong summer sun.

Parterre, view from the first floor.

Trimont,
the garden of the intellectual

General view, garden.

L. Thomassin de Mazaugues, engraving by Cundier.

If the Trimont Pavilion has continued to be the "dream house" in Provence of both the garden-lover and the art historian, it is probably because of its human scale. It was built around 1700, and yet fits well into our contemporary way of living; one can see oneself living there. As soon as one enters, one gets the impression of perfection in how all the elements of the site relate to one another: the harmonious proportions of the façade and garden, that well-measured balance between garden area and dimensions of the house and between rural and urban aspects of the design.

The garden of the pavilion stretches for 60 meters along a stream that flows from the Pinchinats Hills. In the past this long narrow garden was entered at the lower end by and alley along the stream where now tall trees grow. Today it is reached by the kitchen garden and orchard, which was doubtless in its halcyon days a shaded outdoor drawing room. A trellis lines a path that leads to a balustrade at the edge of the terrace; to one side there is a gateway and a few steps to the formal part of the garden. Over a fountain set against the terrace is the coat of arms of the first owner—a curious array of scythes, blades up, all identical, lined up in three rows. It is not known if this device was supposed to refer to the rural origins of the owner or had some other subtle meaning. Did this strange escutcheon motivate its subsequent owner, also a great collector? Louis Thomassin de Mazaugues, president of the parliament in Aix-en-Provence, numismatist and bibliophile, had in his collection the manuscript of Peiresc, that other great collector, and his library enjoyed a well-deserved reputation.

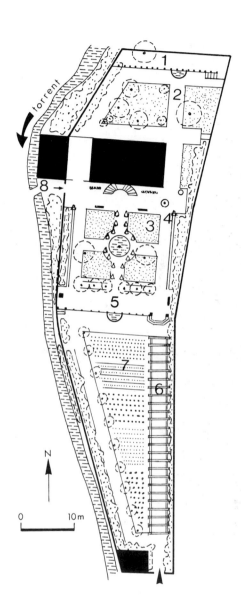

Plan, Trimont summer-house garden, A. Laprade

1. Terrace - 2. *Jardin de fraîcheur* - 3. Parterre - 4. Fountain and channel - 5. Terrace - 6. Arbor - 7. Kitchen garden, orchard - 8. Old mill.

Stone shell: water supply for garden irrigation drains, A. Laprade.

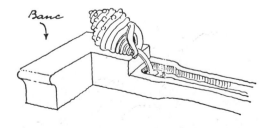

Albertas,
a garden for a garden's sake

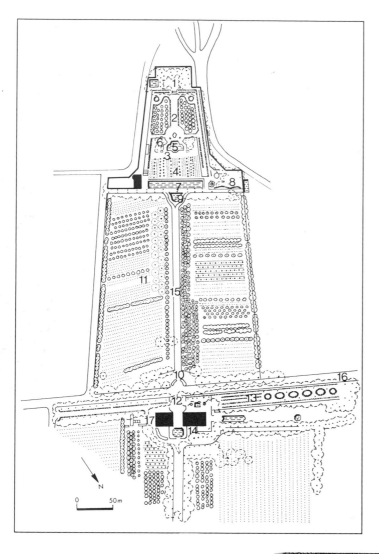

In the eighteen century it was not unusual for a garden to be finished before the foundation stone of the house was laid. The craze for gardens surpassed that for houses.

The Marquis Jean-Baptiste d'Albertas was the first president of the Provence Audit Office; a great garden-lover, he already had a luxurious town house in Aix-en-Provence and a château in Gémenos, when he decided, around 1751, to create the garden at Bouc. His marvelous, though rather remote, garden in Gémenos was not yet completed. Nevertheless, the marquis wanted another garden, no less wonderful but nearer to town. The vast estate at Bouc, inherited from his mother, offered enormous possibilities for the creation of a summer garden. It was situated on a cool north-facing slope overlooking fertile land and meadows, and watered by five abundant springs.

A plan of the site (signed T. C.) was drawn up in 1751. Building yet another expensive residence was of secondary importance, so priority was given to work on the garden. This garden stretched out on either side of the road between Aix-en-Provence and Marseilles. The main or upper garden, centered on the hill and the site of the future château, descended in terraces to the road. On the other side the pavilion garden maintained the central axis toward the north but spread its

Plan
1. Terrace with copse - 2. Terrace with parterre - 3. Terrace with kitchen garden - 4. Former ornamental kitchen garden - 5. Pool with 17 fountains - 6. "Green Room" - 7. Canal - 8. *Salle fraîche* - 9. Pool - 10. Gateway, upper garden - 11. The former meadow - 12. Gateway, lower garden - 13. *Tèse* - 14. Hunting lodge - 15. Main alley - 16. Road from Aix-en-Provence to Marseilles - 17. Outbuildings.

Portrait of J.-B. d'Albertas.

D'Albertas coat of arms over gateway.

General view, pool with seventeen
fountains.

Basin, lower garden.

Triton blowing his conch shell.

charms in a westerly direction along the road to Marseilles. Later on, between 1764 and 1767, the pavilion was transformed by the architect Laurence-Alexandre Vallon.

The marquis, who no doubt, preferred gardens to buildings, continued with the creation of the extraordinary garden of Gémenos, the same garden where he was to be assassinated in 1790. The main house at Albertas, scheduled to top the composition, was therefore never built. Today the layout of the garden remains largely unaltered. The architecture, the terraces, the main composition, the lines of trees, the contours of the box borders, and the works of art are still in place.

This miraculous state of preservation is due to the care bestowed on the garden by the d'Albertas family who still owns this land. And they keep it up as in the past. Although the present Marquis d'Albertas has transformed the former meadow, kitchen garden, and even the parterre into nurseries, he has nevertheless remained faithful to the basic scheme of the garden as conceived by his ancestor more than two hundred years ago.

It would be erroneous to imagine that beds of artichokes are not in keeping with the superb ornamental pool and its seventeen water jets. However, if one looks at the plans made in 1751, one can see that these vegetable beds are true to their eighteenth-century conception: as in the plan they are set out, side by side, just like paints in a box of watercolors and framed by bushes trimmed in the Villandry fashion. On one of the drawings the architect-landscaper has even written "winter artichokes."

Two wrought-iron gates leading into each of the garden parts, face each other across the road. One is prolonged by iron railings in front of the hunting pavilion and the outbuildings, and the other, which opens onto the upper reaches of the garden, indicates where the château would have been built. This beautiful dwelling would have opened out onto a second terrace facing north, known as the terrace of the groves. There, two

Hercules.

Atlantes, pool with seventeen fountains.

groves of twenty-four horse-chestnut trees at either end of an open lawn still frame the view, and they would have invited visitors to approach and admire the parterre. The terrace of the groves is prolonged on each side by two walks bordered by flowers, overlooking the parterre and the kitchen garden. The walks take one to the spot where the best view of the parterre was to be found, the ornamental pool with its seventeen jets and canal, where a trellised arbor drawn on the plan was perhaps constructed. This arbor would have served as a kind of "royal box," a perfect place for admiring the parterre which had been conceived as precisely as a stage set. On the lower terrace the parterre enclosed the kitchen garden and also was bordered by two walks. Its very classical scheme, including U-shaped walks, is continued throughout the rest of the composition, even in the meadow which is lined with young elms and poplars. In the midst of the upper garden six atlantes crowned with leaves support the wall of the ornamental pool which backs onto the parterre terrace. The seventeen water jets from the rim of the pool and from conch shells, into which tritons eternally blow, converged at the center of the pool. The pumps would have to be repaired to deliver the tritons from their despair at having nothing to blow but air. Today the water is absent, and yet it seems present: the expressive force of the tritons sketches out for us its former trajectories. At the edge of the parterre terrace, four statues stand sculpted in the classical tradition; the motivation of their gestures is left to the onlookers' imagination.

Slightly behind the others with his weight shifted to one leg, Hercules gazes at his three neighbors frozen in action: Samson with his fist raised before him, Ephesus with his hand on his sword, and David holding his sling. He too has overcome his adversary, slain the lion Nemeus whose skin he holds in one hand, but already he seems to be thinking about taking a rest. His other hand caresses the symbol of the place, the three apples of the garden of Hesperides. With two green arbors

Samson.

David and Hercules.

planted at the sides of the ornamental pool and, depending on the wind, a fine atomized spray from the seventeen jets, this would have been a deliciously cool place in the summer.

Between the canal closing the upper garden and the pool terminating the double central alley of the meadow, a shaded walk leads to a small octagonal room, a *salle de fraîcheur* which is unfortunately in a very dilapidated state. As is indicated by the term "cascade" marked on the map, the walls of this small room were covered in concretions, concealing a multitude of small pipes that sprayed the visitors with a cold, refreshing shower.

In the meadow a transverse alley of poplars closed off the upper garden with a double curtain. Only the central axis remained open to the other garden. Today a row of plane trees bordering the road fulfills the same role. When one steps through the wrought-iron gateway with its zigzag design, the lower garden still retains its air of mystery. A square courtyard decorated with four corner fountains curving inward leads to the hunting pavilion and the farm. The terrace of the pavilion distinguishes it from the farm and the symmetry of the two buildings prolongs the axis of the upper garden. Behind the hunting pavilion, on a level below it, are two secret-looking places: one can choose either the *tèse*, rather overwhelming with its four alleys planted in box and spindle trees, or an open park shaded by plane trees and horse-chestnut trees and decorated with a large reflecting pool at whose four corners stand large stone urns. As far as I am concerned, I prefer the latter.

. . . .

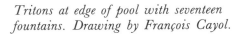
Tritons at edge of pool with seventeen fountains. Drawing by François Cayol.

lined alleys, from which could be seen in the distance all the jets of water coming out of a kind of moat. Beyond this, there was a fine alley of poplars that led to a superb cascade.

Today a few splendid traces of the garden of Gémenos are to be found dispersed about the village, abandoned to the whims of their private owners or to that present-day prince, the municipal authorities, who are installed in the château.

Beyond the brilliant display of the jets, sheets of water, and fountains of Gémenos, the park of Saint-Pons stretches up the valley for three kilometers as far as the riverhead. This is the second part of the show, created for the enthusiasts of the new-style gardens, or more exactly, for those seeking landscapes that would bring them new emotional experiences. Here the prince de Ligne declared: "Imagination does not need mythology for stimulation." In his *Homme champêtre*, the Abbot Delille sang the praises of this "enchanted valley." It's a park for those who have read Rousseau and the *Rêveries du promeneur solitaire*. The Abbot Papon came here to listen to the nightingale, and this plunged him into a "gentle reverie." Rather too impressed by the wildness of the enclosing hills, he reassured himself by listening to the music of "the natural cascades, their tumultuous humming, the burbling of small streams... and the murmuring of the zephyrs." Forty years later, on May, 16, 1838, Stendhal wrote during a visit, "this wood of St Pons that rises up with such dense shadows, the sound of water at the bottom of the ravines, the moaning of the breeze through the branches." However, it is the prince de Ligne who described the vital principle of this park best.

Along the road leading to the riverhead were three corn mills, one olive-oil mill, a workshop making red fezzes for the Turks, two copper workshops, and four paper mills, all presenting "utilitarian" scenes. Next came scenes devised to provoke more emotional reactions: a torrent that could be turned on or off as desired, an aqueduct crossing a river, a gigantic waterfall,

the terrible rumbling of an enormous stream, and finally higher up the stream issuing from "the crack between two rocks that look like the partition of the earth." The prince goes on to say that "all these separate parts are very agreeable taken individually, but yet one has managed to make of them a harmonious whole." Along the path the marquis built bridges, platforms, benches, seats, green bowers, small buildings from which multiple vistas could be enjoyed, a view of all meadowlands, or a close-up view of a strange rock. J.-B. d'Albertas was a man of his time, and he planted in this place a great number of exotic trees and plants. Just forty years after his death, they had so proliferated in the well-exposed places that the Comte de Villeneuve feared for the future of indigenous trees.

At the top of the park the valley widens out, forming a circular expanse of meadowland bordered by trees, bushes, and the ruins of an abbey. Far from discreet eyes, the nuns had led such a debauched existence in this inspired and inspiring place, that the abbey was closed down by order of the bishop in the fifteenth century. It is exactly on this spot, amid the fallen stones of the abbey, that I came upon one evening, basking in the last rays of light, a statue of a rather provocative female sphinx whose breasts had been badly damaged. How did she get up there? Did the marquis have her placed there as though to wink at the history of the abbey? The sphinx has since disappeared, and one will doubtless never know.

The valley has suffered from the parceling out of land in the midnineteenth century. Today the utilitarian scenes of the workshops and the mills, now in ruins, have become picturesque scenes. The park belongs to the *département*, but the higher reaches of Saint-Pons still wield a strange power over the visitor who manages to be there alone. This has become difficult on a Sunday, but two centuries ago, "the merry men of Marseilles and their charming companions" came here by the dozens "to try to sound the *galoubet* [three-holed Provençal flute] that transforms lovers into troubadours."

Cascades in the Saint-Pons valley.

classical temple—all this in a building that was, in principle, a dairy. An inscription, stolen a few years ago, gave a supposed clue to the mystery of this strange construction:

Grandeur is too often followed by ignominy,
From being a temple, I became a church.
I was too proud of this, and was made a stable.
Passerby, you who observe the shame that has
                                 followed my glory,
Learn without protest to yield to destiny.

When the comte had this inscription engraved, did he have a premonition of his own destiny, that of his château and his garden? A troubling conjecture where the past, the present, and the future intermingle.

Apart from the pastoral pleasures that it offered, the garden was also the scene of great festivities. The Comte de Valbelle organized all sorts of games, theatrical productions and concerts. It was also a meeting place for all society.

Like Castille in the Gard and Château Raba near Bordeaux, the park of Tourves is one of the regional examples of neoclassical parks of the eighteenth century, loathed by the common people. Like many places that were destroyed during the Revolution, the semiofficial pillage of the château and the park has left a feeling of communal guilt that tends to make this temple of high culture into something Provence would rather forget.

The numerous ornamental constructions in the landscape parks show how curious eighteenth-century man was about the world he lived in, and how motivated he was by the spirit of the times. Defying and breaking temporal and spiritual laws, the artificial devices in the park were placed there to arouse feelings of surprise, wonder, and curiosity. They made the garden a privileged place where everything was possible.

The sham ruins of the Vacherie invite the visitor to meditate on the passage of time and the ephemeral nature of life, and the pyramid, to reflect upon immortality.

What was the actual plan of the park? No drawing exists, but in 1985 during a campaign to clear the undergrowth, the remains of a star-shaped design in box were found at the foot of the western façade. It was a large parterre, probably contrived at the end of the seventeenth century or at the beginning of the eighteenth. A bit farther toward the west, diagonal alleys led to the various ornamental buildings of the landscape park. Although relics of bygone days have become rare at the site of the château and the park, one does not have to go far to discover unmistakable traces of them. A simple stroll through the streets of the village via the town hall square and the church will quickly enlighten one as to the fate history reserved for them. It is true that the park of Tourves died with the Comte de Valbelle. However, its presence has not fully succumbed to "ignominy." Did the comte strike a bargain with the devil so that the legend of the park would live forever in the hearts of the people of Tourves?

But one cannot take leave of Tourves without calling attention to an amusing story recounted by the villagers. It concerns the tomb of the Comte de Valbelle, which was erected by his mother in 1783 in the church of the convent of Montrieux. Legend has it that the comte, feeling death was not far away, asked the young women of Tourves to pose for statues destined to decorate his monument.

They refused. It is said that he took four of his mistresses, one of whom was Clairon, a tragedienne, and another La Guinen, a dancer with the Paris opera. He organized a fête to unveil the monument. When his mistresses saw themselves portrayed as hired mourners, they were outraged! In Tourves people say that the comte died during the fête at the exact moment that he was simulating his death. But this is only a legend: the comte in fact died in Paris.

After the destruction of the monument during the Revolution, the statues were bougth by the Préfet of the Var, before being distributed in the *département*. Great travelers, these statues, hidden away or exhibited in turn, they have each had a very special destiny and

Dairy.

successive roles ascribed to them. One of them is in a standing pose with a pitcher in her hand. She is now at Fréjus, where she is known as the Vestal. The second, lying down in a languorous pose, was called Hope. She was transferred to the grotto of Sainte-Baume, a presumed refuge of Mary Magdelene, and today represents the saint. A kind of attendant justice, Doctor Fontan admitted that the statue was "none other than the actress Clairon." This change of role provoked some lively reactions from prominent puritans, and in his brochure on the Sainte-Baume, Father Lacordaire voices his indignation: "Inglorious marble statues inhabit the chapel of the Saint, and behind her altar reclines a profane statue utterly unworthy of the majesty of the place where all sacred memories are saddened by its presence."

A third statue was known as Strength, a reference to the Virtus on the emblem of the Valbelle family. Her hand originally held a garland. The garland was replaced by a sword when she was transferred to the Palais de Justice in Draguignan, where she now represents Justice.

The fourth statue is now in Toulon, and is called Provence. She was first placed in the town hall but was later transferred to the rue des Pucelles. Some time later, her hand was broken with stones, and Provence was moved to the town gardens. Finally, she came to rest in a niche in the museum. The four sculptures are attributed to a Marseilles sculptor, Fossati.

. . . .

Village fountain which previously stood in the park.

Mausoleum of the comte de Valbelle after sketch of his tomb.

Châteauneuf-le-Rouge,
in praise of a gardener

Someone once told me that a labyrinth of box exists at Châteauneuf-le-Rouge that is so immense that it takes the gardener the whole year to trim it. He begins at one end and by the time he has finished, a new year with a new circuit is about to begin. On closer investigation, I discovered that the gardener is not an embodiment of Sisyphus, and that trimming the labyrinth really takes him six months. However, the garden he looks after, with so much understanding of its spirit, is in keeping with the legend. Discovering this unusual garden, situated before the terrace of a gigantic *bastide* built on Gallo-Roman foundations, the visitor is immediately attracted by the vista of one of the immense corridors of box, so precisely trimmed that they are like great walls almost three meters high.

Once inside the labyrinth, the explorer is dwarfed by the difference in scale and is irresistibly drawn forward toward the patch of sky visible at the end of the sixty-meter-long corridor. Halfway along the central alley, on an axis with the house, one comes to a circular space, edged by a low ring of box. It is difficult to retrace your path when visiting this garden. The enormous height of the hedges plunges one into semidarkness and cuts off all lateral views and any views of the surrounding countryside. It is only by moving slightly away from the labyrinth toward the meadow which borders it lower down that the visitor discovers the edge of the Cengle plateau, exactly parallel to the walls of box. The superposition of the two borders, one vegetable and the other mineral affects a harmony between the garden and the terrain.

From the windows of the house, the box seems so close that you want to touch it. Looking at the labyrinth, I found myself wondering what giant or what gnome lived there. Had the guardian spirit noiselessly followed us? Or had he gone before us?

Aerial view.

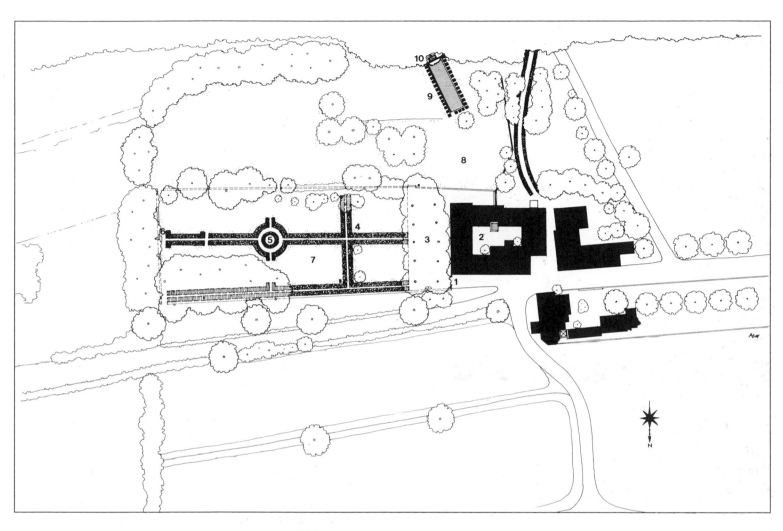

Garden plan, Châteauneuf-le-Rouge.
1. Entrance gateway - 2. Château -
3. Terrace planted with plane trees - 4. Tall, clipped box hedges (3 meters high) -
5. Circular garden path - 6. Bench - 7. Site of former kitchen garden - 8. Large meadow -
9. Basin surrounded by crenellated box walls - 10. Fountain and sculpture (Lion's head).

Main axis, maze, view from above.

Turning in an alley, I came upon a strange contrivance: a combination giant ladder and stepladder with a great number of small wheels that made it look like a tank. It was stationed near a wall of box. I learned later that this was the gardener's secret weapon. Gaby, the gardener who has worked at the château since 1953, uses this mobilized ladder to carry out all the work of trimming the box hedges. From August to January he practically lives in the park. Rushing his adjustable lad-

Main axis, maze, view from château.

Gaby, the gardener.

der before him he begins his work on the box hedges toward the end of August: ''Box should not be trimmed during the height of summer nor during heat waves. If one trims with the sun it gets crown gall, and then dies.'' Every morning during this period Gaby climbs his ladder. Slowly, methodically, with the help of the ladder, the box is trimmed from bottom to top, step by step, without rope or string. Good trimming takes knack, a special kind of motion that Gaby has taught himself: ''You see, when I trim, like that, at the very first glance, just by looking, I see right away the leaf or small branch that has to be cut, so when I get to it, I cut it; it's instinctive, you hold the tool like that ... one learns and learns ... I taught myself everything, it comes naturally to me to do it like that.''[2]

His employers wanted to buy electric clippers for him, but he feels this would be dangerous: ''Hand clippers can't hurt you.'' This is Gaby's technique. He advances continually and never looks back. He goes up the ladder step by step, trimming both sides of the alley from the same step. ''When I get to the end of the alley, I turn back toward the château; afterward there is the big pool and then there's the entrance.'' For six months Gaby reshapes the labyrinth. Each year sees him start off on the same slow journey during which, clippers in hand, he traces and retraces the original design. The circuit is always the same, but each year the amount to be trimmed increases by several millimeters.

Although the gestures made by gardeners may seem merely repetitive, they are not artless. With a movement of ever-growing amplitude, Gaby reclaims contours, and like a patient and inspired builder, he imposes form on a work of art, on strange walls that are alive.[3]

. . . .

Box wall sets off the rivage of the Cengle plateau.

Pool, crenellated boxwood wall.

Saint-Marc-de-Jaumegarde, the garden without a "bastide"

The absence of a *bastide* makes the garden of Saint-Marc a very singular place. It is unusual to come upon a lone garden in the countryside. This garden, although abandoned, imposes its strange presence on the surrounding fields and woods. One can imagine a *bastide* having once been here or invent one, as did J.-L. Vaudoyer,[4] but there is no evidence of a building of any importance inside or near the garden, except for a small pavilion near a nymphaeum. If one thinks of the garden of d'Albertas at Bouc Bel Air, one could suppose that there was a project for a house that was never carried out. However, the garden belongs to a neighboring château on the other side of the road, so it is much more probable that it was created for its own sake, completely independent of any immediate residence. Such a disassociation was quite common in the late eighteenth and early nineteenth centuries, when the park became a place to discover natural wonders during a walk, and not a complement to the house. This phenomenon was much rarer in the mideighteenth century.

At the central point of this garden, in a place usually reserved for the *bastide*, is the nymphaeum, a fine example of classical architecture. This monumental niche

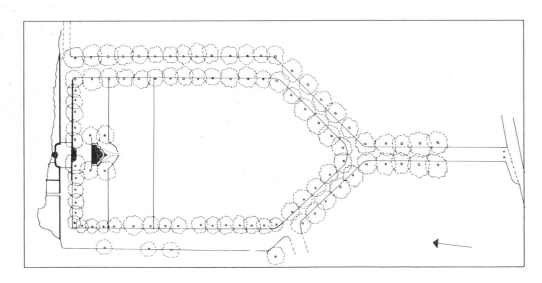

Lower basin of the nymphaeum, detail.

Garden plan, Saint-Marc-de-Jaumegarde.

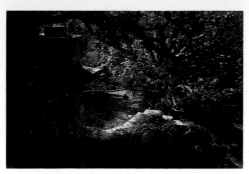

framed by two pairs of pilasters is set against a small cliff where, until quite recently, water gushed out of the mouth of a fish coiled round the foot of a beautifully decorated urn. Although this urn has since disappeared, this place still retains the magical quality evoked by Vaudoyer who wrote at the beginning of this century: "The most beautiful fountain in the country around Aix is probably the one to which Edouard Aude drove me one fine day in May; it is on the left side of the road in the direction of Vauvenargues and is hidden from view by a cluster of tall trees. This beautiful place belongs to the Château St Marc whose four massive towers protect a tiny village, from which one has one of the most exhilerating and breathtaking views of the Mt. Sainte-Victoire. In front of this secret, hidden fountain is a curved meadow, lush, as full of water as a field in Normandy, and its grass swarms with *orties-mouches*, small flowers with brown velvety bodies and two wings, whose mauve tint fades away into the palest of pinks. A walk has been laid along the fringe area at the bottom of a dry wooded slope. Here are found in juxtaposition the two elements of the Provençal countryside: on the one hand, low-growing vegetation, wet, almost marshy ground (the kind of land that did not exist in Provence before the creation of the numerous ingenious irrigation canals) and, on the other hand, a tough vegetation, economical, almost immortal, which seems to be nourished not by water but by fire. Pines, holly oak, and laurels overhang and shade the long wide walk, which would make a fitting setting for the gallant and leisure-loving personages of Watteau or Monticelli. In the center, set against a small artificial mountain, is a simply designed water tower, a large arch that shelters in a high niche a gigantic vase, very ornate, that may have been designed by Chastel of Aix. The nymph of the shrine is not portrayed in person. It is enough to have dedicated this beautiful vase to her, at whose foot she lets flow her crystal blood.

In days gone by, the water continued on its journey around the terraced garden, where the inhabitants of the burning-hot château came to chat in the cool. The outlines of this garden are now almost all effaced. Only the main decorative architectural element still survives and protects the flowered urn. A local stone was used here; the beautiful stone of the quarries of Bibémus not far from St Marc. This stone seems to have been steeped in honey or in oil, in substances born of a long agreeable relationship with the sun. Almost all of Aix is built of this stone which has, in its natural state, the shining tone the Romans tried to obtain and obtained in the coloring of the stuccoed walls of their palaces."[5]

. . . .

Upper view, nymphaeum.

Nymphaeum amid foliage.

The garden of Entrecasteaux, for the view

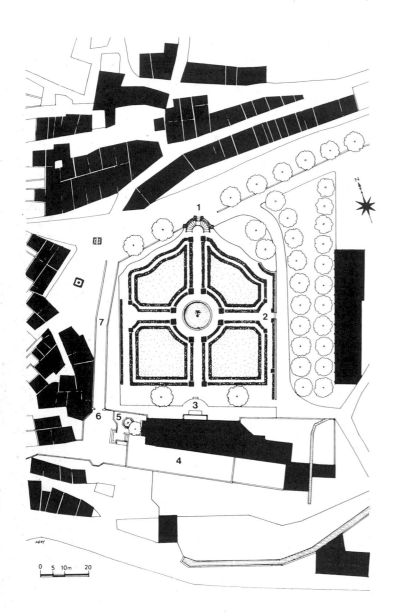

The garden of Entrecasteaux is not difficult to find; it is situated right in the heart of the village just below the château that complements it with its imposing mass. Very little is known about this small garden with its geometric parterre. If the date inscribed on the entrance gate is to be believed, the garden was finished in 1781. A very gripping story attributes it to Le Nôtre who died in 1700. Le Nôtre is said to have executed the plans for it without seeing the site, on the request of the Marquise de Sévigné whose son-in-law M. de Grignan was the owner of the château.

The château, built on a rock, dates from the fifteenth and seventeenth centuries, and it is reached via a ramp from which one can admire the garden. It was occupied in the eighteenth century by J. B. Bruni, marquis of Entrecasteaux and president of the parliament of Aix (his son was the famous Admiral d'Entrecasteaux).

If you also visit the château, you can ask Mrs. McGarvie Munn, owner of the château since 1974 to

Garden plan
1. Entrance gateway and staircase -
2. Clipped box parterre and lawns - 3. Main balcony overlooking garden - 4. Main terrace - 5. Forecourt with octagonal basin
6. Gateway - 7. Access ramp.

Parterre, view from château.

open the central balcony for you, from where you can admire the garden.[6] This is by far the best viewing point. Like many French-style gardens, the garden of Entrecasteaux was designed to be seen from above. Enclosed by the château and by moderately high walls, it fits between the château and the village in a hollow, enabling a bird's-eye view.

The parterre is quite simple. It is composed of four sections of lawn whose contours are formed by a double palisade of box which in the past was about two meters high, and so hid the visitor from sight. In the center, at an alley crossing, there is a round ornamental pool with a water jet. Two box borders follow the curves of the wall between the garden and the village. The garden is reached by a double flight of stairs. In wintertime the bare trees reveal the precise geometry of the box hedge with its dark shadows. In the spring the vista is softened by the rosy blossoms of the Judas tree, and with the approach of summer the garden becomes more and more shaded, finally providing a very pleasant verdant bower extremely restful to the eyes.

Today, this garden seems out of bounds. People rarely venture there, although it is now public property. The suggestion of a past royal presence in this far away locality still seems to keep people at a distance. To go down into the garden means facing the gazes of the château and of the village. The inhabitants of Entrecasteaux still prefer to lean on the enclosing walls and contemplate the garden like birds on a perch than to use the fine stairway which is now theirs. The architecture of the place is stronger than they are.

. . . .

Moulin Blanc, a northern dream

In an arid and rocky countryside the park of the Moulin Blanc enjoys the same "botanical miracle" as the forest of Sainte-Baume. Likewise, it is planted mostly in beech, which is not a common wood in Provence, though it is landscaped in the picturesque style of the midnineteenth century.

In 1850 this wet and richly wooded countryside had all that was necessary to inspire the owner, Alphonse de Saporta, who had completely absorbed fashionable ideas on English gardens and alpine landscaping. For de Drée, a Parisian landscape architect of the picturesque school, this commission in the Midi offered an unhoped for, though not ideal, site.

The park is shaped like a conch, orientated toward the east, and it slopes toward a meadowland steeped in water. An extensive network of canals, a legacy of a former flour mill, crosses the property. Two rivers, the Huveaune and the Péruy, meet at the edge of the park and constitute its boundaries to the north and to the east. A small canal dug between the two rivers turns the meadow into a kind of island, in the center of which is a man-made pond.

On approaching the village of Saint-Zacharie by the highroad, the visitor can see from afar the turrets of the château of the Moulin Blanc emerging among the trees. If you have permission to visit the park, you follow the long avenue lined in plane trees to the château, and on arriving at the terrace you will be able to admire a dream castle, designed by Revoil, the kind flanked with four turrets in which children delight. The park is not visible from the front of the château; this was intended by the landscape architect. Only after advancing to one edge of the terrace is it possible to see it.

A break in the thick woods covering the slope provides a vantage point for viewing the meadowland below. In the midst of this clearing is a clump of giant bamboo soaring skyward, stems swaying in the wind, not far from a pond covered with water lilies. The path zigzagging down to the meadow leads past a grotto overgrown with moss, where the burbly rhythm of the water

Aerial view, Moulin Blanc.

Plan, Moulin Blanc.
1. Entrance gateway - 2. *Bastide* - 3. Dovecot -
4. Grotto - 5. Large meadow - 6. Pond -
7. Massed bamboo - 8. "Small Sainte-
Baume" and its "cliff" - 9. Pine grove -
10. Oratory - 11. Pools-Reservoirs -
12. Waterfall.

in the canal is heard against the rushing sound of the two rivers that surround the park and the roar of the cascade of the Huveaune, where trout are still caught only thirty kilometers from Marseilles. Alphonse de Saporta's first act, on leaving his neighboring property of Montvert for the Moulin Blanc, was to plant an alley of plane trees along the highroad to shelter his property from the dust of cart traffic brought there by the wind.

Next came the work on the park and on the château. Alphonse, and especially his son Gaston—one of the founders of the science of paleobotany—were responsible for extensive plantations of trees. Among these were many varieties of exotic species: many oak trees and maples, different kinds of beeches, conifers, sequoias, bald cypresses, tulip trees, all of which are perfectly at home in this corner of Provence. Certains views of the park are as foreign to a Provençal as the drawings representing the landscapes of the Quaternary Period which the paleobotanist Gaston de Saporta has authored. There is some kind of kinship between the park and those scientific images of the past. You cannot leave the Moulin Blanc without realizing the extent to which it represented a privileged place at the end of the nineteenth century. While in the neighboring towns the *rocaille* builders were trying, with much use of imitation wood and imitation cascades in cement, to make *bastides* sweltering under the sun look like chalets, the Moulin Blanc offered the possibility of realizing the most difficult dream of all in Provence—creating a more northern clime.

. . . .

Sectional view.

Pond.

The garden of Gardini

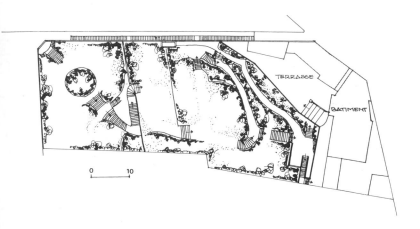

This is a garden of *rocaille*, almost impossible to find. One really has to be born there not to get lost in the labyrinths of Marseilles. Always bordered by walls, the endless stairways are known there as *montées* and the narrow lanes as *traverses*. In Marseilles, one mounts and traverses a great deal, especially in the neighborhood of the Roucas Blanc.

On this hillside overlooking the sea the whiteness of the rocks has finally been covered with pine needles. The *rocaille* builders of the last century were commissioned to create a new decor here, to render more dramatic a nature that already had its own strong personality. Assembling rocks and tons of cement, all of it carried up the hill on men's backs, they redesigned the mountain according to their fancy, constructed grottos, decorated the façades, cooked monstrous cakes in the sun to nourish the weekend dreams of the Marseilles middle class.

Along the crossroads that lead to this exceptional picturesque garden, it is unusual to see another car. Any car that ventures into this maze will be blocked in a dead end, from which the only escape route is one of the stairways that plunge straight down to the sea one hundred meters below. The car that had just passed mine, although I was squeezed into the recess of a gateway, was just on the verge of being swallowed up by the automatic doors of a garage when I asked my now-ritual question, a question that my astonished interlocutors usually asked me to repeat: ''There isn't a *rocaille* garden behind this wall, is there?'' ''Indeed there is, and the most beautiful one in Marseilles!'' replied the owner. The lovely mistress of the garden was not mistaken.

The garden was restored by Tobie Loup de Viane, and so does not have that abandoned look that most *rocaille* gardens have today. If one arrives at the base of the garden via the old gateway, the effect is even more astounding. The *rocaille* builders must have covered at least seven terraces with cement. Recreating the landscape, they made stone boulders, dug out a succession

Garden plan: six terraces laid out with rockwork.

Below the terraces: grottos adorned with peculiar scenes and figures.

Signature of Gaspard Gardini.

of grottos intended to provide shade or to hide the practical jokes of the *rocaille* builder.

On the first terrace there is an enormous sea bass hanging from a rock, which seems to be waiting for someone to carry it up to the kitchens. Inside the first grotto, a craftsman had discreetly slipped his visiting card into a dark corner; there engraved on the rock in Roman characters one can read: *Gaspard Gardini, Rocailleur 1892.*

In the depths of the second grotto, a besieged soldier has retreated behind the ramparts and aims his two cannons at the unfortunate visitor; uneasy in his role of enemy, the visitor continues his explorations and passes under a portico disguised as a rustic grotto. Each arcade is a pretext for some new spectacle. Switching ceaselessly from the gigantic to the minute, the *rocaille* builder quickly transformed the itinerary into something resembling *Gulliver's Travels*. The sensation of feeling

Grottos beneath last terrace.

dwarfed at the entrance to this monumental grotto becomes quickly reversed on going into the dark and discovering little by little a marvelous corner, where a tiny Nativity scene is set out. Awaiting the month of Christmas and the arrival of the *santons*,[7] the windmills turn slowly, driven by water that comes out from the depths of the real rock. In the corner of this scene an amiable monster emerges from a sort of petrified wave, where the visitor can sit and rest before going to the upper terrace—and a new grotto.

On the upper terrace stalactites surround a deep opening that is rather disconcerting because it is closed off with iron bars. It is in fact the ventilation for the cellar. The house is approached from below. Therefore on the upper terrace which is bordered by a balustrade, the imitation rocks are more discreet, and they disappear under the geometric form of the greenhouse. The last swellings of the *rocaille* are sheltered by a small smooth-walled pavilion. The artist-craftsman has placed the best part of his exercise in visual deception as near to the house as possible. At the window, a life-sized Pierrot (a clown) sadly contemplates the rocky landscape. The scale of this is diminished by a few meters in order to show a large expanse of the Roucas Blanc with its hills, where near a small cottage by the sea a fisherman is pulling in his net. The shock comes on turning round and going out into the sunshine. There in front of you is exactly the same scene as the one depicted in the small theater. From above, the successive terraces conceal all the artificial contrivances; the view seems the same as when the Greeks first came here over two thousand years ago. The altitude raises the level of the horizon in this countryside open to the sky and to the sun.

. . . .

The garden of Mas Calendal, an inspiration of a self-taught landscape gardener

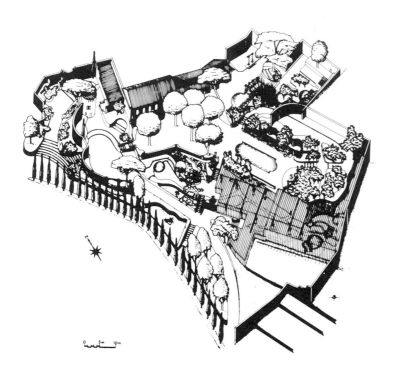

One does not expect to find a garden behind these high surrounding walls that form a veritable rampart against the mistral. However, the gentleness of the contours of the walls following the curve of the land suggests that there is something lovely hidden inside. This garden which faces south has a view of the sea from its western side. It is protected by the southerly slopes of a semi-circular screen of hills and enjoys a privileged local climate.

The garden is situated on terraces built on steeply sloping land, especially around the house, constructed on the northern edge of a slope. The parterres, the reflecting pool, the walls adorned with vases, the pigeon house, the carefully chosen decor, all bear witness to a very real desire to integrate all the outward signs constituting a *bastide*. The creator of this garden was a fervent defender of Provençal traditions, both a vintner and a botanist. All his memories and all the lore he gathered throughout Provence went into the creation of his garden. Born in Marseilles in 1881 of an architect father, he inherited this property from his grandfather and came to live there in 1920. At that time there was only a modest country house, whose main building he renovated and enlarged.

He had the peculiarity of being himself an agriculturist, a sort of Provençal gentleman farmer and also a self-taught landscape gardener. He was among those people who rediscovered the charm of rural houses and

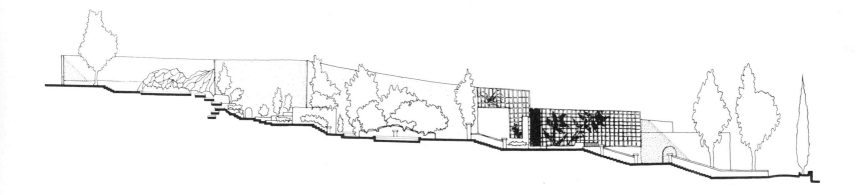

The pool.

changed and improved them by recalling all the remnants of their aristocratic and middle-class culture. How did the manage to finance his daring transformations? ''By selling wine,'' was his reply. Thanks to his white wine which is still world famous, this knowledgeable wintner managed to realize his dream of making a house and a garden exactly as he wanted them and also to indulge in his numerous passions: the collection of plants, birds, objects of art, and, among a host of other things, rocks, some of which are placed in a garden where a *rocaille* grotto, a Japanese garden, and copies of two antique columns from Arles appear side by side. Without a compass or a ruler, without a T square or a plan, completely without a preexisting symmetry, but with an exceptional grasp of the possibilities of the site and a capacity for making concessions to the exigencies of the land, he accommodated the growth of the vegetation by the curves and contours of the stone walls and the greenhouses he built. A lover of exotic plants, he managed in 1924 to plant them in the open ground. In 1952 he began to build the first greenhouse, which was followed by others up until 1956. From being a vintner, he became a botanist. His professional life, which brought him contacts from all over the world, did not interfere with his personal life. On the contrary, it helped it. ''By exporting all over the world, I had contacts with all the oceangoing liners and the maritime agencies and, through them, with the directors of botanical gardens.'' The extraordinary variety of plants found in the garden was therefore due to the old tradition of navigators bringing back plants.

This astute agriculturist-cum-gardener had the gift for creating the necessary climatic conditions where they did not exist, constructing screening walls, glasshouses, and orangeries, and even carrying out a special system of watering for plants accustomed to the monsoon. The garden thus attracted many visitors, among them garden-lovers, researchers, and specialists in the accli-

Greenhouses.

Greenhouse interior.

matization of plants. Apart from the pleasure it gave, this garden was also an experimental research station in its own way, and it supplied the Faculty of Science with a considerable amount of precious material.

The paradoxical nature of this multifaceted personality may seem surprising. On one hand, here was the botanist in his garden trying to introduce exoticism into Provence and, on the other, the fervent lover of all things Provençal, trying to preserve the traditional values of the architecture and language of Provence. He always treated his visitors to a lesson in Provençal, demonstrating its superiority over the French language. Like Frédéric Mistral's hero, the Provençal poet, he fought a solitary battle. His creation is in the image of his contrasting personality, a Provençal Don Quixote, a man of action and a dreamer, a businessman and a storyteller, a vintner and a lover of Provençal lore.

Vestiges.

The parc de Saint-Bernard, a cubist garden

Proudly built over the ruins of a feudal castle which dominated Hyères, the Villa Noailles, its garden, and its park are representative of an epoch: the twenties. The simple volumes of its modern architecture, which respected the existing trees, harmonizes with the medieval ramparts. It looks rather like an ocean liner grafted onto the enormous wall and gives back to the site all its ancient arrogance. During its halcyon days the yellow-and-red flag of the Noailles flew from the observatory tower topping the construction, floating in the vital wind of the gay twenties.

They were both young, sporting, attuned to the artists of their generation—the writers, the Surrealist painters, the musicians, the sculptors, the avant-garde film-

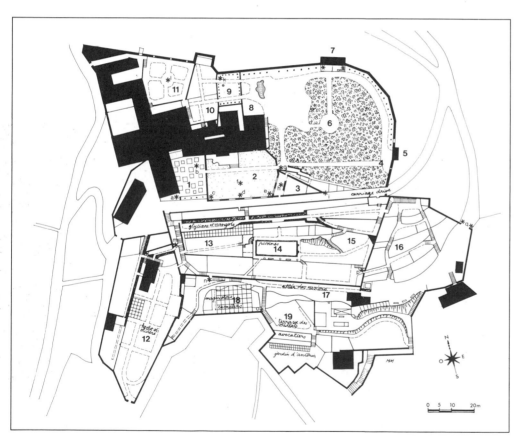

Places.

1. *Cour des pieds carrés* - 2. Green courtyard - 3. The Guévrékian garden - 4. "Pigeons' walk" - 5. Dovecot - 6. Maquis - 7. Salon and vantage point - 8. Enclosure - 9. "Lions' courtyard" - 10. "Parrots' terrace" - 11. Cloister - 12. Currel rock garden - 13. Olive terrace - 14. Peony garden - 15. "Garden of the demoiselles" - 16. "Le Golfe" - 17. The rose walk - 18. Covered terrace - 19. Carnation terrace.

Sculptures.

a) Life-size statue "The White Lady"- b) Ceramic bear - c) M.-L. de Noailles, reclining (wrought iron by Dominguez) - d) The Leaf (ditto) - e) The Cat (ditto) - f) White marble phallus - g) Sculpture in the early twentieth-century style - h) "The Cat," by Dominguez - i) "The Couple," a rotating metal sculpture - j) Two white classical statues in niches - k) Two reclining lions in captivity - l) Bronze of a man's head on a marble pedestal - m) Abstract - n) Two

statues - o) Head of a Gaul - p) "The Pregnant Woman," abstract, rounded block of stone - q) Two white lions, reclining.

The Mallet-Stevens villa.

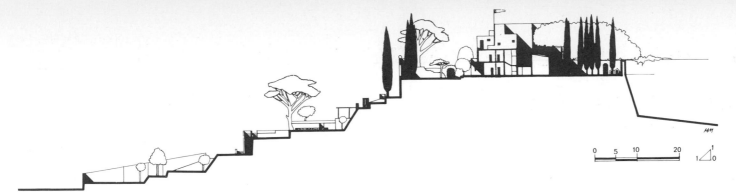

makers. To realize their dream of constructing a winter villa on the Côte d'Azur, Charles and Marie-Laure de Noailles engaged the architect Robert Mallet-Stevens. It was his first commission.

In 1924 Charles de Noailles was already a great gardening enthusiast. The approaches to the villa were as carefully planned as the actual building. Mallet-Stevens also designed the courtyard framing the southern façade, and in 1926 Gabriel Guévrékian was commissioned to design a triangular garden. Meanwhile the master of the house was making his gardening debut in the "cloister" to the north, the grove to the east, and the terrace to the south. In the house, flowers were greatly appreciated. A room that had completely bare walls was especially set aside for flower arranging. The floral compositions were more important to the decor than the numerous pictures by Picasso, Braque, Juan Gris, Chagall, Ernst, and Miró, which were only brought out of the cupboards on special occasions so as not to disturb the purity of the lines.

In 1925 the work on the villa was finished. Just as Charles de Noailles was about to turn to the garden and to the various other exterior amenities, he discovered all the novel creations of the Exposition des Arts Décoratifs in Paris. For sleeping outside and for the siesta, he ordered a bed from Pierre Chareau, to be hung outside on the terrace. It was, however, the "water and light" garden of Gabriel Guévrékian that particularly moved him. He was seduced by the beauty and insolence of this tiny garden whose modernism enraged the critics. He noticed also that its triangular form would fit very nicely into the space in front of the green drawing room, so he instantly ordered a copy.

The young Armenian architect, who was also asked to integrate a mobile by Lipchitz into his design, proposed an appreciably different organization of the available space. The garden-objet d'art could be admired from the terrace above, framed like a picture by the large openings. It was planned to enclose the space by a wall pierced by a bay window; it was opened toward

Sectional view.

Enclosed terrace opening on Guévrékian garden and framed panorama.

Plan of the Guévrékian garden.

The Guévrékian garden, looking outside.

The Guévrékian garden, looking inside.

Model of the Guévrékian garden.

the front. The garden was reached from the green drawing room under the terrace. At one of the vertexes of the triangle the *scachiera*, a checkboard of square ceramic tiles in red, grey, blue, and yellow, sloped up toward the couple of Lipchitz's embracing "Joy of Living," whose provocative motorized contorsions surprised more than one visitor. In front of the windows of the drawing room two dwarf Chinese orange trees, set off by black-and-white paving stones, framed the checker-board; next came a narrow pool prolonged on the axis of the statue by a short path, a *viottolo*, in violet mosaic. At the other two vertexes of the triangle, flower tubs of triangular shape were arranged in zigzags and filled, in the Japanese style, with plants of two contrasting shades of green. Through the use of animated sculpture, reviving the notion of automatism, but above all through the introduction of a succession of triangular espaliers that unfolded like an accordion and through optical tricks, Guévrékian filled his small garden with motion. Unfortunately, the mosaics could not resist the bad weather, and this precious geometric garden had rather an short existence.[8] Later it was obscured by an incongruous porch added by Marie-Laure de Noailles.

Until 1929 the garden was tended by five gardeners. One can imagine the atmosphere of the villa and garden during that fabulous period when all the famous artists assembled there to work under the winter sun. Man Ray gives us an idea of this in his film *Les Mystères du château du dé* which was filmed in the villa and its garden.

In the other parts of the garden Charles de Noailles devoted himself to his love of plants and landscaping. By 1936 he had become so skilled at it that Ernest de

Ganay in *La Gazette illustrée des amateurs de jardins* compared him to the prince de Ligne and at the same time J.-J. Rousseau. "Like the first, he has an insatiable appetite for gardens, and like the second he has a passion for plants and flowers. Luckily he is less uncivilized than the latter, and yet is not ready to lose his freedom as was the former, for he would hate to be a courtier or even a lounge-lizard. In short, he is the most perfect example of a garden-lover that we have ever known."

Antonin Settimione, one of the gardeners, said more or less the same thing in simpler terms: "Monsieur le Vicomte, when he came back, whatever the hour, went round his garden, even at night, before going to bed."

. . . .

Plan, Cour des pieds carrés, *after Charles de Noailles' gardening log.*
1. Judas tree, 1930 - 2. Mulberry tree, 1955 -
3. Parasol pine, 1928 - 4. Cycas, 1929 -
5. Mahón boxwood - 6. Medlar tree -
8. Mulberry tree - 9. Enormous cypress tree -
10. Cypress tree, 1952 - 11. Pear tree -
12. Hybrid buddleia - 13. Magnolia -
14. Wisteria - 15. Medlar tree - 16. Cycas -
17. Wisteria - 18. Mulberry tree - 18. Dead mimosa - 20. Bushy-topped Pittosporum tobira.

A creation in harmony with the Haute Provence

A cultural center was recently constructed in the Haute Provence. The landscaping was carried out by Henri Fisch, and the architecture was designed by Pierre Barbe. Transforming several hundreds of hectares into a landscape garden, where agriculture, the arts, literature, the sciences, and music can in a sense grow on the same tree, is a fantastic challenge. Many years of patience and devotion were necessary to realize this project, to decide on the appropriate design that would be harmonious with the site.

The approach to this center in the Haute Provence immediately gives one a panoramic view of the whole Var forest right up to the blue mountains of the Maures, which rise up in the distance like petrified waves. To get to the building compound, one has first to find a way through a maze of small roads. On entering the property, the visitor's attention is drawn at once to the judicious modifications carried out in the landscape: impeccably restored terraces, areas cleared to prevent fire, double rows of lavender accompanying the lines of olive trees, long hedges of cypress. The cascade, which can be glimpsed on arrival, has not always been there. Around the buildings the land that lacked character or had suffered during the construction work was corrected, reshaped, and replanted according to plants' lighting needs: pine, heather, thyme, rosemary, and cytises in sunny places, and in the shade, the holly oaks and periwinkle.

The Grande Maison is a place for meeting people and for music. To the north, just below a terrace with olive trees, six plantations of box, each trimmed to form a small rectangular maze, whose alleys are paved with tiles from Clausonne, screen the entrance courtyard. Right from the beginning the mood is set: the design of the box commands one to be serious; the fountain asks one to listen. The southern side of the Grande Maison abuts three terraces that descend gently to the garden. The view ricochets from the paved terrace to the "Amphion" by Laurens. A path by the house leads down to a garden where Tuscany and Provence join forces to provide a scenic reverie. To reach the garden, the visitor must first pass through a small green bower with an Italianate fountain where there are planted long rows of cypresses. In the garden one walks along paths edged with the aromatic plants that grow wild in the surrounding countryside and through the traditional plantations of olive trees.

A second house illustrates once again the system adopted for the architecture here: courtyard, patio, and terrace, all impeccably constructed to prolong the exterior of a building, a concept that is both traditional and modern. Sculptures find here their ideal sites. The changing surroundings are a stimulating source of interest whatever one's humor or the time of day. Near the kitchen there is a small herb garden with square beds; diamond-shaped paving stones on the perimeter of the garden separate the garden from the countryside.

"Amphion," by Laurens, against the landscape.

"Capricorn," by Max Ernst, amid the heather.

To the west a water labyrinth provides a patio with coolness. Enclosed by a stone wall underplanted with yellow jasmine and yuccas, the patio is open to the sky and to the tops of pines peeking behind the wall. Beneath the terrace to the south, a royal couple is enthroned among the heather, the "Capricorn" of Max Ernst. Not far from the Grande Maison, a small wood of oak trees shades the Computer Pavilion.

Raised into the sun by Takis like a forest of javelins, strange rock-boring bits anchored firmly in the earth guard the pavilion. At the other end of the property enormous augers planted upright like extraterrestrials' antennas, seem to signal them. Going round them and descending in the direction of the view, the visitor arrives at a small loggia, where he can sit in the setting sun and read the very fine text of Michel Serres engraved on a large copper disc. Above the disc, a work by Takis seems to sum up the spirit of the place. Supported by two invisible wires, two agricultural implements remain suspended in midair, a few centimeters from an enormous black magnet that continually attracts them. One can really sense the tension here between the countryside and the architecture, and between the architecture and the works of art, a tension essential to all creative projects.

· ·

Sculptures by Takis, near the small hut cum loggia.

Rows of lavender between the olive trees.

The alpine garden of the Lautaret, the highest garden in Europe

This pass is not an ordinary one. In a land submerged by the eternal snows of the Meije, the Lautaret affords a passage between the southern and the northern Alps. At an altitude of 2,100 meters, it is still part of the upper forest zone that persists up to 2,300 meters. This site is unique; it is well exposed to the sun but also to the wind, and it is known for its very severe climate.[9] In the garden there is a monument commemorating the fact that the Antarctic expedition team of Captain Robert F. Scott tested its apparatus here during the winter before reaching the South Pole in 1912.

In 1894 M. Lachmann, a professor of botany in Grenoble, developed a project for creating a garden at the Col du Lautaret. Between 1895 and 1908 nearly 2,000 varieties were planted. In 1920 the site was threatened by the construction of a road over the Alps and a new location was proposed by the P.L.M. Company for

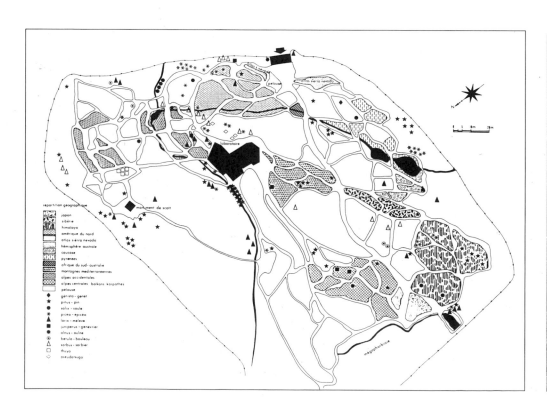

Plan, alpine garden.

Marsh marigold, ranunculus and Caucasian primavera. In the background: the Meije (3987 meters).

creating a model alpine garden. In the 1930s the garden comprised between 3,000 and 4,000 varieties which included an almost complete collection of western alpine plants. In 1950 when R. Ruffier-Lanche was appointed to restore the garden, which had been destroyed in 1944, only about 200 plants remained. Under his direction the garden soon possessed more than 5,000 varieties and enjoyed an international reputation. However, for eight years after Ruffier-Lanche's death the garden was neglected and once again the number of plants was reduced. Since 1981 the collections have been maintained by G. Cadel and J. Lestani. The great mountainous regions of the world are all represented and grouped in rock gardens. Plants from Japan, Siberia, the Himalayas, the Rocky Mountains, the Atlas Mountains, the Sierra Nevada, the Caucasus, the Pyrénées, and the Balkans are all neighbors here with the plants of the western and eastern Alps.[10]

General view, rock gardens of the Caucasus and the Pyrénées.

Marsh marigold, ranunculus, and Caucasian primavera.

Beauregard,
splendid isolation

Toward 1750 Charles de Villeneuve left his princely estates of Tourettes, near Vence, and La Napoule, and went westward to Beauregard where he founded the Villeneuve-Beauregard branch of his family. The château of Beauregard is built in the hills facing a vast forest range that stretches as far as the Esterel and the Maures mountains in the horizon.

Before of the château stretches a parterre of trimmed box. The terrace garden has not been altered since its creation in the seventeenth century, and it is one of three gardens in Europe, the others being the garden of Vignanello near Rome and the garden of Esquelbecq in the north of France, to enjoy this distinction. Beauregard is surrounded by fields and forests. The south-facing terrace is built at the foot of the château which protects it from the east wind; on the northern side it is bordered by a retaining wall followed by an open pavilion, and on the western side by a chapel. The parterre is divided by alleys into four parts around a central circle, thus reproducing the Cross of Malt in memory of an ancestor, Hiéron de Villeneuve, who belonged to that order.

In three centuries the box hedges, which are more than one meter high, have gradually encroached on the space left between the linear plantations that governed the design of the hedges, and today there exists only a very narrow passage between these elements of the original geometry. Here the art of topiary is practiced by a descendant of the Villeneuve-Beauregard family, the Comte de Clarens, who devotedly trims the bushes twice a year. These do not all grow at the same rate; some have had to be replaced, so not all the spheres of box that top the voluminous green shapes have the same size. But at Beauregard you may discover the extraordinary potentialities of sculpting greenery on an monumental scale. One never tires of contemplating them. The light plays on these motionless volumes, inside which the busy sap ceaselessly circulates. The density of this leafy surface is such that one is tempted to touch it.

At the end of the terrace, on the side of the chapel, four rounded arcades of a small gallery rise above the parterre. Was this an orangery? It does not seem ever to have been glazed. Open to the rays of the winter sunshine, less vertical than those of summer, and paved with old terra-cotta tiles, it provides a shelter for old people to sit, or for delicate plants, and a refuge for dreamers and for all the small inhabitants of that place, the furtive lizard or the lazy bumblebee who haunt this venerable garden.

Mountain garden.

Arcades of the small winter garden, set back from parterre.

Box parterre in Maltese cross pattern.

Parterre, detail.

This tower overlooked the sea and later became the Clewses' mausoleum.

"Poet, sculptor, author, chivalrous knight of La Mancha, supreme master of the Humourmystics..." This was the epitaph that Henry Clews wrote for himself. Son of an American banker, he refused to pursue a formal education, preferring to go his own way. He held several exhibitions in New York, but art critics found his work difficult to understand, for it did not correspond to any school with which they were familiar. Clews met Rodin in Paris, and the later's influence is perceptible, but, generally speaking, Clews was less interested in the work of his contemporaries than in the grotesques carved by medieval sculptors and the exotic qualities of Mexican, African, and Indian art. His most characteristic pieces are satirical portraits of decadent individuals. He stalked his subjects in the gambling halls of the casino in Cannes, his favorite hunting grounds. He produced creatures that were half-divine, half-animal, that would have been at home in fantasy tales or science fiction. These bizarre figures were stationed in various parts of the house, to which Mary had successfully imparted a feeling of architectural unity, and all throughout the gardens.

The avenue visible from the entrance leads through a park planted with pine, eucalyptus, and cedar trees during the nineteenth century. In Mary Clew's composition the stately, linear approach to the château courtyard hides other garden areas behind double hedges. On the right there is the Roman garden set in a slight depression, with a rectangular pool and a row of columns in the foreground. On the left stands of trees and outdoor "rooms" carved out of the greenery create a feeling of spaciousness on either side of the avenue called the allée des Sarcophages, or sometimes the "tennis court," after the court that had previously occupied this spot. Between two knolls, a small bridge arching over a stone-paved walk leads toward a colonnade cut out of the surrounding wall to provide a view of the new harbor of La Napoule. Engraved upon a sarcophagus is a poem by Henry Clews celebrating the celestial blue eyes of Mary, his "angel" and "Virgin of la Mancha." Majestic conifers dominate a garden resembling a vast parterre of clipped cypress, laurel, yew, boxwood, holly, and euonymus in which specific shapes are repeated wherever paths and avenues intersect.

Beyond the arcade covered with creepers that leads to the courtyard in front of the château stands a bronze statue of the god of the "Humourmystics" inspired by Don Quixote de la Mancha. Near the cloister gallery stands a small statue, "Glik," an appealing, round-bellied figure perched atop a fountain. There are three other gardens in the grounds: the terraces, the mauso-

Roman garden. Pool restored by Christopher Clews, president of the Foundation, grandson of Henry and Mary Clews.

The pruning of the cypresses.

leum garden, and the garden of the Venetian well. South of the main building, narrow cypress-lined corniches look out over the water. This might be considered a garden of the elements, for the sound of the wind and sea dominate all else. After crossing a "thoroughfare" within the fortified château, one reaches a garden walled in by high, clipped hedges laid out on several levels. This leads to the Tower of la Mancha, the mausoleum, overlooking the rocky shore. At the opposite end of the château, to the east, there is a small, triangular green space containing a monolithic well. This space is planted with boxwood and entirely enclosed by walls, in which a single window opens up on the bay side.

Of the four gardens, the largest has changed the most in the last fifty years. With its basins, fountains, and great domesticated birds—ibis, storks, cranes, white peacocks, and flamingos—strutting across the lawn, the garden must have looked as if it belonged to a crusader back from fighting the infidels. It is not known precisely which flowers were planted there. Mrs. Henri de Vilmorin, who visited the estate in the Clewses' time, remembers a "white garden." Today it is planted with dark evergreens against the russet-red of the porphyry walls, and one can well imagine that a palette of snowy and silvery tones would have lightened this somewhat oppressive atmosphere.

Eventually, an art foundation was set up at La Napoule at the initiative of Mary Clews. In 1950 the Council of the New York State Department of Education gave its stamp of approval to this project. The Château is now a museum in which the work of Henry Clews (1876-1937) and other artists are exhibited. Its avowed objective is to encourage cultural exchanges between the United States, France, and all countries interested in the development of the fine arts. Christopher Clews, Henry's grandson, wishes to restore the gardens with his wife, Noële. They have also decided to organize landscape design courses each summer at the La Napoule Art Foundation.

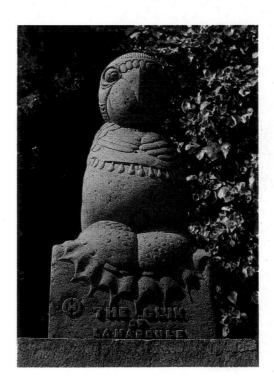

Garden of the Venetian well.

Sculpture by Henry Clews.

Rêve d'Or,
a Greek shipping magnate's
conifer collection

On a height called La Croix des Gardes, where one of Cannes' ancient guard posts stood in the past, there is an arboretum containing close to eighty types of conifer. The name of the park, Rêve d'Or, is also the name of an old-fashioned variety of pale-yellow rose. A botanical particularity of this garden is that its ecosystem has altered tree growth habits over time. Moreover spontaneous hybridization has occurred in several cases. This collection of indeciduous trees was started in 1928 by a tree-lover named Demitrius Stathatos, a Greek shipbuilder. The estate itself changed hands several times before and after his period of ownership. It is thought that the property was purchased in 1917 by a businessman named Paul Girod, who lived there during the construction of a villa on a hilltop on the La Croix des Gardes estate. It was rented by Winston Churchill in 1922 for a season and later belonged to a British officer. In 1965 a forty-apartment condominium was erected where the original house once stood, on the high point of the site, a promontory facing south.

Topiary art.

Woodland scene.

The present gardener-in-chief at Rêve d'Or, Jean-François Plantivaux, recounts that the *cultivars* imported from England and Japan were replanted without any assistance from landscape gardeners. Conifers are everywhere, and their fragrance fills the air. They line narrow terraces descending toward a large semicircular pergola covered with rambler roses. On the west slope some trees were sacrificed when the condominium owners decided to put in tennis courts. The southern slope, also covered with conifers, was ideal for shade-loving shrubs; camellias, rhododendrons, and azaleas provided bright touches of color from February to May. Under the rounded terrace of the old house there still exists an underground passageway. It leads to a building in a garden under the trees, through which run three paths to the swimming pool, one of the first to be built in Cannes. The approach is distinctly theatrical, with its pedimented gateway and large niches. Multi-hued cypress clipped into geometric shapes surround the paved pool area. This pool and garden space overlooking the sea is set on a promontory whose bow-shaped form and proximity to the water must have exerted a strong attraction on the Greek shipbuilder.

. . . .

Swimming pool.

Champfleuri and its gardens from around the world

The route de la Californie goes up a wooded hill to the east of Cannes. In the latter half of the nineteenth century it was a favorite promenade of the Riviera's winter residents, much appreciated for its view of the Iles de Lérins and the peaks of the Esterel. Beginning in 1880, a number of large villas were built on this hill.

These estates changed hands several times, and the largest were subdivided after the First World War. This was a common occurrence in many resort towns along the Riviera. After 1960 apartment buildings began to go up in the place of several houses on the route de la Californie. Fortunately, many parks retained much of their own identity even after the real estate deals were concluded and the buildings erected. The Champfleuri gardens have survived almost intact, although a condominium numbering no fewer than one hundred apartments has been built in their midst.

Champfleuri was planned to be a collection of Japanese, Florentine, Spanish, Provençal, Dutch, and Moorish gardens. Initially, there was an enclosed water garden whose playing fountains adding vital movement to the entire ensemble. Around 1912 Champfleuri was bought by Marino and Danaé Vagliano, a well-known couple who, as a perusal of society columns of the time will confirm, attended all the social events of the Cannes season. Marino Vagliano was an excellent golfer and transmitted his love of the sport to the members of the Mandelieu Club, whereas his wife was an ardent and meticulous amateur gardener. The villa, built some thirty years prior to its purchase by the Vaglianos, was in the Florentine style. The first thing Danaé Vagliano did was have all the palm trees planted by previous owners removed. She had the house totally remodeled and wanted her garden to be different from the gardens to be seen at neighboring estates. For the most part, her neighbors on the route de la Californie had parks with winding paths designed to set off groves of naturalized imported trees underplanted with borders, massed flowers, and flower beds. There was the occasional well-ordered Italianate terraced garden, examples of which

Moorish garden.

Pond in Japanese garden.

Stone lantern.

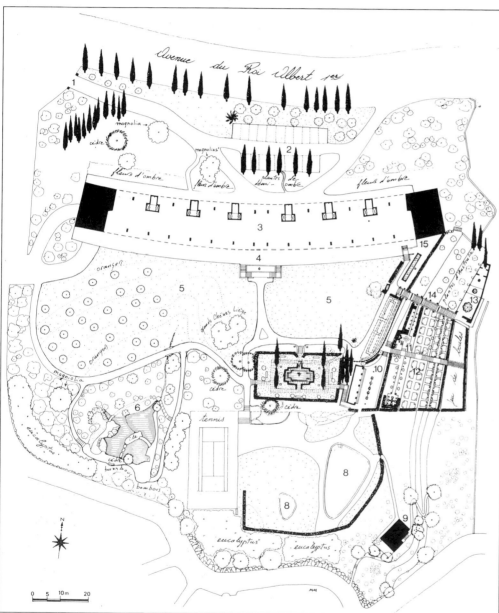

Plan of the today's garden.
1. Entrance - 2. Parking - 3. House on pylons - 4. Terrace under the house - 5. Lawn - 6. Japanese garden - 7. Florentine garden - 8. Swimming pool - 9. Head-gardener's house - 10. Spanish garden - 11. Provençal garden - 12. Dutch garden - 13. Moorish garden - 14. Stairway pergola - 15. Water garden.

Fountain in Spanish garden.

Florentine garden.

Spanish garden.

could be found at the Villa Fiorentina or Villa Isola Bella. Other gardens combined these two styles in any number of different ways. The Vaglianos broke with local habits and planned gardens that borrowed from different cultures and countries. At that time the concept of the "theme garden" was in the air. Countess Ephrussi de Rothschild was just finishing her villa in Cap Ferrat, which was surrounded by gardens of this type. In England, near Bournemouth, Compton Acres was designed somewhat later along similar lines. At the turn of the century Albert Kahn had begun work on a park in Boulogne-Billancourt near Paris which includ-

ed gardens done in a variety of different styles. He later went on to typical regional landscape gardens such as the celebrated Forest of the Vosges.

The Champfleuri entrance is located on the upper edge of the property, on the south side. The condominium stands just about where the terrace of the former house used to be, and the original garden layout has not been greatly disturbed. The new building was erected on bearing pillars, and one passes beneath it to gain access to the garden. In the foreground there is an English lawn which, although less extensive than in the past, still effects the transition between the edifice and the upward slope of the park on which the "collection" of small theme gardens is arranged. Across the lawn stands the grille of the Florentine garden, built to face the villa, which gives onto a rectangular space, bare of flowers and walled in by cypress hedges about six feet high. Bird-shaped boxwood, statues of children, and stone benches are arranged around a pool surmounted by a mossy fountain basin. Farther west, a spring emerges between a stand of large cork oaks and an orange grove and flows downhill in a trickle to a pond in the miniature Japanese garden where Danaé Vagliano used to raise pink flamingos and mandarin ducks. The pond, with its sharply defined, irregular edging, is the centerpiece of this garden set in a small hollow. Here is found an abundance of aquatic plants, bamboo, flowering shrubs, and conifers, not to mention a great number of decorative objects customarily used by western gardeners attempting to strike an Oriental note. A dozen steps lead away from the Florentine garden to the Spanish garden, a small patio featuring terra-cotta paving, a white colonnaded pavilion, and cypress arcades. Water sounds rise up from fountains sunk into the paving. Below the patio one comes across pool and tennis courts in an area which, in the Vaglianos' day, had been reserved for seed pans, cuttings, and greenhouses.

On the east side a walk paved in travertine takes one to two stairways interrupted by landings paved with

Paving, Provençal garden.

Paving, Dutch garden.

cobble-and-marble mosaics set in brickwork, which descend through the terraced Provençal and Dutch gardens. In the sunny corners, earthenware pots, arbor-shaded benches, and tables of the Provençal garden are arranged between two rusticated fountains. The Dutch garden features rows of brick-enclosed compartimented spaces, no doubt intended to evoke, when in bloom, orderly fields on the polders. One of the stairways cutting through these gardens is a pergola, its pillars supporting crossbeams covered with wisteria. In the past it ran along three terraces, each with its own color scheme. Flowers on the first terrace were mauve, followed by orange-tinted blossoms on the second, and variegated ones on the third. Slowly but surely, the steps lead the unsuspecting visitor to the secret Moorish garden. The three fountains that rose up from rose-shaped basins have long since ceased to flow, but the light of early afternoon still plays across the sculpted stucco surface of the garden walls. Above the stairway pergola, just at the foot of the condominium, under arching cypresses, one finds a narrow channel that had belonged to the old *jardin d'eau*; guests must have exclaimed in amazement as they strolled under soaring jets of water without getting the slightest bit wet. In 1928 *Country Life* published an article extolling the Champfleuri gardens. Its author thought it a refreshing change from the other gardens in the Cannes vicinity, which he found wanting in inventiveness and overly reliant on flowers. In his opinion, the small gardens at Champfleuri were well-proportioned, the paving admirable in its simplicity, and plant combinations had been handled with delicacy.

Today the park features nearly five hundred plant species and varieties, making it one of the most remarkable botanical collections on the Riviera. Many of these were introduced by Jean-Noël Demoulliers, whose ultimate ambition is to restore the neglected parts of the gardens to their original state. Perhaps the fountains in the Moorish and Provençal gardens will flow again someday.

. . . .

Red bridge in Japanese garden.

Villa Domergue and its magic soirées

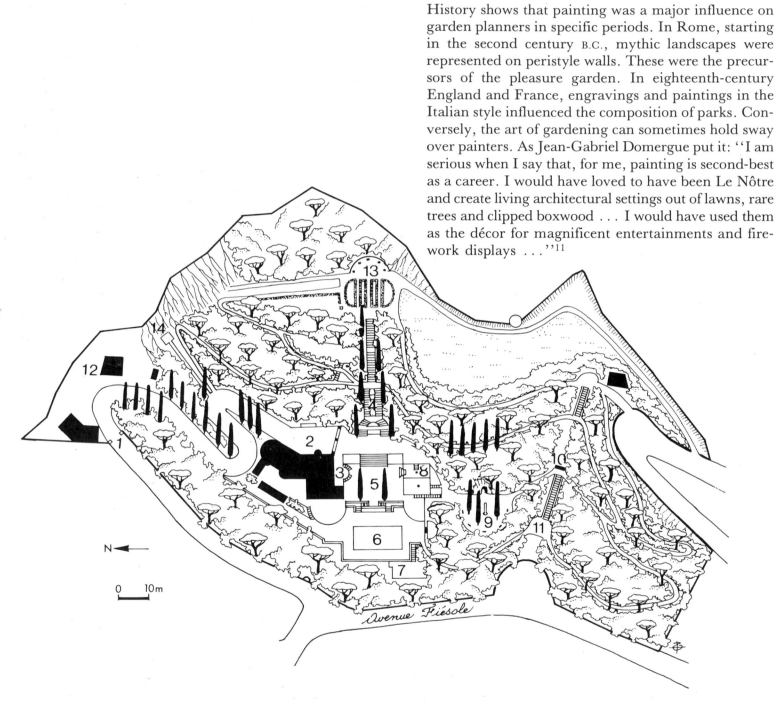

History shows that painting was a major influence on garden planners in specific periods. In Rome, starting in the second century B.C., mythic landscapes were represented on peristyle walls. These were the precursors of the pleasure garden. In eighteenth-century England and France, engravings and paintings in the Italian style influenced the composition of parks. Conversely, the art of gardening can sometimes hold sway over painters. As Jean-Gabriel Domergue put it: "I am serious when I say that, for me, painting is second-best as a career. I would have loved to have been Le Nôtre and create living architectural settings out of lawns, rare trees and clipped boxwood . . . I would have used them as the décor for magnificent entertainments and firework displays . . ."[11]

Plan
1. Entrance gate and housekeeper's lodge -
2. Court of the villa - 3. Villa Domergue -
6. Site of the former reflecting pool -
7. Entrance to the sculpture studio -
8. "Jardin de Priape" - 9. Terrace of the obelisk - 10. Odette and Jean-Gabriel Domergue's mausoleum - 11. Hemicycle of the Four Seasons - 12. Gardener's house -
13. Parterre of fragrant plants - 14. Replica of the mausoleum tomb.

House, view from terrace.

Domergue specialized in the female portrait. He was fascinated by women, their seductiveness and finery, all the outer signs of their feminity, and the way they looked both dressed and undressed. In 1920 he was the guiding spirit behind the first fashion show presenting fashion creations done by artists (Salon de la Mode par les Artistes). He designed settings for an impressive number of balls and gala affairs held in connection with the show. He seized this opportunity to create ephemeral "gardens," although it is stretching accuracy a bit to use the term "garden" for his "fashion garden" and "garden of Greek roots" creations. With his creative instincts thus whetted, Domergue did not waste much time before finding a place where he could give form to his flights of fancy and indulge his desire for playing waters and plants out in the open. In 1926 he purchased land on the steep west slope of la Californie in

Alley crossing waterfall.

Sectional view of the garden.

1. Entrance to the sculpture studio. -
2. Terrace of the former reflecting pool -
3. Archway between courtyard and garden -
4. Waterfall - 13. Upper garden.

Cenotaph stairs.

Cannes. The property had plenty of water to work with and an extended view over the bay of La Napoule. The painter opted to go back through history for inspiration to a style predating the classicism favored by Le Nôtre whom, as we have seen, he so admired. He went back to the Italian Renaissance. In 1918 Domergue had married Odette Maugendre-Villers, a sculptor. They decided that they wanted to build a house in which they could set up studios and receive both clients and friends. The result was the Villa Fiesole, where between 1936 and 1962 they held parties celebrated in Cannes for their gaiety and brilliance, and attended by personalities from the world of entertainment: producers, stars, and their sycophants. The house and garden still stand and bear witness to their days of glory.

On the approach to Villa Domergue, as it was called after Odette Maugendre-Villers left it to the city of

View of the terraces from above.

Sculpture, along the cascade.

Cannes, the painter's studio comes into view. It is a rotunda with high north windows which afforded even lighting all day long. The sculpture studio was located below a wide terrace around the house. An arcade connects courtyard and garden, in which an unexpected sight awaits: a flight of steps moving up the wooded hill toward the sky with, in counterpoint, a cascade tumbling down the center of the stairs. The gushing waters flow over a series of oval landings, cross various basins, and disappear, only to reemerge below as a fountain set in the axis of the wide terrace where there used to be a reflecting pool. This pool has been filled in, but one can imagine its calm waters, a fitting culmination for the playful descending movement of the cascade. The "stone-and-water" staircase is lined by statuary and columns interspersed with clipped cypress. South of the house an avenue leads to a section of the garden containing an arcaded pavilion facing the slope. Against its three arcades is a sculpture representing the Domergues in Etruscan guise executed by Odette Maugendre-Villers. A long stairway connects this cenotaph—the mausoleum is empty—and a hemicycle down the hill that contains four eighteenth-century figures of the seasons. By taking paths winding up the slope through the shade of Aleppo pines, evergreen oaks, cork oaks, arbutus, and mimosa or by way of the cascade stairway, visitors reach a herb garden in the upper part of the estate. Planted on a flat surface about halfway up the hill, the herb garden, whether in bloom or not, is deliciously fragrant all year round. The scents of rosemary, lavender, and santolina permeate the air. As one looks back down the stairs, one's eye follows the gleaming streams of the cascade from the foreground through a sequence of planes to the the Esterel range on the horizon. The waterfall has a plaintive sound in this deserted garden. But when the city of Cannes uses the villa for a reception and its terrace is thronged with guests, the garden awakes and once again works its magic, the legacy of the Domergues.

. . . .

Vista of cascade.

The Grand Jardin in "New Spain"

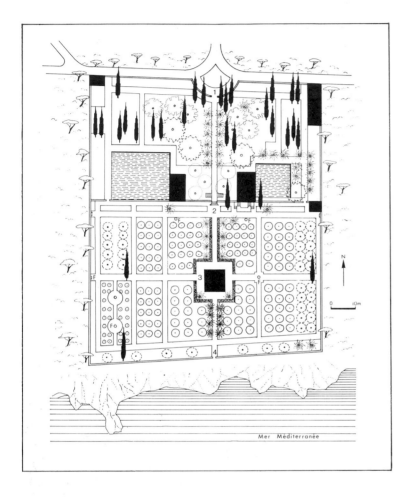

Entirely enclosed by walls, the Grand Jardin is almost perfectly square, with one side set along the shore of Sainte-Marguerite. The pine forest covering these islands surrounds the garden on the three other sides. On the high ground of the walled garden at the end away from the sea, cypress, cedar, and various types of palm tree tower over African plantings, fruit trees, grapevines, aromatic herbs, and flowering plants arranged without much regard for symmetry. The main alley runs toward two houses, each with an irrigation catch basin filled with clear green water. Between these houses an arcade opens onto a second part of the Grand Jardin. Several steps lower, orange trees in regular patterns are flanked by rows of olive trees. At the intersection of slightly raised alleys, bordered by rose bushes and a variety of perennials, rises a two-story monument known locally as the Oubliettes, or the Tower of the Grand Jardin. It looks like a small keep with four identical sides, each with its own door. Beyond this curious structure the central avenue continues to a gateway looking out over a strait separating Sainte-Marguerite from the neighboring island, Saint-Honorat. In 1834 the same year that Lord Brougham set about making the reputation of Cannes, Prosper Mérimée, inspector general of historic monuments, wrote a report in these terms: "The Grand Jardin . . . is said to be the hottest spot in all of Provence. In effect, plants are cultivated out in the open which, elsewhere, must be nurtured with the greatest of care. In the midst of this enclosure stands a bizarre edifice: it is square in shape and its sides face the four points of the compass . . . it has a round-arched door on each side . . . and suggests a central block flanked by four square towers. A terrace in ruins without a balustrade surmounts the structure . . . A square hole known as the 'Oubliettes' is set perpendicular to this platform . . . On the ground floor, one finds a domed room . . . with, in the corners, four empty, round-arched niches positioned several feet above the ground; each niche has a hole pierced in the lower part . . . Some say this was the work of cunning priests who wanted to create the illusion that the statues in the niches could talk. Nobody could give me the slightest information as to who erected this building or for what purpose . . . the walls are not very thick and there are a great many doors, so one can hardly suppose that it was used for defensive purposes. Yet it would be difficult to imagine anything more

Plan
1. Access to the upper garden - 2. Lower garden - 3. Tower of the Grand Jardin - 4. Portal opening to the sea - F. Fountain.

Tower, view from upper garden.

uncomfortable to live in; the upstairs rooms are so low-ceilinged that one can barely stand up . . . its form would seem to rule out a religious function . . . and if it originally had a commercial function, why would it have all those niches in it? . . . The shape of the doors, the construction of the vaulting and, to a certain extent, the type of masonry could all be of the twelfth century, but the three-centered arcades and square upper openings seem to be more typical of the sixteenth century. Monuments are generally dated from their ornamental detail, but here there is absolutely no detail at all.''

The author of Colomba well knew how to "force secrets out of monuments," according to Théodore de Banville[12]. Yet, try as he might, he was unable to establish a date for the tower in the Grand Jardin. Neither could he understand why it was built. He examined the building closely, noticing that its sides faced the cardinal points, but did not mention the overall garden plan.

One of the noteworthy events in the history of the Iles de Lérins was their occupation during the Thirty Years' War by Spanish troops. From 1635 to 1637 the islands, rechristened "New Spain," were inhabited by Spanish troops building fortifications so that Spain might retain possession of these strategically important little islands. Le Grand Jardin was chosen at that moment as a gathering place for Richelieu's army, but the tower was not mentioned in accounts dating from the time. It may have been built later by Jean de Bellon, a noble to whom the Duke of Guise, then governor of Provence and an admiral in the Levant had given the island.[13]

Whatever the origins of the tower and walled garden, the old irrigation system used at the Grand Jardin is reminiscent of the system used in gardens of Moorish influence in southern Spain. The arrangement of rows of olive trees bordering the orange trees—of which a section on the western side is missing—also corresponds to an old Spanish practice typical of the region near Valencia, where olive trees were planted to form a windbreak for the citrus groves.[14] Can one infer that

this was a consequence of the seventeenth-century invasion? Immediately after the conquest of Ile Sainte-Marguerite, the construction of a fort was undertaken on the opposite side of the island. As for the tower, its size and lack of apparent useful purpose seem to indicate a relationship to picturesque monuments, an allusion to associative monuments used to adorn parks in the eighteenth century. Moreover a keep built nearby could have served as the model: the fortified twelfth-century château on the shore of Ile Saint-Honorat where monks from the abbey took refuge in times of danger.

In answer to Prosper Mérimée's question, one might say that the Tower of the Grand Jardin, which was not a fort, residence, or commercial establishment, was simply the product of an architect's imagination. It was not built to "make the statues speak," but as a setting for pleasant moments outside on a terrace listening to the sea or inside, with all four doors opening out onto orange trees in bloom.

Plants typical of African oases.

Channel separating the Lérins islands.

Fortified monastery on the St. Honorat island.

The Villa Noailles:
a return to the source

There are places in Provence between the Esterel range and the Var valley where the landscape has always had a decidedly Tuscan look about it, although now the pace of land development is such that this impression might be but a fleeting one. Moreover this look also depends on the light conditions, the time of day and season of the year, and on the clarity of the air when the mistral wind is blowing or the day after it rains. One part of the countryside strongly reminiscent of Tuscany is the Saint-François district, west of Grasse. Slopes covered with olive trees stretch out beneath a limestone cliff, which provides a grey and ocher background. Here, the Villa Noailles—using the Italian meaning of *villa*, or the estate in its entirety—is perfectly integrated into the landscape of cultivated terraces. From a road on the next hill, it appears to be just another olive grove, yet in this garden lies a suite of compositions born of a true gardener's inspiration, memories, and love of plants. The theme of the garden is water, the vital element springing out of the plateau above, hidden by the cliff. ''Here, everything centers upon water from the source. I was able to put in quite a few fountains,'' said the Viscount of Noailles, almost ninety years old, as he spoke to us of his garden one day in November 1981. ''The avenue lined with Judas trees was the idea of an English landscape gardener,'' he told us. Then, smiling, he indicated his library: ''As you can see, all of my books on plants and gardening were written by English authors.''

It is impossible to describe the personality of Charles de Noailles, now deceased, in a few words. He and his wife, Marie-Laure de Noailles, were deeply interested in the arts in general, and in 1923 they gave expression to their own creative instincts in the construction of their Hyères villa, Le Parc Saint-Bernard,

Glimpse of the peony parterre.

Portrait of Charles de Noailles, by Jean-Claude Janet.

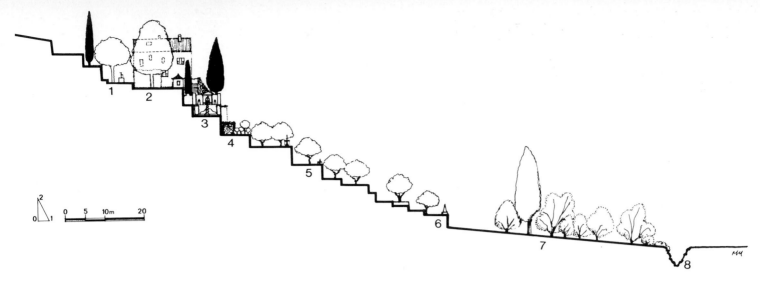

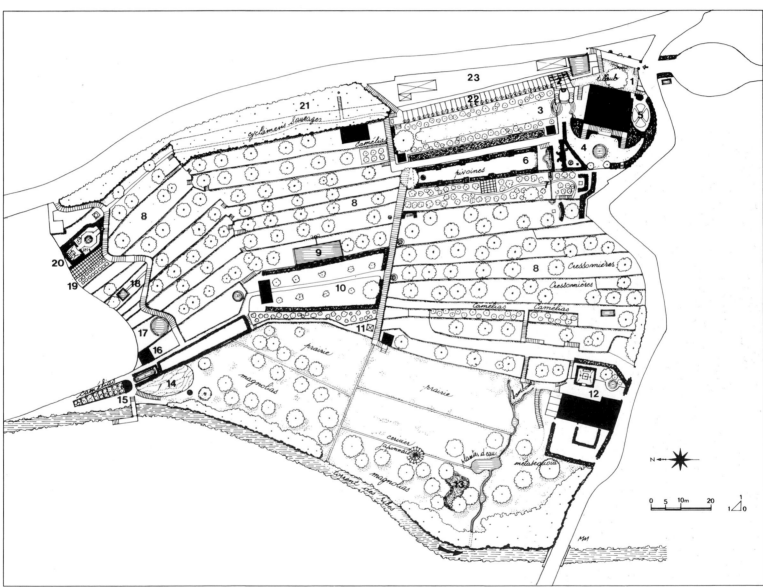

Sectional view

1. Nymphaeum - 2. Lawn-*allée* of the linden tree, matching pavilions - 3. Parterre of the obelisk - 4. Gargoyle - 5. Olive grove - 6. Pyramid - 7. Magnolia meadow - 8. Rushing water.

Plan (after Osamu Shimizu)

1. Nymphaeum - 2. "Petites-fontaines" - 3. Lawn-*allée* of the linden tree, matching pavilions - 4. Former terrace of the *bastide* and double staircase - 5. Oval garden -

6. Parterre of the obelisk - 7. Terrace of the gargoyle - 8. Olive grove - 9. Grand pool - 10. Garden of wisteria pavilion - 11. Pyramid - 12. Farm, old mill - 13. Peat-bog - 14. Rockwork of the mandragora - 15. Pavilion of the camelias - 16. Aviary - 17. Pool of the water-lilies - 18. Shaded pool - 19. Garden of medicinal herbs - 20. Garden of the octagonal pool - 21. Laurel tree grove - 22. Walkway covered by Judas trees - 23. Kitchen garden.

the first of their houses in the south of France. Later the viscount supervised the restoration of a house and garden he had inherited in Fontainebleau that had formerly belonged to the Marquise de Pompadour. In 1947 he moved into his house near Grasse, purchased some twenty years before at a "candlelight auction," as he liked to put it. He was a patron of the arts with a passion for botany. The Belgian landscape gardener

Outdoor "room" near the bastide.

Curving steps.

Terrace and arch.

René Péchère related that the viscount would go several hundred kilometers out of his way to "go see a flower that he had never seen before." With his reputation for knowing everything there was to know about the gardens of Europe, he was elected president of the French Society of Amateur Gardeners. For years he planned trips for this group and must have collected many ideas in this manner to bring home to Villa Noailles. Near the house one senses a strong architectural feeling. But there is a gradual shift in tendency in favor of natural landscaping as one moves outward, and the garden finally melts into the surrounding countryside. Pavilions, pools, and plant collections are deliberately separated to create an element of surprise as the visitor comes across each new garden space. Everywhere one walks, olive branches spread overhead. Before hay is made, paths are cut with scythes in the long waving grass. Stone steps in the walls take one directly from one cultivated terrace (*planche*) to the next, so that one does not have to return to the steep central paved alley—called a *calade* in Provençal—with fountains all along its length. Water is the central and vital theme of the garden. What a difference between this garden and that at Le Parc Saint-Bernard, where finding water for sprinkling was quite a headache! After passing through the entrance, visitors hear a background noise which they eventually perceive to be the tumult of fountains in the courtyard. Each fountain plays on a different register and seems to emanate from one of the statues in the grotto. Ferdinand Bac, who designed Les Colombières in Menton, discovered this *bastide* in 1908, called the Hermitage of Saint-Francis by local people. He was struck by the charm of this fine example of eighteenth-century rural architecture and would have liked to oversee its restoration. Tinted drawings accompanying his article on Mediterranean villas and gardens, published in *L'Illustration* in 1922, show that the gateway, the grotto (whose niches were empty), and the circular basin in front of the *bastide*'s façade already existed at that time. There were only a half-dozen narrow

Fountain.

terraces where later, in the de Noailles garden, there would be a lawn-alley, twin pavilions, and lower down, the peony parterre. When the visitor crosses the grotto courtyard today, he comes to a series of aligned fountains under a long overhang of rambler roses. A few paces further begins a long stretch of lawn set on the axis of an immense lime tree. Other fountains flow and gurgle amid the clipped boxwood but remain invisible. On the terrace of the house, an old round basin, whose urns are filled with the flowers of the season, provides, horizontal counterpoint to the tiled vaulting over the front steps, a more subdued rendering of one of architect Emilio Terry's wild inventions. Farther along, a

spray rising out of the obelisk splashes the peony parterre. In the olive grove a deep rectangular pool shimmers; this is where the viscount bathed. He found the azure-blue pools popular on the coast amusing and had this pool built for his personal use. It is as wide as a cultivated terrace. Like the old, perfectly proportioned rural irrigation basins which can still be encountered here and there in remote areas of the countryside, it gives the impression that it has always existed. As one reaches the lower part of the garden, an olive grove gives way to a meadow irrigated by narrow stone channels. Here, the alluvium has been planted with deciduous magnolias. In March and April each tree becomes a pyramid of waxy white flowers. Under the olive trees the meadow grass is full of narcussi, small anemones, and even fritillarits near the great metasequoia, plant-

Japanese cherry trees and magnolias in bloom.

Mandragora.

Gargoyle.

ed a short time after the "discovery" of these conifers in China in 1941. Then, in succession, one comes across a peat bog, collections of camellias, a rock garden, a herb garden, and a shady path bordered by expanses of wild cyclamen. An alley of espaliered Judas trees returns toward the house which, in April, forms a tunnel of bright pink and white. Or one can walk along the edge of a sweeping lawn bordered with tiers of flowering plants: trees underplanted with shrubs and massed flowers and a herbaceous border. Either way, one returns to the grotto by way of a short trip through space and time, for there are a number of references to other gardens along the path. The aligned fountains and obelisk may be taken to allude to the Villa d'Este and Villa Aldobrandini in Italy, whereas the lawn-alley near the house, with its mixed borders and symmetrical pavilions with their curving roofline and pediment sculpted out of boxwood, is clearly a tribute to Lawrence Johnston's garden composition at Hidcote Manor in England. Johnston came to play squash with Charles de Noailles at the Hyères villa. They had planned a botanical trip through Birmania together in 1938, but their plans had to be canceled as war became imminent.

The garden remained open to the inhabitants of Grasse for thirty years or so, until the summer of 1981. An inscription at the gates to the estate invited visitors to enter the grounds. On Saturdays wedding parties came to have their pictures taken. Yet the peace and quiet remained undisturbed, and "not a single flower was picked," wrote the viscount. The entrance to the garden and grotto, designed to extend the living space of the house, give visitors the feeling of being an invited guest. As George Sand once said of another park,[15] the universe created by Charles de Noailles is "as natural" as the surrounding environment while being "much better composed." It imparts an eloquent serenity. As Charles de Noailles phrased it to us: "I wanted to create a quiet garden, without extravagant effects or grand vistas, which would bring pleasure to the inhabitants of the house and visitors."

. . . .

Garden of the octagonal pool.

Pediment of pruned box.

Villa Victoria,
a Victorian park

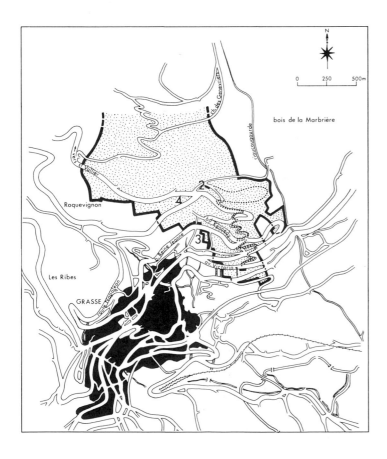

Yellow berets for apprentices, red for workers, navy blue for shift supervisors, and royal blue for foremen. At Villa Victoria, the estate owned by baroness Alice de Rothschild in Grasse prior to her death in 1922, there was a strict coding for headgear worn by gardening staff. Blue and yellow were the colors of the glazed-tile roof on the tea pavilion in the wooded upper part of the estate, and of the silks worn by jockeys riding Rothschild thoroughbreds. Some of the old-timers in Grasse remember seeing squads of gardeners heading home after work, berets firmly in place on their heads. It took a permanent gardening staff of fifty to maintain the vast park and thirty or forty extras to help prepare for the Season — from October to March—and the arrival of La Baronne, as everyone in Grasse called her. Every year acres and acres of land had to be seeded at the end of summer so that, when she finally arrived from England, La Baronne would be satisfied at the sight of impeccable lawns. But once the irascible Alice de Rothschild was settled in her winter residence, this legion of gardeners was supposed to be invisible. All horticultural work had to be laid aside during the hours of her daily promenade.

To the east of the house, the estate sloped up from avenue Victoria (350 meters above sea level) to the Napoléon Plateau (530 meters above sea level). The baroness transformed this site into an extensive pleasure garden. Although it is now subdivided into small lots, one can still imagine how impressive it was in size, because two very distinct buildings mark out the former property lines. The first, the Palais Provençal, is a former luxury hotel turned condominium; it is located where the villa used to stand. The second, the tea pavilion with the blue-and-yellow roof at the opposite end of the park, is visible from the route Napoléon running above the town of Grasse. In 1887 Alice de Rothschild, forty-two years old and unmarried, stayed at the Grand Hotel in Grasse, still in existence today, and for the first time saw the local landscape which would later be visible from the windows of her house. The tran-

The City of Grasse and the Alice de Rothschild estate prior to 1926.
1. Villa Victoria and grand garden - 2. Tea pavilion - 3. Garden of Princess Pauline - 4. Swimming pool.

quility of this little town removed from the bustling cosmopolitan colonies of the Riviera greatly pleased her. The following year she purchased land covered with olive trees on the Nice road. The construction of her villa was entrusted to the same builders who had already erected the Carlton Hotel in Cannes. The terracing was demolished, but most of the olive trees were left untouched. She planned to redefine the contours of the land for a garden containing all the marvelous plants the Mediterranean climate could produce. She had learned to love plants at Waddesdon Manor, the family seat in Buckinghamshire where her brother, Baron Ferdinand de Rothschild, had laid out the gardens which have been appreciated by endless numbers of visitors ever since they were given to the National Trust. She discovered a personal vocation for landscape gardening and launched a campaign to acquire all the land on the slope facing her house. Lot by lot she managed to purchase all the land stretching up to the foot of the mountain, and after several years, she was in possession of approximately 335 acres. The road to Nice, later christened avenue Victoria, cut her estate into two sections, passing behind the house. The first part was called the Grand Jardin, located south of the main house, some portions of which have survived to the present day. On the slope beyond the road, above another garden laid out where olive groves had originally stood, a natural landscape garden blended into a forest of evergreen oak and Aleppo pine. The composition of the Grand Jardin was typical of the day, relying on irregularity of form

Tea pavilion.

to enhance the selection of subtropical plants able to survive the climate of Grasse, cooler than that of the coast. Curving garden paths and avenues, ground level variations outlined by wide expanses of lawn, and a small artificial stream created a general framework into which palm, yucca, agave, large cactus, citrus, and acacia—commonly called mimosa—collections were set. Just off the drawing room was a winter-garden grotto for orchids and flowers forced to bloom out of season. In short, Villa Victoria, which owed its name to a sojourn by the Queen of England in 1891, had an extensive collection of plant varieties that had been introduced on the Riviera in the later half of the nineteenth century. Alice de Rothschild was extremely demanding when it came to landscaping. She took infinite care in studying how to juxtapose different types of foliage for effect, sought to procure the rarest of plants, and had full-size trees transported to her estate if she deemed it necessary. On the other side of the road, next to the Palais Provençal or, as it is known locally, "the Palace," standing where the villa used to be, can still be seen the house with red-and-green glazed-tile roofing where the estate steward used to live (45 avenue Victoria). This was the starting point for the picturesque avenue which wound its way up to the tea pavilion across the Rothschild land. Today this three-kilometer stretch of road still exists under the names of chemin de la Coste d'Or and chemin de la Corniche. From her carriage, the baroness could contemplate the more elaborate parts of the park: a great rock garden providing a suitable decor for cacti, a meadow carpeted with bulb flowers, and plantings of exotic tree specimens. One would pass the Villa Verte, an old *mas* ("farm-house") now called La Verdière; this was where gardener-apprentices of all nationalities lived during their apprenticeship on the estate. There, where the boulevard de la Reine Jeanne is now, stood a wooded fringe in which many clearings were cut to afford a better view for the baroness. Below the woods different rock and plant compositions and a cascade completed a pleasant sequence of landscapes,

Stable where La Baronne *kept her ponies.*

just upon arrival at the tea pavilion. The baroness went for her afternoon airing in a little pony-drawn carriage to the tea pavilion, which had a distinctly Bavarian look with its turrets and pointed roof surmounted with an eagle, its wings outspread. This, no doubt, had some relation to the fact that this branch of the Rothschild family was of Austrian origin. Near the tea pavilion is an old *bastide* surrounded by olive trees and the cedars that the baroness had put in. A tiny stable was built for the ponies, complete with the ceramic tiling typical of local dovecotes. One can still read a nameplate, "Pacha," over a feed trough. This was La Baronne's own "Hamlet," where she could be sure of finding peace and quiet.

From 1914 to 1918 Alice de Rothschild did not come to France, but otherwise the park was carefully tended until 1922, the year of her death. Her heir, baron Edmond de Rothschild, shared her love for gardens—he himself had commissioned Sir Joseph Paxton, who had designed the Crystal Palace in England, to landscape the park around his Château de Boulogne—but did not wish to take over an estate that required such a large staff and so much personal attention. Aware that many people thought that land ownership should not be so concentrated, he turned the estate over to the city of Grasse, stipulating that a portion become a public park. In this manner the township created Square Victoria, the Parc de la Corniche up on heights and the Jardin de la Princesse Pauline, a grove of evergreen oaks looking out over the city, in memory of the occasion upon which Pauline Borghèse, sister to Napoléon Bonaparte, had herself carried in a sedan chair up to this spot during her visit to Grasse in 1811. The great estate was subdivided into three hundred lots. A luxury hotel called the Park Palace was built in 1926 on the site of the villa, which later became a condominium, the Palais Provençal. In 1926, too, the boulevard Alice de Rothschild was built across the hillside. Today many of the original trees may still be seen in individual gardens, but the memory of the largest park in the Alpes-Maritimes *département* would no doubt have entirely faded away by now if it had not been for Marcel Gaucher, member of the Board of the French National Horticulture Society. Son of the last steward at Villa Victoria, he was born in the house with the glazed red-and-green tile roof. In recounting the story of his lifelong passion for gardens, he described the park that brought him to his true calling. His apprenticeship was carried out at Waddesdon Manor, and he went on to manage two other Rothschild estates.

Palms in the park of Alice de Rothschild's estate.

The Château in Gourdon, trees and terraces

Gourdon, a village of Saracen origin, is a well-known locality between Grasse and Vence. It is literally a "high place" of local tourism, for it overlooks the Loup valley from a height of well over 1,500 feet. Its twelfth-century château was built by the Counts of Provence on the site of an earlier fortress, and later belonged to several families of the local aristocracy. A family from Grasse, the Lombards, enlarged it and laid out the terraced garden in the seventeenth century. In 1919 an American named Miss Norris bought and restored the château, in which she hung her collection of primitive paintings. The following proprietors converted the edifice into a museum; in 1950 its doors opened to the public.

In front of the pale stone façade of the château, there is a quiet little square in which the plashing sounds of a fountain can be heard. Through a pedimented gate is a paved, enclosed courtyard, and through the door of the last room is the garden. Stepping into the garden, visitors have the impression of leaving a mineral world behind and entering an organic green space suspended in the air: this is the first terrace, where spherically clip-

ped lime trees look like hot-air balloons and the dark green horseshoe designs of a boxwood parterre have been delineated against the lighter green of the turf. In the center a small stream of water trickles from a round stone tub overgrown with *Iris pseudacorus*, a bright yellow variety of gladiola. Beyond the railing you look out over grey and silver mountains to the west, whereas to the south all is empty blue sky. Four steps up from the esplanade, a small *jardin d'apothicaire* comes up against the south side of the château. Its dwarf hedges are set in a Greek fretwork pattern and enclose varieties of angelica, hyssop, rue, columbine, and old shrub roses. This garden was designed by Tobie Loup de Viane in 1972, who also laid out the part of the garden below the terraces dating from the classic period. Leading away from the esplanade are two, symmetrical staircases on the west side: the first descends toward a round lawn space planted with elders, mahonia, and sumac and edged with a series of tapered boxwoods. This line of boxwoods is set at the base of the château foundations on the north side. The other staircase turns around the southernmost angle of the ramparts and continues down to the Italian garden, another green terrace where boxwoods clipped into round and conical shapes define two rectangular on a green lawn. This garden, laid out in the seventeenth century on top of four enormous barrel vaults visible for miles, looks out onto a bluff and a torrent roaring below. Rock-loving plants such as members of the aubrieta, sempervivum, and sedum families cling to the base of the wall. Facing the symmetrical stairways, a shelf planted with several varieties of the *Magnolias grandiflora* overlooks a deep, narrow ravine which is grassy, but bare of trees. On the opposite slope, myrtle, cistus, and lavender effect a transition between the garden and the indigenous plants of the nearby *garrigue*.

. . . .

Italian garden.

Terrasse d'honneur.

Box parterre.

Bastide du Roy, gardens of a musician

Up in Banksian roses climbing gaily up an olive tree in the yellow-and-white garden, there sits a European warbler. He whistles a few notes and inclines his black-capped head to listen while a tree frog in the cherry-laurel hedge answers. With another trill the warbler flits off between the cypresses to stake out his hunting grounds. It is just after noon. A sea breeze blows through the garden, as it does at this hour every day, and the garden seems to radiate its pleasure. The garden used to belong to a music-lover, Marie-Blanche de Polignac, and just a few years ago it resounded with music and song. There goes the warbler, over near the theater, an octagonally shaped space defined by closely cropped cypress hedges with a green lawn for a stage. In an instant it will reach the box parterre planted with

stock and tulips against a myosotis background, where the dark green frames a rich palette of pale blue, blue, purple, mauve, violet, and carmine.

Countess Marie-Blanche de Polignac was a pianist and had a fine trained voice. Within the circles frequented by the count and countess were many who shared their love of music. Among their friends were composers and concert artists, writers and artists. Many illustrious visitors came to the Bastide du Roy, such as Henri Sauguet, Georges Auric, Igor Markevitch, and Arthur Rubinstein. This is where Francis Poulenc composed a portion of the opera *Dialogues des carmélites*. The count, Jean de Polignac had a deep passion for navigation and took any friends who so desired along with him on board his yacht. On one occasion the pianist Arthur Honeg-

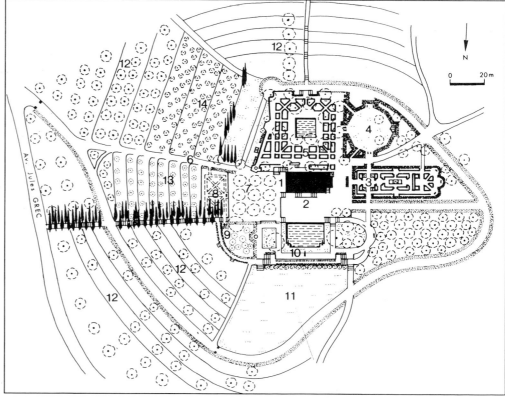

Plan

1. *Bastide* - 2. *Cour d'honneur* - 3. Spanish garden, known as yellow-and-white garden - 4. Theater - 5. Main parterre - 6. Paved pathway - 7. Courtyard planted with lindens - 8. Garden of santolina - 9. Bust of Count Jean de Polignac - 10. Pool of the obelisk - 11. Meadow - 12. Olive trees - 13. Fruit trees - 14. Orange trees.

Blue-and-red garden.

The box parterre in winter.

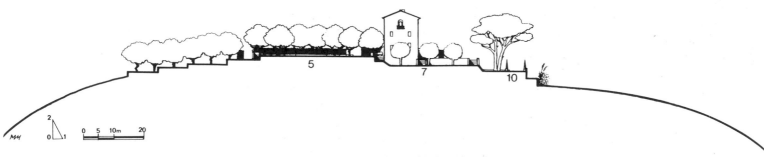

Sectional view of the garden.

Patte d'oie (*"goosefoot"*).

ger reportedly "played while the deck was pitching and awash."[16] In Paris gifted musicians came when they liked to play amid the Impressionist paintings gracing in the drawing rooms of the Countess of Polignac. The shimmering light in Impressionist works is reminiscent of the main parterre at the Bastide du Roy, still planted annually with flowers to obtain a cameo effect. Colette, Louise de Vilmorin, François Mauriac, André Maurois and Jean Cocteau were among the talented guests who sat contentedly in this garden. Originally, as its name indicates, the property belonged to the crown, for it had been acquired by Henri IV in 1608. At the beginning of the eighteenth century, when the *bastide* was built, the estate included farms, mills, a salt house, and a brickyard. Louis XV gave the land to the city of Antibes, and it was sold off to private owners. A guide to Antibes and the surroundings dating from 1877 indicates that an agronomist by the name of Bruery had installed a steam-driven machine to "bring water up under pressure" from the river down in the valley to irrigate the high ground on his property. When the Polignacs purchased the Bastide du Roy in 1927, they commissioned landscape-architect Jean-Claude-Nicolas Forestier to lay out a garden around the house.

Forestier was confronted with a massive edifice atop a terraced knoll; esplanades built on different levels lay to the east and north, whereas the other sides were plant-

The yellow-and-white garden.

Small pool, center of the yellow-and-white garden.

ed with olive trees. The site presented a strong architectural feeling upon which Forestier was able to capitalize in his garden design. He laid out a series of hedge-enclosed gardens under spreading trees. A stairway lined with cypress ran from a terrace featuring brick-paved walks laid out in the form of Saint Andrew's cross down to the bottom of the slope, creating a long vista to the east. North of the *bastide*, ornamental obelisks were put in to decorate the reflecting pool below the main courtyard. The south pool became the focal point of a large boxwood garden, crisscrossed with narrow paths set at regular intervals. Green markers clipped out of the hedges drew attention to the places where borders changed direction. The composition was sober, its most eye-catching effect being the soft shimmer of spring and summer flowers under the olive branches. A cypress screen west of the house hides the entrance to the yellow-and-white garden where the warbler has built its nest. April's marigolds, tulips, arums, and daisies fill the box-enclosed squares. Benches in alcoves carved out of the hedges face a shallow little basin in the center. Between this secret walled garden and the formal parterre, with its view toward the sea, passages through the cypress lead to the theater, a virtually monochromatic circular space in which two olive trees flank a fountain under the high-domed sky.

. . . .

April blooms.

Plantations inside box squares in February.

La Chèvre d'Or, garden and legend

Inland from the shoreline, on a road running by a Roman ruin which local inhabitants claim played a role in the Provençal legend of the Golden-Horned Goat,[17] there is a house surrounded by its own plant world, a garden that seems to represent the ideal Mediterranean garden. This was not the result of a few weeks' work, unlike many gardens today done by local landscaping businesses. It took shape over the course of years, which may partly explain why it is such a successful garden. Its history has always been closely intertwined with the life of its owners. It took them time to settle in here and take up gardening, time to give a new feeling to this peaceful place already inhabited by legend. The garden composition may have been long in the making, but once the plants were in, they grew quickly owing to the very favorable conditions of this site. The garden benefits from a southeast exposure and rises in terraces up a slope which is slightly concave in form. Wooded hills

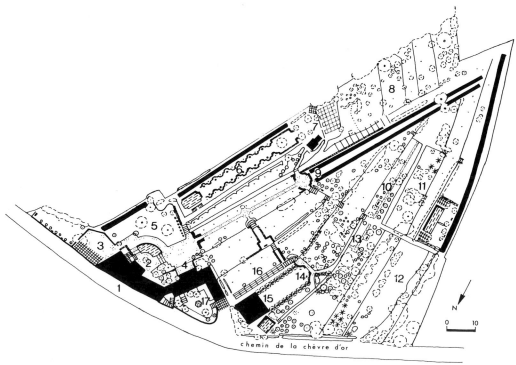

Plan

1. Access - 2. Gate - 3. Pergola with vines - 4. Arcade - 5. Main terrace - 6. Green garden - 7. Wisteria pavilion - 8. Arboretum - 9. Alley of cypress trees - 10. White garden - 11. Pink garden - 12. *Jeu de boules* - 13. Sundial - 14. Border of clipped olive trees - 15. Orange tree grove - 16. Alley of lemon trees - 17. Patio.

Main terrace.

Wild part of the garden.

serve as as screen against the prevailing winds, the cold, dry north and west winds which hinder plant growth.

Around 1940 there was nothing here but two small farms side by side, on the part of the tract facing south. It was decided to connect these farm buildings and make a single house. The main room built on the second floor afforded a view over a neighboring village which Corot no doubt would have gladly painted. The proprietors initially only came south in the summers; therefore they rejected the idea of having to maintain a lawn all year-round. They opted for a cobblestone mosaic design reminding them of paved terraces they had seen in Italy. A convenient way to add color would be to arrange potted flowers on the terrace when they were in residence. Their first potted plants were aromatics such as jasmine and certain types of pelargonium. Later other summer-blooming plants able to withstand heat were added out in the garden: different varieties of datura, hibiscus, plumbago, solanum, desmodium, lagerstroemia, tumbergia, and so forth. Paving then was laid out toward the garden, where they had to decide whether to rebuild the traditional dry stone retaining walls in view of putting in horizontal plantings; this appeared to be the simplest way to organize the available space, which was empty save for an old orange tree and two olive trees near the house. Orange and olive trees were planted on the terraces; these are, along with the cypress most typical of the southern French countryside. Many other species were also included which, although not indigenous, had naturalized successfully and could be used in combination or juxtaposition with native trees. A painterly effect was obtained through the way color values, shades, and textures interacted. The geometric layout used in some parts of the garden effectively offsets the apparently random plantings used elsewhere to create a pleasing overall effect. A cobblestone path leads away from the house and gives way to a cypress-

lined avenue; this transition serves to accelerate the way space is channeled, adding movement and depth to the overall composition. Below there is a double arabesque traced in boxwood interspersed with myrtle-crowned earthenware jars. Here a grassy carpet has been allowed to spring up to furnish a lighter green background. The box parterre catches the eye from the main terrace; in the typical Provençal garden the parterre has

Shrub wisteria.

Japanese wisteria.

Ceanothus

always played a major role, for this is where one finds a corner of freshness.

Above the central alley a series of terraced gardens, including a large agapanthus parterre, leads to a pavilion in the orangery. Spreading out in front of the pavilion is a *tapis vert*, a stretch of lawn, surrounded by clipped olive trees. A little farther on, a collection of different types of white-bloomed cistus, spirea, viburnum, silver-foliaged shrubs, helicrysum, artemsia and teucrium, light up the spring garden. The pink garden on the next terrace flowers later and features cestrum, *Hibiscus mutabilis*, cistus, and wild gladiola. Great attention has been paid to the needs of individual plants, to overall esthetics, and to the way the plants interact or "socialize" among themselves. Some were planted in keeping with recommendations made by Basil Leng, the English botanist, such as the *Teucrium fruticans azureum*, which is of a deeper shade of blue than the most frequently encountered type of teucrium. Leng went

on botanizing expeditions in most countries with a Mediterranean climate and in the Alps. He settled in Antibes and grew rare South African bulbs, orchids, and lotus for his own amusement, as he put it. His main interest was the palette of hues afforded by flowers and foliage. He knew where each plant grew naturally and could gauge its capacity to adjust to the climate of the Côte d'Azur. His friends benefited from his expertise and his collections: many species from his collections found their way into their gardens.

Behind the house olive trees underplanted with shade-loving flowers and shrubs frame a fountain throwing up its slender jet of water. This scene seems to resume the philosophy of this garden to perfection. Here are refined paving materials, the delicate colors of flowering plants grouped together with deliberate artistry to produce a feeling of rightness and proportion. In short, it epitomizes the kind of magic that leaves one spellbound at La Chèvre d'Or.

Orange grove, lawn edged with clipped olive trees.

Tapis vert.

The Thuret Gardens, a nineteenth-century arboretum in constant evolution

In *Lettres d'un voyageur*, published in 1868, George Sand gave her impressions of the Thuret Gardens, which had been planted about twelve years previously, in these terms: "Occupying a strip of land between two bays, the garden offers a view of undulating clusters of trees of all shapes and hues, tall enough to hide the foreground of the surrounding landscape ... One finds oneself in a Garden of Eden at the heart of infinity, with absolutely nothing between that immensity and the wall of foliage closing off the view of the coast with its arid slopes, sad little buildings and countless prosaic details... Did the creator of this lovely garden realize what he was doing? Could he have been aware, as he laid out the garden, that his trees would afford such a strange and unique sight at this stage in their development?... Alas, in several years the trees will obstruct the view of the sea and, eventually, the Alps. There is nothing to do but resign oneself to this prospect, for if one were to cut back the top branches in hopes of protecting the view over the horizon, their supple green festoons would be robbed of their grace and divine freedom of movement. What will remain is a beautiful botanical garden." It is true that the garden here is a green world unto itself, where trees screen the view of the nearby coast, the same coast which George Sand pronounced to be spoiled in 1868. The layout has not been altered since the establishment of the garden in 1856. The house sat at the entrance to grounds that sloped gently, facing north. Alleys circled out from the house and descended to the lower part of the property, defining a large heart-shaped space and, along the boundary lines, other areas reserved for plantings. Tall species towered over lower-growing shrubs planted on lawns over about 7.5 acres. As one strolls through the garden, two categories of plant can be distinguished: those grown to maturity and the new plants. The first have adapted to the climate; it generally takes about twenty-five years for them to strike a metabolic balance. Plants in the second category are not yet advanced enough to show whether they can adapt or not. Degeneration of

Palms and araucarias.

New plantations beneath eucalyptus trees.

these species sets in, on the average, after a period of seventy to one hundred years, but instead of allowing the specimens to complete their life cycle, they are replaced. This is appropriate for this special testing ground, intended to hold as many trees and shrubs as possible with the ultimate purpose of increasing the botanical potential of this region. Moreover, in recent years, it has become clear that the oldest trees spread fungi to neighboring trees and shrubs. As chemical treatments are never used, a radical solution had to be implemented: every three or four years, the entire population of a given section would be pulled out and replaced by new species. A central area, occupied presently by young specimens, was the first section to be repopulated in this manner. The plants are endemic to regions characterized by a Mediterranean-type climate. Each bears an identification tag showing where it comes from: Ethiopia, the Canary Islands, the Southern Himalayas, Java and Sumatra, China, Japan, Australia, Tasmania, New Zealand, California, Mexico, Chili, Argentina, and South Africa. The contemplation of trees from all over the world is a pleasant occupation that allows the imagination to roam and globe-trot at will. This was a practice initiated by Gustave Thuret, among others. Launched on a diplomatic career, he served as an attaché at the French Embassy in Istanbul. Botany attracted him much more than diplomacy, however, and he concentrated on the study of marine algae typical of French coastal areas. In 1855 he settled in Cap d'Antibes and had a villa built to house both living quarters and his laboratory, where he discovered the mode of sexual reproduction of algae. Botanical naturalization experiments were begun right away in the noncalcareous soil in this area of the cape, which supported a wide range of plant types. Gustave Thuret encountered difficulty in obtaining seeds and plants, as intercontinental shipping was still very slow. Furthermore the site had no water, so he was obliged to send a mule cart for it down the road to Antibes. Local people called this stretch of the road the Montée du Parisien,

or Parisian's Hill. By dint of constant effort he managed to put together a large botanical collection. His objective was above all to carry out experiments, and for this reason he declined to pursue any particular line of specialization. After his death in 1875 his heirs decided to turn the property over to the State.

The botanist Charles Naudin, whose genetics research lay the groundwork for Mendel's laws, was the first director of the experimental garden site and established a collection of eucalyptus. His successors continued to experiment with ligneous plants endemic to hot regions which could withstand the protracted drought of the Mediterranean summer. In 1927 the Thuret garden and villa became part of the Institut National de Recherche Agronomique as a botanical and plant pathology study center. The study center corresponds and exchanges seeds with botanical gardens all over the world. Research has been undertaken to identify suitable varieties to replace plants disappearing from the Mediterranean coast in France and neighboring countries due to repeated forest fires and pollution. A second arboretum has been set up on a seventy-five acre tract in the Esterel to conduct reforesting experiments. In the long run these experiments could once again change the face of the Côte d'Azur, whose natural landscape was altered a century ago when the massive introduction of new plant species gave it an entirely new look.

. . . .

Collection of palms.

Le Palais de Marbre, garden at a villa dating from the Second Empire

"You should go and see the Villa Gastaud—its park was famous during the nineteenth century." "Have you been to the garden at the Palais de Marbre [Marble Palace] on the Fabron road?" "One of oldest picturesque parks in Nice belonged to the Villa Les Palmiers." "There is a formal French garden in front of the municipal archives." "Have you visited the Château de Barla? Napoléon III visited it in 1860." "When I was a child, I lived at Les Grands Cèdres with a park around the apartment complex which was wonderful for children." "Did you know that Octave Godard, the landscape architect, designed a garden in the Sainte-Hélène district?"...

When engaged in research on gardens, we heard many things about what turned out to be one and the same place in Nice. It is easy enough to find because one of the Nice access roads, an expressway (*voie rapide*) brings the motorist right alongside the property. One cannot help but notice a series of rock formations emerging from an abundance of green shrubbery, with grottos carved into them and flights of balustraded steps winding in and out, the entire ensemble surmounted by a large apartment complex. As it turns out, this is the interesting subject of garden research mentioned by so many local people. It is indeed located in the Sainte-Hélène quarter off the avenue de Fabron, and a large gateway bears an inscription: Les Grands Cèdres. Cedars rise up among the palms and indeciduous trees growing at the foot of buildings over ten stories high. Proceeding in a southerly direction, the visitor goes through the building complex and follows a pergola to a viewpoint, where there must have been a breathtaking view of the Baie des Anges before the city developed to its present size. Narrow paths descend the bluff—upon inspection, it proves to be man-made—toward grottos where one comes across the dry bed of a water-

Vantage point at the edge of the cliff.

fall, also artificial. Coming back from this rupestrine setting to the top of the stairs with their concrete imitation-branch railings, the visitor unexpectedly sees a canal traversing a formal French parterre toward a classical revival house, Villa Les Palmiers, or the Palais de Marbre, which today houses the city archives. This deliberately ordered garden is enclosed on three sides by apartment buildings. The villa was built on the site of the Villa Gastaud, also known as the Château de Barla, the same house that was rechristened Les Grands Cèdres about sixty years ago. The name has been retained for the present complex comprising the Nice archives, the condominium buildings, the formal parterre and the landscape park.

At the beginning of the nineteenth century, the Gastauds, a prominent family in Nice, owned a fifty-acre country estate on this plateau surrounding the villa-château-palace. Later it was described as a pleasure-house of great interest because of ''its varied landscapes and beautiful conservatories, the attention given to maintenance details and especially all the shaded paths which wind down to the route de France.''[18] In 1860 Napoléon III and the empress came to claim Nice after its population had voted by referendum to become part of France. They visited the Villa Gastaud. In 1871 the estate was subdivided into lots and sold at public auction. The lot on which the main house stood was purchased by Ernest Gambart, an engravings publisher and art dealer of the period known as the Prince of Victorian Art. When he retired from business life, Grambart became the Spanish consul in Nice and commissioned a Nice architect, Sébastien-Marcel Biasini, to design a winter residence for him. He wanted it to be worthy of housing the paintings and art objects which he had collected during his career. In short, he wanted to set up a sort of personal museum. Biasini found inspiration—like many others in 1872—in both the sculpture-enriched façades of specific Italian villas and in eighteenth-century French architecture. He designed a little ''palace'' with two floors and a central project-

Lantern turret and terrace, Villa Gastaud, where the Russian imperial family stayed in 1864.

O. Godard's canal.

Palais de Marbre, general view.

ing loggia surmounted with statuary. The story goes that the marble façade elements were sculpted in Carrara, Italy, and shipped in twenty-seven ships to an unloading dock built specially for the occasion, equipped with a pontoon bridge. Contrary to what one might have expected, the formal French garden does not date from this period. The flat surface in front of the villa was irregular in shape, as was the rest of the park; here, inherited from previous owners, was a conservatory 127 yards long prolonging the winter garden on the east side of the Palais de Marbre. The conservatory extended along the esplanade to the edge of the arrangement of

rock formations. At the foot of the bluff, an avenue lined with *Phoenix canariensis* continues to justify the name of Villa des Palmiers given by Ernest Gambart to his estate.

But the story does not end there. In 1902 the Palais de Marbre was bought by a Russian colonel named Alexandre Faltz-Fein, who made few alterations in the garden. The long conservatory continued to draw exclamations of admiration from guests of the colonel who, in his time, had commanded the corps of pageboys under Czar Nicholas II. His was a very active social life centered on the Russian colony in Nice. He added a ballroom onto the west wing and, on the axis of the façade, had a rotunda erected to contain all the statues that had been dispersed throughout the park. In 1923 the property came into the hands of Edouard Soulas, who finally decided to give the gardens a unified aspect and gave the commission to Octave Godard, who had already created many gardens between Cannes and Menton. Like Ferdinand Bac, Godard was an adversary of the then-fashionable trend in favor of acclimatizing exotics in gardens on the Côte d'Azur. He himself was rediscovering the luxuriousness of empty space and the beauty of geometric designs, and recommended moderation in selecting what kind of plantings to put in. At the Soulas house Octave Godard made a rare exception to his rule that reflecting pools should not be used when the site has a view over the water. He used a canal across the esplanade to recenter the entire composition. He felt that drastic means were necessary to achieve his ends and had the winter garden, conservatory, ballroom, and rotunda razed. In front of the main façade facing south, he composed a parterre in which water, lawns, box clipped in round shapes and garden statuary were placed against a cypress hedge background. This ordered composition presents a strong contrast with the natural landscaping of the park, planted with araucaria, cedar, eucalyptus, and palm trees. On the west side of this formal French-style garden, he added a Roman garden communicating with the espla-

Artificial grotto.

nade via three long terraces adorned with marble figures and terra-cotta urns. One can still see the vestiges of a small cascade with regular basins, niches containing busts, and, at the foot of one of the new buildings, an "antique" bench placed against an openwork wall. A lantern turret rises up in this part of the garden, a typical nineteenth-century piece of picturesque garden furnishing. The uppermost alley is bordered with orange trees and leads toward the sea, to a roundel hedged with cypress in a small evergreen oak grove. Here is an enigmatic statue with an armload of webbed-footed birds. With some vestigal memories of schoolboy Latin, this figure may be identified—after considerable effort—as a *haruspex*, a minor Roman priest, bearing sacrificed birds. Other unidentified representations of equally obscure figures are placed all along the esplanade and terraces. These mythological allusions, the bluff, grotto, cascade, and silent fountains in the park of introduced species, now overgrown in some areas, furnish a perfect setting for children who come here to play and for photographers in search of surrealistic effects.

Neoclassical pool at the foot of an apartment building.

A piece of sculpture in the garden.

Valrose, lyricism and gardening

Valrose is the only park dating from the Second Empire in Nice that has retained its original twenty-five acres and most of its ornamental structures, the picturesque garden monuments of the period. This estate, which stretches out below the Cimiez hillside on the west side, was purchased in 1961 by the City of Nice to build the University of Nice and the Faculty of Sciences there. Inside the gate a narrow valley rises on a gentle upward incline, its slopes populated by a collection of conifers behind which university buildings are scarcely visible. An avenue runs along mowed lawns where palm, cedar, magnolia, and ginkgo trees are ranged separately, as if specimens in a botanical garden, and where, in the old days, a bandstand used to be. The Valrose collection numbers over one hundred varieties of imported trees

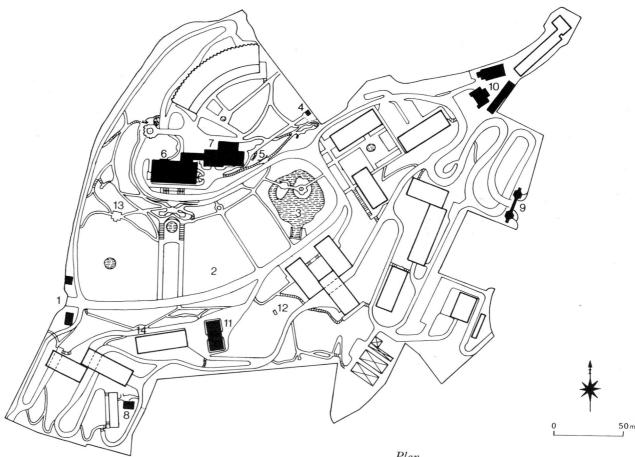

Plan

1. West portal - 2. Arboretum - 3. Pond - 4. Roman temple in ruins - 5. Temple of Bacchus - 6. Château - 7. Outdoor theater - 8. *Isba* - 9. East portal - 10. Small château - 12. Horse's tomb - 13. Grotto of the hydraulic train - 14. Site of the rose garden and greenhouses.

and shrubs. Overlooking the slope facing south, there is a grandiose white stone edifice with a slate roof that recalls the architectural fantasies of Louis II of Bavaria. Yet the real centerpiece of this ''landscape'' is the artificial lake farther uphill, with its islands and boat dock. Quite a scenic effect is produced with this stretch of water surrounded by graceful, uneven contours of the land. Valrose was finished in 1869, the same year that the Buttes-Chaumont park in Paris was completed. One passes three university buildings, which, according to a plan of the original park, occupy the site of the former rose garden and greenhouses, before coming to the drive leading up to the château, punctuated with several picturesque monuments. On one of the islands, a pavilion houses the nymph Amalthea,

Representations of the Seasons hidden by greenery.

Fountain at garden entrance.

after Pradier's statue in Queen's Dairy in Rambouillet. On the adret, or southern slope, grottos are hollowed out of artificial rock overhanging a nineteenth-century version of a Roman temple in ruins. Next to the avenue is a pedimented doorway sculpted, appropriately enough, on the Bacchus theme: here is the entrance to the wine cellars, dug into the hillside. The drive then winds up to the arcaded main courtyard of the château, where a fountain used to play on the upper side, and continues up the hill to one of Valrose's most intriguing features: the theater. Like the theater in Bayreuth, it had an orchestra pit and could accommodate an audience of four hundred. Valrose was built for the Baron Paul von Derwies, who, as builder of the first railway network in Russia, was ennobled by Czar Alexander II for services rendered. Concerts and operas were performed for the baron's own pleasure and for his guests'; occasionally, they were open to the public and attended by the winter residents of Nice. Performances were given by an orchestra of fifty and a chorus of eight. In the 1870s well-known concert artists such as singer Adelina Patti performed about twenty operas at Valrose, and von Derwies even staged an opera of his own composition there, entitled *La Comtesse de Lascaris*. The park also boasted an *isba*, a typical Russian pine dwelling, which had been dismantled in Odessa and shipped to the site. Today it is hidden from view by university buildings on the opposite slope of the small valley. On the east side, at the boulevard de Cimiez entrance, stands a monumental portico flanked by towers. The slope is so steep that a hanging ramp had to be built; it turns back upon itself before coming to two hairpin turns at the edge of the man-made lake, formerly the focal point of the park. In its days of glory water was everywhere at Valrose. A reservoir in the upper part of the property fed small cascades that sang in little grottos where the rocks today are sadly eroded. One of these rocky niches contained a miniature railway the size of a child's toy train, driven by a hydraulic system. The tracks and several gearwheels are still

Island summerhouse on lake.

there. Today the fountains are started up every day to make sure they are in good working order, although they are quickly turned off again. When the moment comes, generally around noon, the fountains come to life and water sounds quietly flow on the air like a Debussy prelude. Some of the statues also seem to have their own voice: at the west gate, a mastiff in stone raises his muzzle to bay an alert, the figure of a little girl lying inert beside him. Elsewhere, a whipper-in restrains his hounds, and one hand cupped at his ear, he attends to the sounds of the hunt. Even today Valrose extends an invitation to listen.

Life-size statue on a horse's tomb.

Time for the fountains to play.

plants and frame a parterre of lavender. Farther up the slope, more rosebushes can be seen near the datura and spirea, and others still over in the corner with the ceanothus and echium. A round-arched gateway called the Narbonne Gate is a passing reference to a pilgrim's resting place. There are many evocative names in this garden: the Temple of Friendship, Staircase of the Sacred Wood, and Astronomers' Circle. Landscaping has been carried out in the meadow and on the lower slopes, for the purpose of restoring an amphitheater effect to the main slope, where fairly horizontal planes alternate with steep sections. Four rows of young olive trees have been planted; for the moment, they can barely be perceived against the meadow background, but once mature, they will provide rhythmical movement across the fan shaped valley. The ground has also been contoured on the hill overlooking the meadow on the east side. Below the *poste aux grives*, a shelter that used to serve as a blind for bird hunters, gently sloping and steep sections again alternate; parasol pines and chestnuts planted here will extend the forest-orchard occupying the upper parts of the property. Lower still is the theater: a few selected columns in the foreground before a shady area at the foot of one of the great sandstone rocks to be found throughout the woods, which were used to build the château. The presence of sandstone and chestnut trees indicate a high silica content in the soil, which explains why, at Vignal, one finds fern, several types of heather and arbutus. The green caterpillar, *Charaxes jasius*, feeds upon arbutus leaves, and sometimes this great brown butterfly is seen fluttering on the breeze, or at rest, artfully displaying the Oriental design patterning the back of his wings.

. . . .

Entrance to green garden.

Le Conseil, *an exedra appropriate for gatherings.*

Miniature garden in a tub.

Esplanade border.

Springs and pool.

Entrance to old rose garden.

La Pallaréa, a solitary retreat in Liguria

Rising up from the Paillon Valley, which in the past afforded access to Turin via the Tende pass, a small road goes up into the hills northeast of Nice, toward the village of Peille.[19] It winds up, down, and around curves through uninhabited pine forests. Travelers one hundred years ago must have driven through a landscape almost identical to this one, except they did not drive a car on a smooth, asphalted road and there were no power lines at the time to mar the view. Near the hamlet of Collet one comes across an oratory at a crossroads. Just past this point a view opens out over the uneven ground of a plateau surrounded by high hills. The plateau culminates at the foot of this small pass, which is where one leaves the car. A great parasol pine and a very old cypress can be seen down farther, marking the location of La Pallaréa,[20] a seventeenth-century *bastide*. From this vantage point an erect stand of century-old trees and the central block in its midst might be taken for a single edifice, placed in the midst of an impressive natural topographical design. This composition would have tempted more than one painter in the classical tradition traveling to and from Italy over the centuries. Nicolas Poussin may very well have seen the Baou de Saint-Jeannet on his way to Rome, for he appears to have used it as a model for the rock profile in his *Paysage avec Polyphème* ("Landscape with Polyphemus").

Walking through the fields toward La Pallaréa, the visitor can perceive a garden located below the great cypress along the south side of the house. It is regular in shape, slightly larger than a tennis court. An alley runs from the house to a fountain built against a retaining wall and divides the garden into equal parts. From above the retaining wall a closer look reveals large oval shapes of clipped yew at the corners of these two squares of turf surrounded by rose beds. It is almost a walled garden in the sense that the site is located in a hollow in the slope and enclosed by walls and buildings. A section of the enclosing walls has been removed, but in the past, a visitor approaching the house, whether on

Yew parterre.

Copse and La Pallaréa.

the avenue coming from the east or the olive-lined road running down the mountain from the pass, could not see the garden until he stood on the threshold to the house. This was not a garden created simply for show. Favorably exposed and protected from the winds, it served as additional living space set in a pleasant outdoor environment. East of the *bastide* there was a terrace bordered with benches of clipped boxwood, where several old lime trees still stand, their foliage so thick that the sun can barely find its way through the leaves. Three steps lead down to a meadow hedged on two sides by slender cypresses planted to form a right angle, back to the north. There is a nicely proportioned reflecting pool set in the grass which, as light plays on its surface across the reflections of trees, recalls specific parts of Italian gardens such as the basin in the meadow at Villa I Collazzi near Florence, or the artificial lake enclosed by cypresses at the Villa Falconierei in Frascati.

A visit to La Pallaréa gives one the feeling of escaping to a calm landscape in the Tuscan or Roman countryside.

. . . .

Fountain in the axis of the arcades.

Reflecting pool.

Garden in June, overflowing with roses.

Les Terrasses de la Darse, a hanging garden

The coast road goes around the ramparts of the citadel in Villefranche and along the harbor, where a rope factory, boat sheds, what used to be the old sailors' home—if one can call galley rowers "sailors"—and one of the old quarantine station buildings still bear witness to former harbor activity. Amid these historic buildings a contemporary garden goes unnoticed by most passersby. Yet one simply has to look up when strolling along the docks, for trees can be spied above the eight white stone entrances to the vaulted boat house erected by Duke Emmanuel-Philibert of Savoy for his men-of-war, today a warehouse for a local shipyard. It is believed that three of his galleys left from here to engage in the Battle of Lepanto (1571). The erstwhile ducal boat house provides a magnificent foundation for the hanging garden, the Terrasses de la Darse, designed by architect Eugène Beaudouin.[21] Beaudouin's plans called for both vertical and horizontal axes: he extended it upward by means of a staircase leading to a patio and laid out a horizontal cement esplanade over the vaulted boat house. A very narrow platform was added to dominate the entire ensemble. His garden thus

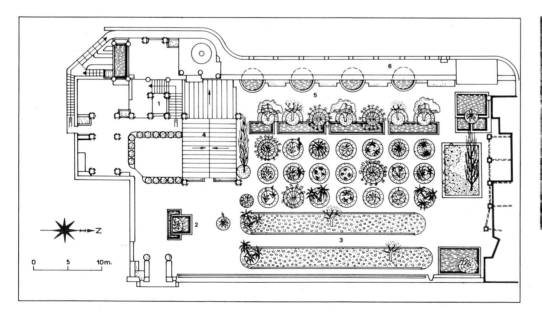

Sixteenth-century vaulting.

Esplanade, view from second terrace.

The winter garden.

included three levels from patio to second terrace, where he used earth and water to bring his design to life. From the arbor one sees that the patio is suspended about midway up the last arcade. Jasmine and yellow bignonia bloom on the wall facing south; a vast, vaulted room gives onto this space. In the nineteenth century this was the kitchen, placed above the arcades; it has now been restored and converted into a study. The stairs continue upward and open out at various levels onto rooms, suggesting troglodytes' caves, set into the second terrace's retaining wall. At the entrance to the esplanade Beaudouin had a dozen or so monumental pillars put in, with ledger strips to support a pergola covered with white grapes and glossy leaved *Vitis Voinieriana*. Between four pillars a glass frame emerges from the terrace floor, the roof of a winter garden built where dynamite blew a hole in the surface in 1944. Between this greenhouse and the old ''capitain's walk'' on the harbor side, sedum and blue marguerite grow on a cement rockwork fountain in a shallow basin. At the opposite end of the terrace a fountain throws up its jet from a cistern, one of several original features on the site which Beaudouin successfully incorporated into his design. Rounded basins were placed at the foot of old half-domed niches in the end wall; facing them are small, rectangular pools that resemble aquatic flower beds for water lilies, butomis, pontederia cattail, and cush-cush. Earth was brought to fill tubs bordering the terrace, and a number of concrete flowerpots are shaped like overturned cones. In the smallest of these, set upon the pergola's pillars,

are found drought-loving plants such as aeniums and aloes. The largest, spaced out at regular intervals along the esplanade, contain rose laurel, Judas trees, yuccas, mimosas, and fruit trees. Lemons, medlars, apricots, and figs can each be picked in season. Plants such as wild thyme, sage, and rosemary, which are both kitchen herbs and aromatics, are mingled in circular beds around the base of each large pot with iris, echium, amaryllis, cineraria maritima, and anthemis, all perennials selected for color and foliage. On either side of an alley overlooking the dockside are long, boat-shaped containers of agapanthus. In the glass-enclosed sunken garden, sunshine and water in a basin provide the heat and humidity needed by tropical plants including *Nephrolepis exaltata*, a large exotic fern, and *Hoya carnosa*, a liana with fragrant white flowers. Above them spreads an enormous-leaved philodendron of the type producing comestible fruit, *Monstera deliciosa*. Two levels up, the second terrace serves as a vantage point to view the esplanade, harbor, and the bay of Villefranche at a single glance. Shade is provided here by the curving tendrils of grapevines and Brazilian passionflower, and both grapes and grenadillas offer themselves at the same time.

Scattered like rugs between benches facing the view and on the landings of the stairs, pebble and terra-cotta mosaics eroded by the sea seem to bring outdoor paving right into the living quarters. These mosaics were the work of the painter, Joséphine Beaudouin, wife of the garden's architect.[22]

. . . .

Pebblework mosaics dominating the Captain's Balcony.

Flowerpot on a pillar.

Villa Ile-de-France, an eternal nautical dream

A late seventeenth-century etching of Cap Ferrat shows it to be a very solitary place, planted in some areas with rows of trees—probably olive trees—and virtually bare of construction except a great fort erected on Saint-Hospice's Point, the pointed roof of the signal station tower up on a height, a house and walled garden, and a windmill on the spine of the raised isthumus connecting peninsula and mainland. Although the Cape is highly built-up today, the chemin des Moulins ("Mill Road") still exists and leads up the hill to a large pink house surrounded by gardens. This is Villa Ile-de-France, which belonged to Mrs. Maurice Ephrussi, Baroness Béatrice Ephrussi de Rothschild. It is now a museum-foundation, housing collections that she left, about fifty years ago, to the Institut de France (a grouping of all French academies of the arts). It might seem strange that Béatrice de Rothschild would select a garden site buffeted by the strongest prevailing winds in the entire area. Yet her choice is more understandable when one sees the view from the villa-museum: it is nothing less than superb, plunging down to the Mediterranean in front and to the bays of Eze and Villefranche on either side of a long garden space created by dint of Herculean earth-moving operations. The villa itself is a 1910 version of a Venician palazzino. South of the house is an esplanade whose parterres and reflecting pools are laid out symmetrically, above a series of small gardens on the western slope which finally disappears into the woods. The visitor cannot gain immediate access to the gardens: first one is required to follow a fixed itinerary through the museum collections of rare objects. One story illustrates just how far Béatrice de Rothschild would go to gain possession of certain rare items: it is reported that for the sake of a single coveted fresco, she bought an entire palace in ruins. She prospected all over the world for rarities. Connoisseurs of Regency, rococo, or Louis XVI furniture, and exceedingly rare tapestries and porcelain will also discover an unexpected collector's item from a window on the ground floor or the second floor loggia: the so-called French garden, actually more Creole in appearance with its palms and agaves clustering around an artifical stream. The stream culminates in a stepped waterfall over which presides a smaller-scale replica of the Trianon's Temple of Love, placed against a pine forest background. Even for visitors out for a stroll, more interested in gardens than collections—and let it be pointed out that the collections are arranged in twelve

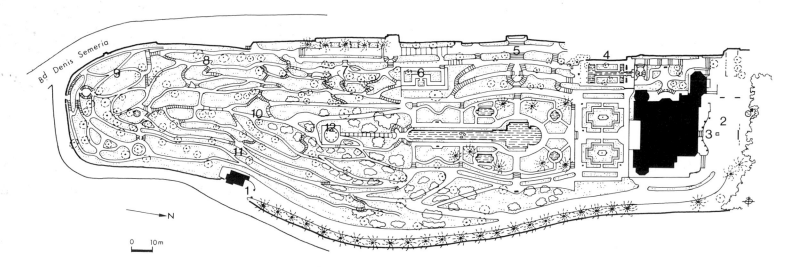

Plan

1. Portal and housekeeper's lodge - 2. *Cour d'honneur* with grotto - 3. Entrance to the museum - 4. Spanish garden - 5. Florentine garden - 6. Lapidiarist garden - 7. Japanese garden - 8. Exotic garden - 9. Former rose garden - 10. Former English garden - 11. Former Provençal garden - 12. Temple - 13. Formal French garden.

Lotuses in the water garden.

Sectional view
1. Road of Saint-Jean-Cap-Ferrat -
2. Florentine garden - 3. French formal
garden - 4. Approach of the villa.

Palazzino, *view from temple.*

drawing rooms, not to mention various assorted galleries, individual collection display rooms, offices, bedrooms, and boudoirs belonging to Béatrice de Rothschild—there is no direct access provided to the spectacular formal garden. An itinerary of small "cosmopolitan" landscape compositions is designed sufficiently well for one to find the way out. A sign on the right at the entrance to the villa-museum indicates the way to the Spanish garden. Several steps lead to a stylized grotto with its echo of trickling water and, farther on, to a patio traversed by a narrow channel full of aquatic plants under pomegranates and large daturas. Then one comes across theme garden terraces. The Florentine garden has horseshoe stairs embracing yet another moist grotto containing a marble ephebus. A shady rock garden features all the pieces of statuary which Mrs. Ephrussi could not fit into her villa, despite the valiant efforts of those whom she considered to be "her" architects.

Mrs. Ephrussi had somewhere between twenty and forty different architects at work drafting projects, for which full-scale models using wooden frameworks and canvas backdrops were sometimes produced; the architect who turned out the highest number of designs for her was Aron Messiah of Nice. Béatrice de Rothschild later tried to get Ferdinand Bac himself to promise to work for her. At the time Bac was staying at Saint-Hospice's Point at the home of the Countess of Beauchamp, working on the interior decoration and garden layout for La Fiorentina. In his memoirs Ferdinand Bac recalls how the "beautiful Madame Ephrussi" arrived one day at his hosts' home, just at mealtime. She would not talk to anyone and took him to the Villa Ile-de-France in her car, saying: "Now, you have to do my estate—it's my turn!" What she had in mind was to have Bac spend a year drafting designs for her. Ferdinand Bac quickly understood her game: she would pretend to accept his designs, discuss them, and then discard them without appeal. Needless to say, he declined the offer. Continuing the fixed itinerary along the garden slope, one sees the vestiges of a miniature Japanese garden before crossing the exotic garden, where steep paths curve around impressive cacti, and the rose garden, which has a slightly Persian air given by four blue faience pieces in a bed of santolina, between a small hexagonal temple and a curious sentry box of marble fretwork. Here the far end of the property is reached. An aloe-lined path brings one back under old olive trees bent by the east wind, reminders of Cap Ferrat's natural landscape and sole survivors of the baroness's gardening dictates. The path crosses sites where the Provençal and English gardens once stood, now unrecognizable even to the discerning eye. A nobly proportioned round-topped temple amid the pines attracts visitors with an appreciation of objects of art. From this monument a vista-in-reverse can be obtained, with the French garden in the foreground and the inevitable mirror image of the rose-colored palazzino in the water. Unfortunately, the cascade, intended to spill exuberantly from the temple base, is now mute, as are the fountains in the basins. The waterworks are only turned on for grand occasions. If one overlooks the conventional flower borders (except the lotus in the lateral basins, visible in June) and the implacably white graveled itinerary, there is a certain oneiric quality to this scene resembling a painted backdrop hanging in the air framed by curtains of water.

Some liken the garden to a ship. This is appropriate in the sense that its collector-proprietor spent much of her time on transatlantic luxury liners. And it is true that the garden-ship theme has been inspiring fertile imaginations for centuries. In the second century B.C., Hiero of Syracuse had ivy- and vine-covered pergolas on board ship. Conversely, the seventeenth-century Villa Isola Bella on Lake Maggiore, with its successive terraces, resembles a galley. It is small wonder that the tempermental Baroness Ephrussi de Rothschild was drawn by the paradoxical interaction of stability and movement.

. . . .

Replica of the Petit Trianon *temple.*

French-style garden.

La Fiorentina and Le Clos, color and fragrance

"A garden is a fascinatingly mobile way of expressing oneself, and all the time new ways of presenting things occur to one." This is how the American writer Roderick Cameron, author of an autobiographical work entitled *The Golden Riviera* and a number of books on his travels, described his feeling toward gardens. In *The Golden Riviera* he recalls settling in Cap Ferrat in the 1940s and his restoration work on two houses, one of which would come to rank among the loveliest villas on the coast. He relates how he learned the art of landscape gardening and how the gardens at *La Fiorentina* and *Le Clos* came into being:

"Quite some time elapsed before, I again visited the South of France . . . My mother had since remarried a Leicestershire hunting squire. Much in love, he had offered her a villa on the Riviera, and in order to look around she had rented La Léopolda. Situated on a hill above Beaulieu, it has sweeping views out over the Mediterranean, and has always been considered one of the show places of the Riviera. Ogden Codman, its builder, a distinguished American architect of the twenties, had based his designs on a combination of two eighteenth-century villas, one the Villa Borelli near Marseilles, and the other the remains of a handsome house he found in Milan. There were few buildings to speak of on the coast at this period, and La Léopolda[23] was Codmans's interpretation if such a thing had existed. Formal gardens with *pièces d'eau* surrounded the house, while a long line of thirty-foot-high cypresses framed the steps leading down to the view. It has been suggested that Edith Wharton helped with the garden, and this might well been the case for the novelist was a friend of Codman's and had collaborated with him on a book dealing with decoration.

"The Codman estate wanted to sell La Léopolda but my mother found the house too formal. Living in it would entail constant entertaining, and my mother, never fond of crowds, wanted something less spectacular and by preference a house with a garden down by the sea. The Agnellis[24] subsequently bought La Léo-

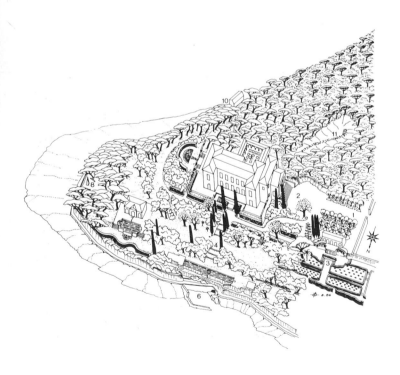

View of seawater pool, La Fiorentina.

Plan

1. Orange tree garden - 2. Courtyard and Palladian façade - 3. Loggia and lavender garden - 4. Grand stairway with grass treads - 5. Swimming pool with sea water - 6. Harbor - 7. New pavilion and swimming pool - 8. Obelisk - 9. Cloister - 10. Library - 11. Japanese garden.

polda and at the time they were exactly the right people for it. Young, handsome and extraordinarily elegant, they had both the taste and the means to live in it properly and it is a great pity they ever left ... It took my mother about six weeks to find her ideal; a house right on the end of Pointe Saint-Hospice with enough land round it to assure complete privacy. Saint-Hospice is a small peninsula jutting out from Cap Ferrat and, bending back like a thumb, faces out across Beaulieu Bay to the mainland, and is about as near to being an island as it is possible to be. Angled east-west, the house faces due south and full out to sea on one façade and to the shelter of a large, open bay on the other. The property was known as La Fiorentina, and as its name suggests, was a Florentine pastiche and was built just after the outbreak of the 1914 war by Comtesse Robert de Beauchamp,[25] and is illustrated, incidentally, in Robert Doré's extremely useful *L'Art en Provence*.

"It was a brave choice and, as it turned out, a very fortunate one. Of course its whole *raison d'être* is the position, its gardens reaching right down to the rocks and the heaving Mediterranean. The end of the point was left wild and grown over with a tangle of stone pines tortured by the wind into weird Rackham-like shapes,[26] and it was for these trees, I believe, that my mother really bought the place."

The purchase was made several months prior to the Second World War in 1939, during which German staff headquarters was installed at La Fiorentina. The house was left in very poor condition as a result, and its restoration represented such a large-scale effort that Cameron moved in to supervise operations.

"Fortunately another house on the property known as Le Clos, the dower house as it were, had escaped completely untouched, and into this charming, late-eighteenth-century building we moved while work on the Fiorentina progressed ... after several false starts we finally decided that it should be rebuilt in the Palladian style, in what one hoped would be a much purer rendering. At least one felt it to be more appropriate

La Fiorentina and Saint-Hospice Point, around 1920.

Cloister designed by F. Bac.

Palladian façade designed by R. Cameron.

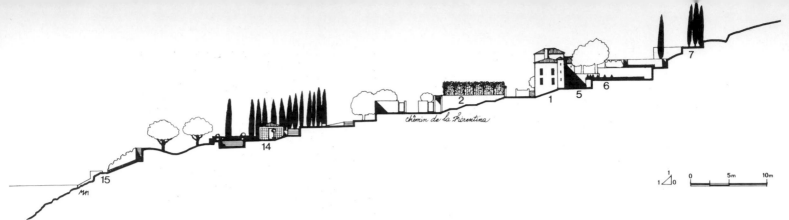

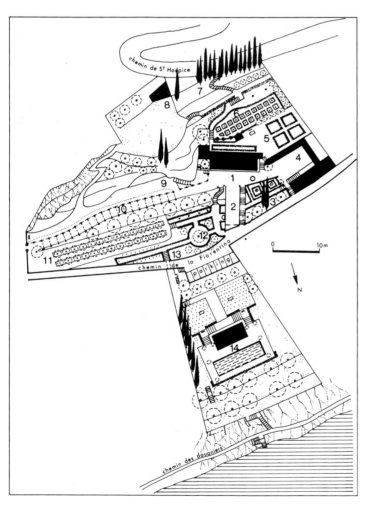

to its position and to the sharp southern light ... It was, after all, in the English tradition to translate the ideas of Veneto's famous architect. They had been transferred to the dunes of Norfolk and the Yorkshire moors, why not to the shore of the Mediterranean. Once having decided what form the garden should take the planning of it was comparatively easy; the bones were already there—the lie of the land dictated the rest. It was to be a predominantly green and white garden, with its main accesses clearly laid out, and within this classical frame a series of separate compartments or rooms walled in behind clipped hedges. From the formal part one was to wander out into the different wild areas— the country's redolent *maquis*. Colours, other than the different blues, silvers and whites, and the dirty pink bloom of the round leafed *Bergenia delavayi*, massed in great carpets under the pines, were to be kept to the minimum. It was not to be the kind of garden in which annuals were bedded out, rather the accent was on shrubs, and not necessarily those which grow only in a temperate zone, but odoriferous ones such as choisya. Those parts of the garden exposed to the prevalent east wind were screened off by twenty-foot hedges of *Pittosporum tobira*, the only fragrant member of that large family; in the spring its small clusters of creamy-white flowers scent the whole garden. The pergola, framing the walk leading from a wild part of our point to the Clos, seemed an ideal place for the different jasmines, while the immediate approach to the house, a large empty rectangle, proved the perfect setting for a plantation of oranges. They were planted four rows deep in lines of ten and the drive swept up the middle. Under the trees we divided the ground up into a geometric pattern of triangles traced in low, clipped box hedges and further delineated, or relieved by alternate spacing of red sand in contrast to the earth; the sand being retained by an edging of tile set edgeways into the ground at the same height as the box. A further detail was the coating of lime with which we daubed the trunks and lower limbs. The citrus growers do this as a protection

Sectional view

1. *Bastide* - 2. Terrace and pergola -
5. Fragrant garden - 6. Boxwood garden -
7. Saint-Hospice's road - 14. Pavilion and swimming pool - 15. Customs agents' path.

Plan

1. *Bastide* - 2. Terrace and pergola overlooking the road - 3. White garden below the terrace - 4. Gardener's house and commons - 5. Fragrant garden - 6. Boxwood garden - 7. Three cultivated terraces - 8. Cabin and meadow - 9. Mossy alley - 10. Wisteria-covered alley - 11. Terrace of mandarin orange trees - 12. Garden of the ebony - 13. Underground passageway - 14. Pavilion and swimming pool - 15. Harbor.

against parasites. We painted them for a purely decorative reason, to give luminosity to the dappled, sub-aqueous light filtering throught the dark leaves. I had never seen this done before and we were rather proud of the result.

"I hesitate to go into too much detail about the garden; amongst the more obvious sweet scented things that we planted were the handsome Carolina *Magnolia grandiflora* and Latin America's graceful trumpet flowered datura, also clumps of the white *Hedychium coronarium* from India. Amongst the less obvious odoriferous plants, and probably the strongest smelling of them all, comes the *Cestrum nocturnum*, the night blooming jasmine, known romantically in the Spanish-speaking countries as *damas de noches*. An inconspicuous shrub with tiny clusters of yellow-green flowers, it is difficult to locate the first time one comes across it in someone else's garden; it is a question of degrees of smell. At the gates we massed a collection of Australian acacia—better known here, of course, as mimosa. At Christmas time they explode in a honey-scented, yellow cloud, to be followed later by the equally sweet-smelling *Coronilla glauca*, indigenous to several parts of southern Europe, and which we had naturalised in the *maquis* under the stone pines.

"Another element in this combination of smells is the terraces of lavender below the loggia, where we always lunched in the summer time. The lavender,

when not in flower, was kept clipped in tidy, clumped balls, treated in exactly the same way as the commercial lavender which one finds planted in great undulating fields in the stony fastnesses of the Var highlands. Clipped hedges of rosemary confined the lavender, the paths in between set, here and there, with flat circular pots of white pelargoniums. Large pots of lemon and bergamot, a species of rough-skinned orange, stood on the corners of the loggia terrace and against the wall behind. Detached from the house the Italianate loggia of honey-coloured tufa constituted one of the few original elements of the property that we kept, modifying it only by adding a terrace framed by curving stairs which lead into the lavender garden. The large terra cotta pots were another legacy, imported from Tuscany where they have been firing them since the times of the Medicis.

"The *pièce de résistance*, the focal point as it were of the garden—and the view most often reproduced in the different gardening books—are the great shallow grass steps leading down to the sea. Falconet[27] sphinxes frame the stairs and along the descent, on both sides, are planted tapering twenty-foot cypress. Below these tight, green columns grow dusty clumps of the Canary Islands blue flowering echium. They advance in waving lines onto the steps, and mixed in with them come a small white flowering convolvulus and the deep blue Corsican rosemary. The last, and seventh step is the

The patio.

Alley of wisteria.

Terra-cotta griffin.

pool spilling over into the glittering Mediterranean, and beyond, across Beaulieu Bay, comes the whole dramatic sweep of the mainland piled in a series of precipitous limestone cliffs rising to a height of nearly two thousand feet before collapsing, in folds of varying pastel shades, into Italy.''

In 1961 or thereabouts, Roderick Cameron found himself living alone at La Fiorentina, which he deemed too large for a single person, and decided to move into Le Clos Fiorentina, the other house which was part of the original estate. He planned a different type of garden here, in keeping with the architectural style of the house.

''I have already explained how we lived in the Clos while rebuilding Fiorentina, which makes the move just a reversal of time. As to the house, it dates from the end of the eighteenth century and is the oldest house on Cap Ferrat, or more exactly the Pointe St-Hospice. It has no pretensions to architecture, but in its simplicity can lay claim to a good deal of charm, and is typical of the country: has red tiled floors and white marble stairs, a Roman tiled roof, green shutters, and pinkish-ochre walls. Directly outside the front door stands the old covered-in well, once the house's only water supply. . . . Both these ground-floor rooms give onto the terrace overlooking Beaulieu Bay and the mainland; a vine shaded area which acts as an outside drawing-room during the hot summer months.

''As regards to the terrace and swimming-pool furniture, I have purposely avoided bright colours. Living in the sun, I find one tends to avoid them, and this, I feel, applies to any of the Mediterranean countries—something to do with the sharpness and quality of the light.

''This question of muted tones is also carried through to the garden, and wherever possible I have kept to a mixture of greens laid out in casual formality. Not actually occupying the house until recently, I have had years to plan the layout. As basic elements, I had the side of a hill buttressed with terraces leading down to

the sea, also the stones from the ruin of an early-seventeenth-century fort to carry on with if any further construction was needed. The fort, as depicted in early drawings, looked a massive affair and was erected as a protection against piratical raids from North Africa. Judged a useless encumbrance by later generations, it was blown up in 1706 by one of Louis XIV's generals, and took two months of concentrated mining to tumble, the walls still bearing the marks where the powder blackened the stones. Along with the terraces, we also inherited some twenty magnificent olives. . . As is usual in this form of cultivation, the olives are planted in rows and are on the same level as the house, centred in a terrace about eight feet wide along which I have clumped great cushions of grey-green echium, a handsome contrast to the grey of the olives when they burst out with their blue candle-like flowers in the spring. Another feature of the garden is a walk of mandarin trees with their trunks daubed in whitewash. Under them, confined by a low border of box hedging, I have planted double rows of arums and it looks very effective when the lilies are out, their white chalices catching the light filtered throught the mandarins' pointed leaves. In one place, copying the Italians, I have massed a bed of aspidistra and on the terraces to the left of the house, where the rocks begin to obtrude and the soil is thin, I have naturalised broad drifts of the wild tulips from Greece and Turkey, also a collection of dwarf narcissi, a native of stony reaches in the Alpilles. The steep banks behind are anchored with a solid flank of Judas trees with, under them, blue drifts of *anemone blanda* alternating with clumps of the pale *iris stylosa*.

''A garden is a fascinatingly mobile way of expressing oneself, and all the time new ways of presenting things occur to one. The idea, for instance, for the topiary work behInd the house came to me while on a flight to Cape Town, the whole terrace, quite broad in this instance, being divided up with squares of box and in the centre of each square a tapering cone of the same plant—nothing spectacularly original but just the right accent,

Orange trees and arums.

Topiary of clipped box.

to my mind, at this particular point of the garden. From here stairs railed in a Chinese Chippendale design mount to a further terrace backed by cypress with an underplanting of agapanthus.

"The terracing, of course, has played a major role in dictating the character of the garden. It has imposed a strict architectural setting, a frame into which I have tried to work a mixture of loose and tight plants. By varying from light to dark and changing from narrow to broad, I have been able to create an illusion of space, the garden appearing much larger than it actually is . . . Fiorentina, as far as concerns my life, is miles away, the Clos being altogether another idiom, small and intimate, surrounded by its hidden gardens."

Roderick Cameron's account is very revealing: it not only shows how important he considered the relationship between garden style and the architecture of the house, but it throws light on his personal life and frame of mind at that particular time. These gardens are located on adjacent sites on Saint-Hospice's Point, but their exposure and topography differ considerably. La Fiorentina is set at the tip of the peninsula like a lighthouse, with different garden areas off each side of the house.

Le Clos, a *bastide*, faces the terracing which is so typical of the region, rising up from the sea. The garden plans derived from physical considerations relating to the spatial arrangement of the architecture and from Cameron's own vision, heavily charged with references (models, memories, and fantasies). Roderick Cameron took advantage of this opportunity to give expression to his personal code of values and principles; at the same time he analyzed the inherent quality of the site. His basic concept was to plan a garden in terms of the feeling he had for the site. An interesting comparison can be made of his treatment of alleys in these two gardens. At La Fiorentina, broad avenues lend themselves to gather and provide access to large spaces well-suited to social functions. Paths and stairways at Le Clos, however, are so narrow that two people cannot walk side by side. La Fiorentina's "green rooms" extend interior space out into the open air and effect a transition with the pine forest and rocky shore, whereas at Le Clos, Cameron capitalized on the agricultural theme embodied by the terraces, using it to create a place where one can sense an underlying concept of time, conducive to quiet reflection.

. . . .

Mossy alley.

Fantastic shapes of pines.

The Jardin Exotique in Monaco, a garden on gigantic rock

Located on a site with an exceptional microclimate, this botanical garden contains cactus specimens as large as those in their native countries (Mexico, Argentina, etc.), measuring well over a dozen yards in height and weighing two to three tons. But many other "succulent plants"—formerly referred to as "fleshy plants"—have grown to considerable dimensions, including the branched euphorbia, aloes from the African continent, and agaves from the Americas.

This garden of xerophytes (plants adapted to drought conditions), planted on an immense rock unique in the world, attracts over half a million visitors annually. These strange-looking plants, which look as if they came from another planet, grow at an astonishing rate. This is due to the garden's proximity to the heat-storing Mediterranean, the protection afforded by a screen of high mountains and its location on a 45-degree slope that allows rapid surface drainage and maximum heat accumulation in winter.

In 1897 the gardener-in-chief at Monaco's Saint-Martin gardens began to collect varieties of cacti, which were novelties at the time. Prince Albert I, who was not only a great naturalist but one of the first oceanographers, took an interest in these plants; plans were made to create a special garden for them along the newly completed route de la Moyenne Corniche. The site was prepared by Louis Notari, civil engineer and deputy mayor, who met the awesome challenge of shaping the cliff site with talent. Walks were laid out, artificial boulders put up, footbridges built, enormous blocks of rock moved, and the necessary amount of earth was brought in. Finally, on February 13, 1933, the doors opened to the public, and since then, 14 million people have come to admire this unusual garden, the product of human resourcefulness combined with the great natural adaptability of plants. The landscape is unlike anything to be found elsewhere, for here, on less than three acres of principality land, one finds, as Paul Guth phrased it, "a tropical fairyland at the same parallel of latitude as Vladivostok." On this rock face prosper speci-

Construction site, 1933.

The garden today.

mens of the typical flora of Madagascar, Chili, Africa, and the Americas.

For about thirty years the Centre Botanique de Monaco has specialized in plant biology, and its collections of succulents is one of the most prestigious in the world. Many species are grown in greenhouses and shelters and are studied by researchers of all nationalities. Every year interns come here to study the particularities of these plants. The collection continues to expand by means of exchanges and botanizing expeditions. Lectures on succulent plants are held, and the garden takes part in international flower shows and exhibitions. The Jardin Exotique, like the principality of Monaco, may be small in size but its sphere of influence is truly international.

. . . .

Large cacti.

In bloom.

Torre Clementina and Villa Cypris, on the gender of gardens

Until 1861, the year in which the Roquebrune area including Cap Martin became a part of France, Cap Martin was officially part of the hunting grounds of the princes of Monaco. Up until the eighteenth century, the princes held stag hunts there. In 1886 a director of a Monte Carlo pigeon-shooting establishment set up a paying hunting area where one could bag rabbit and pheasant. But, scarcely two years later, a real estate company acquired and subdivided most of the peninsula and built a grand hotel out on the point. Stephen Liégeard described the latter in his book, *La Côte d'Azur* (1894): "An English-style hotel called the Cap Martin Hôtel rose rapidly from the earth, a leviathan of luxury and comfort. One cannot think of anything it does not have: telegraph, telephone, areas to park carriages in, and even carriage service to Monte Carlo." Its clientele included grand dukes, princes, and sovereigns and the nearby luxury development area still known today as the *Domaine* was soon covered with elegant and grandiose mansions. Empress Eugénie had her celebrated Villa Cyrnos erected there. On the western shore of the cape there were two parallel rows of turn-of-the-century houses characterized by the eclecticism of the age. Among these were Torre Clementina and Villa Cypris, both built just after 1900. The gardens of both houses were created by the same Italian artist, Raffaële Mainella, summoned by the proprietor of Torre Clementina from Venice, where he lived and showed his paintings. Relatively little is known about Mainella who was, according to Ferdinand Bac's thumbnail sketch, an "architect from Naples and former hairdresser's assistant who landscaped the gardens at Torre Clementina and Villa Cypris." He was a native of Benevento in Campania, where Roman, Byzantine, Arabic, and medieval architectural influences coexisted. All of these influences, with an added Venetian touch, emerged in Raffaële Mainella's interior decorating and landscaping. He chose brick as the primary material and stone for sculpted elements, which were either imported from Italy or, as in the case of repeated motifs, executed on site. With regard to Torre Clementina, Ferdinand Bac wrote: "Decorative elements taken from China and Japan, Syracuse and Trebizond, not to mention the translucent marbles of San Miniato, are used in an attempt to create an air of mystery; there is an accumulation of effects lacking nothing but a degree of Christian restraint . . ." Ernesta Stern was a woman

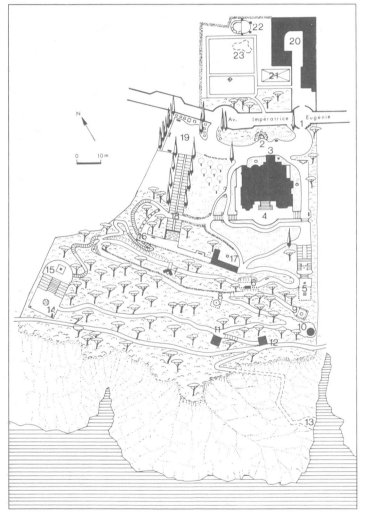

Plan
1. Portal with arch - 2. Grotto-fountain - 3. Torre Clementina - 4. Paved terrace - 5. Garden of the sundial - 6. Bridge with belltower over the old waterfall - 7. Arch - 8. Fountain - 9. Urn under pergola - 10. *Poivrière* - 11. Kiosk - 12. Portal in style of ruins - 13. Harbor - 14. Terrace and fountain - 15. Monument - 16. Colonnade - 17. Gallery - 18. Alley with grass treads - 19. Square courtyard - 20. Commons - 21. Greenhouse - 22. Greek theater - 23. Site of the Japanese garden.

The Stern House.

of letters who had published tales and poetry under the pseudonym of Maria Star. She was a central figure in Riviera social life and, at her own social events, tended to mix ''seers and Russian dancers with heads of state, singers with bishops.'' Ferdinand Bac went on to say that she ''took herself to be a reincarnation of Semiramis, held seances, interrogated Moses, Jesus Christ, and Napoléon who miraculously answered to the call of gifted mediums.'' Mrs. Stern demanded that the setting be worthy of her esoteric practices. Architect Lucien Hesse designed a keep for her in the Romantic style, complete with wooden corbeling, small coupled columns, and sculpture in low relief. Its masonry was a rather baroque combination of stone, brick, and pebblework. As for the interior décor, Mainella went back to the Ottoman Empire for inspiration. Mosaics, sculpted stucco, translucent onyx panels, and a central domed salon rising up on four pillars created a suitably Ottoman ambiance. In the small entrance hall there was a fountain in front of a window giving directly onto another fountain outdoors, playing under the trees. The garden contained two or three symmetrical compositions, but aside from these, it was a picturesque park typical of the early twentieth century. Greater emphasis was placed on original site features than in the previous century. An alley wound down from house to water through the pines, punctuated with colonnades, galleries, arches, fountains, and kiosks where the visitor could stop off along the way and indulge in a little mystical meditation. A Japanese garden—miniature, of course—with the inevitable small lake and pagoda, and a tiny Greek theater, both set at some distance from the main path, provided the finishing touches to this ensemble. According to the Countess Sanjust di Teulada, granddaughter of Ernesta Stern, one of the preferred amusements of the brilliant Riviera society was *pétanque*, the French bowling game usually practiced by the ''common people.'' The countess reported that silver basins were placed on either side of the bowling green so that the players might wash their hands once

Sectional view
1. Sea level - 2. Sea-side path - 3. Kiosk -
4. Fountain - 5. Galerie - 6. Torre
Clementina - 7. Avenue Impératrice
Eugénie - 8. Site of the japanese garden -
9. Greek theater.

Grassy steps toward the hill.

Arch over garden path.

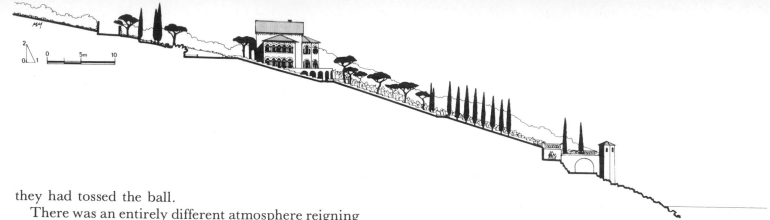

they had tossed the ball.

There was an entirely different atmosphere reigning at Villa Cypris, a reference to one of the names of Aphrodite. The proprietor was Mrs. Robert Douine, mother of Virginie Hériot, who, in the 1920s, broke many a yachting record on her sailboat, "Ailé 4." Here we find a more Byzantine look. By the time Raffaële Mainella had finished with Torre Clementina next door, the house had already been built by Edouard Arnaud, architect-in-chief at the French national institute for civil buildings and palaces. The stonecutter from Menton who worked on Villa Cypris had a son, Pascal Molinari, who later became a professor at the prestigious engineering school in Paris, the Ecole Centrale. For years Molinari referred to the Villa Cypris in his course on construction. Generations of graduates went off into the world with the plan, section, and elevation drawings of this villa in their heads. Inside, one is confronted once again with Mainella's excessive penchant for columns,

mosaics, and translucent panels. The obvious landscaping solution was chosen: the space was left wide open, facing the sea, allowing Mrs. Douine to keep track of her daughter as she tacked backward and forward on the water. A tour of the garden first takes you around the house to a rounded terrace where the view of the water hits one—almost literally—right in the face. Two flights of stairs finally meet in the middle of the slope and continue down to a transversal gallery open to the water, whose surface is agitated at this particular spot by currents moving in opposite directions. To the right and left of this podium, which is open on all sides, are pines; sinuous paths, rocks, and dwarf palms are dispersed throughout the pinewoods. One cannot help but notice that this part of the garden was laid out according to a strictly controlled plan. Other, less imposing elements hidden in the trees finish off the design.

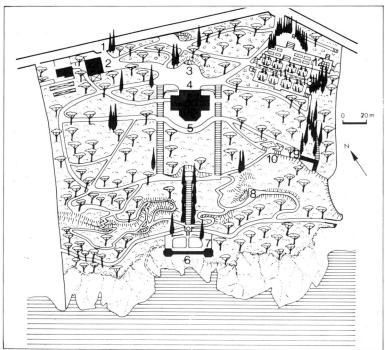

Sectional view, Villa Cypris.

Galerie entrance.

Plan
1. Portal with arch and belltower -
2. Housekeeper's lodge - 3. Fountain -
4. Villa Cypris - 5. Terraces with
plantations - 6. Gallery and colonnades -
7. Mosaic chapel - 8. Pond in the rocks -
9. Venetian temple - 10. Bridge above former
waterfall - 11. Moorish garden with canal and
water stairway - 12. Pergola mosque in ruins
- 13. Waterfall.

Stairs and galerie at water's edge.

Beyond a little bridge under the pines is a second colonnaded gallery, known as the Venetian Temple, set on a bluff over the water. Farther up is a Moorish garden with, amid the dark green of large yew trees, a long pool and lavender parterres set in brickwork. On a terrace, planted with orange trees overlooking the canal, is to be found one of the most surprising garden monuments on the entire coast: sixteen marble pillars surmounted by brick arches covered with climbing roses. This pergola is reminiscent of a "mosque in ruins," wrote Gertrude Jeckyll, writer and landscape architect, in a description dating from about 1920.

As the gardens at Torre Clementina and Villa Cypris are adjacent to one another, garden-lovers may visit them on the same day. There seems to be a curious relationship between these gardens, done by the same landscape architect for two very different women. In the words of Ferdinand Bac, Ernesta Stern was "enamored of grandeur, of power through the ages, and her white,

massive form was adorned with Merovingian pendants." Yet along the serpentine garden path at Torre Clementina, garden monuments and pavilions are revealed amid the pines like feminine charms unveiled one by one. On the other hand, the story goes that Mrs. Douine started out as a salesgirl before she married the director of several Parisian department stores, who gave her a free rein with his fortune. The strictly controlled composition of the Villa Cypris seems to echo with the sound of a command and emanates virile authority. Both gardens, independently of gender-related attributes which one can imagine in them, give onto the immense horizon of the Mediterranean. But at Torre Clementina the view is revealed through stands of trees, whereas it is served up in brutal fashion at Villa Cypris. The Semiramis in Ernesta Stern would have been more at home at Villa Cypris, whereas if one tried to express Mrs. Douine's nature in garden terms, one might very well have come up with something like the tortuous alley at Torre Clementina. One is very tempted to think that Raffaële Mainella switched the designs around, executing the perfect garden for each woman but at the neighbor's house.

. . . .

Venetian temple.

Pergola, Moorish garden.

La Serre de la Madone, the second garden planned by Lawrence Johnston

As we drove down the Gorbio to Menton early in March, we came across the entrance to a garden sloping up the west side of the valley, located just outside the Menton city limits. We stopped and could see, just inside the gate, large Asian magnolias whose blossoms were just coming out. A gardener busily trimming a hedge informed us that the grounds had been landscaped by an English gardener. This was none other than the Mediterranean garden of Lawrence Johnston, who had previously designed one of the most celebrated gardens in England: Hidcote Manor.

Several days later, with the authorization of the owner, we entered the gate, strolled along alleys, and admired the water garden and shade garden. We went past the house and patio up through the tangle of native *maquis*, taking paths running along the slope from top to bottom. At La Serre de la Madone, there is a haunting atmosphere which is easier to describe than forget. ''The rhythm of our life here is different from other people's,'' says Mrs. Jean-Claude Bottin, daughter of Lawrence Johnston's former butler, who spent her entire childhood on the grounds and now lives at the edge of the property. ''It is a world apart.'' No doubt this impression is due to the way indigenous and imported species grow in such a tangle. The erstwhile botanical garden, formerly maintained by a gardening staff of twelve, has been completely overwhelmed by its main component: the plants themselves, which have been left largely untouched for years. This is a quite a fascinating sight. One can measure the effects of time at La Serre de la Madone and the impact the latter has on visitors. ''At twilight, in this garden, I have had the feeling that time comes to a halt,'' said one friend accompanying us. Another declared: ''When I went to Venice, I felt Venetian, but here I feel at home.'' Lawrence Johnston planted trees thickly, guided by his typically English love of woodlands. Could he anticipate that this garden would become a perfect place for introspection? This was his winter house, a sort of botanical midway point between the far-flung tropics encountered

Glasshouse entrance.

Shade garden.

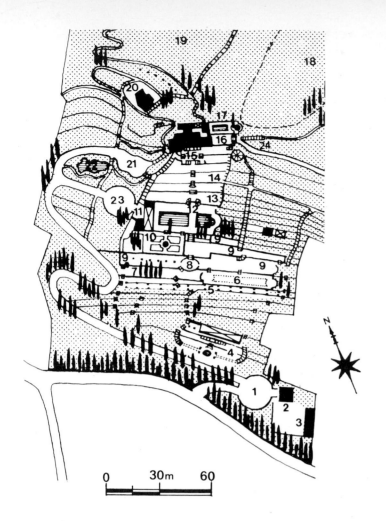

in his travels and his English estate. A review of the events in his life enables one to better understand how the Menton estate fit into his life-style.

Lawrence Johnston was born in Paris in 1871. His parents were American, and his tutor French. In 1894 he went to England and spent three years reading history at Cambridge, where there was an experimental botanical garden. Johnston opted to become a British citizen and took part in the Boer War. He traveled throughout South Africa, the place of origin of many plants which have been introduced in Europe. After several years on a Scottish farm, in 1907 he bought Hidcote Manor in the Cotswolds, in the west of England. Here, he planned a garden which would become a landmark in English gardening history and a mine of ideas for future generations. After the First World War, he turned to his second garden-planning endeavor near Menton, where he intended to put into application his ideas about the Mediterranean garden. Johnston went

Plan

1. Entrance courtyard - 2. Gardener's house - 3. Garages - 4. Garden and cold frames - 5. Alley covered with clematis - 6. Green room - 7. Terrace of the succulents - 8. Rondel of boxwood - 9. Terraces on axis overlooking fountains or sculptures - 10. Shaded garden - 11. Winter garden - 12. Water garden - 13. Lawn - 14. Four terraces with plantations - 15. Main house - 16. Paved courtyard of the mandarine orange tree - 17. Patio - 18. Former aviary -

19. Wood of pines, green-oaks, and imported species - 20. Casa Rocca, guest house - 21. Recent parking - 22. Recent Japanese garden.

Reflecting pools and orange grove.

on several botanizing trips for the purpose of personally selecting the specimens he wanted for his garden. At La Serre de la Madone, plants from the southern regions of Africa and China unable to survive in the Cotswolds could be readily naturalized. Hidcote Manor is often considered exemplary of a new school of twentieth-century English gardening. Its design shows a return to formalism without excluding natural areas. Johnston placed great emphasis on the use of small garden spaces and on color effects, which he obtained by combining familiar plants with the latest botanical and horticultural discoveries. Although these tendencies had surfaced before 1907, Johnston perfected their application. It must be recalled that nineteenth-century English society had undergone profound changes which were mirrored in English gardens. Vast landscaped parks fell into decline, neglected by owners who could not or would not spend what was necessary for their maintenance. New owners settled in the countryside and wished to landscape their own gardens. In England, a reaction set in against the conventionalism of Victorian parterres and the systematically eclectic gardens of the period. Typical of this movement was the Surrey School, named for the county where its members met, whose members included architects, painters, journal-

ists, horticulturalists, and landscape gardeners. The most prominent figures were Gertrude Jekyll and William Robinson, who were both landscape gardeners and writers. The Surrey School advocated the expression of a theme when designing a garden closely related to the architectural style of a house. It sought new sources of inspiration in small, village gardens and in the ways plants ''associate'' in nature. It rediscovered the Elizabethan garden, in which individual garden scenes could be multiplied by compartmentalizing overall available space. Lawrence Johnston subscribed to these precepts but had the additional benefit of his own expertise, described by author Harold Nicolson as ''a calculated alternation between the elements of expectation and surprise.'' At Hidcote Manor one discovers, one after the other, about twenty small gardens of interlocking design within an overall plan organized along two, unequal axes forming a tee. These spaces are highly differentiated outdoor ''rooms'' surrounded by hedges, each centered on a given focal point: a fountain or pool, the volume of specific plantings, a color combination or the unique coloring of foliage and blossoms. Lawrence Johnston's ''tapestry'' hedge (planted with several greens) and his borders associating plants of varying sizes showed a sense of daring and simplicity which was

Patio.

new in the gardening world. This spirit of invention marked an entire generation of gardens in England; the best known is Sissinghurst, in Kent, planned around 1930 by Vita Sackville-West and Sir Harold Nicolson. La Serre de la Madone is much less complex in design than Hidcote Manor, whose country setting imposed no particular topographical restrictions. The Gorbio valley estate obliged Johnston to reckon with the configuration of the site and its narrow cultivated terraces. Terraces lie below the house, halfway up the hill; across them run a straight walk and two narrow lateral paths. In the center a series of landings and single or double staircases flank fountains in niches and evoke the layout of several Renaissance gardens in Italy. Both ends of most of the transversal paths were originally marked by a piece of statuary, but the formal part of the garden was not separated from more freely planted areas stretching up the slopes on either side and above the house. Johnston took the existing farmhouse and added on to it; he had bought a dozen small properties to piece together his estate. Accompanied by a pack of dachshunds and a house staff of ten—not to mention his gardeners—he used to spend winters here before the war. His life-style, like his manner of arranging plants, may have been nonconformist but that did not prevent him from receiving the *crème de la crème* of international society on the Riviera. In 1936 Ernest de Ganay described him in an article in the *Gazette illustrée des amateurs de jardins*, a French garden magazine : "He comes to greet you in corduroys straight from his terraces, with dirt on his hands like a gardener, or like a hunter with his dogs behind him. Yes, he is a hunter, a skilled plant hunter, but he gives life to plants instead of taking animals' lives."

At La Serre de la Madone, there are few small exotics, because the accent was placed on tall varieties.

Terraces, center of garden.

One can still see the *Mahonia lomariifolia* and the *Mahonia siamensis*, brought as seeds from China, which successfully naturalized on the site. When Lawrence Johnston died in 1958 at La Serre de la Madone, his heir Nancy Lindsay placed the rare plants on the estate at the disposition of the botanical garden in Cambridge, where Johnston had spent his university years, and had the statuary and large earthenware pieces in the orangery removed. Trees that were too large to transport remained on the grounds. Miss Lindsay was a botanist and a variety of hybrid mahonia bears her name. Around 1960 the third proprietor of the estate altered one of the reflecting pools in order to put in a swimming-pool and an incongruous miniature Japanese garden near the house. Then La Serre de la Madone passed into the hands of Count Jacques de Wurstemberger, who preserved the compositions of Lawrence Johnston during his period of ownership, but the estate has once again just been sold.

Sculpture, water garden.

Bust of Antinoüs.

Circular arbor.

There are two approaches to this garden. Like the colonel's visitors, one can come up the winding avenue to the large circle and cross the orangery to the heart of the formal garden, or from the entrance courtyard, one can take the path interrupted by stairs that goes under the trees to the left of the first greenhouse. The latter enables one to come up the slope and make a gradual discovery of what each terrace has to offer: the alley covered with clematis, the outdoor "room," the terrace planted with succulents, and so on. When one reaches the *jardin d'eau*, a splash of lovely ocher halfway up the hill draws one's attention. This is the façade of the house, whose color is in perfect harmony with the green of its surroundings. Its upper floors give directly onto southern terraces. The first is extended by a narrow staircase going up a steep wooded slope comparable to the mount in the Villa Medici garden in Rome; the second might be called a hanging patio, with its arcades, basin, and fountains. Continuing one's upward path via either of these paths, crossing the collections of exotic acacias and yuccas, half-buried under oaks and pine, one may come across one of the gates, now in ruins, of the immense aviary that covered a section of the forest. Mrs. Bottin still remembers the names of the birds that lived there, among them seven big macaws, a hornbill, a crowned crane and a demoiselle crane, peacocks, ibis, golden and silver pheasants.

....

The house, view from water garden.

Val Rahmeh, a botanical nest

Val Rahmeh, the botanical garden belonging to the Museum of Natural History in Menton, offers visitors a sumptuous autumn flowering, whereas gardens elsewhere on the coast cannot muster up more than a few rearguard rose-laurels and plumbagos, roses coming back into bloom, yuccas from Guatemala raising their white spikes, strelitzias, asters, and cyclamens. The majority of plants opening up at this time of year come from the Southern Hemisphere, where the spring season coincides with our fall, and the remainder are plants indigenous to hot or temperate climates above the equator which open their corollas after a period of rest in summer.

In October, on the terrace at Val Rahmeh, the blues of *Thunbergia grandiflora* and *Solanum ratonnettii*, the delicate pink of *Podranea brycei*, a bignonia from southern parts of Africa, the reds and oranges of the hibiscus and lantana, and the pink-tinged whites and pale yellows of the tropical datura provide a range of hues perfectly orchestrated around the yellow-ocher patina of the house. Entwined in the terrace balustrade is *Rosa laevigata*, originally from China, its five pure white petals balanced around the golden bouquet of its stamens. Farther down, in the terraced garden set above a small valley, the amaryllis holds up its pink trumpets without showing a single leaf, the sulfur-colored hemerocallis glistens beneath orange trees laden with green fruit, and varieties of Mexican sage form a mauve and blue mist. From Mexico too comes a variety of bamboo, *Arundinaria longifolia*, with leaves as finely textured as feathers. Dahlias from American tropical climates begin to open their enormous dark red flowers, and the vibrant deep blue of *Salvia caerulea*, a type of South African sage, borders the path leading up to the house. In 1925 Lord Percy Radcliffe, governor of Malta, built this neo-Mediterranean house with its many loggias and arcades and had the first exotic plants put in the garden. Miss Campbell, the next-to-last owner of the property, started to make Val Rahmeh into a true botanical garden in 1950.

Villa terrace.

Behind the house, a small olive grove spared by the pruning knife is underplanted with shrubs, some of which are autumn-flowering: white-blossomed olearia, phylica and malvaviscus, and *Osmanthus fragrans*, whose minute flowers give off a heady fragrance. On the east slope of the valley lies a wilder part of the garden; chamaerops, arecastrum, yucca, dasylirion, cordyline, and cycas present sundry stiff and supple forms and a wide range of dark, blue-tinted, and bright greens. A trickle of water can be heard down in the deep valley long before Humphrey Waterfield's fountain comes into sight. Miss Campbell asked Waterfield, an English landscape architect who owned a garden in Menton, to design a fountain for the spring on this spot. Today, at the foot of a tangle of filao, old olive trees, and brachychiton, waters from the spring gently fall into a moss-covered basin in which ferns have made their home.

. . . .

Spiny trunks of Brazilian Chorisia speciosa.

Palm-lined drive.

Fontana Rosa,
on novelists and gardens

On the road to the Pian olive grove, which is one of Menton's parks, a gateway with an unusual iconography attracts the attention of passersby. Painted ceramic portraits of Balzac, Cervantes, and Dickens figure above the inscriptions: *Fontana Rosa, Le Jardin de Romanciers, El Jardin de los Novelistas* ("The Novelists' Garden"). If one glances through the fence, one can see small buildings decorated with polychromed faience, a colonnaded hemicycle, and empty pedestals among the trees. This was the garden of Spanish author Vicente Blasco Ibañez.

This novelist, born in Valencia in 1867, defined himself as a political agitator. His life was full of adventures of all sorts. He wrote pamphlets opposing the monarchy and fought duels with ideological adversaries; the wrath of justice finally came down on his head with several convictions and exile. Yet it was he who founded *El Pueblo*, the first socialist newspaper in Spain, and served for ten consecutive years as an elected Republican representative from Valencia. During this period he continued to write and publish novels describing various social environments with which he was familiar.

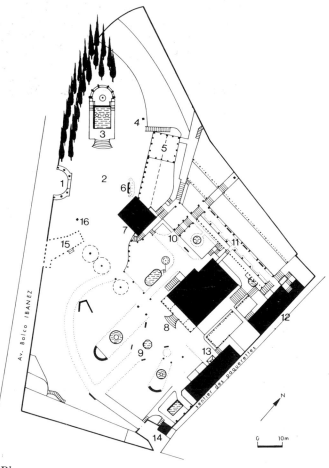

Plan

1. Entrance with the novelists' portraits -
2. Entrance courtyard - 3. Cervantes' rotunda - 4. Pedestal of the bust of Dickens -
5. Aquarium - 6. Monument of V. Blasco Ibañez, former site of the bust of Flaubert -
7. Gardener's house - 8. Villa Fontana Rosa, site - 9. Pools and benches covered with ceramic tiles - 10. Staircases of Balzac -
11. Pergola leading to the cinema -
12. Cinema - 13. Library - 14. Belvedere -
15. Pergola above the former garage -
16. Site of the bust of Dostoïevski.

Portraits on the gateway.

The Cervantes rotunda.

V. Blasco Ibañez in his garden.

0   10m   20

In 1908, weary of political life, Blasco Ibañez embark-
ed for the New World. He gave lecture tours on Spanish literature in the United States and South America
and decided to try his hand as a planter in Argentina.
Within four years, he had established two villages:
Nueva Valencia in northern Argentina and Colonia
Cervantes in Patagonia. [28] He then returned to
Europe, but still in exile, he lived in Paris, and then
on the Riviera. When he came to Menton, the rights
to many of his works had just been purchased by American film studios. At Fontana Rosa a projection room
was built especially to show films based on his novels:
The Blood of the Arena, Mare Nostrum, and *The Four Horsemen of the Apocalypse*. The Menton property purchased
by Blasco Ibañez in 1922 was made up of narrow
cultivated terraces planted initially with olive and lemon
trees, subsequently with cutting flowers. They overlooked flat ground where a villa dating from 1880 stood,
in a garden full of large ficus and palm trees. Two small
houses adjoined the main house, where Blasco Ibañez
and his second wife, Elena, moved in. One of these
houses was occupied by Ramon, the gardener who had

Sectional view of the garden.

Ceramics and rambler roses.

Rotunda colonnade.

come with him from Spain, and the other was converted into a library building with five individual book-lined library rooms inside. Over the preceding years the author had lived episodically in Valencia, Madrid, Buenos Aires, Paris, and Nice, and his books were scattered among his various places of residence. With all fifty thousand or so of his books conveniently near at hand, he felt a desire to live among three-dimensional representations of the literary "figures" who meant the most to him. He commissioned a Russian sculptor, Leopold Bernstamm[29] to execute a series of bronze busts, starting with Flaubert, and continuing with an ever-longer list of names: Balzac, Boccacio, whose humor he appreciated, Cervantes, who to him was a cult figure, Dickens, Dostoyevsky, Goethe, Hugo, Poe, Stendhal, Tolstoy, Zola, and, finally, Beethoven. Some of the pieces were placed in his study, but the majority were set upon pedestals and put out in the garden. This pantheon required worthy surroundings; Blasco Ibañez had architectural elements put in that recalled city parks in Spain. The imposing entrance to Fontana Rosa, symmetrically placed columns and vases, and

vine-covered pergolas could be compared to El Jardin de Monforte in Valencia, whereas the fountains and colorful ceramics bring to mind the *alcazars* and patios of southern Spain. Long, curved benches covered with faience were placed in the shade; the faience either came from Manises, a small town near Valencia which has specialized in ceramics ever since the Moorish occupation, or from a craftsman in Menton. These hospitable benches were well-placed for one to listen to the playing of the fountains, whose music seemed to vary as one changed seats. This was a perfect place to relax with friends or follow the example of Blasco Ibañez, because "not a day went by but that three or four hours were spent reading." This was not the first garden of this type: in Seville, the summerhouse by the Alvarez Quintero brothers in the Maria-Luisa Park served at that time as an open-air reading room. It formed a fronton containing bookshelves on either side of a fountain, extending out to a reflecting pool surrounded by benches for readers. The hemicycle at Fontana Rosa, dedicated to Cervantes, also gives onto a pool, and the bench-back following the curve of the colonnade is adorned with one hundred enameled scenes from *Don Quixote*. This was Blasco Ibañez's tribute to the man whom he considered to be the father of the Spanish school of realism. The story goes that he himself planted the cypress hemicycle surrounding the exedra.[30]

Benches and fountain in front of the house.

The nearby classical revival building contained the fish ponds in which Blasco Ibañez raised fish brought back from his travels. At the opposite end of the garden, a tower bearing the name of the estate was erected to provide a ''view of the water''; an elevator was to have been installed to give access to the top terrace, but the installation was never completed.

The inscription, *Fontana Rosa*, written on the gable over the projection room, could be read from the railroad track edging the property; visitors leaving the train at the Menton-Garavan station were thus informed of where this celebrity lived.

In 1928 a few years after creating his small literary world, Blasco Ibañez died. Ten years later, his wife returned to Chile, land of her birth. Before leaving Menton, she commissioned Bernstamm to execute a monument that she intended to present to the city: a bust of the writer, placed on a sort of altar surfaced in ceramic, to be erected at the bottom of the old town. However, a newly elected mayor who looked askance at the ideology of the author of *The Four Horsemen of the Apocalypse*, and the official inauguration ceremony was postponed a number of times. Elena wrote from Chile to request that the monument be moved to Fontana Rosa; the bust of Blasco Ibañez replaced Flaubert in the place of honor, facing the gateway. During the Second World War the Novelists' Garden was damaged during borderline skirmishes: the house was pillaged, the estate neglected completely. Sigfrido, son of Blasco Ibañez, sold the upper part of the grounds in 1970, and a line of four-story buildings went up. He gave the garden and houses to the city of Menton. The busts and remaining books were put in storage. The gardener's house was occupied once again, but the garden itself was abandoned for several years. Grass eventually masked paved alley motifs and ivy grew over benches and fountains. Recent plans, however, aim to restore the Novelists' Garden. The main house no longer exists, but the library building still stands. When Blasco Ibañez's books once again line the shelves, the busts replaced on their pedestals, the fountains turned on, and the faience surfaces restored, Fontana Rosa may rediscover its literary purpose, as Menton's ''reading room'' garden.

Bench under the pergola.

The adventures of Don Quixote: one hundred pictures adorning hemicyclic bench-back.

Curved bench.

Bench near the house.

Les Colombières, the garden of Ulysses

From the seaside promenade by the old town, one can see Les Colombières, a red house with a symmetrical roofline against a cypress screen, above the white tide of recent apartment buildings on the Italian border end of the walk. On either side of the *corniche* that takes you to the Les Colombières driveway are Swiss chalets and a varied assortment of turrets, verandas, and ceramic balustrades, but once past the gate, another world opens up. On a façade which has almost no openings, an inscription reads: *Inveni portum, spes et fortuna valete, sat me lusistis, ludite nuc alios* (''I have found my harbor, Hope and Luck farewell, all too often you have used me as your plaything, now find others to toy with''). Indeed, the themes worked out by Ferdinand Bac at Les Colombières were the farewell to the traveler's existence and travel memories. This is the only one of Bac's gardens to have survived intact. Bac, a repented caricaturist and chronic traveler, had been all through Europe and the Mediterranean region, writing and painting as he went. Halfway through his long life, he set about landscaping

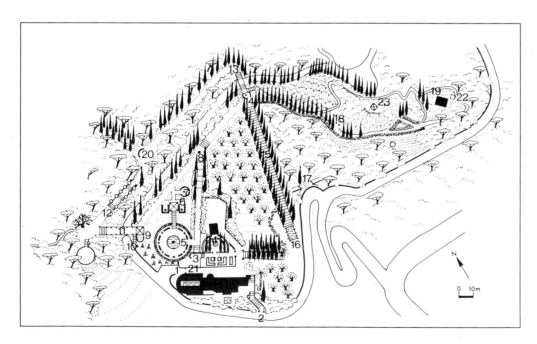

Plan

1. Main entrance - 2. Small entrance and fountain of the doves - 3. Alley of the fountain of Nausicaa - 4. *Trompe-l'œil* garden - 5. Rotunda and obelisk - 6. Palladio's casino - 7. Slave with necklace - 8. Child with butterfly - 9. The Bella Vista - 10. Staircases of the Philosopher - 11. Bridge leading to the carob tree - 12. *Allée des Jarres* - 13. Bridge of the quarry - 14. Nymph, by Jean Goujon - 15. Alley of the quarry - 16. Spanish pool - 17. Red bridge - 18. Rock of Orpheus - 19. Gardener's house - 20. Roman bench - 21. Homer's garden - 22. Altar of Niké - 23. Mausoleum.

Rotunda cypresses, view from garden entrance.

his friends' gardens. He would move in and work on the design until he felt it was right; sometimes his design would call for alterations in the architecture of the house. In 1919 Mr. and Mrs. Emile Ladan-Bockariry asked him to landscape their Menton estate, giving him a free hand and allowing him to take as long as he deemed necessary. Bac became so attached to this garden that he never left his friends again: a mausoleum for all three of them was erected on a rock overlooking the garden. In the foreword to his book, *Les Colombières, ses jardins et ses décors*, Ferdinand Bac elucidates his overall conception: "The idea was to start with small private spaces near the house, with a secret garden enclosed by walls and gates, and gradually, as one moved away from the dwelling, amplify the movement with more daring geometrical arrangements; these would in turn defer to a rebellious natural topography full of grandeur, full of hills and dales, ravines and rocks, stubbornly resisting any attempt to tame it; the culmination would be reached with a promontory and view of the last Alpine range plunging towards an infinite horizon . . . On this rather small property, we worked up the scale from small spaces up to infinity, taking care to leave most of the grounds in their natural state. In these areas, we set isolated, small garden monuments, placed with respect for the character of this natural setting which, if truth be said, required no help from human hands."

Bac's humility was superfluous. He knew full well that he had succeeded in his celebration of the site and fulfilled his own artistic potential. The monuments are of course what provide the elements of surprise. Several of these pavilions, colonnades, and bridges, and even the mausoleum, "frame" vistas over the old town, nearby capes, and rocky peaks. This *camera oscura* effect seems to bring distant landscapes into closer view and convert them into decorative elements of the garden: "To make the best of topographical features, we chose precise parts of distant landscapes and invited them up into the orchard, where their beauty could provide us with the finishing touch."

Carob tree (drawing by François Cayol).

Thus he used architectural elements in the garden to frame these *tableaux*. This was standard practice in classical gardens to create a focus on a pool or statue, but here, in several cases, Bac selected trees as the focal point: "One day, as I was thinking about how to place a colonnade . . . my gaze fell upon the *Arbre-Dieu* and a feeling came over me that I had to bridge the gap between myself and that tree . . . And so, this bridge was built, leading to the carob tree."

Other trees were also given special treatment: "Below this cliff, we placed a red bridge. Must I admit that it was not needed to get from one bank to the other? What had been lacking was a bridge to the cypress." The red bridge has since collapsed, but perhaps some day it will be rebuilt to preserve Bac's original intentions. Cypresses are a major component in this garden. There are cypress hedges, cypress avenues, and cypress arches in between columnlike trunks of isolated trees. At Villa Croisset, in Grasse, Bac had patterned the use of immense "dark pyramids shading the patios" after the cloister in Assisi. He felt that the cypress and the "eternal and twice sacred" olive allowed him to approach the natural structure of this landscape with sensitivity and nostalgia, and ultimately to restore its Latin character. In 1922 he wrote the following in *L'Illustration*:

"Ever since the day that Lord Brougham planted the first palm tree in his garden in Cannes, nearly a century ago, the Mediterranean villa has been treated in an exotic manner that defies all natural traditions of the land."

Nymphaeum.

Mausoleum.

In the same article Ferdinand Bac justifies his choice of deep colors for architectural elements: "For over forty-five years, local builders, no doubt hypnotized by building habits current over in Algiers, have idolized enamel paint and introduced raw white for all exterior building surfaces. By using enamel paint, plaster, and stucco finishes, they have produced chalky, blinding monotony in a landscape that was made for color. White may be ideal as a complementary color, useful in creating a tidy effect, but it has been employed excessively, to the point of becoming a veritable optical affliction. All those white cubes reflecting the scintillating glare off the water are intolerable to look at ... I wanted to distill the wisdom and transparent logic produced by a millennium of rural living on the land. In Carpaccio's paintings, I finally found this red surface finish; when I held it up against the background of olive trees covering the mountains in Grasse, I knew that I had discovered their complementary color."

Umbrian cypresses and Venetian red were not the only elements that Bac brought back from his travels. In his book on Les Colombières, he enumerated his sources of inspiration; there are many of them, but common threads can be detected: "When traveling among the things which the forces of progress had left standing on the shores of the Mediterranean, I never left one of those special places, where genius had left its mark over generations, without wishing to take souvenirs with me ... But is the fetishism of the past any less ridiculous than that of the present? Each is absurd in its own way. My fundamental objective was to choose native Mediterranean forms, detect the special characteristics defining them as a product of a specific period, religion, or reign, and effect a synthesis clear enough to pinpoint the ancestral sign unifying them all into a single family bound by geography, climate and cultural origins ... On this small estate, the sole element I worked with was, precisely, this bond, born of a thousand years' common existence on the shores of the Mediterranean, a bond that holds all of us ... Thus, at Les Colom-

Bridge leading to the carob tree.

bières, we find hints of Tuscany mingling with visions of Venice, Attic Greece and the Spain of Philip IV, and one half expects an Argonaut or two to pop up in the garden.''

Of course these reminiscences, expressed in keeping with the surroundings and the scale, constituted an interpretation by the author; this interpretation can be called his style. Elsewhere in this book Ferdinand Bac defines the criteria that guided him at Les Colombières to produce a composition that, although it borrows heavily from the past, is a sort of manifesto for the contemporary period: ''When, in 1913, I first attempted to renovate the Mediterranean garden by means of simplicity, the idea clearly came to me that we should go back to *simple* materials, those used in indigenous rural architecture. A garden can only be based on two principles: that of a genuinely natural setting, rendered accessible, or else a deliberately geometric design, which allows one's imagination to range freely within a disciplined framework. In my imagination, I could see a garden like a cathedral, in which the idea of infinity would be the first thing to establish by means of spacious vistas; the chapels would be 'secret chambers' in which one could find privacy, solitude with a book and harmony in reflection. Above all, Mediterranean art should fit the landscape and satisfy climatic needs which have always existed . . . We thus eliminated decorative detail

Allée des jarres.

Homer's garden.

which, over the centuries, had accumulated to the point of nausea, cluttering up forms embodying reason itself. In esthetics, what seem to be the most revolutionary acts are often no more than a return to an ancient ideal that men had sacrificed to their need to add to forms already sufficient in themselves. This undertaking required the collaboration of sun and sea in a plot against 19th-century routine, guilty of denaturing one of the loveliest natural areas in France. The clouds, rocks, and olive trees must all help make a garden out of a place which, without them, would be nothing more than dry rock.''

''Use of topographical features,'' ''simple materials,'' ''rural architecture''—Ferdinand Bac's terms have a singularly modern ring to them. As for the collaboration of sun and sea, one would think that Bac had second sight when one observes the cult of the sun practiced

Bella Vista.

on beaches today. This man, full of nostalgia, could look forward as well as backward. Yet he could not have predicted that one day anyone could gain access to shores which, in the 1930s, were off-limits except to the select few, whose names often appeared in the society column of the daily paper, *L'Eclaireur de Nice*. In 1952 Mrs. Edouard Ladan-Bockairy, daughter-in-law of Ferdinand Bac's friends, decided to turn Les Colombières into a hotel and open up the garden to the public. After a tour of the grounds, one can enter the house, admire the murals executed by Ferdinand Bac, and have a cup of tea. It is pleasant to sit and look through the dining room windows, whose small columns frame a view of the old town like a precious painting, or to go and sit by the pool in Homer's Garden, a patio dedicated to the heros and gods in the Odyssey.

Le Clos du Peyronnet, reflections on water gardens

Set between vertical cliffs and the sea, the Garavan district lies right at the Italian border. Etymologically, "Garavan" is thought to derive from *gardé-du-vent* or "sheltered from the wind," which may well be true, as this spot benefits from an exceptionally mild microclimate due to its location. Before the invasion of winter residents, this south-facing slope was terraced, planted with olive, orange, and lemon trees, especially the latter; its produce was shipped by water to various destinations such as Hamburg or New York. Today avocado and specific types of banana mature in villa gardens on the site of the former citrus groves. Here, in this sheltered area Mr. and Mrs. Derick Waterfield purchased a small, terraced tract where several olive trees and cypresses stood around Le Clos Peyronnet, a house built in the late nineteenth century. Like so many English people at the time, they found the Riviera a congenial place to winter. In the garden a marble plaque on an ivy-covered wall bears this inscription:

Derick and Barbara Waterfield
and later their son Humphrey
created and loved this garden
1915 - 1971

Today, Humphrey's nephew, William Waterfield, lives in the house and has introduced a number of rare plants in the garden, including about thirty varieties of sage. One can no longer get an idea of what Barbara and Derick Waterfield's garden originally looked like, for their son, Humphrey, painter and landscape gardener, subsequently left his own mark upon its design. Around 1950 Humphrey Waterfield relandscaped the park at Abbots Ripton in Huntingdonshire, England, and his own Hill Pasture estate in Essex, in addition to others, less well known. In Menton, he designed the fountain down in the valley at Val Rahmeh.

Le Clos du Peyronnet is an Italianate villa typical of the *Belle Epoque* with its projecting roof and ground-floor colonnade. A *Wisteria sinensus* has taken over the façade and crushed a peristyle column in its tentacular embrace; its tendrils are now laying siege to the second-

Long pool.

Epigraph.

Water lily pond.

floor terrace. Out in front the drive arrives at the house, veering around a clump of American trees—oreopanax, catalpa, and jacaranda—underplanted with shrubbery and adorned with large Anduze urns. On the right, when facing the façade, a wet grotto overgrown with ferns, whose fronds fall vertically from the vaulted ceiling, anticipates the *jardin d'eau* east of the house. On the other side a round basin, part of the white grottos, is hidden by architectural cypress forms; wide curbing is used to outline the basin and emphasize its value in this small, walled-in enclosure. Behind the house rise three terraces planted with heat-loving plants, including varieties of hibiscus and solanum, as well as fruit trees and a small kitchen garden. At either end of the middle terrace is a stone arcade, used by people crossing the terrace as an entrance and exit. This idea may have come from a terrace treated comparably at the Hanbury Botanical Gardens near La Mortola, a village about two miles from Menton on the Italian coast. Plants at Le Clos du Peyronnet sometimes came from this garden, La Serre de la Madone, Val Rahmeh, or from Villa Roquebrune at Cap Martin. The proprietors of these gardens were often only too delighted to offer other garden-lovers specimens of rare or newly introduced plants. The eastern area at Le Clos is composed of a water garden, which provides a transition between the exuberant upper garden area and the geometric spaces on the same level as the house. In the water garden surfaces of water and lawn are traversed by a walk whose stone steps lead directly down into one of the pools. From the top of the garden the sky is mirrored in six stretches of water, as if parts of a reverie. An aquatic staircase is formed by three small fountain basins, going up to the two basins in the heart of the garden. The vista culminates in the Mediterranean, affording a perfect example of the technique of fitting a view of the water into the overall design.

. . . .

Stone pinecone.

Solanum wendlandii.

Peristyle.

Alley between arcades.

Water staircase.

Notes

About Paradise

1. Plutarch, *Decline of the Oracles*.
2. Vitruvius, Book VII, Chapter 1.
3. Stendhal [Marie Henri Beyle], *Mémoires d'un touriste*, 1838.
4. Gustave Flaubert, *Voyages*, 1845.
5. Flaubert, *Voyages*, p. 441.
6. Stendhal, *Mémoires d'un touriste*, p. 271.
7. The Comtat comprises the part of the *département* of the Vaucluse that belonged to the popes from 1274 to 1791.
8. Pliny the Elder, *Natural History*, Books XVI and XVII.
9. Michèle Salmona, *Jardins maraîchers et littoral varois*, C.O.R.D.E.S., 1980.
10. Louis Roussel, *Photologie forestière*, Masson, Paris, 1972.
11. Henri Bosco, *Un rameau de la nuit*, Gallimard, 1950.
12. M. Salmona, *Jardins maraîchers*.
13. Marie Mauron, *Lorsque la vie était la vie*.
14. Jules Michelet, "Tableau de la France," *Histoire de France*, 1861.
15. Anonymous.
16. Jean-Robert Pitte, *Histoire du paysage français*, Taillandier, Paris, 1983.
17. N. Coulet, "Pour une histoire des jardins-vergers et potagers à Aix-en-Provence," *Le Moyen Age* n° 2, 1967.
18. Ibid.
19. N. Coulet, "Pour une histoire des jardins," *Le Moyen Age* n° 2, 1967.
20. From Sylvain Gagnière, *Le Palais des papes d'Avignon*, Les Amis du Palais du Roure, 1983.
21. Petrarch, *Familiari*, Book VIII, Chapter 3.
22. Ibid.
23. Noël Coulet, Alice Planche, and Françoise Robin, *Le Roi René, le prince, le musicien, l'écrivain, le mythe*, Édisud, 1983.
24. Michel Darluc, *Histoire naturelle de la Provence*, 1786.
25. J.-J. Gloton, *Renaissance et Baroque à Aix-en-Provence*, École française de Rome, 1979.
26. Elizabeth Sauze, "Du potager au jardin botanique: l'environnement des châteaux du pays d'Aigues du XVIᵉ au XVIIIᵉ siècle," *Colloque du C.I.R.C.A.* 9/12 July, 1980.
27. Darluc, *Histoire naturelle*.
28. Elizabeth Sauze, "Du potager au jardin botanique."
29. Darluc, *Histoire naturelle*.
30. Dora Wiebeson, *The Picturesque Garden in France*, Princeton University Press, 1978.
31. Cl. Watelet, *Essai sur les jardins*, 1774, Minkoff reprint, Geneva.
32. Ibid.
33. Ibid.
34. J.-Ch. Krafft, *Plans of the Most Beautiful Gardens in France, England and Germany*, 1809.
35. Comte de Villeneuve, *Statistique des Bouches-du-Rhône*, vol. 4, 1827.
36. G. Gromort, *L'Art des jardins*, Vincent Fréal, 1934.
37. Stendhal [Marie Henri Beyle], *Mémoires d'un touriste*.
38. Joseph Méry, *Marseille et les Marseillais*, 1860.
39. Michel Racine, *Architecture rustique des rocailleurs*, Le Moniteur, Paris, 1981.
40. Patrice Gouy, *Pérégrinations des "Barcelonnettes" au Mexique*, Presses universitaires de Grenoble, 1980.
41. Comte de Villeneuve, *Encyclopédie des Bouches-du-Rhône*, Vol. 4, 1827.
42. Ibid.
43. Ibid.

The invention of the Côte d'Azur: travelers' gardens

1. Quoted by H. O. Reichard, *Conseils aux touristes de 1793*.
2. Other geographical features contribute to the singular climate protected from abrupt variations along the coast from Toulon to Menton. The temperature is regulated by the sea, which accumulates heat during the day and cools only slowly at night. Seasonal high and low pressure systems are responsible for the mild winter, wet spring and fall, and dry summer. After A. L. Giuglaris, *De l'acclimatation des végétaux exotiques dans le midi de la France*, 1940.
3. Michel de l'Hospital, First letter to Jacques du Faur. Quoted by A. Merquiol, *La Côte d'Azur dans la littérature française*, 1949.
4. Quoted by A. Denis, *Promenades pittoresques à Hyères*, 1853.
5. Antoine Godeau (1605-1672) was a member of the literary salon of the Marquise de Rambouillet before taking holy orders.
6. Marquise de Sévigné, Letter to Coulanges dated April 10, 1691.
7. Quoted by A. Merquiol, *La Côte d'Azur dans la littérature française*, 1949.
8. A. Durbec, "Les vieux bourgs de Provence," *Annales de la société scientifique et littéraire de Cannes et de l'arrondissement de Grasse*, 1953.
9. A. Risso, *Histoire naturelle des orangers*, 1818.
10. J. P. Papon, *Voyage littéraire de Provence*, 1780.
11. T. G. Smollett, *Travels through France and Italy*, London, 1766.
12. R. Latouche, *Histoire de Nice*, Vol. 1, 1951.
13. A. L. Thomas, *Œuvres complètes*, Letter to Jean-François

Ducis dated December 28, 1782.

14. S. Liégeard, *La Côte d'Azur*, 1887.

15. F.E. Fodéré, *Voyages aux Alpes-Maritimes*, 1821.

16. Quoted by A. Denis, *Promenades pittoresques à Hyères*, 1853.

17. Arch. Nationales, Fos 443, 434. Capatti. Quoted by A. L. Giuglaris, *De l'Acclimatation des végétaux exotiques dans le Midi de la France*. 1940.

18. Eucalyptus lanceolata, Melaleuca myrtifolia, Eucalyptus baconis, Melaleuca stricta, Fabricia loevigata, Leptospernum pubescens, Fabricia leptospernum, Metrosideros pinifolia, Metrosideros citrinella.

19. A. Beaumont, *Voyage historique et pittoresque du Comté de Nice*, 1787.

20. One of these residences can still be seen today, surrounded by a part of its garden, at 61 promenade des Anglais. It was built in 1787 by Lady Rivers, and inhabited by Pauline Bonaparte in 1808. Today, it is the Villa Furtado-Heine.

21. T. G. Smollett, *Travels through France and Italy*, London, 1966.

22. Cape Noli, between Imperia and Savone.

23. Prosper Mérimée, Letters of December 17 and 18, 1856.

24. Michel Marié, *Un territoire sans nom*, C.N.R.S., 1982.

25. Individuals had been sending parcels of flowers for decades. In Letters from Nice, 1763-1765, T. G. Smollett wrote: "In winter, gifts of carnations are sent by mail from here to Turin and Paris, and even all the way to London."

26. Guy de Maupassant, *Sur l'Eau*.

27. E. André, *Traité général de la composition des parcs et jardins*, 1879.

28. With Barillet-Deschamps, André created the Buttes-Chaumont Park in Paris; he also did Sefton Park in Liverpool.

29. Philippe Moisset, *Notes de voyage au Pays de Galles*, 1983.

30. Dr. Valcourt, *Cannes, son climat et ses promenades*, Paris, 1878.

31. F. Honoré, "Les jardins et les architectures de la Villa Maryland," *L'Illustration*, March, 1922.

32. After data recently collected by Mr. Graeme Moore, landscape gardener at Colchester, Essex.

33. E. de Ganay, *La Gazette illustrée des amateurs de jardins*, 1936-1937.

34. J. L. Graverend, *Les Jardins méditerranéens*, Bilbliothèque d'horticulture pratique, J. B. Baillère et Fils, 1959.

35. Mrs. Philip Martineau, *Gardening in Sunny Lands: the Riviera, California, Australia*, 1924.

36. Viscount C. de Noailles, *Gazette illustrée des amateurs de jardins*. 1976.

37. André Véra, *Les Jardins*, Emile Paul, Paris, 1919.

38. F. Bac, "Villas et jardins méditerranéens," *L'Illustration*, Christmas, 1922.

39. J.-C.-N. Forestier, *Jardins. Carnets de plans et de dessins*, Emile Paul, Paris, 1920.

40. Octave Godard, "L'art des jardins dans le Midi," *Bulletin de la Société d'Horticulture pratique de Nice et des Alpes-Maritimes*, 1921.

41. "L'Art des Jardins à l'Exposition des Arts Décoratifs," *L'Illustration*, August, 1925.

42. J.N. Durand, *Précis de leçons d'architecture*, Paris, 1809.

43. Quoted by André Véra, *L'homme et le Jardin*, Plon, 1950.

44. Ferdinand Bac, "L'art des jardins à l'Exposition des Arts Décoratifs," *L'Illustration*, August, 1925.

45. Name given by Guillaume Apollinaire to painting practiced by Robert Delaunay, Francis Picabia, Fernand Léger and Marcel Duchamp, then midway between Cubism and abstract art.

46. See the chapter on the Villa Noailles in Grasse.

47. Russell Page, *The Education of a Gardener*, Random House, New York, 1962.

48. Anita Pereire and Gabrielle Van Zuylen, *Jardins privés en France*, Artaud, 1983.

Gardens of Provence and the French Riviera

1. A.C. Thibaudeau, *Reflet des Bouches-du-Rhône, Mémoires*.

2. Interview at Châteauneuf-le-Rouge, October 1982.

3. Shortly after this visit, the garden was sold to the municipality, and it is no longer maintained in the same manner.

4. J.-L. Vaudoyer, *Beautés de la Provence. L'ombre et l'eau*, Bernard Grasset, 1926.

5. Ibid.

6. Mrs. Tubbs and Mr. McGarvie Munn have undertaken the restoration of the château at their own expense.

7. *Santons* are small clay figurines representing not only the customary Holy Family of the Nativity but also local Provençals: the garlic seller, the fisher, the fishwife, the spinner, among others.

8. It will soon be reconstructed as part of a current project to rehabilitate the villa.

9. The average low temperature in July is 4.4°C, and the average high is 16.2°C; in January the average low temperature is −8.9°C, and the average high is −0.3°C.

10. The plants of the western Alps, the region that most interests the general public, are grouped in four kinds of rock gardens:
- A natural habitat garden which shows how the different varieties live together under similar conditions.

- A medicinal plant garden.
- A garden grouping varieties belonging to the same family or to the same species: Astragalus, Pediculars, Lilies, Campanulas, Ericaceous, Yellow Compositae, etc.
- A regional rocks garden: Queyras, Western Italian Alps, Maritime Alps.

11. Gérard-Louis Soyer, *Jean-Gabriel Domergue, l'art et la mode.*
12. T. de Banville, *La mer de Nice*, 1861.
13. C. Bianchi, "Contribution à l'histoire de l'île Sainte-Marguerite." *Annals of the Société scientifique et littéraire de Cannes et de l'arrondissement de Grasse*, 1977.
14. J. H. Bennett, *La Méditerranée, la Rivière de Gênes à Menton*, 1880.
15. George Sand, *Le château des Désertes*, 1866.
16. *Revue de Nice*, March, 1861.
17. The goat with the golden horns was said to guard a secret treasure hidden at the bottom of an *oubliette* or cave. Legend has it that she appears from time to time and that greedy treasure hunters who manage to find her den vanish, never to be seen again.
18. *Revue de Nice*, March, 1861.
19. Turin was first capital of the Duchy of Savoy (1418), then the kingdom of Sardinia (1720). The county of Nice belonged to both of these states at various different times before becoming part of France in 1860.
20. Pallaréa comes from the Latin words *palla*, the tool used to level the ground, and *area*, "cleared space," or place where the grain was threshed.
21. Eugène Beaudouin (1898-1983), French architect, recipient of the Grand Prix of Rome, architect-in-chief at the French administration of public buildings and national palaces, member of the French Institute, and professor at the Ecole Nationale des Beaux-Arts in Paris. He is also known for city planning projects which he executed in several cities in France and other countries.
22. When Joséphine Beaudoin shows her work, she calls them *marmorées*: marmoreal figurative pieces inspired by the veining of the marbles she uses.
23. Edith Wharton (1862-1937), American novelist and garden-lover, author of *Italian Villas and Their Gardens*. She had houses in Ile-de-France, Saint-Brice, and in the south of France, an estate in Hyères (Sainte-Claire-le-Château) near that of Viscount Charles de Noailles.
24. The Agnellis were the son and daughter-in-law of Giovanni Agnelli, who started Fiat in 1899.
25. *La Fiorentina*, designed by Aron Messiah, was built in 1919 for Countess Robert de Beauchamp. Ferdinand Bac also worked on the architectural plan (cloister, loggia,) and designed the interior and the garden.
26. Arthur Rackham (1867-1939), English illustrator of fantasy tales.
27. Etienne Falconet (1716-1791), French sculptor and protégé of the Marquise de Pompadour.
28. Camille Pitollet, *Blasco Ibañez, ses romans et le roman de sa vie.*
29. Léopold Bernstamm, sculptor born in 1859 in Riga, (Latvia) died in France around 1925. He showed his works at the *Salon des Artistes Français* around 1888 and was awarded the gold medal at the Paris Exhibition of 1900. He was curator at the Grévin Museum in Paris.
30. Sarah Safir-Lichnevsky, *Les fantômes de Fontana Rosa.*

Bibliography

General works

ADAMS, W.H., *Les jardins en France, 1500-1800*, l'Equerre, 1979.

ANDRE, E., *Traité général de la composition des parcs et jardins*, Paris, J.B. Baillère et Fils, 1959.

Atlas historique de Provence-Alpes-Côte d'Azur.

BLONDEL, Jacques-François, *De la distribution des maisons de plaisance*, Paris, 1737.

BLONDEL, Jacques-François, *Cours d'architecture*, Paris, 1771-1777.

CAILLOIS, R., *Pierres réfléchies*, Paris, Gallimard, 1975.

CAISSE NATIONALE DES MONUMENTS HISTORIQUES, *Jardins en France, 1760-1820.*

CAMERON R., *The Golden Riviera*, Paris, Gallimard, 1975.

CHARMAISON, Raymond, *Les jardins précieux*, Meynial, 1919.

COMITE DE L'ART DES JARDINS DE LA SOCIETE NATIONALE D'HORTICULTURE DE FRANCE, *Jardins d'aujourd'hui*, Paris, Vie à la campagne, 1932.

DEZALLIER D'ARGENVILLE, A.J., *La théorie et la pratique du jardinage où l'on traite à fond des beaux jardins*, 1709-1749.

DUBOST, Françoise, *Côté Jardins*, Scarabée et Compagnie, Paris, 1984.

DUVILLIERS, F., *Les parcs et jardins*, Paris, 1871.

EITEL, E.J., *Feng-Shui or the rudiment of Natural Science in China*, London, 1873, 1973.

FLAUBERT, G., *Voyages*, Seuil.

FORESTIER, J.C.N., *Jardins, Carnets de plans et de dessins*, Paris, Emile Paul, 1920.

FREGNAC, C., *Merveilles des Châteaux de Provence*, Paris, Hachette, 1965.

GANAY, Ernest de, *Les Jardins à la française en France au XVIIIᵉ siècle*, Van Oest, 1943.

GANAY, Ernest de, *Les jardins de France et leur décor*, Paris, Larousse, 1949.

GIRARDIN, R. L. de, *De la composition des paysages*, Paris, reprint Champ Urbain, 1979.

GORSE, Georges L., Genoese Renaissance Villas: a typological introduction, *Journal of Garden History*, December 1983.

GRAVEREND, J.L., *Les jardins méditerranéens*, Bibliothèque d'horticulture pratique, J.B. Baillère et Fils, 1959.

GROMORT, G., *L'art des jardins*, Paris, Vincent Fréal, 1934.

HAFFNER, J.J., *Composition des jardins*, Paris, 1931.

HUGO, J., *Le regard de la mémoire*, Le Paradou, Actes Sud, 1983.

HUNT, John Dixon, WILLIS, Peter, *The Genius of The Place*, London, Paul Elek, 1975.

LAPRADE, Albert, *Croquis III*, Paris, Vincent Fréal, 1950.

LASSURE, C., Architecture rurale en pierre sèche, *Revue de l'architecture populaire anonyme*, 1977-1978-1979.

LEQUENNE, F., *La vie d'Olivier de Serres*, Paris, Julliard, 1945.

LIEGEARD, S., *La Côte d'Azur*, Imprimeries réunies, 1887.

LIVET, R., Habitat rural et structures agraires en basse Provence, *Annales de la Faculté de Lettres*, Aix-en-Provence, 1962.

MAC DOUGALL, Elisabeth B., HAMILTON HAZLEHURST, F., KARLING, Sten, STRANDBERG, Runar, GOLLWITZER, Gerda, *The French formal garden*, Dumbarton Oaks Trustees for Harvard University, Washington D.C., 1974.

MERIMEE, P., *Notes de voyage*, 1834.

MICHELL, John, *L'esprit de la terre ou le génie du lieu*, Paris, Seuil, 1975.

MILLIN, Aubin-Louis, *Voyages dans les départements du Midi de la France*, Paris, Tourneisen, 5 volumes, 1807-1811.

MOLLET, André, *Le jardin de plaisir*, 1651. Reprint Le Moniteur, 1981.

PAGE, Russel, *Gardens in the south of France*, Garden Chron., 1962.

PAPON, J.P., *Voyage littéraire de Provence*, Paris, Barrois, 1780.

PEAN, P., *Jardins de France*, Paris, 1925.

PEROUSE de MONCLOS, J.M., De la Villa rustique d'Italie au pavillon de banlieue, *Revue de l'Art*, n. 32, 1976.

PITTE, Jean-Robert, *Histoire du paysage français*, Paris, Taillandier, 1983.

PLINE L'ANCIEN, *Histoire naturelle*, Books XVI and XVII.

PLUTARQUE, *Le déclin des oracles.*

RACINE, Michel, *Architecture rustique des rocailleurs*, Paris, Le Moniteur, 1981.

ROUSSEL, Louis, *Photologie forestière*, Paris, Masson et Cie, 1972.

SERRES, Olivier de, *Théâtre de l'Agriculture et Mesnage des Champs*, Paris, 1600.

STENDHAL, *Mémoires d'un touriste*, Paris, 1838.

TANIZAKI JUNICHIRO, *Eloge de l'ombre*, 1933, Paris, Publications orientalistes de France, 1983.

TESTU, C., *Les roses anciennes*, Paris, Flammarion, La maison rustique, 1984.

THACKER, C., *Histoire des jardins*, Paris, Denoël, 1981.

THOMAS, A.L., *Œuvres complètes*, Paris, 1822.

TONGIOGI TOMASI, Lucia, Projects of Botanical and Other Gardens: a sixteenth-Century Manual, March 1983, *Journal of Garden History.*

TOURNEFORT, P. de, *Eléments de botanique ou méthode pour connaître les plantes*, Paris, 1694.

VAN ZUYLEN, Gabrielle, PEREIRE, Anita, *Jardins privés en France*, Artaud, 1983.

VERA, André, *Les Jardins*, Paris, Emile Paul, 1919.

VERA, André, *L'homme et le jardin*, Plon, 1950.

WATELET, *Essai sur les jardins*, Genève, Minkoff reprint, 1774.

WIEBENSON, D., *The picturesque garden in France*, Princeton University Press, 1978.

WOODBRIDGE, K., *Princely Gardens, the origins and development of the French formal style*, Thames and Hudson, 1986.

WOODBRIDGE, K., *The nomenclature of style in Garden History*, The Colonial Williamsburg Foundation, Williamsburg, 1979.

Provence

ALGOUD, H., *Mas et bastides de Provence*, Marseille, Detaille, 1927.

ARTAYD, A., *Un amateur marseillais*, Paris, Georges Roux, 1980.

BORRICAND, René, *Châteaux et Bastides du pays d'Aix*, Borricand, 1979.

BOSCO, Henri, *Un rameau de la nuit*, Gallimard, 1980.

BOYSSET, Bertrand, *Traité d'arpentage*, Bibliothèque Inguebertine, Carpentras.

BRIOLLE, C., FUZIBET, A., *Jardin de Guévrékian, Projet de reconstitution*, 1982.

BROMBERGER, LACROIX, RAULIN, *L'architecture rurale en Provence*, Paris, Berger-Levrault, 1980.

BRUN, Lucienne, *Provence*, Horizons de France, 1974.

BUNEL, J.-P., *Promenades pittoresques descriptives et historiques dans le Var*, Draguignan, 1853.

CONARD, S., Tourves, fabriques et géométrie, *Monuments historiques*, n. 5, 1976.

CORDOLEANI, M. and C., *Les éléments linéaires du paysage à Auriol*, DRAE, Aix-en-Provence, 1983.

COULET, N., Pour une histoire des jardins. Vergers et potagers à Aix-en-Provence: 1350-1450, *Le Moyen Age*, n. 2, 1967.

COULET, N., *La naissance de la bastide,*

Géographie historique du village et de la maison rurale, CNRS, 1979.

COULET, N., PLANCHE, A., ROBIN, F., *Le roi René, le prince, le musicien, l'écrivain, le mythe*, Aix-en-Provence, Edisud, 1983.

CROZE-MAGNAN, *Essai sur les jardins pittoresque convenables au territoire de Marseille*, Académie de Marseille, 1813.

DARDE, R., *L'habitation provençale*, 3 volumes, Paris, 1786.

DARLUC, M., *Histoire naturelle de la Provence*, Avignon, 1786.

DE GERIN-RICARD, *Les antiquités de la Vallée de l'Arc*, Aix-en-Provence, 1907.

DELOBEAU, P., *Une famille parlementaire aixoise au XVIIIᵉ siècle, les Barrigues de Montvallon*, 1943.

DESHAIRS, Léon, Une villa moderne à Hyères, *Art et décoration*, December 8, 1928.

DETAILLE, Albert, *Provence des Mas et Bastides*, Marseille, Detaille, 1792.

DIRECTION DES ESPACES VERTS DE MARSEILLE, *Les botanistes à Marseille et en Provence du XVIᵉ au XIXᵉ siècle*, Marseille, 1982.

DOBLER, H., *Le cadre de la vie mondaine à Aix-en-Provence aux XVIᵉ et XIXᵉ siècles*, Aix-en-Provence, Marseille, 1928.

DORE, R., *L'art en Provence*, Paris, Editions des Beaux-Arts, 1929.

FEISSART, *Des plantations*, Marseille, 1828.

FUSTIER-DAUTIER, N., *Les bastides de Provence et leurs jardins*, Paris, Serg, 1977.

GLOTON, J.-J., *Renaissance et baroque à Aix-en-Provence*, Ecole française de Rome, 1979.

GUEIDAN, *Manuel des cultures pour la Provence*, Marseille, 1895.

GUIGUE, Julien, *La fontaine Vaucluse.*

HENSELING, L., *Zig-zags dans le Var*, 1939.

LIGNE, Prince de, *Coup d'œil sur Bel Œil*

et sur une grande partie des jardins de l'Europe, 1781.

MARCHAIS, A., *Les jardins dans la région de l'oranger*, Antibes, 1884.

MARIE, Michel, *Un territoire sans nom*, CNRS, 1982.

MARRAST, *Jardins*, Paris, Charles Moreau, 1926.

MASSON, P., La Provence au XVIIIᵉ siècle, *Annales de la faculté des lettres*, Aix-en-Provence, 1935-1936.

MASSOT, J.-L., *Maisons rurales et vie paysanne en Provence*, Paris, Serg, 1979.

MAURON, Marie, *Lorsque la vie était la vie.*

Merveilles des Châteaux de Provence, Paris, Hachette, 1965.

MERY, J., *Marseille et les Marseillais*, 1860.

MILLIN, Aubin Louis, *Voyage dans les départements du midi de la France*, 1807.

MORTON SHAND, P., An essay in the adroit: at the villa of the Vicomte de Noailles, *Architectural Review*, April 1929.

MOSSER, Monique, *Jardins et peinture*, CNDP, 1984.

NOAILLES, vicomte de, *My Garden*, Botanical diary.

NOAILLES, vicomte de, LANCASTER, R., *Plantes de jardins méditerranéens*, Larousse.

PEIRESC, Claude Nicolas Fabri de, *Correspondance.*

PEZET, M., *Châteaux des Bouches-du-Rhône*, Paris, Nouvelles éditions latines.

PUECH, Laurent, *Mémoires romancés du Comte de Valbelle*, Grasset, 1986.

RACINE, Michel, Les jardins de la vallée de l'Huveaune, *Monuments historiques*, n. 133, June, 1984.

SAUREL, Alfred, *Notice historique sur St-Jean de Garguier, l'Abbaye de St-Pons et Gémenos*, Marseille, 1863.

THIBAUDEAU, A.-C., *Reflet des Bouches-du-Rhône, Mémoires d'un Préfet sous l'Empire.*

VAUDOYER, J.-L., *Beautés de la Provence*, Paris, Bernard Grasset, 1926.

VILLENEUVE, Comte de, *Statistique du département des Bouches-du-Rhône*, Marseille, Ricard, 1821.

VITRUVE, *De Architectura*, VII, 1.

VOVELLE, M., *L'irrésistible ascension de Joseph Sec, bourgeois d'Aix*, Aix-en-Provence, Edisud, 1975.

ZOLA, E., *La faute de l'abbé Mouret*, Ernest Flammarion.

French Riviera

ALFORD, H., *The Riviera: pen and pencil sketches from Cannes to Genoa*, col. plts, 1870.

BAC, F., *Jardins enchantés, Un Romancero*, Paris, Conard, 1925.

BAC, F., Villas et jardins méditerranéens, *L'Illustration*, Christmas, 1922.

BAC, F., L'art des jardins à l'exposition des arts décoratifs, *L'Illustration*, August, 1925.

BAC, F., *Les Colombières, ses jardins et ses décors*, Paris, Conard, 1925.

BANVILLE, T. de, *La mer de Nice*, Paris, Poulet-Malassis, 1861.

BEAUMONT, A., *Voyage historique et pittoresque du Comté de Nice*, Genève, Bardin, 1787.

BENNET, J.-H., *La Méditerranée, la rivière de Gênes et Menton, comme climats d'hiver et de printemps*, Paris, Asselin et Cie, 1861-1880.

BERENGUIER, R., *Châteaux des Alpes-Maritimes*, Paris, Editions Latines, 1962.

BERNER, L., *Introduction dans le Midi des plantes grasses*.

BOURELY, général, *Les perles de la Côte d'Azur*, Paris, H. Laurens, 1900.

BOURSIER-MOUGENOT, E.-J.-P., RACINE, M., Les jardins néo-méditerranéens, deux jardins nés du voyage, *Monuments historiques*, February, 1983.

BOURSIER-MOUGENOT, E.-J.-P, Les jardins de la Côte à l'époque du comté de Nice, *Monuments historiques*, June, 1985.

BRESSON, J., *La fabuleuse histoire de Cannes*, Monaco, Editions du Rocher, 1981.

BUTTURA, A., *L'hiver à Cannes et au Cannet*, J.-B. Baillière et fils, 1883.

CAMERON, R., *The Golden Riviera*, London, Weidenfeld and Nicolson, 1975.

CHABAUD, B., *Végétaux exotiques cultivés en plein air dans la région des orangers*, Toulon, 1871.

CHABAUD, B., *Les jardins dans la région de l'oranger*, Antibes, 1885.

CHABAUD, P., *Les jardins de la Côte d'Azur*, Toulon, 1910.

CHOUARD, P., *Jardins botaniques de la Côte d'Azur*, La Terre et la Vie, 1932.

DENIS, A., *Promenades pittoresques à Hyères*, Toulon, 1853.

DIESBACH, G. de, *Ferdinand Bac*, Paris, 1979.

DORE, R., *L'Art en Provence*, Paris, Editions des Beaux-Arts, 1929.

DURBEC, J.-A., *Biot, un vieux village provençal*, Mandelieu, 1978.

FODERE, F.-E., *Voyage aux Alpes-Maritimes*, Paris, Levrault, 1821.

GALANTE, P., GALL, A. et M., *Les années américaines*, J.-C. Lattès, 1985.

GANAY, Ernest de, *La Gazette illustrée des amateurs de jardins*, 1936-1937.

GAUCHER, M., *Les Rothschild, côté jardins*, Paris, Gilbert, 1982.

GAUCHER, M., *Les jardins de la fortune*, Paris, Hermé, 1985.

GIRARD, J.-B., *Cannes et ses environs*, Paris, Garnier Frères, 1859.

GIUGLARIS, A.-L., *De l'acclimatation des végétaux exotiques dans le Midi de la France*, Nice, 1940.

GODARD, Octave, L'art des jardins dans le Midi, *Bulletin de la Société d'horticulture pratique de Nice et des Alpes-Maritimes*, 1921.

GODARD, Octave, *Les jardins de la Côte d'Azur*, Paris, Massin, 1927.

GUIGOU, Mgr., *Histoire de Cannes et de son canton*, Cannes, H. Vidal, 1878.

HONORE, F., Les jardins et les architectures de la villa Maryland, *L'Illustration*, March 25, 1922.

L'Illustration, Le jardin, May 28, 1932.

ISNARD, G., HUSSEY, C., Garden ornament, *Country Life*, 1927.

JEKYLL, G., HUSSEY, C., *Garden ornament*, Londres, Country Life, 1921.

JOHNSON, H., *L'art des jardins*, Paris, Nathan, 1980.

KARR, Alphonse, *Voyage autour de mon jardin*, Paris, Dumont, 1845.

KING, R., *Les paradis terrestres*, Paris, Albin Michel, 1980.

LATOUCHE, R., *Histoire de Nice*, 1951.

LENTHERIC, C., *The Riviera ancient and modern*, London, 1895.

LIEGEARD, S., *La Côte d'Azur*, Paris, 1887.

MALCOM, A., *Letters of an Invalid from South of France, 1826-1828*, London, 1897.

MARTINEAU, Mrs P., *Gardening in sunny lands*, London, Cobden-Sanderson, 1924.

MAUROIS, A., COLTON, D.-J., *Le monde étrange de Henry Clews*, Paris, Joseph Foret, 1959.

MAYRARGUES, H., *De villa en villa*, Nice, Malvano, 1877.

MAUPASSANT, G. de, *Sur l'eau*, Paris, 1888.

MEIFFRET, J.-B., *Guide d'Antibes et de ses campagnes*, Nice, 1877.

MERIMEE, P., *Notes d'un voyage dans le midi de la France*, 1835.

MERQUIOL, André, *La Côte d'Azur dans la littérature française*, Nice, Jacques Dervyl, 1949.

MILLER, W., *Wintering in the Riviera*, 1879.

MILLIET-MONDON, C., *Cannes 1835-1914, villégiatures, urbanisme, architectures*, Nice, Serre, 1986.

MILLIN, A.-L., *Voyage en Savoie, Piémont et à Nice*, Paris, Wasserman, 1816.

MONTAUT, H. de, *Voyage au pays enchanté, Cannes, Nice, Monaco, Mentone*, 1880.

NOAILLES, vicomte de, LANCASTER, R., *op. cit.*

ORGEAS, J., *L'hiver à Cannes, Saint-Raphaël et Antibes*, Cannes, 1889.

PAGE, R., *The education of a gardener*, London, Collins, 1962.

PAPON, J.-P., *Voyage littéraire de Provence*, Paris, Barrois, 1780.

PAPON, S., *Voyage dans le département des Alpes-Maritimes*, Paris, 1804.

PEREIRE, A., VAN ZUYLEN, G., *op. cit.*

PIERRUGUES, J.-J., *Cannes à travers les âges*, Cannes, 1931.

PITOLLET, C., *Vicente Blasco Ibañez, ses romans et le roman de sa vie*, Paris, Calmann-Lévy, 1921.

POLIGNAC, H. de, *Les Polignac*, Paris, Fasquelle, 1960.

Revue de Nice, Promenades, Nice, February, 1861.

RISSO, A., POINTEAU, A., *Histoire naturelle des orangers*, 1818.

SAFIR-LICHNEVSKY, S., *Les fantômes de Fontana Rosa*, Menton, 1978.

SAND, G., Lettres d'un voyageur. A propos de botanique, Paris, *Revue des Deux-Mondes*, 1868.

SAND, G., *Le château des Désertes*, Paris, Michel Lévy Frères, 1866.

SACKVILLE-WEST, V., *Hidcote Manor Garden*, The National Trust.

SARTY, L., *Nice d'antan*, Nice, Isnard, 1921.

SAUVAIGO, E., Les plantes exotiques introduites sur le littoral méditerranéen, *Revue des sciences naturelles appliquées*, June, 1982.

SMOLLETT, T.-G., *Lettres de Nice sur Nice et ses environs*, translated by E. Pilatte, Nice, 1919. (*Travels through France and Italy*, London, 1766.)

SOUZA, R. de, *Nice, capitale d'hiver*, Paris, Berger-Levrault, 1913.

SOYER, J.-L., *Jean-Gabriel Domergue, l'art et la mode*, Editions Sous le vent, 1984.

TESTU, C., *Les roses anciennes*, Paris, Flammarion, 1984.

THOMAS, A.-L., *Œuvres complètes*, Paris, 1822.

VALCOURT, Dr., *Cannes, son climat et ses promenades*, 1878.

VARILLE, M., *Bonheur des jardins*, Paris, Marius Audin, 1944.

WHITE, C., BUTOR, M., ALLARY, D., BINE-MULLER, N., *Rêveuse Riviera*, Paris, Herscher, 1983.

WITSEY, F., A Hilltop garden in the maquis, *Country Life*, May 25, 1978.

WITSEY, F., Idealised Riviera garden in reality, *Country Life*, June 16, 1977.

WITSEY, F., Flattery at every angle, *Country Life*, July 15, 1982.

WITSEY, F., No need for greenhouses, *Country Life*, May, 1979.

WITSEY, F., Provençal passion for plants, *Country Life*, January 8, 1976.

WITSEY, F., Exuberance on a Southern hillside, *Country Life*, July 10, 1980.

Plants

The Alpine garden, plants

Rockworks "Western Alps," and off rockwork

Gentiana Lutea L. - Leontopodium alpinum L. - Artemisia umbelliformis Lam. - Eryngium alpinum L. - Larix decidua L. - Pinus uncinata Miller - Pinus mugo Turra - Veratum album L. - Gentiana acaulis L. - Pinguicula alpina L. - Pinus cembra L. - Gentiana punctata L. - Polygonum alpinum All. - Juniperus nana Willd. - Globularia cordifolia L. - Hieracium villosum Jacq x H. lanatum Vill. - Campanula thyrsoides L. - Paeonia officinalis L. - Papaver Rhaeticum Ler. - Sorbus aucuparia L. - Cirsium x purpureum All. - Salix herbacea L. - Saponaria ocymoides L. - Arctostaphylos uva-ursi L. - Dryas octopetala L. - Potentilla valderia L. - Gentiana asclepiadea L. - Alnus viridis D.C. - Rhododendron ferrugineum L. - Vaccinium uliginosum L. - Pedicularis rostrato-spicata Crantz. - Rhodiola rosea L. - Potentilla delphinensis Gren. - Salix helvetica Vill. - Senecio incanus L. -Dianthus pavonius Tausch. - Campanula barbata L. - Salix serpyllifolio Scop. - Geranium argenteum L. - Geum reptans L. - Ranunculus glacialis L. - Saussurea depressa Gren. - Saxifraga aizoides L. - Silene acaulis Jacq. - Xanthoria elegans Th. Fr. - Picea abies L.

Rockworks "Central and Eastern Alps," "Karpates," and "Balkan Mountains"

Wulfenia carinthiaca Jacq. - Achillea clavennae L. - Senecio abrotanifolius L. - Geum coccineum Sibth. et Sm. - Potentilla nitida L. - Erica herbacea L. - Rhododendron hirsutum L.

Off rockwork

Betula nana L. - Salix lapponum L. - Cytisus purgans Bossi.

Rockworks "Pyrenees" and "Massif Central"

Carduus carlinoides Gouan. - Senecio leucophyllus DC. - Ramonda myconi Reich. - Iberis sempervirens L. - Geranium cinereum Cav. - Polemonium caeruleum L. - Iris latifolia Voss. - Erodium manescavi Coss. - Lilium pyrenaicum Gouan. - Potentilla fruticosa L.

Rockwork "Sierra Nevada," and "Atlas"

Ptilotrichum spinosum Boiss.

Rockworks "Caucase"

Caltha polypetala Hochst. - Veronica gentianoides Vahl. - Geranium ibericum Cav. var. Platypetalum F. et M. - Cephalaria brevipalea Litwin. - Nepeta grandiflora Bieb. - Centaurea macrocephala Puschk.

Several places in the garden

Papaver lateritium C. Koch. - Papaver nudicaule L.

Rockwork "Siberia"

Ligularia macrophylla DC. - Dracocephalum grandiflorum L. - Iris sibirica L. - Nepeta sibirica L.

Rockworks "Himalaya Mountains," "China," and "Japan"

Potentilla atrosanguinea Lod. - Pseudomertensia primuloides C.B. Clarke. - Delphinium vestitum Wallich.

Rockwork "North America"

Lewisia x cotyledon Rob. - Polenium pulcherrimum Hooker. - Penstemon scouleri Dougl. - Castilleja. - Phlox subulata L. - Lupinus polyphyllus Lindl.

Off rockwork

Picea pungens Engelm.

Some medicinal cultivated plants

Achillea erba-rotta All. - Achillea millefolium L. - Acinos alpinus L. - Aconitum napellus L. - Aconitum vulparia Reich. - Alchemilla alpina L. - Allium schoenoprasum L. - Antennaria dioica (L.) Gaertn - Anthyllis vulneraria L. - Arctostaphylos uva-ursi L. - Arnica montana L. - Artemisia absinthium L. - Artemisia umbelliformis Lam. - Carlina acaulis L. - Chenopodium bonus-henricus L. - Colchicum autumnale L. - Delpinium elatum L. - Digitalis lutea L. - Dryas octopetala L. - Epilobium angustifolium L. - Filipendula ulmaria (L.) Maxim. - Gentiana lutea L. - Heracleum spondylium L. - Hieracium pilosella L. - Hyssopus officinalis L. - Levisticum officinale Koch. - Meum athamanticum Jacq. - Myrrhis odorata (L.) Scop. - Narcissus poeticus L. - Narcissus pseudo-narcissus L. - Peucedanum ostruthium (L.) Koch. - Plantago lanceolata L. - Plantago media L. - Polygonum bistorta L. - Sanguisorba officinalis L. -Saponaria officinalis L. - Sedum acre L. - Stachys officinalis (L.) Trev. - Tanacetum vulgare L. - Thymus serpyllum L. - Tussilago farfara L. - Urtica dioica L. - Vaccinium myrtillus L. - Valeriana officinalis L. - Veratrum album L. - Viola calcarata L.

Moulin Blanc park, Var

List of remarkable trees published in 1951 and executed after the survey of Mr. Laurent, the museum curator in 1934.

Abies Nordmanniana. - Abies pectinata. - Abies Pinsapo. - Cèdre de l'Himalaya. - Cedrus Atlantica. - Cryptomeria Japonica. - Cupressus Bentani. - Fagus Tricolor. - Gingko Biloba. - Hêtre à feuilles laciniées. - Hêtre pleureur. - Liquidembar Orientalis. - Maclura. - Parrotia Persica. - Pavia. - Pinus Austriaca. - Pinus Sabiniana. - Platanus Orientalis. - Platanus Orientalis Digitata. - Populus Lasiocarpia. - Quercus Lusitanica. - Quercus Mirbeckii. - Sequoia Gigantea. - Sequoia Sempervirens. - Taxodium Dymtichum. - Thuya Dolobrata. - Thuya Gigantea. - Tillolus Silvestris. - Torrulosa. - Tulipier de Virginie. - Ulmus Effusa América

Mas Calendal, Cassis

Botanical survey by Mr. André Jullien, 1966.

Abelia floribunda. - Abies Nordmannia. - Abies Pinsapo. - Acacia. - Acer. - Agapanthus umbellatus. - Agapanthus africanus. - Agave americana. - Agave potatorum. - Albizzia Julibrissin. - Aloes. - Amorphophallus Rivieri. - Ampelopsis Henryana. - Anemone. - Aralia Sieboldii. - Araucaria araucana. - Araucaria Bidwilli. - Araucaria imbricata. - Arbustus andrachne. - Arum dracnculus. - Aspidistra eliator. - Aucuba Japonica. - Azalea indica. - Bignonia. - Billbergia mutans. - Bougainvillea spectabilis. - Brachychiton populneum. - Brahea Roezlii. - Broussonetia papyrifera. - Buddleia Asiatica. - Buxux Baleriaca. - Callitris quadrivalvis. - Camellia Japonica. - Campsis Chinensis. - Canna indica. - Cassa floribunda. - Casuarina equisetifolia. - Cedrus Atlantica. - Cedrus Deodora. - Cedrus Libani. - Cedrus Libanotica. - Cephalotaxus Fortunei. - Ceratonia siliqua. - Ceratostigma. - Cestrum elegans. - Cestrum Parqui. - Chaenomcles Japonica. - Chamaccyparis Cachomiriana. - Chamaerops. - Choisya ternata. - Cineraria platanifolia. - Cinnamomum camphora. - Cistus congiloïdes. - Cistus villosus. - Citrus. - Clerodendron foetidum. - Cocculus laurifolius. - Coccos Yatay = pulposa. - Colletia cruciata. - Convolvulus Mauritanicus. - Cordyline indivisa. - Cotoneaster horizontalis. - Cryptomeria Japonica. - Cupressus Arizonica. - Cycas circinalis. - Cycas revoluta. - Cyphomandra betacea. - Cytisus Laburnum. - Danae racemosa. - Daphne odora. - Dasylirion acrotichum. - Datura arborea. -Datura sanguinea. - Diospyros Kaki. - Dracunculus vulgaris. - Erythaea armata. - Eucalyptus. - Fatsia Japonica. - Feijoa sellowiana. - Ficus repens =pumila. - Firmiana platanifolia. - Fuschia. - Gazania Splendens. - Genista monosperma. - Ginkgo biloba. - Gledistehia triacanthos. - Grevillea. - Habrothamnus elegans. - Hedera helix Marengo. - Hemerocallis fulva. - Heyderia decurrens. - Hypophae rhamnoïdes. - Hydrangea Hortansia. - Illicium anisatum. - Indigofera Gerardiana. - Ipomea purpurea. - Jacobinia pauciflora. - Jasminum primulum. - Jusbea spectabilis. - Justicia rosea. - Kleinia. - Koelreuteria paniculata. - Lagerstroemia indica. - Laurus camphora. - Lavendula dentata. - Libocedrus decurrens. - Ligularia stenocephala. - Liquidambar. - Liriodendron tulipifera. - Livistonia australis. - Loropetalum Chinensis. - Lycium barbarum. - Maclura aurantiaca. - Magnolia grandiflora. - Mahoberberis Neuberti. - Mahonia. - Melia Azedarach. - Mesembryanthemum. - Metasequoia. - Mucuna imbricata. - Musa Japonica. - Myrtus communis. - Nandina domestica. - Nolina recurvata. - Ophiopogon japonicus. - Opuntia. - Oxalis. - Paeonia Moutan. - Parkinsonia aculeata. - Parthenocissus Henryana. -Pelargonium zonale. - Penstemon. - Persea gratissima. - Phlox. -Phœnix. - Phormium tenax. - Photinia serrulata. - Phyllostachys. -Phytolacca. - Picca Parryana. - Pinus. - Pittosporum Mayi. - Pittosporum Tobira. - Plumbago Capensis. - Plumbago Larpentae. - Podocarpus neriifolia. - Poinciana Gilliesi. - Poinciana pulcherima. - Pritchardia filifera. - Prunus. - Punica granatum nana. - Retinospora ovata. - Rhododendron ponticum. - Richardia Elliottiana. - Ruscus. - Sabal Adamsoni. - Salvia. - Schinus molle. - Sciadopitys verticillata. - Sechium edule. - Senecio petasitis. - Solanum. - Sterculia acerifolia. - Styrax officinalis. - Taxodium distichum. - Tecoma. - Tetraclinis articulata. - Viburnum. - Washingtonia filifera. - Wistaria sinensis. - Yucca. - Zantedeschia Elliotiana. - Zauschneria california.

Champfleuri

Data by J.-N. Demoulliers, head-gardener

Abelia grandiflora. - Abutilon striatum. - Abutilon hybridum. - Acacia floribunda. - Acacia verticillata. - Acacia longifolia. - Acacia cultriformis. - Acacia howittii. - Acacia hanburyana. - Acanthus mollis. - Acer palmatum dissectum atropurpurem. - Acer negundo elegans guichardi. - Acer palmatum septemlobum. - Acer japonicum aureum. - Acer platanoides schwedleri nigra. - Actinidia hinensis. - Aesculus hippocastanum. - Aesculus parviflora. - Agapanthus umbellatus. - Albizzia julibrissin. - Albizzia cathartica. - Aloe arborescens. - Alsophila australis. - Antholyza aethiopica. - Agave americana marginata. - Arbutus unedo. - Arctotis hybrida venidio. - Araujia exelsa. - Arundinaria sp. - Asystasia sp. - Astilbe arendsii 'Rheinland'. - Astilbe arendsii 'Fanal'. - Asparagus sprengeri. - Asplenium sp. - Aspidistra eliator. - Aucuba japonica. - Aucuba crotonoides. - Bauhinia grandiflora. - Bergenia crassifolia. - Bergenia purpurascens. - Berberis thunbergii atropurpurea. - Begonia sp. - Beloperone guttata. - Beschorneria yuccoides. - Bletia hyacinthina. - Bougainvillea glabra. - Bougainvillea spectabilis 'Killy Campbell'. - Bougainvillea buttiana 'Sandiego Red'. - Brachychiton populneum. - Brachychiton acerifolium. - Brunfelsia hoppeana. - Brunfelsia calycina. - Brunfelsia latifolia. - Buddleja davidii. - Buxus sempervirens. - Caladium hortulanum. - Colacedrus decurrens aurea variegata. - Colacedrus decurrens. - Carissa grandiflora. - Carex pendula. - Capparis spinosa. - Caesalpinia gilliesii. - Caesalpinia tinctoria. - Callistemon citrinus. - Callistemon viminalis. - Caryopteris clandonensis 'Kew Bleu'. - Carpobrotus acinaciformis. - Cassia corymbosa. - Canna hybrida. - Cedrus libani. - Cedrus atlantica. - Cedrus atlantica glauca. - Cedrus atlantica glauca pendula. - Cedrus deodora. - Cercis siliquastrum. - Ceratonia siliqua. - Cestrum elegans. - Cestrum newellii. - Ceanothus thysiflorus repens. - Ceanothus austromontanus. - Cereus peruvianus. - Chamaerops humilis. - Chaenomeles lagenaria. - Chlorophytum elatum. - Chamaecyparis obtusa nana gracilis. - Chamaecyparis nootkatensis pendula. - Chamaecyparis pisifera nana. - Chamaecyparis pisifera filifera aurea. - Cinnamomum camphora. - Citrus aurantium. - Citrus limon. - Citrus sinensis. - Citrus nobilis. - Citrus mitis. - Clematis armandii. - Clivia miniata. - Clerodendrum thomsonae. - Colocasia antiquorum. - Cordyline australis purpurea. - Coronilla glauca. - Cotoneaster horizontalis. - Cotoneaster microphyllus. - Cotoneaster lacteus. - Cotoneaster franchetii. - Cotinus coggygria atropurpureus. - Cocos nucifera. - Corylus avellana contorta. - Corokia virgata. - Crataegus oxyacanthoides. - Cupressus sempervirens. - Cupressus arizonica glauca. - Cycas revoluta. - Cyperus papyrus. - Cyperus alternifolius gracilis. - Cyphomandra betaceae. - Choisya ternata. - Convolvulus cneorum. - Casimiroa edulis. - Dalhia arborea. - Dalhia imperialis. - Dasylirion acrotichum. - Datura sanguinea. - Datura arborea. - Datura cornigera. - Datura versicolor. - Desmodium penduliflorum. - Diervilla grandiflora. - Dombeya burgessiae. - Dryopteris filix-mas. - Duranta plumieri. - Echium « fastuosum. - Echinocactus grusonii. - Elaeagnus pungens. - Elaeagnus pungens variegata. - Erythrina crista-galli. - Eriobotrya japonica. - Erythea armata. - Escallonia macrantha. - Euonymus japonica. - Ephedra altissima. - Euphorbia pulcherrima. - Euphorbia pugniformis cristata. - Euphorbia splendens. - Euphorbia myrsinites. - Eucalyptus ficifolia. - Eucalyptus globulus. - Fatsia japonica. - Feijoa. - Felicia amelloides. - Ficus pumila. - Ficus carica. - Fittonia verschaffeltii. - Forsythia viridissima intermedia. - Fortunella margarita. - Freesia refracta. - Fuchsia triphylla. - Gardenia jasminoides. -Gazania splendens. - Gerbera jamesonii. - Ginkgo biloba. - Grevillea sp.. - Greyia sutherlandii. - Gunnera scabra. - Hedera canariensis. - Hedera helix. - Hedychium coronarium. - Hedychium gardnerianum. - Heliotropium peruvianum. - Helleborus niger. - Heracleum mantegazzianum. - Hibiscus rosa sinensis.

- Holmskioldia sanguinea. - Hosta sieboldiana. - Hosta fortunei aurea. - Hosta undulata medio-variegata. - Howea fosteriana. - Hydrangea macrophylla. - Hydrangea aspera. - Hypericum olympicum. - Hypericum calycinum. - Hypericum patulum. - Ilex crenata 'Golden Gem'. -Iochroma ·ubulosum. - Iris hybrida. - Iris unguicularis. - Jacobinia suberecta. - Jacaranda ovalifolia. - Jasminum officinalis. - Jasminum revolutum. - Jasminum primulinum. - Jasminum sambac. - Juniperus scopulorum. - Juniperus chinensis aurea. -Juniperus chinensis glauca. - Kalanchoe blossfeldiana. - Kerria japonica. - Kniphofia sp. - Laburnum vulgare. - Lantana camara. - Lagerstroemia indica. - Laurus nobilis. - Lavandula x burnati. - Lavandula dentata. - Leonitis leonorus. - Leptospernum laevigatum. - Ligularia sp. - Ligustrum japonicum. - Liquidambar sp. - Mahonia japonica. - Magnolia grandiflora. - Magnolia soulangeana. - Malus floribunda. - Malus 'Everest Perpetu'. - Malva sylvestris. - Malvaviscus arboreus. - Mammillaria vaupelii cristata. - Medicago arborea. - Melaleuca armillaris. - Melia azedarach. - Melianthus major. - Mimulus glutinosus. - Monstera deliciosa. - Musa ensete. - Musa sapientum. - Nandina domestica. - Narcissus pseudonarcissus. - Nelumbo speciosum. - Nerium oleander. - Nolina recurvata. - Nymphaea odorata. - Nymphaea chromatella. - Olea europaea. - Olearia macrodonta. - Ophiopogon japonicus. - Opuntia microdasys. - Opuntia ficus indica. - Osmanthus fragrans. - Osteospermum ecklonis. - Pandorea jasminoides. - Passiflora edulis. - Passiflora quandrangularis. - Pelargonium capitatum. - Pelargonium grandiflorum. - Pelargonium hederaefolium. - Peperomia obtusifolia. - Persea gratissima. - Picea glauca nana. - Picea smithiana. - Pieris jamonica. - Pistacia lentiscus. - Pittosporum tobira. - Pittosporum tenuifolium mayi variegatum. - Phalaris arundinacea picta. - Philadelphus 'Snow Flake'. - Philodendron cordifolium. - Philodendron selloum. - Phlomis fruticosa. - Phœnix canariensis. - Phormium tenax variegatum. - Phylica ericoides. - Phyllitis scolopendrium. - Phyllostachys nigra. - Platycerium alcicorne. - Plumbago capensis. - Podocarpus sp. - Polianthes tuberosa. - Polystichum falcatum. - Pontederia cordata. - Potentilla sp. - Prunus laurocerasus. - Prunus cerasifera pissardii. - Prunus lusitanica. - Punica granatum. - Punica granatum nana. - Pyrostegia venusta. - Quercus ilex. - Quercus suber. - Ranonculus lingua. - Raphiolepis umbellata. - Raphiolepis indica. - Reinwardtia trigyna. - Rosmarinus officinalis. - Ruscus aculeatus. - Russelia juncea. - Ruta sp.. - Sagittaria. - Salvia greggii. - Salvia aurea. - Salvia fulgens. - Salvia guaranitica. - Salvia discolor. - Santolina chamaecyparissus. - Sarcococca hookeriana humilis. - Schefflera sp. - Schinus molle. - Sedum morganianum. - Senecio petasitis. - Skimmia jamonica. - Solandra hartwegii. - Solanum rantonnettii. - Solanum aviculare. - Sorbaria sorbifolia. - Sparmannia africana. - Spathiphyllum wallisii. - Spiraea cantoniensis. - Stachys lanata. - Stapelia hirsuta. - Stranvaesia nussia. - Strelitzia reginae prolifera. - Strelitzia angusta. - Streptosolen jamesonii. - Taxus baccata cuspidata. - Taxus baccata hibernica fastigiata. - Templetonia retusa. - Teucrium fruticans. - Teucrium chamaedrys. - Thunbergia sp. - Thunbergia grandiflora. - Thunbergia alata. - Thuja sp. - Tipuana tipu. - Trachycarpus fortunei. - Ulex europaeus. - Veronica speciosa. - Viburnum tinus. - Viburnum opulus 'Sterile'. - Viburnum rhytidophyllum. - Viburnum carlesii. - Villarsia nymphoïdes. - Vinca minor. - Vitis voinieriana. - Washingtonia filifera. - Wistaria sinensis. - Yucca sp. - Yucca gloriosa. - Zantedeschia aethiopica.

Villa Noailles

Main collections:

Magnolia grandiflora. - Magnolia virginiana. - Magnolia thompsoniana. - Magnolia Loebneri. - Magnolia x soulangeana 'Delaunay'. - Magnolia griselinia littoralis. - Magnolia Lennei Minier. - Magnolia purpurea. - Magnolia stellata. - Magnolia x soulangeana 'Lennei'. - Magnolia Yulan Genest-Barge. - Magnolia alba. - Magnolia kobus moorheim. - Magnolia verbanica. - Magnolia rustica rosea. - Magnolia x soulangeana 'Alexandrina'. - Magnolia rustica rubra. - Magnolia mollicomata Hillier. - Camellia 'Lavigna Maggi'. - Camellia 'Margarita Caleoni'. - Camellia 'Victor Emmanuel'. - Camellia 'Fred Sanders'. - Camellia 'Souvenir de Bahuau Litou'. - Camellia 'Valtevaredo'. - Camellia 'Hippolithe Thobi'. - Camellia 'Bahaud Litou'. - Camellia 'Giardino Schmitz'. - Camellia 'Martin Cachet'. - Camellia 'Frau Mina Seidel'. - Camellia 'La Pace'. - Camellia 'Festiva'. - Camellia 'Kenny'. - Camellia 'Mathotiana Alba'. - Camellia 'Lalarouk'. - Camellia 'Elegans Chandleri'. - Camellia 'Princesse Baciochi'. - Camellia 'Adolphe Audusson'. - Camellia 'Kelwingtonia'. - Camellia 'Donklari Eugène Lise'. - Camellia 'Imperator'. - Camellia 'Princesse Clotilde'. - Camellia 'Mathotiana'. - Camellia 'Comtesse Lavigna Maggi'. - Camellia 'Honskongonensis'. - Camellia 'Auguste Haerens'. - Camellia 'Victor de Bischop'. - Camellia 'Lady Clarc'. - Camellia 'Auguste Desfossé'. - Camellia 'Drama Girl'. - Camellia 'Alba Plena'. - Camellia 'Alba Simplex'. - Camellia 'Director Loutil'. -Viburnum carlesii. - Viburnum bitchuanense. - Viburnum x carlcephalum. - Viburnum chenaulti. - Viburnum x bodnantense. - Viburnum x burkwoodii. - Viburnum judii. - Viburnum tomentosum mariesi. - Viburnum awafuki. - Viburnum fragrans 'Basil Leng'. - Viburnum opulus Sterile.

Shrub peonies and narcissi

Valrose

Pinus mughus. - Pinus canariensis. - Cedrus atlantica. - Cedrus deodora. - Cedrus libani. - Abies pinsapo. - Picea pungens. - Araucaria bidwillii. - Araucaria heterophylla. - Cupressus macrocarpa. - Cupressus lusitanica. - Taxus baccata. - Ginkgo biloba. - Phœnix canariensis. - Phœnix dactylifera. - Trachycarpus fortunei. - Chamaerops humilis. - Washingtonia filifera. - Yucca australis. - Cordyline australis. - Casuarina cunninghamianna. - Ficus macrophylla. - Magnolia grandiflora. - Persea gratissima. - Cocculus laurifolius. - Brachychiton populneum. - Citrus aurantium. - Acer oblungum. - Eriobotrya japonica. - Photinia serrulata. - Acacia dealbata. - Acacia retinodes. - Acacia verticillata. - Eucalyptus globulus. - Celtis australis. - Liriodendron tulipifera. - Ailanthus altissima. - Aesculus hippocastanus. - Platanus x acerifolia. - Prunus 'Pissardii'. - Prunus lauro-cerasus. - Albizzia julibrissin. - Albizzia lophata. - Cercis siliquastrum. - Sophora japonica. - Robinia pseudacacia. - Catalpa bignonioides.

Thuya orientalis. - Chamaecyparisus lawsoniana. - Cephalotaxus drupacea. - Ephedra sp. - Cycas revoluta. - Yucca gloriosa. - Furcraea gigantea. - Dasylirion acrotrichum. - Asparagus sprengeri. - Hakea saligna. - Hakea oleifolia. - Grevillea hilliana. - Ficus pumila. - Muehlenbeckia complexa. - Bougainvillea spectabilis. - Opuntia ficus-indica. - Paeonia arborea. - Berberis stenophylla. - Berberis thunbergii. - Mahonia aquifolium. - Iberis semperflorens. - Hypericum calycinum. - Camellia japonica. - Euonymus japonicus. - Rhamnus californica. - Colletia spinosa. - Colletia cruciata. -Parthenocissus quinquifolia. - Parthenocissus tricuspidata. - Sedum praealtum. - Escallonia rubra. - Pittosporum tobira. - Spiraea vanhouttei. - Spiraea cantoniensis. - Spiraea japonica. - Pyracantha coccinea. - Cotoneaster pannosa. - Cotoneaster salicifolia. - Cotoneaster horizontalis. - Retama monosperma. - Cytisus sessilifolius. - Medicago arborea. - Wistaria sinensis. - Lagertroemia indica. -Callistemon speciosus. - Callistemon salignus. - Callistemon rigidus. -Melaleuca thymifolia. - Fatsia japonica. - Aucuba japonica. - Forsythia suspensa. - Jasminum nudiflorum. - Arauja albens. - Buddleia davidii. -Echium x fastuosum. - Nicotiana glauca. - Datura arborea. - Hebe speciosa. - Adhatoda vasica. - Tecomaria capensis. - Phlomis fruticosa. - Lonicera nitida. - Lonicera pileata. - Eupatorium micranthum. - Senecio deltoideus. - Senecio grandifolius. - Senecio cineraria. - Eriocephalus africanus.

Villa Ile-de-France

Arecastrum romanzoffianum. - Butia eriospatha. - Erythea armata. - Acacia cultriformis. - Acacia howitti. - Acacia eburnea. - Acacia cyanophylla. - Casimiroa edulis. - Cyphomandra betacea. - Aberia caffra. - Eugenia sp. - Araucaria bidwillii. - Araucaria heterophylla. - Umbellularia californica. - Cinnamomum camphora. - Schafflera actinophylla. - Meryta sinclairii. - Oreopanax nymphaefolia. - Bauhinia sp. - Caesalpinia tinctorra. - Brachychiton discolor. - Brachychiton heterophyllum. - Brachychiton populneum. - Ephedra altissima. - Buddleja madagascariensis. - Vitis voinieriana. - Ficus grandiflora. - Yucca guatemalayensis. - Manihot carthagenense. - Alectryon coriaceum. - Grevillea robusta. - Phytolacca dioica. - Iochroma grandiflora. - Senecio antheuphorbium. - Senecio grandifolius. - Senecio petasitis. - Strelitzia angusta. - Sparmannia africana. - Eriocephalus africans. - Homalocladum platyphyllum. - Thunbergia grandiflora. - Portulacca affra. - Cestrum purpureum. - Datura sanguinea. - Eupatorium atrorubens. - Salvia leucantha. - Salvia involucrata.

Clos du Peyronnet

Access

Schinus molle. - Ligustrum sinense. - Adhatoda vasica. - Orepanax nymphaeifolium. - Buddleja sp. - Sterculia diversifolia. - Jaracanda ovalifolia. - Philadelphus sp. - Plumbago capensis. - Agapanthus sp. - Urginea maritima. - Daphne odorata. - Bergenia sp. - Sternbergia lutea. - Amaryllis belladonna. - Catalpa sp. - Washingtonia filifera. - Acacia arabica longifolia. - Acacia Howittii 'Clair de lune'. - Acacia verticillata. - Acacia retinodes 'Quatre saisons'. - Acacia cyanophylla. - Rosa radziwill. - Rosa 'Bengal Rose'. - Rosa banksiae. - Salvia guaranitica. - Salvia elegans. - Salvia confertiflora. - Salvia semiatrata. - Salvia microphylla. - Salvia interrupta. - Salvia broussonetii. - Salvia canariensis. - Salvia involucrata. - Salvia mexicana minor. - Salvia somaliensis. - Salvia coccinea. - Salvia mellifera. - Salvia gesneriiflora. - Salvia uliginosa. - Callistemon speciosum. - Ballota pseudodictamnus. - Canarina campanula. - Venidium sp. - Senecio cineraria. - Jacobinia suberecta. - Lantana camara. - Calliandra sp. - Albizzia julibrissin. - Hydrangea quercifolia. - Solanum warczewiczii. - Heteromorpha arborescens. - Tecomaria capensis. - Arbutus unedo. - Streptosolen jamesonii. - Polygala myrtifolia. - Nerium oleander. - Podranea ricasoliana. - Escallonia sp. - Cobea scandens. - Veronica sp. - Eleagnus pungens. - Phœnix canariensis. - Wigandia caracasana. - Abutilon striatum. - Montanoa bipinnatifida. - Cestrum cultum. - Clerodendrum fragrans. - Buddleja officinalis. - Buddleja asiatica. - Spiraea sp. - Iochroma grandiflora. - Passiflora quadrangularis. - Solandra hartwegii. - Aloe hanburiana. - Aloe ciliaris. - Ceratonia siliqua. - Yucca gloriosa. - Nolina recurvata. - Opuntia tunicata. - Senecio anteuphorbium. - Aegle marmelos. - Furcreae sp. - Nolina sp. - Dasylirion acrotrichum. - Nepeta cataria. - Fremontodendron mexicanum. - Erythrina crista-galli. - Solanum jasminoides. - Datura arborea. - Grevillea rosmarinifolia. - Eleagnus pungens. - Crinum sp. - Geranium palmatum. - Macleaya bella. - Bletilla striata. - Ochna serrulata. - Acanthus arboreus. - Brodiaea uniflora. - Felicia sp. - Scilla hispanica

Lower garden

Jasminum stephanense. - Jasminum polyanthum. - Jasminum primulinum. - Lampranthus sp. - Setcreasea purpurea. - Fuchsia fulgens. - Solanum aviculare. - Jacobinia carnea. - Ligularia sp.. - Begonia 'Cleopatra'. -Gerbera sp. - Lotus bertholettii. - Arctotis sp. - Pavonia hastata. - Citrus limon. - Lonicera sp. - Stauntonia hexaphylla. - Brachychiton populneum. - Phaedranthus buccinatorius. - Sparmannia africana. - Antholyza ringens. - Saxifraga stolonifera. - Buddleja 'Royal Purple'. - Gomphocarpus physocarpa. - Sedum rubrotinctum. - Hoya carnosa. - Thunbergia gibsonii. - Hypoestes sanguinolenta. - Hedychium coronarium. - Hedychium gardnerianum. - Hedychium greeni. - Chasmanthe aethiopica. - Brunnera macrophylla. - Cardiospermum grandiflorum. - Tradescantia sp. - Buddleja. - Oxypetalum ceruleum. - Datura 'Grand Marnier'. - Abutilon yellow. - Gynura sarmentosa. - Iris germanica. - Fremontia californica. - Wistaria sinensis. - Pandorea jasminoides. - Heliotropium peruvianum. - Raphiolepis sp. - Acidanthera bicolor. - Tuberose 'The Pearl'. - Ornithogalum arabica. - Gaura lindheimeri. - Barleria cristata. - Protaea cynaroides. - Bocconia fruticosa. - Salvia lavandulifolia. - Salvia argentea. - Salvia farinacea. - Salvia splendens. - Salvia chamaedryoides. - Methysticodendron amesianum. - Polygonum capitatum. - Echeveria sp. - Geranium 'Galilée'. - Gladiolus colvillei. - Gladiolus nanus 'Amanda Mahy'. - Papaver atlanticum. - Watsonia fulgens. - Watsonia bulbillifera. - Grevillea acanthifolia. - Yucca desmetiana. - Centaurea ragusina. - Raphiolepis sp. - Paeonia sp. - Lycopodium sp. - Asparagus sprengeri. - Convolvulus mauritanicus. - Pelargonium 'Black Knight'. - Rosa laevigata. - Camelia 'Gloire de Nantes'. - Romneya coulteri. - Chamaelaucium uncinatum. -

Anigozanthus. - Cestrum nocturnum. - Kennedya comptoniana. - Phornium tenax. - Monstera sellowa. - Chlorophytum elatum. - Hosta sp. - Philodendron selloum. - Streptocarpus sp. - Leptospermum pubescens. - Prostanthera rotundifolia. - Ficus pumila. - Hebe hulkeana. - Ophiopogon japonicus. - Crassula lactea. - Nephrolepis sp. - Laurus nobilis. - Fascicularia sp. - Aloe vera. - Yucca guatemalensis. - Pteris cretica. - Pittosporum tobira. - Spiraea vanhouttei. - Osmanthus fragrans. - Yucca aloifolia. - Danae racemosa. - Mackaya bella. - Fuschia arborescens. - Camellia magnoliflora. - Embothrium coccineum. - Rhodendron 'Anna Rose Whitusy'. - Digitalis sp. - Ruscus hypoglossum. - Muehlenbeckia axillaris. - Rosa 'Dorothy Perkins'. - Escallonia sp. - Berberis sp. - Hermodactylus tuberosus. - Ophrys fusca. - Asphodelus microcarpus. -Tropaelum sp. - Clematis armandii. - Aponogeton distachyus. - Allium subhirsutum. - Tulbaghia alliacea. - Homeria breyniana. - Papyrus sp. - Aristolochia elegans. - Periploca graeca. - Ampelopsis brevipedunculata. - Sollya heterophylla. - Mina lobata. - Clerodendron fragrans. - Iris pallida variegata. - Nelumbo lotus. - Callistemon viminalis. - Puschkinia scilloides. - Sparaxis grandiflora. - Leucojum autumnale. - Ornithogalum arabicum. - Ornithogalum thyrsoides. - Ornithogalum umbellatum. - Brodiaea sp. - Muscari moschatum. - Muscari comosum. - Allium ostrowskanium. - Iris bucharica. - Iris danfordiae. - Iris graeberana. - Iris hoogiana. - Iris aucheri. - Iris magnifica. - Iris suziana. - Romulea ramiflora. - Crocus angustifolius. - Crocus sativus. - Crocus ochroleucus. - Crocus nudiflorus. - Fritillaria persica 'Adivaman'. - Fritillaria meleagris. - Tulipa ingens. - Tulipa lanata. - Tulipa altaica. - Tulipa didieri. - Tulipa clusiana. - Tulipa saxatilis. - Scilla messenaica. - Scilla cilicica. - Scilla italica. - Scilla sibirica. - Scilla hyacinthoides. - Scilla peruviana. - Colchicum autumnale. - Ixiolirion macracanthum. - Anemone blanda. - Anemone coronaria. - Narcissus juncifolius. - Narcissus bulbocodium. - Narcissus elegans. - Galanthus corcyrensis. - Galanthus reginae olgae. - Eranthis cilicica. - Moraea villosa. - Moraea glaucopsis. - Bellevalia pycnantha. - Veltheimia sp.

Long pool

Strelitzia reginae. - Yucca gloriosa. - Citrus 'Jérusalem orange'. - Citrus myrtifolia. - Vitis voineriana. - Hydroleys commensonii. - Zephyranthes candida. - Zephyranthes citrina. - Zephyranthes grandiflora. - Zephyranthes rosea. - Canna lilies. - Fortunella margarita. - Pontederia cordata. - Lobelia cardinalis. - Sisyrynchium striatum. - Iris japonica. - Iris ochroleuca. - Teucrium fruticans.

Terrace of the Magnolia

Firmania Simplex. - Pittosporum tobira. - Pittosporum undulatum. - Russellia juncea. - Tulipa maximowiczii. - Acokanthera venenata. - Salvia blepharophylla. - Salvia blancoana. - Salvia cacaliaefolia. - Salvia interrupta. - Salvia coccinea. - Salvia officinalis. - Sarcococca ruscifolia. - Methysticodendron amesianum. - Oxypetalum caeruleum. - Tradescantia sp. - Chlidanthus sp. - Nectaroscordum siculum. - Campanula rupestris. - Allium stipita' ım. - Tritoma sp. - Sisyrynchium brachypus. - Sisyrynchium bermudiana. - Sisyrynchium convolutum. - Tubalghia alliacea. - Iris graminea. - Iris unguicularis. - Belamcanda chinensis. - Catha edulis. - Eucomis sp. - Ismene festalis. - Melianthus major. - Asclepias curassavica. - Helichrysum argenteum. - Libertia elegans. - Vitex agnus-castus. - Bocconia fruticosa. - Dianella tasmanica. - Acidanthera murieliae. - Hypoxis sp. - Abutilon megapotamicum. - Thermopsis caroliniana. - Campanula portenschlagiana. - Hedychium gardnerianum. - Arthropodium cirrhatum. - Brunnera macrophylla. - Saxifraga stolonifera. - Asphodelus lusitanicus. - Eucomis bicolor. - Convallaria majalis. - Solanum aviculare. - Ruta graveolens. - Heuchera sanguinea. - Tigridia sp. - Ixia sp. - Alchemilla mollis. - Vellozia elegans. - Macleaya sp. -Rudbeckia sp. - Nerine bowdenii. - Glechoma hederacea. - Plectranthus australis. - Neomarica gracilis.

Center garden

Abelia floribunda. - Streptosolen jamesonii. - Magnolia x soulangeana. - Euphorbia myrsinites. - Euphorbia dendroides. - Euphorbia characias. - Senecio petasitis. - Lavatera maritima. - Prunus campanulata. - Solanum rantonettii. - Solanum auriculatum. - Punica granatum. - Polygala myrtifolia. - Kniphophia sp. - Phlomis viscosa. - Cistus albidus. - Cistus salviaefolius. - Cistus villosus. - Cistus florentinus. - Cistus delilei. - Cistus nigricans. - Cistus canescens. - Cistus gargerotae. - Cistus symphitifolius. - Cistus monspeliensis. - Cistus ladanifer. - Cistus heterophyllus. - Cistus crispus. - Cassia sp. - Cassia corymbosa. - Cassia tomentosum. - Cassia macranthera. - Duranta plumieri. - Scabiosa cretica. - Anthyllis barba-jovis. - Malvastrum latericium. - Lavatera sp. - Prunus cerasus campanulata. - Solanum rantonettii. - Polygala myrtiflora. - Sparmannia africana. - Clematis armandii. - Kniphophia sp. - Solanum auriculatum. -

Zanthoxylum planispinum. - Myoporum sp. - Dietes grandiflora. - Chlidanthus sp. - Chamaelaucium uncinatum. - Anigozanthus sp. - Phyllostachys sulphurea. - Ferula communis. - Echium simplex. - Haemanthus kalbreyeri. - Asphodeline lutea. - Scilla tubergeniana. - Scilla hyacinthoides. - Scilla autumnalis. - Yucca aloifolia variegata. - Agave ferox. - Alpinia speciosa. - Dracaena sp. - Asphodelus sp. - Pitcairnia sp. - Iris versicolor. - Iris reticulata. - Iris spuria. - Zauschneria californica. - Aphyllanthus monspeliensis. - Schizostylis coccinea. - Tulipa turkestanica. - Tulipa tarda. - Furcraea selloa. - Sisyrinchium convolutum. - Sisyrinchium coelestis. - Eremerus bungei. - Graptopetalum paraguayense. - Gasteria verrucosa. - Ornithogalum narbonense. - Ornithogalum pyramidale. - Asphodelus fistulosus. - Othonnopsis cheirifolia. - Malvaviscus mollis. - Correa speciosa. - Abutilon megapotamicum. - Abutilon avicennae. - Decumaria racémosa. - Sphaeralcea umbellata. - Cordyline indivisa. - Cordyline australis. - Agave americana. - Agave attenuata. - Agave rigida. - Kleinia gregorii. - Tetrapanax papyriferus. - Clianthus puniceus. - Dracaena draco. - Eupatorium riparium. - Hebeclinum ianthinum. - Yucca guatemalensis. - Beschorneria yuccoides. - Antholyza sp.. - Watsonia rosea. - Watsonia meriana. - Hebe hulkeana. - Tulipa kaufmanniana 'Berlioz'. - Tulipa 'The first'. - Tulipa gregiolensis. - Tulipa praecox. - Tulipa linifolia. - Aloe comosa. - Raphiolepis sp. - Allium stipitatum. - Allium giganteum. - Erythrina corallodendron. - Pancratium illyricum. - Asphodeline liburnica. - Asphodeline lutea. - Bowkeria gerardiana. - Bauhinia variegata. - Watsonia sp. - Gladiolus papilio. - Eryngium agavifolium. - Amaryllis bella donna 'Alba'. - Clivia sp. - Aristea sp. - Kniphophia sp. - Bletilla striata. - Abutilon darwinii. - Phlomis viscosa. - Narcissus 'Paperwhite'. - Phormium tenax. - Furcraea sp. - Plectranthus sp. - Leonotis leonurus. - Reinwardtia trigyna. - Eichhornia crassipes. - Coreopsis sp. - Aloe sp. - Senecio mandraiiscae. - Crassula argentea. - Sedum compressum. - Aeonium sp. - Escallonia sp. - Homeria breyniana. - Mahonia lomariifolia. - Mahonia aquifolium. - Berberis darwinii. - Genista lydia. - Cuphea hyssopifolia. - Pyracantha sp. - Santolina neapolitana. - Podocarpus macrophyllus. - Salvia discolor. - Salvia buchananii. - Salvia chamaeleagnea. - Salvia aurea. - Salvia sessei. - Salvia greggii. - Diospyros sp. - Campanula vidalii. - Vernonia neocorymbosa. - Odontospermum sericeum. - Geranium tomentosum. - Geranium odoratissimum. - Pavonia hastata. - Dyckia sp. - Cortaderia sp. - Pittosporum tobira variegata. - Exochorda grandiflora. - Semele androgyna. - Sarcococca ruscifolia. - Muehlenbeckia platyclados. - Artemisia incarnata. - Helichrysym sp. - Monstera deliciosa. - Cupressus arizonica. - Tibouchina semidecandra. - Passiflora violacea. - Chasmanthe aethiopica. - Sophora tetraptera. - Carpenteria californica. - Dahlia imperialis. - Forsythia sp. - Tradescantia sp. - Akebia quinata. - Convolvulus mauritanicus. - Buddleja sp. - Eriobotrya japonica. - Ceanothus 'Gloire de Versailles'. - Rosa bracteata. - Rosa 'Alberic Barbier'. - Rosa 'Michèle Meilland'. - Rosa 'Perle d'Or'. - Philadelphus incanus. - Aster uliginosum. - Aster fruticosus. - Kleinia gregori. - Mimulus sp. - Russelia juncea. - Hebe sp. - Campsis radicans. - Murraya exotica. - Jasminum revolutum. - Lagerstroemia indica. - Bougainvillea sp. -Malvastrum latericium. - Lopezia lineata. - Lavatera maritima. - Lavandula sp. - Lantana sellowiana. - Haemanthus coccineus.

Upper garden

Correa alba. - Brachychiton. - Spirea trilobata. - Cercis siliquastrum. - Spartium junceum. - Malus lemoinei. - Rhus typhina. - Iris douglasiana. - Iris fimbriata. - Trachelium campanuloides. - Dianella tasmanica. - Doryanthes palmeri. - Kolkwitzia amabilis. - Hyacinthus sp. - Buddleja colvillei. - Buddleja crispa. - Sutherlandia frutescens. - Heracleum mantegazzianum. - Ranunculus sp. - Anemone sp. - Polygala virgata speciosa. - Sphaeralcea umbellata. - Cytisus battandierii. - Cassia nemophylla. - Caryopteris sp. - Allium neapolitanum. - Hyacinthoides hispanica. - Leptospermum sp. - Helleborus corsicus. - Helleborus foetidus. - Helleborus niger. - Helleborus orientalis. - Coronilla sp. - Zizyphus jujuba. - Rosa brunonii. - Erythrina crista-galli. - Melia azederach. - Datura sanguinea. - Juniperus rigida. - Laurus nobilis. - Eccremocarpus scaber. - Asparagus falcatus. - Vitis coignetiae. - Callistemon phœniceus. - Callistemon paludosus. - Callistemon roseus. - Cupressus kashmeriana. - Berberis wilsoniae. - Viburnum tinus. - Barosma lanceolata. - Muscari sp. - Eucomis sp. - Allium fragrans. - Cotoneaster sp. - Linum sp. - Galtonia candicans. - Brodiaea sp. - Brodiaea laxa. - Bursaria spinosa. - Cistus x aguiilari. - Cistus x pulverentus 'Sunset'. - Galanthus elwesii. - Gingko biloba. - Melaleuca pauperiflora. - Lagunaria patersonii. - Antholyza ringens. - Exochorda grandiflora. - Crataegus. - Photinia arbutifolia. - Mahonia siamensis. - Photinia serrulata. - Cantua buxifolia. - Dombeya natalensis. - Xanthoceras sorbifolia. - Lavandula dentata. - Syringa tomentella. - Philadelphus sp. - Forsythia sp. - Tritonia sp. - Libertia elegans. - Chimonanthus praecox. - Buddleja fallowiana. - Buddleja Davidii. - Scilla peruviana. - Iris foetidissima. - Iris innominata. - Cestrum psitticanum. - Erica cucullata. - Convolvulus cneorum. - Lycium barbarum. - Nandina domestica. - Viburnum rhytidophyllum. - Stranvaesia davidiana. - Lonicera periclymenum. - Lonicera standishii. - Phylica ericoides. - Hedera nepalensis. - Lytanthus salicinus. - Medicago arborea. - Cneorum tricoccum. - Freylinia lanceolata. - Rhus ciliata. - Rhamnus prinoides. - Bupleurum fruticosum. - Dononea viscosa purpurea. - Eucalyptus vernicosa. - Melaleuca halmaturorum. - Melaleuca retusa. - Acacia podalyriifolia. - Tecoma stans. - Mackaya bella. - Dovyalis hebecarpa. - Hippophaea rhamnoides. - Aeonium arboreum. - Hibiscus syriacus. - Feijoa sellowiana. - Eugenia guabiju. - Rosa 'Mermaid'. - Sedum morganianum. - Podocarpus sp. - Caesalpina horrida. - Euphorbia pulcherrima. - Persea gratissima. - Jasminum grandiflorum. - Retama monosperma. - Lobelia laxiflora. - Podranea ricasoliana. - Hibiscus lasiacarpus. - Mimosa pudica. - Lippia citrodoria. - Passiflora sp. - Cynanchum acuminatifolium. - Hunnemannia fumariifolia. - Annona cherimola. - Barleria cristata. - Eschsholzia californica. - Nerine undulatum. - Setcreasea purpurea. - Citrus medica. - Brachychiton acerifolium. - Anemopaegma subunculata. - Solanum jasminoides. - Cyphomandra betacea. - Ismene festalis. - Gladiolus natalensis. - Aristea sp. - Physalis peruviana. - Arum dracunculus. - Campanula portenschlagiana. - Anemone japonica. - Ageratum sp. - Crinum sp. - Agapanthus umbellatus. - Vica acutiloba. - Malvaviscus conzattii. - Gladiolus ignescens. - Lavandula multifida. - Pelargonium odoratissimum. - Zantedeschia aethiopica. - Rosa 'Albertine'. - Rosa 'Caroline Testout'. - Passiflora edulis. - Actinidia chinensis. - Musa sp. - Helianthus tuberosus. - Citrus arantiifolia. - Caesalpinia sepiaria. - Hibiscus rosa-sinensis. - Hedera colchica. - Lantana camara. - Solanum wendlandii. - Sechium edule. - Fortunella sp. - Eriocephalus africanus. - Kalanchoe fedtschenkoi. - Aloe aristata. - Chrysantemum haradjanii. - Echeveria derenbergii. - Mammillaria elongata. - Opuntia microdasys. - Trachycereus pingei.

Sources, and photographic credit

Except those mentioned below, all photographs were executed by the authors.
Aerial views: S.T.U., Ministery of Equipment.
La Gaude, Lenfant, Albertas, Saint-Marc-de-Jaumegarde: plans after Nerte Fustier-Dautier, *Les bastides de Provence et leurs jardins*, Serg.

P. 7, La Violaine: Jean Bernard. P. 14, The G. Grammont Garden, by J.-A. Constantin: Réunion des Musées nationaux. P. 16, Map of the Atlantis: Bibliothèque nationale. P. 17, The grotto of the Sainte-Baume: Musée de la Chambre de Commerce et d'Industrie de Marseille. P. 18, The Vaucluse fountain, by Constantin: Musée des Beaux-Arts, Marseilles. P. 22, Greenhouses, Vallauris: Sylvie Tubiana. P. 23, The Grilles district, Lauris: Inventaire général Roucaute-Heller. P. 24, Greenhouses, Antibes: Sylvie Tubiana. - Drawing by F. Cayol: Jean Bernard. P. 26, Drawing by F. Cayol: Jean Bernard. P. 27, Water-deviner: Musée dauphinois, Grenoble. P. 29, Borély park: Etienne Revault. P. 32, "The fish Pond:" Daspet. Pp. 34, 36, *Treatise on surveying* by B. Boysset: the National Library in Vienna, Austria. P. 37, Emilie in the garden: the National Library in Vienna, Austria. P. 38, "The Bird Charmer:" Daspet. Pp. 40, 41, Petrarch, King René: Musée Arbaud, Aix-en-Provence. P. 45, Belgentier: Musée Arbaud, Aix-en-Provence. Portrait of Peiresc: Bibliothèque Méjanes, Aix-en-Provence. P. 49, Plan, Beaulieu estate: private collection. P. 51, La Torse: Inventaire général Roucaute-Heller. P. 55, Le Verger : Jean-Luc Gille. P. 60, Parterre, Albertas: private collection. P. 61, Summer palace of Monseigneur du Bellay: Bibliothèque de Draguignan. P. 62, *Tèse*, Beaulieu: private collection. *Tèse*, la Tour-d'Aygues: Bibliothèque Méjanes, Aix-en-Provence. P. 66, Plan, la Tour-d'Aygues: Bibliothèque Méjanes, Aix-en-Provence. P. 69, View of St. Pons by Constantin: Musée des Beaux-Arts, Marseilles. P. 71, Member of the Marseilles bourgeoisie: Musée du Vieux-Marseille, Marseilles. P. 76, Mexican villa: Hélène Homps. P. 79, Villa Winslow and Château des Tours: in Stephen Liégeard, *La Côte d'Azur*. P. 82, The Antibes harbor, and View of Antibes: Bibliothèque de Cessole, Musée Masséna, Nice. P. 83, St. Hospice bay: Bibliothèque de Cessole, Musée Masséna, Nice. P. 86, Orange and fig picking: Bibliothèque de Cessole, Musée Masséna, Nice. P. 87, Parterre, Hôtel Théas de Thorenc: Photothèque de Grasse. P. 93, Stained-glass window, Beausoleil: Sylvie Réol. P. 97, Colonnade from the Tuileries: Sylvie Réol. P. 98, Rockwork, Villa Rothschild: Camille Milliet-Mondon. P. 99, Park, Villa Rothschild: Marc Heller. P. 100, Summerhouse, Villa La Cava: Camille Milliet-Mondon. P. 101, Araucarias and palms: Marc Heller. P. 102, Harold Peto in his garden: *Country Life*. Villa Maryland: Service photographique de la Ville de Nice. P. 109, Terra-cotta jar: Service photographique de la bibliothèque nationale. P. 114, Villa Croisset: Service photographique de la Ville de Nice. P. 115, Spanish garden, Villa Croisset: Alain Sabatier. P. 119, Stairway-pergola, Champfleuri: Jacques Repiquet. P. 120, Staircase, Fontana Rosa: Catherine Baïs. P. 121, Watercourse, Harold Peto: Giletta, Spadem. P. 122, The *allée rouge*: service photographique de la Bibliothèque nationale. P. 123, *Jardin d'eau et de lumière*: black, Gabriel Guévrékian; color, Richard Wesley. P. 159, Sauveur Mignard: private collection. P. 162, Simon Lenfant: Bibliothèque Méjanes, Aix-en-Provence. P. 164, Thomassin de Mazaugues: Bibliothèque Méjanes, Aix-en-Provence. P. 166, J.-B. d'Albertas: Bibliothèque Méjanes, Aix-en-Provence. P. 169, Atlantes: Jean Bernard. P. 171, Drawing by F. Cayol: Jean Bernard. P. 172, Gémenos, garden plan: Archives départementales des Bouches-du-Rhône. Géménos garden: Musée de la Chambre de Commerce et d'Industrie de Marseille. P. 176, Park, Tourves: Serge Conard. P. 177, Park, Tourves: Serge Conard. Drawing by F. Cayol: Jean Bernard. P. 188, Drawing by F. Cayol: Jean Bernard. P. 199, Mallet-Stevens villa, terrace: Etienne Revault. P. 200, all photographs by Gabriel Guévrékian. P. 220, Villa Domergue, sculpture at the edge of water fall: Etienne Revault. P. 226, Staircases, villa Noailles: Sylvie Réol. P. 228, Japanese cherry-tree and magnolias: Jean Bernard. P. 254, Château du Vignal: Etienne Revault. P. 255, Drawing by F. Cayol: Jean Bernard. P. 267, La Fiorentina and St. Hospice point: Service photographique de la Ville de Nice. P. 272, Construction site: Jardin botanique de Monaco. P. 273, Large cacti, in bloom: Jardin botanique de Monaco. P. 288, Blasco Ibanez in his garden: Service photographique de la Ville de Nice. P. 293, Drawing by F. Cayol: Jean Bernard. P. 296, Homer's garden: Sylvie Réol.

This book was set in Baskerville by Le vent se lève..., Chalais, France, and printed by Tardy Quercy, Bourges, France.